METHODS IN MOLECULAR BIOLOGY™

Series Editor
John M. Walker
School of Life Sciences
University of Hertfordshire
Hatfield, Hertfordshire, AL10 9AB, UK

For further volumes:
http://www.springer.com/series/7651

Ribozymes

Methods and Protocols

Edited by

Jörg S. Hartig

Department of Chemistry, University of Konstanz, Konstanz, Germany

☀ Humana Press

Editor
Jörg S. Hartig, PhD
Department of Chemistry
University of Konstanz
Konstanz, Germany

ISSN 1064-3745   e-ISSN 1940-6029
ISBN 978-1-61779-544-2   e-ISBN 978-1-61779-545-9
DOI 10.1007/978-1-61779-545-9
Springer New York Dordrecht Heidelberg London

Library of Congress Control Number: 2012930264

Printed on acid-free paper

Humana Press is part of Springer Science+Business Media (www.springer.com)

Preface

Since the first reports of the fascinating phenomenon of catalytically active RNAs in the 1980s, ribozyme research has become a major field in the molecular and biological sciences. Following the initial, ground-breaking discoveries, in the 1990s phosphodiester-cleaving ribozymes have been studied intensively both as basic tools as well as for therapeutic applications in order to suppress target gene expression. The past decade has then seen a variety of fundamental insights into ribozymes which have profoundly changed our view on RNA catalysis: The increasing biochemical and structural knowledge of ribosomal protein synthesis revealed that the fundamental peptidyl-transfer reaction is catalyzed by rRNA. In addition, the discovery of extended motifs of the hammerhead ribozyme and its subsequent high-resolution structure solved a long-standing discrepancy in understanding hammerhead catalysis. The discovery of riboswitches as wide-spread mechanisms for the regulation of gene expression widened our view on functional roles of RNAs in modern organisms. The glmS riboswitch was discovered as the first ligand-dependent, ribozyme-based genetic switch. Apart from these highlights, an increasing number of catalytically active ribozymes are discovered in a variety of organisms, although in most cases their functional role is still unknown. In addition to these fascinating discoveries in nature, in vitro selection techniques in combination with latest developments in the fields of RNA chemical and synthetic biology allowed for the development of a variety of engineered ribozymes utilized for various tasks. Taken together, ribozymes have been discovered in many places and show wide-spread functions and applications.

The present Methods in Molecular Biology collection starts with an introduction by Eric Westhof, followed by protocols for investigating the catalysis of individual ribozymes. Then, a series of chapters covers specific methods useful for studying different aspects of ribozymes. The third part of the collection highlights techniques for specific applications of ribozymes. Hence the present collection covers both protocols and techniques for the characterization of ribozymes as well as introduces a variety of novel applications for catalytic RNAs. It is aimed both at the biochemist interested in the investigation of mechanistic aspects of ribozyme catalysis as well as the biomedicine researcher and life scientist interested in the application of ribozymes as tools for manipulating cellular functions. I would like to thank all authors for contributing to the present methods collection and hope that it will help fostering ribozyme research in all of the above-mentioned aspects.

Konstanz, Germany   *Jörg S. Hartig*

Contents

Contributors

SERGIO D. AGUIRRE • *Department of Biochemistry and Biomedical Sciences and Department of Chemistry and Chemical Biology, McMaster University, Ontario, Canada*

M. MONSUR ALI • *Department of Biochemistry and Biomedical Sciences and Department of Chemistry and Chemical Biology, McMaster University, Ontario, Canada*

BETTINA APPEL • *Institut für Biochemie, Universität Greifswald, Greifswald, Germany*

YONG BAI • *Division of Infectious Diseases and Vaccinology, School of Public Health, University of California, Berkeley, CA, USA*

SOUMITRA BASU • *Department of Chemistry and Biochemistry, Kent State University, Kent, OH, USA*

PHILIP C. BEVILACQUA • *Department of Chemistry, The Pennsylvania State University, University Park, PA, USA*

KRISTA TRAPPL • *Innsbruck Biocenter, Divison of Genomics and R Nomics, Medical University of Innsbruck, Innsbruck, Austria*

RONALD R. BREAKER • *Department of Molecular, Cellular, and Developmental Biology, Department of Molecular Biophysics and Biochemistry, Howard Hughes Medical Institute and Yale University, New Haven, CT, USA*

BROŇA BREJOVÁ • *Natural Sciences II, University of California, Irvine, CA, USA*

JANINA BUCK • *Institute of Organic Chemistry and Chemical Biology, Johann Wolfgang Goethe-University, Frankfurt am Main, Germany*

LUCIA CARDO • *Institute of Inorganic Chemistry, University of Zurich, Zurich, Switzerland*

ANDREA L. CERRONE-SZAKAL • *Institute for Bioscience and Biotechnology Research, NIST, Rockville, MD, USA*

DURGA M. CHADALAVADA • *Department of Chemistry, The Pennsylvania State University, University Park, PA, USA*

LORENZO CITTI • *Institute of Clinical Physiology, National Research Council, Pisa, Italy*

DANIELA DONGHI • *Institute of Inorganic Chemistry, University of Zurich, Zurich, Switzerland*

P. PATRICK DOTSON II • *Department of Physiology, University of Kentucky, Lexington, KY, USA*

MATTHIAS D. ERLACHER • *Division of Genomics and R Nomics, Innsbruck Biocenter, Medical University of Innsbruck, Innsbruck, Austria*

OLGA FEDOROVA • *Howard Hughes Medical Institute and Department of Molecular, Cellular and Developmental Biology, Yale University, New Haven, CT, USA*

BORIS FÜRTIG • *Institute of Organic Chemistry and Chemical Biology, Johann Wolfgang Goethe-University, Frankfurt am Main, Germany*

ANNA-SKROLLAN GEIERMANN • *Institute of Organic Chemistry, Center for Molecular Biosciences (CMBI), University of Innsbruck, Innsbruck, Austria*

MARKUS GÖSSRINGER • *Institut für Pharmazeutische Chemie, Philipps-Universität Marburg, Marburg, Germany*

YUKI GOTO • *Department of Chemistry, Graduate School of Science, The University of Tokyo, Tokyo, Japan; PRESTO, Japan Science and Technology, Kawaguchi, Saitama, Japan*

DAGMAR GRABER • *Institute of Organic Chemistry, Center for Molecular Biosciences (CMBI), University of Innsbruck, Innsbruck, Austria*

CHRISTIAN HAMMANN • *Heisenberg Research Group Ribogenetics, Technical University of Darmstadt, Darmstadt, Germany*

JONATHAN HART • *Department of Chemistry, University of Kentucky, Lexington, KY, USA*

JÖRG S. HARTIG • *Department of Chemistry, University of Konstanz, Konstanz, Germany*

ROLAND K. HARTMANN • *Institut für Pharmazeutische Chemie, Philipps-Universität Marburg, Marburg, Germany*

DOMINIK HELMECKE • *Institut für Pharmazeutische Chemie, Philipps-Universität Marburg, Marburg, Germany*

HANNAH E. IHMS • *Department of Chemistry, University of Illinois at Urbana-Champaign, Urbana, IL, USA*

JERMAINE L. JENKINS • *Department of Biochemistry and Biophysics, University of Rochester, Rochester, NY, USA*

RANDI M. JIMENEZ • *Department of Pharmaceutical Sciences, University of California, Irvine, CA, USA*

ANNE KALWEIT • *Heisenberg Research Group Ribogenetics, Technical University of Darmstadt, Darmstadt, Germany*

KRISHANTHI S. KARUNATILAKA • *Wayne State University, Detroit, MI, USA*

BENEDIKT KLAUSER • *Department of Chemistry, University of Konstanz, Konstanz, Germany*

JOLANTA KRUCINSKA • *Department of Biochemistry and Biophysics, University of Rochester, Rochester, NY, USA*

CATERINA LANDE • *Institute of Clinical Physiology, National Research Council, and Human Morphology and Applied Biology Department, Pisa, Italy*

MICHEL V. LÉVESQUE • *Département de Biochimie, Université de Sherbrooke, Sherbrooke, QC, Canada*

YINGFU LI • *Department of Biochemistry and Biomedical Sciences and Department of Chemistry and Chemical Biology, McMaster University, Ontario, Canada*

JOE C. LIANG • *Division of Chemistry and Chemical Engineering, California Institute of Technology, Pasadena, CA, USA*

JOSEPH A. LIBERMAN • *Department of Biochemistry and Biophysics, University of Rochester, Rochester, NY, USA*

GEOFFREY M. LIPPA • *Department of Biochemistry and Biophysics, University of Rochester, Rochester, NY, USA*

FENYONG LIU • *Division of Infectious Diseases and Vaccinology, School of Public Health,*

and Program in Comparative Biochemistry University of California,
Berkeley, CA, USA

YI LU • *Department of Chemistry, University of Illinois at Urbana-Champaign,*
Urbana, IL, USA

ANDREJ LUPTÁK • *Department of Pharmaceutical Sciences, University of California,*
Irvine, CA, USA

JOANNE MACDONALD • *Department of Medicine, Division of Experimental*
Therapeutics, Columbia University, New York, NY, USA

THOMAS MARSCHALL • *Institut für Biochemie, Universität Greifswald,*
Greifswald, Germany

PHILLIP J. MCCOWN • *Department of Molecular, Cellular, and Developmental Biology,*
Yale University, New Haven, CT, USA

ALBERTO MERCATANTI • *Institute of Clinical Physiology, National Research Council,*
Pisa, Italy

RONALD MICURA • *Institute of Organic Chemistry, Center for Molecular*
Biosciences (CMBI), University of Innsbruck, Innsbruck, Austria

WENDY W.K. MOK • *Department of Biochemistry and Biomedical Sciences*
and Department of Chemistry and Chemical Biology, McMaster University,
Ontario, Canada

HOLGER MORODER • *Institute of Organic Chemistry, Center for Molecular*
Biosciences (CMBI), University of Innsbruck, Innsbruck, Austria

MARK J. MORRIS • *Department of Chemistry and Biochemistry, Kent State University,*
Kent, OH, USA

SABINE MÜLLER • *Institut für Biochemie, Universität Greifswald, Greifswald, Germany*

HENRIK NIELSEN • *Department of Cellular and Molecular Medicine,*
The Panum Institute, University of Copenhagen, Copenhagen, Denmark

CHRISTOPHER NOE • *Department of Chemistry, University of Kentucky,*
Lexington, KY, USA

CATHERINE PAZSINT • *Department of Chemistry and Biochemistry,*
Kent State University, Kent, OH, USA

RENJUN PEI • *Department of Medicine, Division of Experimental Therapeutics,*
New York, NY, USA

MARCOS DE LA PEÑA • *IBMCP (CSIC-UPV), Universidad Politécnica de Valencia,*
Valencia, Spain

JEAN-PIERRE PERREAULT • *Département de Biochimie, Université de Sherbrooke,*
Sherbrooke, QC, Canada

NICOLAS PIGANEAU • *Institut für Biochemie und Molekularbiologie,*
Universität Hamburg, Hamburg, Germany

NORBERT POLACEK • *Department of Chemistry and Biochemistry, University of Bern,*
Bern, Switzerland

RITA PRZYBILSKI • *Department of Microbiology and Immunology, University of Otago,*
Otago, New Zealand

LADISLAV RAMPÁŠEK • *Faculty of Mathematics, Physics and Informatics, Comenius*
University, Bratislava, Slovakia, USA

CHRISTIAN RICHTER • *Institute of Organic Chemistry and Chemical Biology,*
Johann Wolfgang Goethe-University, Frankfurt am Main, Germany

LUKAS RIGGER • *Institute of Organic Chemistry, Center for Molecular Biosciences (CMBI), University of Innsbruck, Innsbruck, Austria*

JOHN J. ROSSI • *Department of Molecular and Cellular Biology, Beckman Research Institute of the City of Hope medical Center, Duarte, CA, USA*

DAVID RUEDA • *Wayne State University, Detroit, MI, USA*

MOHAMMAD SALIM • *Department of Biochemistry and Biophysics, University of Rochester, Rochester, NY, USA*

ATHANASIOS SARAGLIADIS • *Department of Chemistry, University of Konstanz, Konstanz, Germany*

HARALD SCHWALBE • *Institute of Organic Chemistry and Chemical Biology, Johann Wolfgang Goethe-University, Frankfurt am Main, Germany*

CARSTEN SEEHAFER • *Department of Biology, Technical University of Darmstadt, Darmstadt, Germany*

NATHAN A. SIEGFRIED • *Department of Chemistry, University of North Carolina, Chapel Hill, NC, USA*

ROLAND K.O. SIGEL • *Institute of Inorganic Chemistry, University of Zurich, Zurich, Switzerland*

CHRISTINA D. SMOLKE • *Bioengineering Department, Stanford University, Stanford, CA, USA*

JESSICA STEGER • *Institute of Organic Chemistry, Center for Molecular Biosciences (CMBI), University of Innsbruck, Innsbruck, Austria*

MILAN N. STOJANOVIC • *Department of Medicine, Division of Experimental Therapeutics, Columbia University, New York, NY, USA; Department of Biomedical Engineering, Columbia University, New York, NY, USA*

ANNE STRAHL • *Institut für Biochemie, Universität Greifswald, Greifswald, Germany*

HIROAKI SUGA • *Department of Chemistry, Graduate School of Science, The University of Tokyo, Tokyo, Japan*

NARESH SUNKARA • *Division of Infectious Diseases and Vaccinology, University of California, Berkeley, CA, USA*

STEPHEN M. TESTA • *Department of Physiology, University of Kentucky, Lexington, KY, USA*

HOSHANG J. UNWALLA • *Department of Medicine, Division of Pulmonary Critical Care and Sleep Medicine, University of Miami, Miami, FL, USA*

TOMÁŠ VINAŘ • *Faculty of Mathematics, Physics and Informatics, Comenius University, Bratislava, Slovakia, USA*

JOSEPH E. WEDEKIND • *Department of Biochemistry and Biophysics, University of Rochester, Rochester, NY, USA*

ERIC WESTHOF • *Architecture et Réactivité de l'ARN, Université de Strasbourg, Institut de biologie moléculaire et cellulaire du CNRS, Strasbourg, France*

JENNIFER L. WILCOX • *Department of Chemistry, The Pennsylvania State University, University Park, PA, USA*

WADE C. WINKLER • *Department of Biochemistry, The University of Texas Southwestern Medical Center, Dallas, TX, USA*

Chapter 1

Introduction

Eric Westhof

Abstract

It is now about 30 years since ribozymes, catalytically active RNA molecules, were discovered. Although the chemical versatility of RNA does not come close to that of proteins, the chemical properties of nucleic acid systems are nevertheless thoroughly exploited in biological systems, leading to diverse ways of accelerating chemical reactions. Ribozymes are truly fascinating biological molecules. After all, is catalytic RNA an accident of life or, instead, is life an accident of catalytic RNA?

Key words: RNA catalysis, Ribozyme, Hammerhead, Overview

It is now about 30 years since ribozymes, catalytically active RNA molecules, were discovered in two different systems by Thomas Cech and Sydney Altman. During that period, our understanding of ribozyme structure and function has progressed quickly and steadily. Gone are the days where catalytic efficiencies of ribozymes were considered to be very poor compared to the protein enzymes. Gone also are the days where ribozyme action would be limited to restricted fossil-like enzymatic catalysis. Although the chemical versatility of RNA does not come close to that of proteins, the chemical properties of nucleic acid systems are nevertheless thoroughly exploited in biological systems, leading to diverse ways of accelerating chemical reactions.

Ribozymes are truly fascinating biological molecules. After all, is catalytic RNA an accident of life or, instead, is life an accident of catalytic RNA? Indeed, ribozymes constitute the physicochemical basis of the RNA World hypothesis of the origins of life, and they are still deeply embedded in several biological systems. The ribozyme sequences, consequently, carry on them marks of the long history of biological evolution. Thus, a complete understanding of ribozymes requires knowledge of the chemistry of catalysis, of the physics of three-dimensional structure folding and organization, and of the bioinformatics of sequence analysis and annotations.

Jörg S. Hartig (ed.), *Ribozymes: Methods and Protocols*, Methods in Molecular Biology, vol. 848,
DOI 10.1007/978-1-61779-545-9_1, © Springer Science+Business Media, LLC 2012

As often in science, the progress in our understanding of ribozymes is due to new developments and technologies. At the same time, the fascination ribozymes exercised on many resulted in a strong push for new experimental approaches and tools. Among those, the establishment of fast and reliable chemical synthesis of RNA played a central role. This, coupled with technological advances in X-ray crystallography (cryo-crystallography, availability of synchrotron sources, automatic crystallogenesis robots) and major developments in NMR technologies, led to a wealth of structural and dynamical data on all major ribozyme families. In the meantime, new ribozyme types have been either discovered or produced by in vitro selection. Bioinformatic searches led to detection of various types of ribozymes in genomes where they were not expected, sometimes in large numbers. Finally, ribozymes are applied in biotechnology and nanotechnology.

Nowadays, it is fully appreciated that metal ions, especially magnesium ions, play often a role but also that acid–base catalysis involving the nucleotide bases is an intrinsic component of the catalytic strategy of many ribozymes. We have learnt further that apparently minor structural changes can lead to changes in the type of chemical reaction catalyzed (e.g., 2′,5′ linkage formation instead of 3′,5′ linkage).

A recurrent problem in the study and analysis of ribozyme action is to obtain a clear separation between the effects observed on catalysis due to chemistry and those due to constraints on RNA folding. Our understanding of RNA architecture and of the folding processes leading to native architectures has enormously benefited from the analyses of ribozymes, especially those on the paradigmatic *Tetrahymena thermophila* group I intron. Group I introns are self-splicing RNA molecules that must adopt complex folded architectures in order to be active. Thus, the extent of splicing activity is a direct measure of the folded state of the ribozyme. Consequently, the effects on the folding extent can be measured precisely as a function of various physicochemical parameters, like temperature, ions, types of ions, and site-directed mutations, by following the extent of splicing activity. Further, the folded state of the molecule can be monitored independently using chemical and enzymatic probing methods, allowing a clear separation between the chemistry of catalysis and the folding processes.

Using such methods, it became apparent that regions far away from the catalytic site in the secondary structure were important for optimal activity. The exploitation of this property led to the discovery of some key RNA recognition modules (especially the roles of GNRA tetraloops in RNA–RNA interactions). At the same time, several mutations could be introduced or domains removed, swapped, or altered without much effect. All these observations pointed to the critical role of the folding process itself on the final amount of molecules in an active architecture.

Most importantly, it became clear that the folding process itself and the number of states accessible to the molecule are both dictated by the local topology of the secondary structure and the resulting possible coaxial stacking of helices. Coaxial stacking of helices restricts the number of positions in space available to the molecule and, if the correct choice of helices has been made, leads to further, and often long-range, stabilizing RNA–RNA interactions with the ensuing selection of the native architecture and optimal catalytic activity.

Secondary structure in RNA is the set of Watson–Crick paired helices that can form by self-folding of the sequence. Within a helix, any base pair can usually be replaced by any one of the common Watson–Crick pairs (A-U, U-A, G=C, or C=G) as well as the two wobble pairs (GoU or UoG). The number of possible sequences yielding a given topology (e.g., a three-way junction where three helices meet) is therefore large. However, depending on the particular sequence, a proportion of oligonucleotide strands will present several or a manifold of alternative structures, either prohibiting the adoption of the native choice of helices or limiting the number of molecules able to reach that choice at any moment of time. Clearly, this diversity of states leads to incorrect evaluations of catalytic rates. Further, sequence mutations influence in unsuspected ways the distribution of states and, thus, the effects of mutations are extremely difficult to monitor.

All these difficulties are compounded by the intricacies of the three-dimensional architectures of folded RNAs. Indeed, in order to obtain a three-dimensional architecture, some non-Watson–Crick pairs involving nucleotides in single strands have to form. The structural and evolutionary constraints on the non-Watson–Crick base pairs are not the same as those for the Watson–Crick pairs. In a nutshell, the isosteric set of Watson–Crick pairs coevolve (both nucleotides have to change), while in each family of non-Watson–Crick pairs the isosteric set of base pairs tends to form neutral networks (another functional pair is obtained if only one base changes). Such constraints on the sequence blur the understanding of the mutational effects. In addition, RNA modules, defined as an ensemble of non-Watson–Crick pairs, display sequence conservations, still not well understood, but most probably resulting from optimal combinations of the formation of covalent linkages compatible with nucleic acid stereochemistry together with good stacking and hydrogen bonding geometries. The final effect is the appearance of consensus sequences to which biological function can be wrongly assigned. In this respect, the hammerhead story is highly relevant. The high conservation observed in the core of the hammerhead ribozyme is the result of the presence of a stabilizing module for the three-way junction that is coupled with a particular nucleotide environment for binding a magnesium ion in proper position to the cleavable phosphate group for catalysis. However, the three-way junction is further stabilized in the active conformation by additional long-range contacts that are highly variable in positions and types. Thus, when those long-range and diverse contacts are absent, the number of states accessible to the RNA is large, the catalytic efficiency low, and the interpretation of the data highly difficult and risky.

The great lesson is again and again to restrict consensus sequences to a mnemotechnic help and to stress the importance of analyzing sequences fully with their whole range of diversity and in their full biological context. Natural selection is not "commitment to excellence" (Al Davis, Raiders, Los Angeles) and biological sequences result from various historical contingencies. Thus, beyond accidents of terrestrial history, what is striking in the chemistry of life is the enormous potential for overall structural and energetic neutrality: nucleotides can exchange while still maintaining structural integrity and similar stability through alternative interactions and contacts. This chemical neutrality allows for the presence of local suboptimal choices that do not destabilize the global architecture. In other words, the architectures and functions are robust to variations in the sequence. However, the resulting effects pave the way for the appearance of alternative or new catalytic mechanisms or functions within a similar architectural fold. Just with four letters, ribozymes are among nucleic acids really amazing molecular objects.

Chapter 2

Characterization of Hammerhead Ribozyme Reactions

Anne Kalweit, Rita Przybilski, Carsten Seehafer, Marcos de la Peña, and Christian Hammann

Abstract

Hammerhead ribozymes are small catalytic RNA motifs ubiquitously present in a large variety of genomes. The reactions catalyzed by these motifs are both their self-scission and the reverse ligation reaction. Here, we describe methods for the generation of DNA templates for the subsequent in vitro transcription of hammerhead ribozymes. This is followed by a description of the preparation of suitable RNA molecules for both reaction types, and their kinetic analysis.

Key words: Hammerhead ribozyme, Catalytic RNA, Kinetic analysis, Cleavage reaction, Ligation reaction, Recursive PCR

1. Introduction

The hammerhead ribozyme belongs to the family of small endonucleolytic ribozymes (1). It consists of a catalytic core of conserved nucleotides that are flanked by three helical arms (Fig. 1), of which arms I and II engage in tertiary interactions. Only if these interactions take place can the catalytic activity of the hammerhead motif be observed under physiological conditions (2–5). The ribozyme catalyzes the establishment of an equilibrium of both self-cleavage and self-ligation and thus, the reaction is reversible (Fig. 1). The cleavage reaction produces two RNA fragments, of which the 3′ product features a 5′ hydroxyl group, while the 5′ product features a 2′, 3′ cyclic phosphate. Originally, the hammerhead motif was found in satellite RNAs (6, 7), circular single-stranded RNA molecules, which can accompany specific plant viruses, aggravating the virus' symptoms on the plant. These satellite RNA molecules replicate by a rolling circle mechanism (8), in which the hammerhead ribozyme cleaves and ligates in cis: it cuts multimeric linear forms into the monomers, which in turn could get their ends joined to

Jörg S. Hartig (ed.), *Ribozymes: Methods and Protocols*, Methods in Molecular Biology, vol. 848,
DOI 10.1007/978-1-61779-545-9_2, © Springer Science+Business Media, LLC 2012

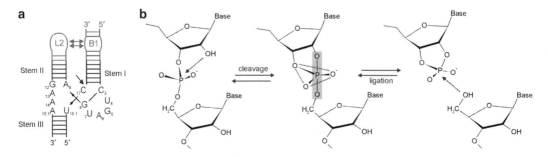

Fig. 1. The hammerhead ribozyme. (a) Shows the motif consisting of helical stems I, II, and III that surround the catalytic core with the indicated essential nucleotides. *Double-headed arrows* indicate tertiary interactions between elements in L2 and B1. The *arrow* shows the position of cleavage. Numbering is according to convention (9). (b) Shows the S_N2 mechanism, in which the 2′ hydroxyl attacks the adjacent 3′,5′ phosphodiester bond. In the transition state, a trigonal bipyramidal arrangement is reached (boxed in *grey*), that is resolved to yield the two products of the cleavage reaction. The 5′ product features a 2′, 3′ cyclic phosphate and the 3′ product a 5′ hydroxyl. These are also the substrates for the ligation reaction.

the circular monomeric form by the ribozyme motif. The latter reaction, however, has not been proven experimentally in vitro.

In recent years, the number of genomic positions, in which hammerhead ribozymes are found, has been dramatically extended (10, 11). After their discovery in various amphibians, schistosomes and cave crickets, hammerhead motifs were also found in a plant genome ((12), and references therein) or as a split version in various mammals (13). Most recently, we have shown the existence of this specific catalytic RNA as ultraconserved intronic motifs in all amniotes, including humans genome (14), as well as in a vast number of organisms, ranging from alpaca to zebrafish, indicating that the hammerhead ribozyme is ubiquitously present along the tree of life (15, 16). The biological function of these novel ribozyme motifs is presently under investigation in our laboratories, and a first, important step is the analysis of the cleavage and ligation reactions for each identified motif.

We provide here protocols for the generation of DNA templates, the preparation of the relevant RNA species, and the kinetic analysis of their reactions. The cleavage reaction of a conventional hammerhead ribozyme can be readily studied in cis, where it follows a first-order reaction kinetic. While the subviral plant pathogens may also circularize through a ligation reaction in cis, when they circularize, this is hard to reconstitute in vitro, as an access to molecules with both required ends, i.e., a 5′ hydroxyl group and a 2′, 3′ cyclic phosphate is limited. Because of this, the ligation reaction is best approximated in vitro by incubating the two reaction products of the cleavage reaction in a second-order reaction. While the in vitro ligation reaction thus proceeds in trans, its kinetic analysis is facilitated by using a large excess of one of the substrates. In this setup, the concentration of the latter substrate remains virtually unchanged, and the reaction can be analyzed according to

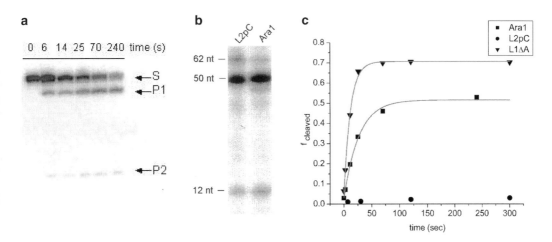

Fig. 2. Cleavage reactions of minimal and tertiary stabilized hammerhead ribozymes. (**a**) Shows a representative time course of a cleavage reaction of the tertiary stabilized *A. thaliana* hammerhead ribozyme Ara1. S, P1, and P2 denote substrate and products 1 and 2. (**b**) Complete cleavage of the 62 nucleotide (nt) long RNA in the two fragments of 50 and 12 nt is observed, at elevated Mg^{2+} concentrations in the course of an in vitro transcription, both for Ara1 and the variant L2pC. This variant cannot form tertiary interactions, which can be inferred from (**c**), a comparison of the kinetic analysis of the cleavage reactions under physiological Mg^{2+} concentrations. The cleaved fraction ($f_{cleaved}$) is plotted against time for the tertiary stabilized molecules Ara1 and L1ΔA, and the L2pC variant, which is inactive under these conditions. Reproduced from Ref. (12) with permission (www.plantcell.org; Copyright American Society of Plant Biologists).

pseudo-first order. In either case, it is essential that the reactions are performed in the presence of physiological concentrations of divalent metal ions like Mg^{2+}. Only then can the reaction of a true hammerhead ribozyme be monitored (Fig. 2), that relies on tertiary interactions between stems I and II (2, 3), which are also essential for the folding of this catalytic RNA (17).

2. Materials

All solutions should be prepared wearing gloves to avoid contamination with RNases, using deionized ultrapure water and analytical grade reagents (see Note 1).

2.1. Recursive PCR

1. Desalted DNA oligonucleotides are ordered from any suitable suppliers. Prepare stock solutions with a concentration of 100 μM. For the overlapping oligonucleotides they are diluted to working concentrations of 5 μM. Stock solutions of the *outer* primers are used undiluted.

2. For PCR, Taq polymerase is used together with the 10× buffer containing 10–20 mM $MgCl_2$, provided by the supplier.

3. The dNTP solution contains 2 mM of each dATP, dCTP, dGTP, and dTTP.

4. A PCR thermocycler.

5. A Stock of 10× TBE is prepared by dissolving 890 mmol Tris base and 890 mmol boric acid with 40 mL 0.5 M EDTA (pH 8.0) in 1 L H$_2$O. After autoclaving for 20 min, 100 mL are diluted 1:10 to yield 1 L 1× TBE.

6. Agarose.

7. Prepare a 10 mg/mL ethidium bromide solution.

8. DNA size standard with fragments in the range 50–200 bp, from the supplier of your choice.

9. The 6× DNA loading dye contains 10 mM Tris–HCl, pH 8.0, 60 mM Na$_2$EDTA, 60% (w/V) glycerol, 0.03% (w/V) bromophenol blue, and 0.03% (w/V) xylene cyanol.

10. If required, a gel extraction kit.

11. Any cloning vector lacking the T7 promoter sequence can be used to clone the PCR fragment of recursive PCR. We normally use the pJET1/blunt that Fermentas (now Thermo Fisher Scientific) has prepared upon request.

12. Competent *E. coli* cells, prepared using standard procedures or purchased commercially. LB medium and antibiotics depending on the resistance encoded in the used cloning vector.

13. For analytical DNA restriction, enzymes and their incubation buffers are required. Which enzyme to use depends on the cloning vector's features.

14. A gel documentation system with UV table suitable to record ethidium bromide stained gels.

15. DNA size marker with fragments in the kb range.

16. PCI mixture consisting of phenol: chloroform: iso-amyl alcohol (25:24:1; V:V:V). Work with PCI requires appropriate protective measures and should be performed under a fume hood.

17. A solution of 3 M NaOAc (pH 5.3) and ethanol, at concentrations of 100 and 70% (V/V).

18. A UV spectrophotometer.

2.2. PCR on Genomic DNA

1. PCR primers to amplify genomic ribozyme motifs.

2. Genomic DNA of the organism under investigation.

3. Material according to Subheading 2.1, steps 2–18.

2.3. In Vitro Transcription

1. Protective equipment to work with radio-labeled material.

2. Ingredients for the 10× transcription buffer are 400 mM Tris-HCl (pH 8.0), 200 mM MgCl$_2$, 20 mM spermidine, and 0.1% TritonX-100; ACG solution (5 mM of each ATP, CTP, and GTP), UTP solution (1 mM), RNase inhibitor (10 U/μL),

(α-^{32}P) UTP and T7 RNA Polymerase (20 U/μL), 0.5 M EDTA (pH 8.0).

3. An rNTPs mixture of 5 mM each.

4. A DNA oligonucleotide that is complementary to the core of the individual hammerhead motif.

5. Plastic syringes (1 mL).

6. Sterile glass wool.

7. Several 15-mL Falcon tubes and 1.5-mL reaction tubes with screw caps.

8. For gel filtration, 20 g of Sephadex G-50 fine are mixed in a bottle with 200 mL 1× TE containing 10 mM Tris–HCl and 1 mM EDTA (pH 8.0) and the mixture is boiled in a microwave for 30 s. Upon setting of the sephadex bed in the bottle, the buffer is poured off, and replaced by fresh buffer. Bring again to the boil and store after cooling at 4°C.

2.4. Gel Purification of RNA Species to Study Cleavage and Ligation Reactions

1. A vertical gel electrophoresis apparatus for polyacrylamide gels, with a power supply generating a voltage of 600 V.

2. Glass plates (25 × 20 cm), 0.5-mm thick spacers and comb.

3. A 12% polyacrylamide:bisacrylamide (19:1) solution with 7 M urea and 1× TBE.

4. Tetramethylethylendiamin (TEMED).

5. Ammoniumpersulfate (APS), dissolved in water to a concentration of 20% (w/V).

6. For the gel run 1× TBE.

7. A 10- or 20-mL syringe with injection needle.

8. Denaturing RNA loading buffer consisting of 95% formamide, 50 mM EDTA (pH 8.0), 0.03% (w/v) bromophenol blue, and 0.03% (w/v) xylene cyanol.

9. Heating block.

10. An RNA size marker.

11. Saran wrap.

12. A phosphorimager.

13. Sterile scalpels.

14. Gel extraction solution of 40% formamide (V/V), 0.7% (w/V) SDS in 1× TE.

15. A shaker for reaction tubes.

16. A solution of ethidium bromide with a concentration of 1 μg/mL 1× TBE.

17. Orbital shaker for trays.

<table>
<tr><td>

2.5. Kinetic Analysis of the In Vitro Cleavage Reaction

</td><td>

1. Reaction buffer (10×) containing 100 mM Tris–HCl (pH 8.0), 100 mM NaCl and 1 mM EDTA (pH 8.0).
2. Start solution containing 6 mM $MgCl_2$ in 10 mM Tris–HCl (pH 8.0).
3. At least 10 reaction tubes with 10 μL denaturing RNA loading buffer (see Subheading 2.4, step 8), consecutively numbered and placed on ice.
4. Material for PAGE as described above (see Subheading 2.4, steps 1–12).
5. Computer software, like Prism to plot and fit the quantified kinetic reaction data.

</td></tr>
<tr><td>

2.6. Kinetic Analysis of the In Vitro Ligation Reactions

</td><td>

1. Reaction buffer (10×) containing 100 mM Tris–HCl (pH 8.0), 100 mM NaCl, and 1 mM EDTA (pH 8.0).
2. Start solution containing 25 mM $MgCl_2$ in 10 mM Tris–HCl (pH 8.0).
3. Material identical to Subheading 2.5, steps 3–5.

</td></tr>
</table>

3. Methods

3.1. Recursive PCR

1. The DNA template contains the T7 RNA polymerase promoter sequence (TAATACGACTCACTATA), to which the sequence GGG, GCG, or GGC is added, depending on structural features of the resulting RNA (see Note 2). These triplets ensure high transcription yields (18). After either triplet, the sequence of the hammerhead ribozyme motif under investigation is added (Fig. 3). The resulting DNA template is therefore 20 nucleotides (nt) longer than the original ribozyme sequence.

2. The DNA template is then split up in an even number of DNA oligonucleotides that partially overlap (Fig. 4). It is crucial that all overlapping sequences for one DNA template have a similar predicted melting temperature (Tm) within 1°C. Tm values can be determined by a wide range of programs accessible via the internet (see Note 3). The size of the template determines the number of DNA oligonucleotides that are suitable to cover

Fig. 3. Design of the DNA template for in vitro transcription. In front of the DNA sequence encoding the hammerhead ribozyme, the sequence of the T7 promoter is added, followed by one of the indicated triplets.

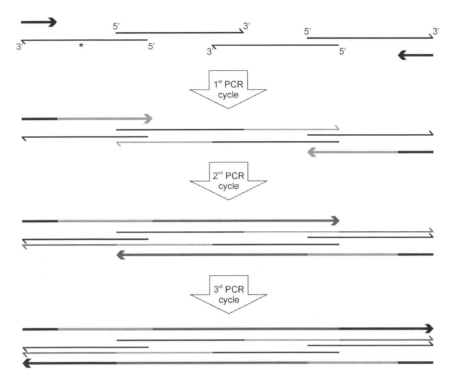

Fig. 4. Principle of recursive PCRs. An even number of overlapping DNA oligonucleotides (*thin arrows*) is designed in a way that the overlapping stretches have the same melting temperature (Tm). To these, two additional *outer* oligonucleotides with identical Tm to that of the overlapping stretches are added in excess (*thick black arrows*). In a first PCR round, the oligonucleotides are extended as indicated (*light grey lines*). In the second round, DNA strands get further extended (*dark grey lines*). In the third round and thereafter, only the *outer* oligonucleotides are extended to cover the entire sequence. The DNA oligonucleotide marked with an *asterisk* can be used in subsequent in vitro transcription reactions to prevent self-cleavage of the ribozyme.

the entire sequence. Currently, the efficient synthesis rate for oligonucleotides can easily yield 50–60 nt. On the basis of this and an average overlapping sequence of about 20 nt (depending on the sequence), the number of required DNA molecules (N) can be estimated according to (1) (see Note 4), in which L is the length of the template:

$$L = 30 \text{ nt} * N + 20 \text{ nt} \tag{1}$$

Additionally to these DNA oligonucleotides, two *outer* primers are required that will amplify the template sequence eventually (Fig. 4). Their sequences are defined by the requirement to have a similar Tm value, as the overlapping sequences of the other DNA oligonucleotides have.

3. For PCR, a mixture is set up containing 1 µL of each overlapping DNA oligonucleotide (5 µM; see Note 5), 1 µL of each *outer* primer (100 µM), 2.5 µL 10× PCR buffer, 2.5 µL dNTP solution, and 0.5 µL Taq-Polymerase in a total reaction volume of 25 µL H$_2$O. This mixture is transferred to a PCR tube

and the PCR is started in a thermocycler with the following parameters:

First step	1×	30 s @ 95°C
Second step	30×	10 s @ 95°C 10 s @ T_M of the used DNA oligonucleotides 10 s @ 72°C
Third step	1×	60 s @ 72°C

4. Prepare a 1.5% standard agarose gel by dissolving 1.5 g agarose in 100 mL 1× TBE in a microwave oven. After the solution has cooled to about 60°C, add 5 μL ethidium bromide solution, cast the gel in a horizontal gel chamber, and insert a comb. After the gel has set, overlay with 1× TBE. Mix 10 μL of each PCR reaction with 2 μL loading dye and load next to a DNA standard covering the size range of the expected PCR product. Connect to a power supply and perform gel electrophoresis at a field strength of 10 V/cm.

5. Make sure to wear appropriate protective gear against the radiation, when analyzing the result by visualization on a UV table. Document the result by a gel documentation system. If no clear PCR product of the expected size is observed, the reaction can be repeated using a reduced amount of overlapping DNA oligonucleotides (see Note 5). As soon as a clear band is obtained use any cloning vector lacking the T7 promoter to clone the PCR product. Follow the manufacturer's instructions for the cloning vector to ligate the PCR fragment in the cloning vector. Transform in competent *E. coli* cells and on the next day, perform Plasmid mini-preparations according to the lab's standard protocols, or as described (19).

6. Use an analytical digest of the DNA of 5–10 prepared plasmids to identify clones with inserts of the expected size. Sequence 2–3 plasmids at any company that offers commercial sequencing. Upon identification of a clone with the correct DNA template (Fig. 3), prepare a 100 mL overnight culture of the original *E. coli* strain and prepare the plasmid DNA, according to your lab's standard protocols. This cell culture volume is expected to yield 100 μL solution of 1 μg/μL plasmid. Use a restriction enzyme that cleaves near after the last nucleotides encoding the ribozyme sequence (see Note 6) and linearize 10 μg of the plasmid preparation. To assess the completeness of the digest, analyze 1/100 (V/V) on a 1% agarose gel, using an equivalent amount of undigested plasmid for comparison and a conventional DNA size marker in the kb range. Proceed as above (step 4) and analyze the gel using a gel documentation system. Upon complete digest (see Note 7) purify the linear DNA by phenol extraction. For this purpose, mix equal amounts of the linear DNA solution and PCI in a 1.5-mL reaction tube,

vortex for 30 s and separate phases at room temperature in a table centrifuge at maximum speed for 5 min. Carefully remove the upper aqueous phase and transfer to a fresh 1.5-mL reaction tube. Add to the linear plasmid solution 1/10 (V/V) 3 M NaOAc, pH 5.3 and 3 (V/V) parts ethanol and incubate 20 min at –20°C. Place the tube in a cooling table centrifuge and start a run at maximum speed for 30 min at 4°C. Remove the supernatant, add 1 mL 70% ethanol and repeat centrifugation for 10 min. After removal of the liquid, dry the pellet by placing the open reaction tube in a heating block set to 50°C (see Note 8), and redissolve subsequently in 100 µL H_2O. Determine the solution's DNA concentration in a UV spectrometer.

3.2. PCR on Genomic DNA

1. If the ribozyme motif and its surrounding sequences in the genomic location under question are known, conventional forward and reverse PCR primers are designed. In front of the sequence of the forward primer, the sequence of the T7 RNA polymerase promoter and either triplet GGG, GCG, or GGC are added (Fig. 5), in analogy to the situation described in Subheading 3.1, step 1 (see Note 2). Perform a standard PCR on genomic DNA using the reaction components as indicated in Subheading 2.1, steps 2 and 3, using 5 µM primer solutions and at the PCR elongation temperature defined by the primers.

2. Analyze and clone the PCR product as described in Subheading 3.1, steps 4–6.

3.3. In Vitro Transcription

1. To generate the hammerhead ribozyme transcripts, linearized plasmid DNA serves as template that was generated either by recursive PCR (Subheading 3.1) or by PCR on genomic DNA (Subheading 3.2). Since the in vitro transcription is performed in the presence of $(\alpha\text{-}^{32}P)$ UTP, appropriate protective measures against radioactivity must be in place in the laboratory and implemented for all work described in the subsequent sections. Mix by pipetting in a total reaction volume of 50 µL the

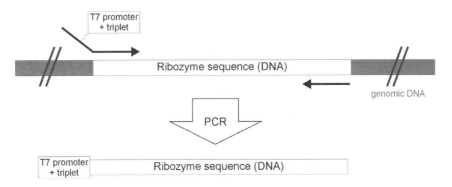

Fig. 5. Amplification of hammerhead ribozyme motifs from genomic DNA. To the forward PCR primer, the sequence of the T7 promoter and a transcription-promoting triplet (GGG; GGC or GCG) is added, yielding the indicated PCR product.

following ingredients: 200 ng linearized plasmid DNA, 5 μL 10× transcription buffer, 5 μL ACG solution, 5 μL UTP solution, 1 μL RNase inhibitor, 1 μL (α-^{32}P) UTP, and start the reaction by the addition of 1 μL T7 RNA Polymerase. Incubate for 30 min at 37°C.

2. In a separate reaction, add a DNA oligonucleotide that is antisense to the core of the hammerhead ribozyme, at a concentration of 10 μM (see Note 9). This will disrupt the folding of the transcript in the hammerhead ribozyme structure and thus allows for preparation of full-length RNA for subsequent kinetic analysis.

3. For kinetic analysis of the in vitro ligation reaction, one of the cleavage products is required in a nonradio-labeled version. To generate this, perform a third in vitro transcription reaction as stated above (Subheading 3.3), however in the absence of (α-^{32}P) UTP, and at a uniform rNTP concentration (see Note 10).

4. Stop all reactions after 30 min by the addition of 50 μL EDTA solution, 100 μL H$_2$O, and 200 μL PCI. Perform the phenol extraction as described above (Subheading 3.1, step 6).

5. After removing the upper aqueous phase, place it on a sephadex G-50 column (Fig. 6) to separate unincorporated nucleotides

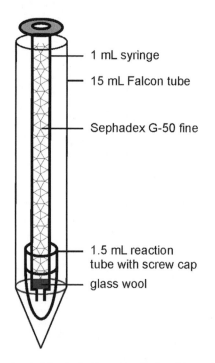

1 mL syringe

15 mL Falcon tube

Sephadex G-50 fine

1.5 mL reaction tube with screw cap

glass wool

Fig. 6. Gel filtration column. A 1-mL plastic syringe is closed by pushing in sterile glass wool using the syringe's forcer. This serves as support for the gel bed. Place the syringe in an empty 15-mL Falcon tube. The syringe is then filled by pipetting Sephadex G-50 fine emulsion in the syringe. Let the material set by gravity flow. The column is set by centrifuging the syringe for 2 min at 1,000 × g within the Falcon tube. Discard the flow through, and place a 1.5-mL reaction tube (without the screw cap) in the 15-mL Falcon tube, into which the syringe is placed again, as shown here.

from the transcript by gel filtration. The sephadex column is prepared as described in the legend to Fig. 6. Add the aqueous phase after PCI treatment on top of the sephadex column and centrifuge the syringe within the 15-mL Falcon tube again for 2 min at $1,000 \times g$. For the radio-labeled RNA, carefully discard the syringe in the radioactive waste and remove the 1.5-mL tube from the Falcon tube by using long forceps (see Note 11).

6. Add 1/10 volumes 3 M NaOAc (pH 5.3), and fill the reaction tube with 3 volumes ethanol. Precipitate RNA as described above (Subheading 3.1, step 6) and redissolve in 20 µL H_2O.

3.4. Gel Purification of RNA Species to Study Cleavage and Ligation Reactions

The transcription buffer contains Mg^{2+} and this is sufficient to initiate the hammerhead ribozyme self-cleavage reaction, resulting in a mixture of the full-length RNA and the two cleavage products after transcription. The addition of an inhibitory DNA oligonucleotide in the transcription reaction (see Subheading 3.3) allows, however, to increase the fraction of uncleaved full-length RNA (see Note 9). For subsequent kinetic analysis, it is necessary to purify the full-length RNA to study the cleavage reaction, and the two cleavage products, to investigate the ligation reaction.

1. Carefully clean the glass plates of ca. 25×20 cm, and suitable spacers and comb, and set them up to cast a vertical gel.

2. Add 150 µL APS solution (see Note 12) and 60 µL TEMED to 60 mL polyacrylamide solution and cast the gel and immediately insert the comb.

3. After the gel has polymerized, remove the comb and flush the slots with 1× TBE, by using a syringe with an injection needle.

4. Place the glass plates with the gel in the electrophoresis chamber, fill the upper and lower tanks with 1× TBE, and connect to a power supply. Check the connections by pre-running the gel at a constant current of 40 mA.

5. Mix equal amounts (10–25 µL) of the RNA and the denaturing loading buffer and denature the RNA by placing the mixture on a heating block set to 95°C. After 2 min, snap cool by placing on ice.

6. Load the mixture in a freshly flushed gel pocket, next to an RNA size marker.

7. Separate the RNA species electrophoretically using a constant current of 40 mA. When the dyes of the loading buffer indicate good separation, after ca. 2 h, switch off the power supply. Carefully lift one gel plate.

8. For the radio-labeled transcripts, mix a small amount of (older) (α-^{32}P) UTP, or any other β-emitter with the loading dye. Use this to draw with a micropipette an asymmetric pattern on those areas of the gel, where no RNA has been loaded.

Carefully cover the gel with saran wrap and expose to a phosphorimager screen for 10–30 min. Read the exposed screen and identify the bands of interest. Make a 1:1 print out of the gel area and place this under the glass plate with the gel and use the drawn pattern to overlay (see Note 13).

9. The gel with nonradio-labeled transcripts is displaced from both glass plates and incubated in a tray with ethidium bromide solution (see Note 14). After slow orbital shaking for 20 min, the gel is covered in saran wrap, and bands are visualized on the UV table of the gel documentation system.

10. Use a sterile scalpel to excise the bands of interest and move carefully to a fresh 1.5-mL reaction tube. Overlay the gel slice with gel extraction solution and place the closed tube in a shaker.

11. After shaking overnight, carefully pipette the liquid in a fresh 1.5-mL reaction tube. Perform a PCI treatment and precipitate RNA as described (see Subheading 3.1, step 6). Dissolve the RNA in 20–40 μL H_2O. Determine the concentration of the nonradio-labeled RNA spectrophotometrically.

3.5. Kinetic Analysis of the In Vitro Cleavage Reaction

1. Mix 10 μL of the radio-labeled full-length hammerhead ribozyme transcript with 80 μL H_2O and 10 μL 10× reaction buffer.

2. Allow the RNA to refold by placing the tube for 1 min at 80°C, followed by snap cooling on ice.

3. Remove 10 μL to a reaction tube with denaturing RNA loading buffer as time 0 control.

4. Place the tube in a heating block set to 25°C and initiate the reaction by adding 10 μL start solution, resulting in an effective Mg^{2+} concentration of 0.5 mM (see Note 15).

5. After suitable time points, e.g., 30 s, and 1, 2, 4, 6, 9, 12, 15 and 20 min, remove 10 μL aliquots from the reaction to prepared reaction tubes with denaturing RNA loading buffer (see Note 16).

6. Heat all 10 samples for 2 min at 95°C, snap cool on ice and load the entire content on a denaturing polyacrylamide gel. Perform PAGE to separate the reaction products (see Subheading 3.4).

7. After reading the phosphorimager plate, quantify the three relevant RNAs (FL: full-length; P_1: product 1 and P_2: product 2) using the phosphorimager's software.

8. Determine the fraction of cleaved RNA $f_{cleaved}$, according to (2):

$$f_{cleaved} = \frac{P_1 + P_2}{FL + P_1 + P_2} \qquad (2)$$

9. Use a computer program to plot $f_{cleaved}$ against time. To obtain the end point of the cleavage reaction $f_{\infty,c}$ and first-order kinetic constant k_{cis}, fit the data according to (3), in which f_0 denotes the small fraction of transcripts that might have cleaved already before the addition of magnesium.

$$f_{cleaved} = f_0 + f_{\infty,c} * (1 - e^{-k^*t}) \qquad (3)$$

10. The constant k_{cis} has the unit 1/s or 1/min. It allows together with the end point of the cleavage reaction $f_{\infty,c}$ to compare different hammerhead ribozymes.

3.6. Kinetic Analysis of the In Vitro Ligation Reaction

1. Mix 10 μL of the radio-labeled small product (S*) of the hammerhead ribozyme self-cleavage reaction with 30 μL H_2O and 10 μL 10× reaction buffer.

2. Mix 10 μL of a 2 μM solution of the nonradio-labeled large product (L) of the hammerhead ribozyme self-cleavage reaction with 30 μL H_2O and 10 μL 10× reaction buffer.

3. Allow the RNA solutions to refold separately by placing the tubes for 1 min at 80°C, followed by snap cooling on ice.

4. Mix the two solutions and remove 10 μL to a reaction tube with denaturing RNA loading buffer as time 0 control.

5. Place the tube in a heating block set to 25°C and initiate the reaction by adding 10 μL start solution, resulting in an effective Mg^{2+} concentration of 2.5 mM (see Note 17).

6. After suitable time points, e.g., 1, 2, 4, 6, 9, 12, 15, 20, and 30 min, remove 10 μL aliquots from the reaction to prepared reaction tubes with denaturing RNA loading buffer.

7. Heat all 10 samples for 2 min at 95°C, snap cool on ice and load the entire content on a denaturing polyacrylamide. Perform PAGE to separate the reaction products (see Subheading 3.4).

8. After reading the phosphorimager plate, quantify the two relevant RNAs that are visible on the screen (FL: full-length; S*: small product) using the phosphorimager's software.

9. Determine the fraction of cleaved RNA $f_{ligated}$, according to (4)

$$f_{ligated} = \frac{FL}{FL + S^*} \qquad (4)$$

10. Use a computer program to plot $f_{ligated}$ against time. To obtain the end point of the ligation reaction $f_{\infty,L}$ and pseudo-first-order kinetic constant k_{obs}, fit the data according to (5), in which f_0 denotes the small fraction of transcripts that might have ligated already before the addition of magnesium.

$$f_{cleaved} = f_0 + f_{\infty,L} * (1 - e^{-k^*t}) \qquad (5)$$

11. The constant k_{cis} has the unit 1/min. It allows together with the end point of the ligation reaction $f_{\infty,L}$ to compare different hammerhead ribozymes.

4. Notes

1. We do not use Diethylpyrocarbonate (DEPC) that can be applied to alkylate and thus inactivate RNases, as we find it dispensable if normal lab standards are maintained.

2. The addition of the triplet might interfere with folding of the RNA into the hammerhead structure. It therefore is essential to compare the motif's predicted secondary structure with and without the triplet by using Mfold (20) or any other RNA folding program. In case that the presence of the optimal GGG changes the folding, GGC or GCG can be used alternatively (18).

3. Tm values determined with a given program are comparable with another, but not necessarily between programs. One example calculator is available at http://mbcf.dfci.harvard.edu/docs/oligocalc.html.

4. Since N has to be an even number, one can round the calculated value. For example, 100 nt is covered by $N = 8/3 = 2.66$, i.e., two DNA oligonucleotides; for a template sequence of 180 nt, the number comes down to $N = 16/3 = 5.33$, i.e., six DNA oligonucleotides will be required.

5. Since the overlapping DNA oligonucleotides are required only for the sake of providing the backbone from which the entire sequence is generated, one can dilute them manifold (1:5, 1:10, 1:20…). This is important when additional bands appear after PCR that have another size than the expected. Presumably these unwanted products are derived from primer dimers or the like. In case dilution of the overlapping DNA oligonucleotides does not yield a uniform PCR product of expected size, one needs to gel purify the desired PCR product using a gel purification kit. For that purpose, bands are visualized on a UV table, excised with a sterile scalpel, and processed according to the instructions provided by the manufacturer of the gel purification kit.

6. For linearization of plasmid DNA that is to be used for subsequent in vitro run off transcription, restriction enzymes producing blunt ends or 5′ overhangs are better suited than those

generating 3′ overhangs, as these might lead to extended RNA transcripts.

7. Occasionally, the digest remains incomplete. In that case it is advisable to gel purify the linearized plasmid, as described (see Note 5).

8. Alternatively, a speed vac can be used.

9. Frequently, the addition of an inhibitory oligonucleotide reduces the yield of transcript, including the full-length RNA dramatically, and additional shorter RNAs are observed. In this case, gel purification of the hammerhead ribozyme is required. In case the template has been generated by recursive PCR, the DNA oligonucleotide marked by a star (asterisk, Fig. 4) can be used. For templates from genomic DNA, a separate oligonucleotide has to be ordered.

10. In the generation of the cleavage products as substrates for ligation reactions do not include the inhibitory oligonucleotide (asterisk, Fig. 4), as in this case, a high fraction of self-cleaved material is desirable.

11. The success of the in vitro transcription of radio-labeled RNA can be immediately assessed by measuring the fraction of radionuclides in the column compared to the screw cap reaction tube, using a Geiger–Müller counter.

12. The APS solution must be stored at 4°C, and should be prepared freshly every month.

13. If no "old" radio source is available, take a small amount of the (α-^{32}P) UTP.

14. The ethidium bromide solution might be reused if stored in a dark bottle.

15. It is essential to apply Mg^{2+} concentrations in the low mM, or better sub-mM range. Only if the hammerhead ribozyme under investigation cleaves at these concentrations, it is a full, tertiary stabilized motif (2, 3, 12, 21). Depending on the organism, from which the ribozyme stems, the reactions might also be carried out at 37°C.

16. Depending on the idiosyncracies of the investigated motif, times have to be adjusted (22).

17. The preferred Mg^{2+} concentrations used here is in the low mM range. This is attributable to the elevated RNA concentration that is present due to the pseudo-first-order reaction setup. Depending on the organism, from which the ribozyme stems, the reactions might also be carried out at 37°C.

References

1. Fedor, M.J. (2009) Comparative enzymology and structural biology of RNA self-cleavage. *Annu Rev Biophys.* **38**, 271–99.

2. De la Peña, M., S. Gago, and R. Flores (2003) Peripheral regions of natural hammerhead ribozymes greatly increase their self-cleavage activity. *EMBO J.* **22**, 5561–70.

3. Khvorova, A., A. Lescoute, E. Westhof, and S.D. Jayasena (2003) Sequence elements outside the hammerhead ribozyme catalytic core enable intracellular activity. *Nat Struct Biol.* **10**, 708–12.

4. Martick, M. and W.G. Scott (2006) Tertiary contacts distant from the active site prime a ribozyme for catalysis. *Cell.* **126**, 309–20.

5. Przybilski, R. and C. Hammann (2006) The Hammerhead Ribozyme Structure Brought in Line. *ChemBioChem.* **7**, 1641–1644.

6. Hutchins, C.J., P.D. Rathjen, A.C. Forster, and R.H. Symons (1986) Self-cleavage of plus and minus RNA transcripts of avocado sunblotch viroid. *Nucleic Acids Res.* **14**, 3627–3640.

7. Prody, G.A., J.T. Bakos, J.M. Buzayan, I.R. Schneider, and G. Bruening (1986) Autolytic processing of dimeric plant virus satellite RNA. *Science.* **231**, 1577–1580.

8. Branch, A.D. and H.D. Robertson (1984) A replication cycle for viroids and other small infectious RNAs. *Science.* **223**, 450–455.

9. Hertel, K.J., A. Pardi, O.C. Uhlenbeck, M. Koizumi, E. Ohtsuka, S. Uesugi, et al. (1992) Numbering system for the hammerhead. *Nucleic Acids Res.* **20**, 3252.

10. Jimenez and Luptak, Structure-Based Search and In Vitro Analysis of Self-Cleaving Ribozymes, Chapter 9, this volume.

11. Jimenez et al., Discovery of RNA Motifs Using a Computational Pipeline that Allows Insertion in Paired Regions and Filtering of Candidate Sequences, Chapter 10, this volume

12. Przybilski, R., S. Gräf, A. Lescoute, W. Nellen, E. Westhof, G. Steger, et al. (2005) Functional Hammerhead Ribozymes Naturally Encoded in the Genome of Arabidopsis thaliana. *Plant Cell.* **17**, 1877–1885.

13. Martick, M., L.H. Horan, H.F. Noller, and W.G. Scott (2008) A discontinuous hammerhead ribozyme embedded in a mammalian messenger RNA. *Nature.* **454**, 899–902.

14. De la Peña, M. and I. Garcia-Robles (2010) Intronic hammerhead ribozymes are ultraconserved in the human genome. *EMBO Rep.*

15. De la Peña, M. and I. Garcia-Robles (2010) Ubiquitous presence of the hammerhead ribozyme motif along the tree of life. *RNA.* **16**, 1943–1950.

16. Seehafer, C., A. Kalweit, G. Steger, S. Gräf, and C. Hammann (2011) From alpaca to zebrafish: hammerhead ribozymes wherever you look. *RNA.* **17**, 21–26.

17. Penedo, J.C., T.J. Wilson, S.D. Jayasena, A. Khvorova, and D.M. Lilley (2004) Folding of the natural hammerhead ribozyme is enhanced by interaction of auxiliary elements. *RNA.* **10**, 880–8.

18. Weber, U. and H.J. Gross, *In vitro RNAs*, in *Antisense Technology, A Practical Approach*, C. Lichtenstein and N. W., Editors. 1997, Oxford University Press: Oxford. p. 75–92.

19. Sambrook, J., E.F. Fritsch, and T. Maniatis, *Molecular Cloning: a laboratory manual.* 2 nd ed. 1989, New York: Cold Spring Harbor Press.

20. Zuker, M. (2003) Mfold web server for nucleic acid folding and hybridization prediction. *Nucleic Acids Res.* **31**, 3406–15.

21. Przybilski, R. and C. Hammann (2007) The tolerance to exchanges of the Watson/Crick basepair in the hammerhead ribozyme core is determined by surrounding elements. *RNA.* **13**, 1625–1630.

22. Przybilski, R. and C. Hammann (2007) Idiosyncratic cleavage and ligation activity of individual hammerhead ribozymes and core sequence variants thereof. *Biological Chemistry.* **388**, 737–741.

Chapter 3

Mechanistic Analysis of the Hepatitis Delta Virus (HDV) Ribozyme: Methods for RNA Preparation, Structure Mapping, Solvent Isotope Effects, and Co-transcriptional Cleavage

Durga M. Chadalavada, Andrea L. Cerrone-Szakal, Jennifer L. Wilcox, Nathan A. Siegfried, and Philip C. Bevilacqua

Abstract

Small ribozymes such as the hairpin, hammerhead, VS, glm S, and hepatitis delta virus (HDV) are self-cleaving RNAs that are typically characterized by kinetics and structural methods. Working with these RNAs requires attention to numerous experimental details. In this chapter we focus on four different experimental aspects of ribozyme studies: preparing the RNA, mapping its structure with reverse transcription and end-labeled techniques, solvent isotope experiments, and co-transcriptional cleavage assays. Although the focus of these methods is the HDV ribozyme, the methods should be applicable to other ribozymes.

Key words: Co-transcriptional RNA folding, pK_a, Proton inventory, Sequencing, Solvent isotope

1. Introduction

There has been deep interest in the mechanisms of catalytic RNA (1–4). Our lab has focused on the structure and function of the hepatitis delta virus (HDV). This chapter summarizes some of the basic protocols used in our lab for studying the HDV ribozyme, with a strong focus on structure mapping, solvent isotope effect, and co-transcriptional cleavage experiments, which are experimentally complex. The four specific methods described here in detail are as follows: (1) cloning of DNA and preparation of RNA (5), (2) analysis of RNA sequence and structure by reverse transcription and end-labeled structure mapping (6–8), (3) characterization of phosphodiester bond cleavage by solvent isotope experiments,

Jörg S. Hartig (ed.), *Ribozymes: Methods and Protocols*, Methods in Molecular Biology, vol. 848,
DOI 10.1007/978-1-61779-545-9_3, © Springer Science+Business Media, LLC 2012

including proton inventories (9–11), and (4) analysis of ribozyme cleavage during transcription (12, 13). Basic kinetic equations relevant to methods 3 and 4 are provided in the text and are described in some detail.

The major procedures described in this chapter include the design and preparation of DNA templates for transcription of any RNA of interest, performance of reverse transcription and enzymatic probing of RNA, design of solvent isotope experiments including preparation of buffers and determination of true pH, and procedures for analyzing cleavage of RNA as it is being transcribed. We make a special effort to include details of procedures that are especially important or tricky both in Subheadings 2 and 3, as well as Subheading 4 at the end.

2. Materials

All solutions are prepared using ultrapure deionized water (Barnstead NANOpure DIamond Water Purification Systems, 18 MΩ) and analytical grade reagents. To assure no RNases are present in the water, we periodically incubate various radiolabeled RNAs in the (buffered) water at 37°C for 24 h and test for absence of degradation, analyzing samples by polyacrylamide gel electrophoresis (PAGE) analysis. In addition, we test for trace amounts of certain metal ions (Li^+, K^+, Na^+, Mg^{2+}, Ca^{2+}) by atomic absorption (AA), and note their concentrations, which are typically found to be 1 μM or less. We have described standard methodologies and precautions used in our laboratory for working with RNA previously (5); researchers new to working with RNA may wish to consult that article. In general, precautions include wearing gloves and using RNase-free plasticware and glassware. Unless otherwise indicated, carry out all procedures on ice.

2.1. Cloning

1. *Overlap extension to generate insert*: Reconstitute (Integrated DNA Technologies, Coralville, IA, USA) (IDT) in TE (10 mM Tris–HCl (pH 7.5), 1 mM EDTA) and (refer to Section 2.2.3). A typical 50 μL overlap extension reaction uses each primer at a concentration of 0.4 μM, 1 mM dNTP mix, and 1 μL of Taq DNA polymerase (New England Biolabs, Ipswich, MA, USA) (NEB), in 1× buffer provided by the enzyme manufacturer. A 12-cycle (3 step) temperature cycling program is used for overlap extension (see Note 1). In step 1, denature primers at 94°C for 5 min. In step 2, cycle reaction mixture 12 times through the temperatures 94°C (denaturation), 58°C (annealing), and 72°C (extension) for 1 min at each temperature (see Note 2). Step 3 is the final extension, and should be carried out at 72°C for 10 min.

2. *Agarose gel conditions*: Run 1.5% agarose gels in 1× TAE gel running buffer (40 mM Tris–HCl-acetate, 1 mM EDTA) at 80 mA for 2 h.

3. *Ligation of insert into vector to generate recombinant plasmid*: Digest both the purified overlap extension product and the vector chosen for cloning, which have been gel purified, with the two appropriate restriction enzymes. Dephosphorylate the vector (e.g., pUC19) using calf intestinal phosphatase (CIP) from NEB using standard protocols (see Note 3). For the ligation reaction, use vector to insert ratios of 1:3 and/or 1:10 to find the best conditions (i.e., those that give approximately 50 colonies). Use the vendor-provided reaction buffer along with 200 units of T4 DNA ligase (NEB) in a 10 µL reaction (see Note 4). Carry out reactions at 16°C for 16–18 h.

4. *Transformation protocol*: Transform ligated plasmid samples into chemically competent DH-5α cells. For each transformation, add 4 µL of ligation reaction to a tube (~50 µL) of competent cells that has been thawed on ice. Leave tubes on ice for an additional 10 min, followed by a 1 min heat shock at 37°C, and incubation on ice for 2 min. Revive cells by addition of 800 µL LB media and constant shaking for 1 h at 37°C. Plate 100 µL of this sample on LB plates containing the appropriate antibiotic and incubate overnight at 37°C (see Note 5).

5. *Plasmid purification*: Obtain purified plasmids from a few select colonies using Qiagen (Valencia, CA, USA) mini- or maxiprep kits. Use this plasmid for sequence analysis, and once the correct sequence is confirmed, use this plasmid for run-off digestion and co-transcriptional assays.

2.2. Reverse Transcription and Structure Mapping

1. *Primer design*: Design a 20–25 nucleotide DNA primer to be complementary to the 3′-end of the 1/99 HDV ribozyme.

2. *5′ ^{32}P labeling of the DNA*: Obtain DNA primer from IDT and use without prior purification. Dissolve primer in 1× TE and quantify by UV absorbance at 260 nm. 5′-end-label the primer with [γ-^{32}P] ATP (Perkin Elmer, Waltham, MA, USA) at 25 µM each, along with T4 polynucleotide kinase and 1× buffer (NEB) and perform PAGE purification (5).

3. *RNA preparation by in vitro transcription reactions*: Digest purified –54/99 plasmid DNA (20–50 µg) (6) with *Bfa* I restriction endonuclease at a concentration of 2 units/µg of DNA in the appropriate buffer supplied by NEB. Use a typical reaction volume of 200 µL, and allow reaction to proceed for 2 h at 37°C. Purify the digested DNA by phenol/chloroform extraction and ethanol precipitation/70% ethanol wash (ice cold). Resuspend precipitated DNA in 1× TE to give a final

concentration of 1 μg/μL. The –54/99 RNA is transcribed using phage T7 polymerase and the MEGAscript high yield transcription kit (Applied Biosystems/Life Technologies, Carlsbad, CA, USA). During transcription, this RNA self-cleaves to generate the –54/–1 upstream RNA and 1/99 downstream HDV RNAs, where 1/99 contains the catalytic core of the ribozyme. Carry out transcription reactions for 4 h at 37°C and quench by addition of EDTA/95% formamide loading buffer (see Note 6). Fractionate the RNA on a denaturing PAGE gel, visualize by UV shadowing, and excise the 1/99 RNA band from the gel. Elute the RNA into 1× TE containing 250 mM NaCl by soaking the gel piece in the buffer overnight with agitation (see Note 7). Isolate the RNA from the buffer by ethanol precipitation and dissolve the resulting pellet in 1× TE.

4. *2.5× Annealing mix*: The following 2.5× annealing mix is used in item 6. This mix contains each dNTP at 10 mM, 2 μM HDV RNA, and 0.04 μM ^{32}P-labeled DNA primer. Set up a 12 μL reaction 2.5× annealing mix to carry out one complete sequencing reaction. Renature the 2.5× annealing mix at 75°C for 3 min, then slow cool at room temperature for 10 min.

5. *2.5× Enzyme buffer mix*: Prepare a 2.5× Reverse Transcription (RT) buffer enzyme mix using 1.5 μL M-MuLV reverse transcriptase and 3 μL of the 10× RT buffer (both from NEB) at a final concentration of 2.5×.

6. *Sequencing reactions*: Prior to setting up the sequencing reactions dilute each ddNTP (Affymetrix, CA, USA) (USB) from a stock of 10 mM to a 5× stock of 0.5 mM. Set up individual reactions for each ddNTP. Each sequencing reaction should contain 2 μL of the 2.5× annealing mix (with RNA, primer, and dNTPs), 2 μL of the 2.5× enzyme buffer mix, and 1 μL of the appropriate 5× ddNTP. Incubate the reactions for 1 h at 42°C. In addition to the four sequencing reactions, set up an additional blank reaction that substitutes water for ddNTP; this allows the monitoring of any structural stops that may be observed in the RNA.

7. *RNA preparation for enzymatic structure mapping*: Carry out transcription reactions as described above. For enzymatic structure mapping, end-label the RNA. For 5'-end-labeling, dephosphorylate the RNA using CIP (NEB) and label with [γ-^{32}P] ATP and T4 polynucleotide kinase (NEB); both steps are described above (see Note 8). Repurify the labeled RNA on a denaturing gel, ethanol precipitate, wash in ice cold 70% ethanol, and dissolve the pellet in 1× TE.

8. *Structure mapping experiments*: Use the 5'-end-labeled RNA generated above for structure mapping in conjunction with several commercially available RNases (purchased from Applied

Biosystems/Ambion, Austin, TX, USA). Typically, complete structure mapping requires the use of several single- and double-strand specific RNases such as RNase T1 and A (ss-specific), and RNase V1 (ds-specific). Dilute RNases with 50% (v/v) glycerol and use a wide range of concentrations to search for optimal, single-hit kinetics (see Note 9). Prior to structure mapping, renature the RNA in 0.5 mM Tris–HCl (pH 7.5), 0.05 mM EDTA, 100 mM KCl, and 10 mM MgCl$_2$ at 55°C for 5 min and slow cool at room temperature for 10 min. After addition of RNases, incubate the reaction mixture for 15 min at 37°C, then add 95% formamide loading buffer with 100 mM EDTA, and immediately freeze in dry ice (see Note 10). Prepare G sequencing lanes by limited digestion with RNase T1 (1 unit/μL) under denaturing conditions and incubate at 50°C for 15 min (see Note 11). Freshly prepare an alkaline hydrolysis ladder by heating the RNA in hydrolysis buffer (50 mM Na$_2$CO$_3$/NaHCO$_3$ (pH 9.0)) and 1 mM EDTA at 90°C for 5 min (see Note 12).

9. *Denaturing polyacrylamide gel*: 20% acrylamide/bis solution (29:1) in 1× TBE (100 mM Tris–HCl, 83 mM boric acid, 1 mM EDTA)/8.4 M urea (see Note 13). To make sure the gel is extremely denaturing, load samples immediately after boiling them and blowing out the wells. Make solutions from 40% polyacrylamide and 10× TBE stocks, filter through a 0.2-μm sterile filter, and store at 4°C in a dark bottle for up to 6 months (see Note 14). Gels of a desired percentage are prepared by dilutions with 1× TBE/8.4 M urea.

10. *Ammonium persulfate*: Prepare a 10% (w/v) solution in deionized water and filter through a 0.2-μM sterile filter. Store this solution at 4°C for no longer than 2 weeks.

11. *10× TBE running buffer*: The 10× solution is 1 M Tris–HCl, 0.83 M boric acid, 10 mM EDTA. Filter the solution through a 0.2-μm sterile filter. Prior to use, dilute it with nine parts deionized water and use as a 1× solution.

12. *2× Formamide loading buffer*: The 2× solution is 95% (v/v) formamide, 10 mM EDTA, 0.05% (w/v) bromophenol blue, and 0.05% (w/v) xylene cyanol. Store the solution at 4°C. Discard unused solution after 2 months in hazardous waste.

2.3. Solvent Isotope Effects in the Absence of Divalent Ions

1. *General notes*: Reagents specific for solvent isotope effects are described here, whereas other general reagents are as described in the previous sections, including the details on DNA preparation (see Subheading 2.1), RNA preparation and end-labeling (see Subheading 2.2), and denaturing PAGE (see Subheading 2.2).

2. *Heavy water*: Use deuterium oxide (D$_2$O) (D, 99.9%) (Cambridge Isotope Laboratories, Andover, MA, USA) for all sample preparations that involve D$_2$O.

3. *Preparation of run-off transcription template by plasmid DNA digestion*: Digest purified plasmid DNA (20–50 μg) with *Bfa* I at a concentration of 2 units/μg of DNA in the appropriate buffer. Use a typical reaction volume of 200 μL and allow reactions to proceed at 37°C for 2 h. Purify the DNA by phenol/chloroform extraction and ethanol precipitation. Resuspend the precipitated DNA in 1× TE to give a final concentration of 1 μg/μL.

4. *RNA construct used in this study*: Transcribe the –30/99 genomic HDV RNA from pT7 –30/99 using phage T7 polymerase and the MEGAscript high yield transcription kit (Applied Biosystems) (11). Generate all mutant plasmids from pT7 –30/99 using the QuikChange kit (Stratagene/Agilent Technologies, Santa Clara, CA, USA). Confirm sequences by dideoxy sequencing after both minipreps and maxipreps (Qiagen). All transcripts contained a G11C mutation that biases the equilibrium between Alt P1 and P1 toward the native fold (6). We studied the ribozyme with only the G11C change and refer to this construct as wild-type (WT). We also studied a ribozyme containing a double mutation (C44U:G73A) in the base quadruple motif and refer to this construct as the double mutant (DM).

5. *Antisense DNA oligomer*: Use a synthetic DNA antisense oligonucleotide, AS (–30/–7), to facilitate the reaction by releasing the Alt 1 mispairing (6). Use this oligonucleotide without further purification, dissolve in H_2O, and split into two aliquots. Dry one aliquot, resuspend in D_2O, and repeat to ensure that all H_2O has been removed.

6. *End-labeling of the transcribed RNA*: End-label the RNA as described above with the following exceptions: Store the 5′-end-labeled RNA in H_2O, then split into two aliquots. Dry one aliquot and resuspend in D_2O, and repeat to ensure that all H_2O has been removed.

7. *Reaction buffers*: We studied the ribozyme reaction as a function of pL over the pL range of ~5.5–9.5 and collected data at ~12 pL values depending on the ribozyme construct and the solvent. Use 25 mM MES for experiments at pL~5–7 and 25 mM HEPES for experiments at pL~7–10, where L=H or D. Prepare the following stock solutions:

(a) 250 mM MES, 250 mM HEPES

(b) 2× ME (50 mM MES/200 mM Na_2EDTA)

(c) 2× HE (50 mM HEPES/200 mM Na_2EDTA)

(d) 1.2× MEN (25 mM MES/100 mM Na_2EDTA/1.2 M NaCl), and 1.2× HEN (25 mM HEPES/100 mM Na_2EDTA/1.2 M NaCl) (see Note 15). Choose the 12 pL values you want to work at; for each H_2O buffer also make a D_2O buffer at the same pL. Start by preparing 250 mM MES in both H_2O and D_2O and 250 mM HEPES in both H_2O and D_2O. For example, to make 36 mL of 250 mM MES in H_2O, dissolve 1.76 g of MES (acid form) in enough H_2O to give 36 mL, and to make 36 mL of 250 mM HEPES in H_2O, dissolve 2.15 g of HEPES (acid form) in enough H_2O to give 36 mL. Then prepare the 2× and 1.2× buffers described above. For example, to make 10 mL of 2× ME at six different pH (or pD) values, mix 12 mL 250 mM MES, 4.46 g $Na_2EDTA\cdot2H_2O$, and 35 mL H_2O (or D_2O). Heat and stir until the solids dissolve. Adjust the volume to 54 mL with H_2O (or D_2O). Split into 6×9 mL aliquots. Adjust each aliquot to the desired pH (or pD) value using 1–4 M solutions of HCl or NaOH prepared with H_2O (or D_2O) (see next step). Adjust the volume to 10 mL with H_2O (or D_2O). Filter (0.2 μm).

8. *pH measurements*: Set up a 37°C (the temperature used for the ribozyme reactions) water bath by the pH meter. Incubate the pH standards and buffers to be measured in the water bath until they reach 37°C. Calibrate the pH meter and measure the pH of the buffers with the solutions at 37°C. Using temperature compensation (ATC or manual) and calibrating the electrode at the same temperature as the solutions to be measured will eliminate most of the temperature-dependent measurement error from the electrode. Since ionic strength, which is also varied in most ribozyme reactions, affects both the activity of H^+ in solution and electrode response, meter readings need to be converted to true pH values. For the reactions in D_2O, the pD is determined by adding 0.4 to the pH reading (14).

9. *Stock solutions for pH calibration curves*: Prepare a series of solutions of known H^+ concentration and variable salt conditions and measure the meter response to these (15). Prepare fresh solutions of HCl and NaOH or KOH to minimize neutralization by CO_2 from air.

10. *Calculate true pH values*: Convert concentrations of H^+ known from preparations in the previous step to $-\log a_{H+}$ (=pH) using the equations for activity coefficients presented in Subheading 3 (15, 16).

11. *Meter readings*: We obtained the meter reading of each stock using a stainless steel electrode (electrode model PH47-SS and meter model IQ150 (IQ Scientific Instruments, Loveland, CO, USA)).

12. *Calibration curve*: Construct a calibration curve of pH (item 10) vs. meter reading (item 11) at each ionic strength of interest, and use this curve to convert reaction buffer meter readings to true pH values.

13. *Description of terms used in this section*: OL^-, lyoxide ion, which is either OH^- or OD^-, depending on solvent composition; pK_a, K_a is an acid dissociation constant; k_{obs}, observed rate constant for self-cleavage; KSIE, kinetic solvent isotope effect; n, the mole fraction of D_2O in solution; nt, nucleotide; pL, pH or pD; WT, wild-type ribozyme, which is in the background of the fast-reacting, monophasic G11C change; ϕ^E, equilibrium fractionation factor, which allows the equilibrium isotope effect of K_H/K_D to be calculated from $K_H/K_D = 1/\phi^E$; ϕ^T, transition state fractionation factor, which describes the deuterium preference of a particular site relative to that of an average bulk water molecule.

2.4. Co-transcriptional Assays

1. *Run-off digestion*: Digest purified plasmid DNA (20–50 μg) with *Bfa* I as described. Extract the digested DNA with phenol/chloroform and ethanol precipitate. Resuspend DNA in 1× TE to give a final concentration of 1 μg/μL.

2. *Co-transcriptional assays*: Set up reactions containing 1 μg digested DNA, 600 μM each NTP (see Note 16), 20 mM Tris–HCl (pH 8.0), 100 mM KCl, 10 mM $MgCl_2$, 1 mM DTT, 100 μg/mL acetylated bovine serum albumin (BSA), and 10 μCi [α-^{32}P] GTP (Perkin Elmer, Waltham, MA, USA) in a 50 μL reaction. Incubate at 37°C for 2 min, and initiate reaction by the addition of 5% T7 RNA polymerase. For most reactions, time points ranged from 15 s to 1 h. At specific time intervals, withdraw 3 μL aliquots and quench with 3 μL of formamide loading buffer containing 0.1 M EDTA. To ensure complete termination of the reaction, transfer tubes to dry ice until all time points have been collected. Fractionate quenched reactions on a denaturing PAGE gel that is run at 25W in 1× TBE for ~2 h at room temperature. The gel image a is scanned using a PhosphorImager Typhoon 9410 (GE, Piscataway, NJ, USA).

3. *Denaturing polyacrylamide gel*: 20% acrylamide/bis solution (29:1) in 1× TBE/7 M urea. Solutions are prepared as described in 2.2.9, and filtered through a 0.2-μm sterile filter and stored at 4° C.

3. Methods

Unless otherwise indicated, carry out all procedures on ice. Also, to protect against RNase contamination, wear gloves and use RNase-free plasticware and glassware.

3.1. Cloning

1. *Overview*: In order to generate RNA for structural and kinetic analysis, the first step is to design a DNA version of the RNA

of interest. This dsDNA is then established in a plasmid vector using routine cloning. Following are the steps we use to design the DNA and obtain this vector.

2. If the DNA insert sequence is ~100 bp long, order the top and bottom strands corresponding to the digested version of the sequence directly from the supplier, which obviates the need to digest the insert (see Note 17). However, if the sequence is longer, generate a double stranded version of the insert by overlap extension (see Subheading 2.1). In the case of the HDV ribozyme, we generated the –54/99 ribozyme by overlap extension using two ~110 nt oligonucleotides with a 20 base pair overlap (6). Fractionate this double-stranded overlap extension on an agarose gel and purify using a gel extraction kit (Qiagen).

3. Introduce unique restriction sites at the 5′- (*Eco* RI) and 3′-ends (*Bam* HI) of the dsDNA fragment. Start and end the double-stranded insert with a fixed 4-nucleotide region (GGCC) to aid efficiency of the restriction endonuclease digestion.

4. Include a T7 promoter sequence (5′TAATACGACTCACTATA′) directly upstream of the insert, which should start with a few G's for optimal yields (17).

5. Lastly, include an inverted *Bsa* I site (5′(N)$_5$GAGACC3′) to ensure that the transcribed RNA does not contain additional vector sequences. Place immediately downstream of the insert to serve as a run-off site (8). In the case of the HDV ribozyme, a naturally present *Bfa* I site was exploited for run-off digestion (6). To summarize, the above steps result in a typical primer composed of the following elements: 5′ N4—upstream cloning site—T7 promoter—DNA sequence—*Bsa* I run-off site—downstream cloning site—N4 3′.

6. Once the dsDNA insert has been generated, clone it into any potential high-copy number vector such as pUC19. First, digest both the insert and vector with the appropriate restriction enzymes, then ligate and transform into *Escherichia coli*; run a 'no-added insert' control; and ideally there should be no (or very few) colonies on this plate (see Note 18). Purify and sequence the plasmid from the resulting clones, after both minipreps *and* maxipreps (Qiagen) to control for revertants. Generate a glycerol stock (15% glycerol) of the selected clone and keep frozen at –80°C; this recombinant-containing cell stock thereby serves as a reservoir of template DNA.

3.2. Structural Analysis of HDV by Enzymatic Digestions and RT

1. *Reaction termination*: After conducting both RT sequencing and enzymatic structure mapping reactions, quench with an equal volume of 2× formamide loading buffer (95% formamide, 0.05% xylene cyanol, 0.05% bromophenol blue tracking dyes, and 100 mM EDTA) and either store the reactions overnight at 4°C or use immediately.

2. *Gel electrophoresis*: Dilute the stock acrylamide/bis solution as appropriate and use to pour 8% (for RT reactions) and 12% (for structure mapping) denaturing sequencing gels. Pre-run the gels on a sequencing gel apparatus (Kodak BioMax STS 45i) at 100 V for 1 h to achieve very denaturing conditions. Blow settled urea out of the wells both before pre-running and before loading samples by squirting 1× TBE running buffer from the gel apparatus into the wells with a syringe.

3. *Sample loading*: Prior to loading the gel, heat the samples to 90°C for 3 min to denature, and load immediately. Electrophorese at 100 V for approximately 1 h, or until the bottom dye (bromophenol blue) has traveled most of the way down the gel.

3.3. Mechanistic Analysis of HDV Self-Cleavage Using Solvent Isotope Effects in the Absence of Divalent Ions

1. *Ribozyme kinetics in the absence of added divalent metal ions*: A typical self-cleavage reaction contains 2 nM 5′-end-labeled RNA, 25 mM buffer, 100 mM Na_2EDTA, 10 µM AS (−30/−7), and 1 M NaCl (11) in a 50 µL reaction volume. Renature the RNA at 55°C for 10 min in the presence of AS (−30/−7) and H_2O or D_2O as appropriate, then cool at room temperature for 10 min. Dilute renatured RNA with an equal volume of 2× ME or 2× HE buffer, and incubate at 37°C for 2 min. Remove a zero time-point (2 µL) and initiate self-cleavage by the addition of an appropriate volume of 1.2× MEN or 1.2× HEN buffer. Quench all time points by mixing with an equal volume of 95% (v/v) formamide loading buffer (no EDTA) and immediately placing on dry ice. Under these conditions, the reactions proceed at rates on the order of $10^{-3}–10^{-5}$/min (11), and so all reactions are performed for 48 h. At the end of each reaction in H_2O, check the pH of the mixture with pH paper to confirm that the correct pH is maintained throughout the reaction. Fractionate time points on a denaturing 10% (w/v) polyacrylamide gel. Dry gels and visualize using a PhosphorImager (Molecular Dynamics).

2. *Data fitting*: Construct plots of fraction product vs. time. Most of the WT ribozyme reactions could be fit to the single-exponential equation (see Eq. 1), where f is the fraction of ribozyme cleaved, A is the fraction of ribozyme cleaved at completion (usually ~80%), $A + B$ is the burst fraction (in all cases, $A + B \approx 0$), k_{obs} is the observed first-order rate constant, and t is time.

$$f = A + Be^{-k_{obs}t} \qquad (1)$$

Obtain parameters using nonlinear least-squares fitting by KaleidaGraph (Synergy Software). Typically, collect 3–6 half-lives of data. For the WT ribozyme reactions at pL~9–10 and most of the DM ribozyme reactions, plots of fraction product vs. time were linear, representing the early portion of an exponential time course. At early time, Eq. 1 reduces down to Eq. 2:

$$f_{early} = Ak_{obs}t \tag{2}$$

Such plots are fit to Eq. 3 where m, the slope of the line, is Ak_{obs} and b is the y-intercept:

$$f_{early} = mt + b \tag{3}$$

Obtain k_{obs} by dividing m by A, which is assumed to be 80% on the basis of time courses that could be fit using Eq. 1. The pL~5.5, 6.5, 7.5, 8.7, and 9.7 data points in each rate-pL profile are performed in duplicate and on different days, and nearly the same rates are obtained in all cases, providing confidence in the rate-pL profiles reported (11). In general, the equations used to fit the rate-pL profiles are derived from schemes involving the minimal number of protonation or deprotonation events necessary to arrive at the active ribozyme species. The log rate-pH profile for WT is fit according to appropriate equations derived from reactions schemes, as described elsewhere (11).

3. *Proton inventory experiments*: The reactions are performed largely as described above. Experiments are performed in the plateau regions of the log rate-pL profiles (pL 6.6 for WT and pL 9.7 for both WT and DM). Appropriate volumes of RNA, AS ($-30/-7$), and buffers in H_2O or D_2O are mixed to obtain the desired mole fraction (n) of D_2O. Choose at least eight n-values to work at for each pL.

4. *Sample calculations for proton inventories*: First calculate n for the buffers made with D_2O (i.e., "100%" D_2O, which turns out to be less than 100% owing to protons from buffer and EDTA). For our studies, we assumed the entire volume of buffer (10 mL) to be D_2O and calculated the moles of D_2O using the density of D_2O (1.107 g/mL) and the molecular weight of D_2O (20.03 g/mol). A more accurate way would be to measure the mass of D_2O used, but we did not take that approach because the masses of solids dissolved are small. We accounted for the exchangeable protons that come from the buffer (1 from MES and 2 from HEPES) and $Na_2EDTA{\cdot}2H_2O$ (6 from $Na_2EDTA{\cdot}2H_2O$). We wrote out the calculations in terms of mole fraction D (atom), which is mathematically equivalent to mole fraction D_2O. For example, *for 2× ME*:

D_2O: 10 mL buffer (1.107 g D_2O/mL) (1 mol D_2O/20.03 g) (2 mol D/1 mol D_2O) = 1.105 mol D

MES: 10 mL buffer (1 L/10^3 mL) (50 mmol MES/L) (1 mol/10^3 mmol) (1 mol H/mol MES) = 0.0005 mol H

$Na_2EDTA \cdot 2H_2O$: 10 mL buffer (1 L/10^3 mL) (200 mmol $Na_2EDTA \cdot 2H_2O$/L) (1 mol/10^3 mmol) (6 mol H/mol $Na_2EDTA \cdot 2H_2O$) = 0.012 mol H

$$n = 1.105 / (1.105 + 0.0005 + 0.012) = 0.989 \text{ in } 2 \times ME$$

Similarly, for *1.2× MEN*:

D_2O: 10 mL buffer (1.107 g D_2O/mL) (1 mol D_2O/20.03 g) (2 mol D/1 mol D_2O) = 1.105 mol D

MES: 10 mL buffer (1 L/10^3 mL) (25 mmol MES/L) (1 mol/10^3 mmol) (1 mol H/mol MES) = 0.00025 mol H

$Na_2EDTA \cdot 2H_2O$: 10 mL buffer (1 L/10^3 mL) (100 mmol $Na_2EDTA \cdot 2H_2O$/L) (1 mol/10^3 mmol) (6 mol H/mol $Na_2ED\text{-}TA \cdot 2H_2O$) = 0.006 mol H

$$n = 1.105 / (1.105 + 0.00025 + 0.006) = 0.994 \text{ in } 1.2 \times MEN$$

For a 50 μL reaction in "100%" D_2O MES buffer:

5 μL 100 μM AS(−30/−7) in 99.9% D_2O ($n = 0.999$)

1 μL RNA in 99.9% D_2O (ignore volume contribution)

5 μL 2× ME ($n = 0.989$)

At this point,

$$n = ((5 \text{ μL})(n = 0.999) + (5 \text{ μL})(n = 0.989)) / 10 \text{ μL or } n = 0.994$$

Remove zero time-point (2 μL), leaving 8 μL (with $n = 0.994$).

Next, we need 42 μL 1.2× MEN ($n = 0.994$).

For the final reaction mix,

$n = ((8 \text{ μL})(0.994) + (42 \text{ μL})(0.994))/50 \text{ μL} = 0.994$. This is the reaction for the $n = 0.994$ point on the proton inventory curve.

Second, consider steps to prepare other points on the proton inventory curve. Prepare 1.2× buffer at the desired n value by mixing appropriate volumes of 1.2× H_2O and 1.2× D_2O buffers at a particular pL (technically, the buffers do not have to be at the same pL value, but they have to be at pL values in the plateau region of the log rate-pL profile).

For example, to make 500 μL of 1.2× MEN at $n = 0.1$:

$$0.1 = x(0.994) / 500 \text{ μL}, \quad x = 50.3 \text{ μL}$$

So mix 50.3 μL 1.2× MEN (D_2O) with 449.7 μL 1.2× MEN (H_2O).

Before adding the 42 µL of 1.2× buffer at $n=0.1$ to the 8 µL of reaction mix (see above reaction), the 8 µL of reaction mix should be at the desired n value.

For example, to make the 8 µL of reaction mix at $n=0.1$:

5 µL 100 µM AS(–30/–7) in H_2O ($n=0$)

1 µL RNA in H_2O (ignore volume contribution)

5 µL 2× buffer in H_2O/D_2O ($n=\times$)

$$n \text{ for 8 µL of reaction mix} = 0.1 = ((5\,\mu L)(n=0)$$
$$+(5\,\mu L)(n=\times))\,/\,10\,\mu L, \times = 0.2$$

So you need to make 2× buffer at $n=0.2$.

For example, to make 500 µL of 2× ME at $n=0.2$:

$$0.2 = x(0.989)\,/\,500\,\mu L, \quad x = 101.1\,\mu L$$

So mix 101.1 µL 2×ME (D_2O) with 398.9 µL 2× ME (H_2O). Now you are set to initiate the reaction by adding 42 µL of 1.2× MEN ($n=0.1$) with 8 µL of RNA and AS oligonucleotide in 1×ME ($n=0.1$).

Proton inventory data at low pL is fit using the standard Gross–Butler equation. This equation represents the simplest mathematical treatment of proton inventories possible. Derivations of four forms of this equation from simple physical models have been derived elsewhere (11), including a more complex population-weighted form developed especially for the system at hand, as well as its simpler non-population-weighted one- and two-proton transfer limits. For the case of two proton transfers, A and B, with nonequivalent transition state fractionation factors (ϕ^T) this equation can be written as (18):

$$k_n\,/\,k_0 = (1-n+n\phi_A^T)(1-n+n\phi_B^T) \tag{4}$$

where k_0 is the observed reaction rate in H_2O and k_n is the observed reaction rate in a solution with mole fraction (n) of D_2O. Because of limits in accuracy of data, one often assumes equivalent transition state fractionation factors (10, 14, 19), which causes Eq. 4 to reduce to

$$k_n\,/\,k_0 = (1-n+n\phi^T)^m \tag{5}$$

with $m=2$ for two proton transfers and $m=1$ for one proton transfer.

The standard graphical predictions for experimental implementation of Eq. 5 are as follows. For a one-proton transfer system, a plot of k_n/k_0 vs. n will be linear with a slope of ϕ^T-1 and an intercept of unity, and for a two-proton transfer system a plot of $(k_n/k_0)^{1/2}$ vs. n will be linear with the slope and intercepts given above. A plot of k_n/k_0 vs. n for a two-proton system will be concave-up or "bowl-shaped."

Because the plateau at high pL is due to a compensation of concentrations of the functional forms of an acid and base, proton inventory data at this pL can be fit to a population-weighted Gross–Butler equation (see Eq. 6), as well as the standard Gross–Butler equation (see Eq. 5):

$$k_n / k_0 = (1-n)^2 + \phi^{T}_{C75}n(1-n)10^{\Delta pK_{C75}} + \phi^{E}_{OL^-}n(1-n)$$
$$10^{-\Delta pK_{LOL}} + \phi^{T}_{C75}\phi^{E}_{OL^-}n^2 10^{\Delta pK_{C75}-\Delta pK_{LOL}} \qquad (6)$$

where ribozymes having an acid and a base with HH, DH, HD (this is done to have a correction with equation 6), and DD compositions are represented by each of the four respective terms in Eq. 6, and $\phi^{E} = K_{D}/K_{H}$ is termed an "equilibrium frac-tionation factor" and is defined as a ratio of ground-state frac-tionation factors and provides a WT inverse equilibrium isotope effect of K_{H}/K_{D} of 0.43 (11) (see Note 19). The ΔpK terms are for either ionization of C75 or autoprotolysis of solvent, and each is equal to the value in D_2O minus the value in H_2O. Specifically, ΔpK_{C75} was found to be +0.44 (=7.92–7.48), whereas ΔpK_{LOL} was +0.85 (=14.23–13.38) (11).

5. *Correction of meter readings to pH values.* Construct a plot of pH vs. meter reading at each ionic strength of interest using the custom calibration standards and pH meter methods described in Subheading 2. The following procedure accounts for both the activity of H^+ in solution and the electrode response. First, calculate the true pH using the following equa-tions. The pH of a solution is defined as

$$pH = -\log a_{H^+} \qquad (7)$$

The relationship between a_{H^+} and c_{H^+} is given by

$$a_{H^+} = c_{H^+} f_{H^+} \qquad (8)$$

and f_{H^+} can be estimated by an approximate form of the Debye–Hückel equation (at 25°C), where I is ionic strength of the solution (16),

$$-\log f_{H^+} = \frac{0.51\sqrt{I}}{1+\sqrt{I}} - 0.1I \qquad (9)$$

According to Eqs. 7–9, pH at a given ionic strength can be calculated from a known c_{H^+} and ionic strength by the following relationship,

$$pH = -\log c_{H^+} - \log f_{H^+} \qquad (10)$$

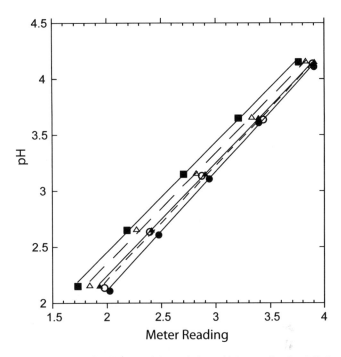

Fig. 1. Calibration curve for Na⁺-containing solutions. Meter reading is plotted on the x-axis and pH on the y-axis. Each *line* represents a different NaCl concentration: (*filled circles*) is 0.1 M, (*open circles*) is 0.2 M; (*filled triangles*) is 0.33 M; (*open triangles*) is 0.5 M; and (*open squares*) is 1.0 M. Reproduced from Siegfried (15) with permission from Nathan Siegfried.

As shown in Fig. 1, the deviations between pH and meter reading are greater for higher ionic strength in NaCl. It is noteworthy that such deviations are even greater at high pH than low (15), although all deviations are smaller in K⁺ than Na⁺ ion, presumably due to the sodium effect on meter response wherein Na⁺ is more similar in size to H⁺ (20).

3.4. Kinetic Analysis of HDV Self-Cleavage Using Co-transcriptional Assays

1. *Run-off digestion*: Linearize the plasmid with *Bfa* I to generate a run-off site and resuspend in 1× TE at an approximate concentration of 1 μg/μL for co-transcriptional assays.

2. *Co-transcriptional assays*: A typical co-transcriptional assay is carried out at 37°C in a total reaction volume of 50 μL. Reaction kinetics are followed during transcription by including trace amounts of [α-³²P]GTP (see Note 20). In the case of the HDV ribozyme, this allows visualization of three RNA species: the uncleaved full-length RNA (–54/99), the upstream cleavage fragment (–54/–1), and the downstream cleaved ribozyme (1/99) (see Fig. 2a). Dry, scan, and quantify gels to determine the extent of transcription and cleavage for the ribozyme.

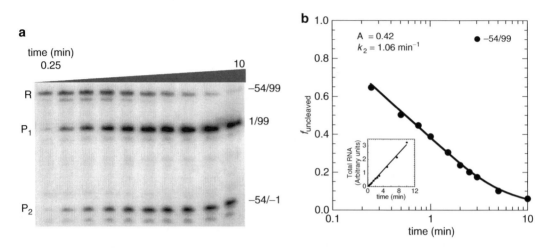

Fig. 2. Co-transcriptional self-cleavage reaction of −54/99 HDV ribozyme. (a) Denaturing gel electrophoresis of the coupled transcription and self-cleavage reaction. The full-length −54/99 RNA (R) is observed to decrease over time, with almost no reactant RNA remaining at 10 min (*last lane*). On the other hand, the product RNAs (P₁ and P₂) accumulate during the course of the reaction as seen by an increase in the intensity of the bands labeled 1/99 and −54/99. The RNA is uniformly labeled with [α-³²P]GTP during transcription. Time points for this reaction range from 15 s to 10 min. (b) A plot of fraction uncleaved vs. time is fit to Eq. 12, with $A = 0.42$, and $k_2 = 1.06$/min. Time points range from 15 s to 10 min and are an average of at least two trials. The *inset* shows that the rate of transcription is constant throughout the reaction, as demanded by Eq. 11. Adapted from Chadalavada et al. (6) with permission from Elsevier.

From these data, plots of fraction uncleaved vs. time can be constructed and used to obtain rate constants for the self-cleavage reaction (see Fig. 2b). The rate of transcription during the reaction can be monitored by generating "RNA transcribed" vs. "time" plots. Fit the data to standard co-transcriptional cleavage equations (see below), which requires the rate of transcription to be constant throughout the time course of the experiment (see Fig. 2b inset) (see Note 21).

3. *Data fitting*: The equation used to fit co-transcriptional data is based on an expression derived for the case where production of a full-length transcript is an intermediate in a two-step reaction; that of transcription followed by cleavage, in which there is no burst fraction and all of the RNA reacts (21).

$$f_{un} = \left(\frac{1 - e^{-k_{obs}t}}{k_{obs}t} \right) \tag{11}$$

where f_{un} is the fraction of precursor ribozyme remaining uncleaved; k_{obs} is the observed first-order rate constant for ribozyme cleavage; and t is time.

When kinetic profiles exhibit clear burst and non-burst fractions, the data are fit to the following modified equation,

$$f_{un} = A\left(\frac{1-e^{-k_1 t}}{k_1 t}\right) + (1-A)\left(\frac{1-e^{-k_2 t}}{k_2 t}\right) \qquad (12)$$

where k_1 and k_2 are the observed first-order rate constants for the burst and non-burst phases, respectively; t is time; and A is the burst fraction. For most of the HDV constructs, the burst phase rates could not be accurately determined because a significant fraction of the RNA was reacted by the first time point at 15 s.

4. Notes

1. Fewer cycles should suffice since there is no amplification here. If problems arise, especially for longer inserts, try fewer cycles since misannealing could occur if there is no extension preceding it. This is also true if there is a random region in the center of the insert.

2. An efficient way to determine the optimum annealing temperature is to use a thermal gradient cycler in which small volume (10 µL) reactions are carried out varying the annealing temperature every few degrees over a temperature range. If a gradient cycler is unavailable, reactions can be run every 5°C in annealing temperature in a regular cycler.

3. It is critical that CIP be inactivated before the next step because if it is left active it will remove the phosphates from the insert too, which will prevent ligation. We take extra steps to inactivate the CIP by using a two-step process: heat at 75°C for 10 min then perform a phenol-chloroform extraction to remove CIP.

4. We typically supplement this reaction with 1 mM 10 mM ATP solution (NEB), to offset ATP degradation in the reaction buffer over time.

5. If there is difficulty getting the colonies to grow, or they grow but with mutations, try carrying out all steps of the transformation and growing at 30°C. This has on occasion allowed us to obtain a recalcitrant mutant. Once established, such plasmids can typically be handled at 37°C.

6. A small fraction of the transcription reaction can be run on an analytical denaturing PAGE to look for desired completion of the self-cleavage reaction. If not complete, further self-cleavage can be encouraged by incubating the large transcription mixture at slightly higher temperatures (e.g., 42°C) for 1 h, with supplementation of another 5 mM of Mg^{2+} from a stock.

7. Increased yields of RNA are obtained if the soak step is repeated (both after transcription and kinase steps).

8. Use 10% DMSO for RNAs that are structured at the 5′-end.

9. We typically test a range of three or more log units of concentration, using serial dilutions to prepare enzyme stocks. These are all tested at the same incubation time (~15 min) at 37°C. This provides a simple way to hone in single-hit kinetics conditions, in which about 90% of the starting material is uncleaved. Single-hit kinetics are important because an RNA could refold after the first cleavage, causing further cleavage events to report on a potentially erroneous fold.

10. Some RNases continue to work under denaturing conditions, so it is vital that these samples be kept on ice to prevent the enzyme from probing the denatured form of the RNA. As a precaution, the RNase could be removed by phenol/chloroform extraction and the RNA concentrated by ethanol precipitation prior to fractionation on the gel.

11. It is not essential that these cleavage events be single-hit; only that every G be revealed.

12. We use a homemade apparatus consisting of two round metal plates with eight holes drilled to accommodate microcentrifuge tubes. The apparatus has a central threaded bolt with a wingnut to tighten the plates. This apparatus is placed in a beaker of boiling water. This setup serves to keep the tops of the eppendorf tubes from popping open during boiling, and allows the tops of the tube to be at the same temperature as the bottoms, which prevents concentrating of the samples due to condensation on the lids of the tubes.

13. If insufficient resolution is obtained, the comb can be altered to widen the lanes to apply the sample over a greater area sample. We use combs with 8×5-mm wells and load between 3 and 5 μL of sample at a time to allow for optimal separation and resolution of bands.

14. To avoid unnecessary handling of solid acrylamide, which is a neurotoxin, we obtain our polyacrylamide as a 40% (29:1) stock from EMD Chemicals (Gibbstown, NJ, USA).

15. The $1.2 \times$ MEN and HEN buffers are actually $1 \times$ in buffer and EDTA and $1.2 \times$ in NaCl.

16. In some cases where lower rates of transcription are desired, the concentration of all NTPs except GTP can be dropped as low as 5 μM (12).

17. Many restriction endonucleases leave a 5′-phosphate, so that should be included in the sequence as well, and can be added during the chemical synthesis of the DNA for a nominal fee

from IDT, or could be added with ATP and polynucleotide kinase.

18. If there is a high background in the no-insert control, the vector can be dephosphorylated prior to the ligation step using NEB CIP and inactivation as described.

19. A few assumptions used in arriving at Eqs. 4–6 are noted. First, we ignored ground-state fractionation factors in Eqs. 4 and 5, as is typically done for proteins since most functional groups involved in acid-based chemistry appear to have fractionation factors near unity (22). (An exception is Eq. 6, involving lyoxide, since its ground-state fractionation factor is significantly less than 1.) Second, we assumed absence of so-called "medium effects," which describe contributions of enzyme protonations at sites other than those involved in the reaction (22). This is the normal assumption in proteins and seems to hold well in the absence of special circumstances (e.g., participation of RS⁻) (22); also, medium effects are often manifested as "dome-shaped" proton inventories, which are not found herein. The influence of solvent on base pairing (23) does not appear to account for the inventories in this study either (11).

20. Co-transcriptional reactions are initiated by addition of T7 RNA polymerase; therefore, accurate and reproducible data could only be obtained 15 s after time zero, and in some cases a significant fraction of RNA is already cleaved.

21. The co-transcriptional equations provided assume that the rate of transcription is constant with time. To assure that this is the case, we measure the level of total RNA transcribed and plot it vs. time to test for a straight line. At any given time point, the total level of RNA is simply the sum of all RNA bands because the RNA is body-labeled.

Acknowledgments

We thank NSF MCB-0527102 for supporting this research.

References

1. Narlikar, G.J. and Herschlag, D. (1997) Mechanistic aspects of enzymatic catalysis: lessons from comparison of RNA and protein enzymes. *Annu Rev Biochem* 66, 19–59.

2. Doudna, J.A. and Cech, T.R. (2002) The chemical repertoire of natural ribozymes. *Nature* 418, 222–228.

3. Bevilacqua, P.C. (2008) Proton Transfer in Ribozyme Catalysis. In: Lilley, D. M., and Eckstein, F. (ed), Proton Transfer in Ribozyme Catalysis. edition edn. Royal Society of Chemistry, Cambridge

4. Fedor, M.J. (2009) Comparative enzymology and structural biology of RNA self-cleavage. *Annu Rev Biophys* 38, 271–299.

5. Bevilacqua, P.C., Brown, T.S., Chadalavada, D., Parente, A.D., and Yajima, R. (2003) Kinetic Analysis of Ribozyme Cleavage. Oxford University Press, Oxford.

6. Chadalavada, D.M., Knudsen, S.M., Nakano, S., and Bevilacqua, P.C. (2000) A role for upstream RNA structure in facilitating the catalytic fold of the genomic hepatitis delta virus ribozyme. *J Mol Biol* 301, 349–367.

7. Brown, T.S., Chadalavada, D.M., and Bevilacqua, P.C. (2004) Design of a highly reactive HDV ribozyme sequence uncovers facilitation of RNA folding by alternative pairings and physiological ionic strength. *J Mol Biol* 341, 695–712.

8. Chadalavada, D.M., Gratton, E.A., and Bevilacqua, P.C. (2010) The human HDV-like *CPEB3* ribozyme is intrinsically fast-reacting. *Biochemistry* 49, 5321–5330.

9. Nakano, S., Chadalavada, D.M., and Bevilacqua, P.C. (2000) General acid-base catalysis in the mechanism of a hepatitis delta virus ribozyme. *Science* 287, 1493–1497.

10. Nakano, S. and Bevilacqua, P.C. (2001) Proton inventory of the genomic HDV ribozyme in Mg^{2+}-containing solutions. *J Am Chem Soc* 123, 11333–11334.

11. Cerrone-Szakal, A.L., Siegfried, N.A., and Bevilacqua, P.C. (2008) Mechanistic characterization of the HDV genomic ribozyme: solvent isotope effects and proton inventories in the absence of divalent metal ions support C75 as the general acid. *J Am Chem Soc* 130, 14504–14520.

12. Diegelman-Parente, A. and Bevilacqua, P.C. (2002) A mechanistic framework for co-transcriptional folding of the HDV genomic ribozyme in the presence of downstream sequence. *J Mol Biol* 324, 1–16.

13. Chadalavada, D.M., Cerrone-Szakal, A.L., and Bevilacqua, P.C. (2007) Wild-type is the optimal sequence of the HDV ribozyme under cotranscriptional conditions. *RNA* 13, 2189–2201.

14. Schowen, K.B. and Schowen, R.L. (1982) Solvent isotope effects of enzyme systems. *Methods Enzymol* 87, 551–606.

15. Siegfried, N.A., Thesis: Folding cooperativity and its influence on pKa shifting in nucleic acids, The Pennsylvania State University, University Park, PA, 2009.

16. Perrin, D.D. and Dempsey, B. (1974) Buffers for pH and Metal Ion Control. In: (ed), Buffers for pH and Metal Ion Control. edition edn. Chapman and Hall, London

17. Milligan, J.F. and Uhlenbeck, O.C. (1989) Synthesis of Small RNAs Using T7 RNA Polymerase. *Methods Enzymol* 180, 51–62.

18. Schowen, K.B. (1978) Solvent hydrogen isotope effects. In: (ed), Solvent hydrogen isotope effects. edition edn. Plenum Press, New York

19. Smith, M.D. and Collins, R.A. (2007) Evidence for proton transfer in the rate-limiting step of a fast-cleaving Varkud satellite ribozyme. *Proc Natl Acad Sci USA* 104, 5818–5823.

20. Bates, R.G. (1973) Determination of pH. Theory and practice. Wiley, New York

21. Long, D.M. and Uhlenbeck, O.C. (1994) Kinetic characterization of intramolecular and intermolecular hammerhead RNAs with stem II deletions. *Proc Natl Acad Sci USA* 91, 6977–6981.

22. Quinn, D.M. and Sutton, L.D. (1991) Theoretical Basis and Mechanistic Utility of Solvent Isotope Effects. In: Cook, P. F. (ed), Theoretical Basis and Mechanistic Utility of Solvent Isotope Effects. edition edn. CRC Press, Inc., Boca Raton, FL.

23. Tinsley, R.A., Harris, D.A., and Walter, N.G. (2003) Significant kinetic solvent isotope effects in folding of the catalytic RNA from the hepatitis delta virus. *J Am Chem Soc* 125, 13972–13973.

Chapter 4

Kinetic Characterization of Hairpin Ribozyme Variants

Bettina Appel, Thomas Marschall, Anne Strahl, and Sabine Müller

Abstract

Kinetic analysis of ribozyme reactions is a common method to evaluate and compare activities of catalytic RNAs. The hairpin ribozyme catalyzes the reversible cleavage of a suitable RNA substrate at a specific site. Hairpin ribozyme variants as an allosteric ribozyme responsive to flavine mononucleotide and a hairpin-derived twin ribozyme that catalyzes two cleavage reactions and two ligation events with the result of a fragment exchange have been developed by rational design and were kinetically characterized. Herein, protocols for preparation of ribozymes and dye-labeled substrates as well as for analysis of cleavage, ligation, and fragment exchange reactions are provided.

Key words: Hairpin ribozyme, Twin ribozyme, Flavine mononucleotide, Allosteric activation, RNA, Cleavage, Ligation, Kinetics, Chemical synthesis, In vitro transcription

1. Introduction

Among the many catalytic RNAs that have been discovered to date, small ribozymes as the hairpin or the hammerhead ribozyme play an important role as tools for specific RNA destruction in biological and medicinal studies (1), or as reporter ribozymes in the context of bioanalytics (2). The hairpin ribozyme (Fig. 1a) belongs to the group of small catalytic RNAs that catalyze the reversible cleavage of a specific phosphodiester bond generating characteristic fragments with 5'-OH and a 2',3'-cyclic phosphate as terminal groups (3). It is derived from the negative strand of the satellite RNA of tobacco ringspot virus (4, 5). In a number of studies, the hairpin ribozyme (Fig. 1a) was reengineered, for example into an allosteric ribozyme (Fig. 1b) with flavine mononucleotide (FMN) as cofactor (6, 7) or into a twin ribozyme (Fig. 1c) that catalyzes an RNA fragment exchange reaction (8–10). Activity of

Jörg S. Hartig (ed.), *Ribozymes: Methods and Protocols*, Methods in Molecular Biology, vol. 848,
DOI 10.1007/978-1-61779-545-9_4, © Springer Science+Business Media, LLC 2012

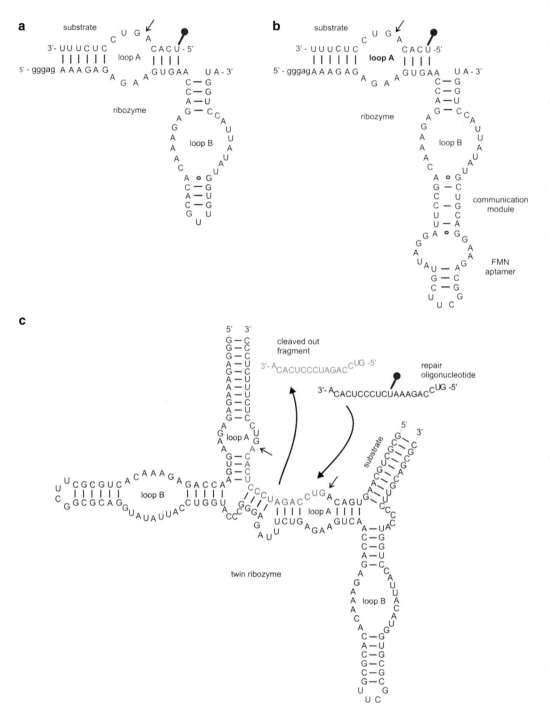

Fig. 1. Hairpin ribozyme variants. Cleavage sites are marked by *arrows*. The substrates are labeled with the fluorescent dye ATTO680 (indicated by the *red dot*). (**a**) Hairpin ribozyme, (**b**) FMN-responsive ribozyme, (**c**) Twin ribozyme.

the FMN-responsive ribozyme (Fig. 1b) can be reversibly switched on and off by control of the redox state of FMN. FMN in its oxidized state binds to the aptamer and thus stabilizes the catalytically active conformation of the aptazyme. Upon reduction, the molecular shape of FMN is dramatically changed, leading to loss of its binding properties. As a result, activity of the aptazyme is significantly decreased (6). The twin ribozyme is derived from tandem duplication of the hairpin ribozyme. In a site-directed and specifically controlled reaction, it promotes the exchange of sequence patches in arbitrarily chosen substrate RNAs by catalyzing two cleavage reactions and two ligation events (8–10).

Reaction analysis for all hairpin ribozyme variants shown in Fig. 1 can be carried out by following the formation of cleavage/ligation products with time and determination of kinetic parameters at substrate saturation (multiple turnover) or at ribozyme saturation (single turnover). For this purpose, labeling of the substrate is required, which alternatively to the attachment of a radioactive phosphorus usually at the 5′-terminus, can be achieved by conjugation of a suitable fluorescent dye to the 5′- or 3′-end of cleavage or ligation substrates, or to an internal position of the repair oligonucleotide in the twin ribozyme reaction.

2. Materials

General information

Chemicals were obtained from commercial suppliers (Merck, Aldrich, Sigma, Acros, Fluka, and VWR) and used without further purification unless otherwise noted. All buffers for PAGE and HPLC purification were filtered through a 0.45-μm membrane. All buffers for HPLC purification were degassed in an ultra-sonic bath before use. Buffers used in RNA purification and labeling protocols were prepared from autoclaved micro-pore water with a resistivity of 18.2 MΩ/cm and were filtered through a 0.20-μm membrane.

Apparatus

1. Gene Assembler Special, Pharmacia.

2. Äkta Purifier, Amersham Biosciences.

3. Anion exchange column: DNAPac PA-100 22 × 50 mm, Guard and DNAPac PA-100 22 × 250 mm, PrepScale, Dionex, BioLC.

4. Reversed-phase column: VarioPrep Guard 50/10 Nucleodur 100-5C18ec and VarioPrep 250/10 Nucleodur 100-5C18ec, Macherey-Nagel.

5. Centrifuge 5804 R and 5804, Eppendorf.

6. UV transilluminator: Chemi Smart 2000, Vilber Lourmat.

7. NanoDrop ND-1000 Spectrophotometer, PeqLab.

8. DNA sequencer 4200L and 4300, LICOR.

9. Table centrifuge, Roth.

2.1. In Vitro Transcription

1. HEPES-buffer (5×): 400 mM HEPES-Na, 60 mM MgCl₂, 10 mM spermidine, 200 mM DTT, pH 7.5.

2. NTP mixture: Prepare a stock solution containing 20 mM of each NTP in water.

3. dsDNA template.

4. T7-RNA polymerase (20 U/μL, Fermentas).

5. DNase I (1 U/μL, Fermentas).

6. Ethanol abs.

7. 0.3 M sodium acetate, pH 7.0.

8. TBE buffer (10×), 0.89 M Tris–HCl, 0.89 M boric acid, 0.02 M EDTA-Na, pH 8.3. Use 1× TBE buffer as running buffer as well as to dilute the acrylamide/bisacrylamide solution.

9. 20 Vol% acrylamide/bisacrylamide (19:1) in 1× TBE, 7 M urea.

10. N,N,N′,N′-tetramethylethylenediamine (TEMED).

11. Ammonium persulfate: Prepare 10% solution in water and store in the dark at 4°C.

12. Denaturing loading buffer: 98 Vol% formamide, 2 Vol% 0.5 M EDTA-Na.

2.2. RNA Preparation

2.2.1. RNA Synthesis

1. Phosphoramidite building blocks: 5′-(4,4′-Dimethoxytrityl)-N-phenoxyacetyl-adenosine-2′-O-TBDMS-3′-[(2-cyanoethyl)-(N,N-diisopropyl)]-phosphoramidite (link technologies), 5′-(4,4′-Dimethoxytrityl)-N-acetyl-cytidine-2′-O-TBDMS-3′-[(2-cyanoethyl)-(N,N-diiso-propyl)]-phosphoramidite (link technologies), 5′-(4,4′-Dimethoxytrityl)-N-(4-isopropyl-phenoxyacetyl)-guanosine-2′-O-TBDMS-3′-[(2-cyanoethyl)-(N,N-diisopropyl)]-phosphor-amidite (link technologies), 5′-(4,4′-Dimethoxytrityl)-uridine-2′-O-TBDMS-3′-[(2-cyanoethyl)-(N,N-diisopropyl)]-phosphoramidite (link technologies), Amino Modifier Uridine (C-6) CED phosphoramidite (ChemGenes) for internal labeling, TFA-Amino C-6 CED phosphoramidite (ChemGenes) for 5′-labeling.

2. Acetonitrile (99.9% extra dry over molecular sieves from ACROS or anhydrous with water content under 20 ppm from emp Biotech).

3. Solid phase support as CPG (controlled pore glass) functionalized with A, C, G, and U: Pac-rA SynBase, CPG 1000/110, 37 μmol/g (link technologies), Ac-rCSynBase, CPG 1000/110, 36 μmol/g (link technologies), iPr-Pac-rGSynbase, CPG

1000/110, 23 μmol/g (link technologies), U RNA Synbase, CPG 1000/110, 44 μmol/g (link technologies), 3'-Amino Modifier TFA Amino C-6 lcaa CPG 45 μmol/g (ChemGenes).

4. Oxidation solution: 0.01 M iodine, 260 mL acetonitrile (HPLC grade), 24 mL trimethylpyridine, 120 mL water.

5. Capping A: 20% *N*-methylimidazole in acetonitrile: 15.86 mL *N*-methylimidazole, 84.14 mL acetonitrile (HPLC grade).

6. Capping B: 50 mL acetonitrile (HPLC grade), 20 mL acetic anhydride, 30 mL trimethylpyridine.

7. Detritylation: 970 mL 1,2-dichloroethane, 30 mL dichloroacetic acid.

8. Activator: 0.3 M in acetonitrile: 2.3 g BMT (5-benzylmercaptotetrazole from emp Biotech) in 40 mL acetonitrile (99.9% extra dry over molecular sieves from ACROS or anhydrous with water content under 20 ppm from emp Biotech).

9. Molecular sieves for amidite solution and activator solution: 3 Ångström, activated (Roth).

2.2.2. Deprotection

1. Deprotection solution: 1:1 (v/v) mixture of 8 M ethanolic methylamine and concentrated NH_3 (30%).

2. Desilylation reagent: 3:1 (v/v) mixture of $NEt_3 \times 3$ HF and DMF (dry 99.8% ACROS Extra Dry over AcroSeal).

3. *n*-Butanol: analytical grade.

4. 0.05 M TEAA buffer: 0.05 M triethylammoniumacetate, pH 7.0.

2.2.3. Purification by Denaturing PAGE

1. TBE (1×): 0.1 M Tris, 1 mM EDTA, 85 mM boric acid, pH 8.3.

2. Gel solution: 20% nonfluorescent acrylamide/bisacrylamide 19:1 (Roth), 1× TBE, 7 M urea.

3. Ammonium persulfate: 10% solution in water.

4. *N,N,N',N'*-tetramethyl-ethane-1,2-diamine (TEMED).

5. Loading buffer: 98:2 (v/v) mixture of formamide and 0.5 M EDTA.

6. Eluting buffer: 0.5 M LiOAc.

2.2.4. Purification by Anion Exchange HPLC

1. Columns named in Subheading 2.

2. Buffer A: 10 mM Tris, 10 mM $NaClO_4$, 6 M Urea, pH 7.0.

3. Buffer B: 10 mM Tris, 500 mM $NaClO_4$, 6 M Urea, pH 7.0.

2.3. Desalination

2.3.1. Ethanol Precipitation

1. 20 Vol% 50 mM $MgCl_2$.

2. 10 Vol% 3 M NaOAc (pH 5.3).

3. 250 Vol% ethanol analytical grade.

2.3.2. Ethanol/Acetone Precipitation	1. Ethanol analytical grade. 2. Acetone analytical grade.
2.3.3. RP Chromatography	1. Reversed-phase column (Sep-Pak Vac 12 cc (2 g), C18, Waters). 2. Acetonitrile (HPLC grade). 3. 100 mM TBK buffer pH 8.5 (triethylammonium bicarbonate buffer, Fluka). 4. Acetonitrile/H_2O 3:2 (v/v).
2.3.4. Gel Filtration	1. 1 g Sephadex G25 fine (GE Healthcare). 2. Autoclaved micro-pore water.

2.4. Post-synthetic Dye Labeling

2.4.1. Labeling Reaction

1. 10 nmol Amino-modified RNA.
2. 50 µL 200 mM Na_2CO_3, pH 8.5.
3. 100 µg ATTO680-NHS in 50 µL dry DMF.

2.4.2. Purification of Labeled RNA by RP-HPLC

1. Columns named in Subheading 2.
2. Buffer A: 0.1 M TEAA, 5% acetonitrile.
3. Buffer B: 0.1 M TEAA, 30% acetonitrile.

2.5. Cleavage Reaction

1. Reaction buffer: 100 mM Tris–HCl, pH 7.5.
2. 100 mM $MgCl_2$.
3. Ribozyme.
4. ATTO680-labeled substrate RNA.
5. Stop mix: 7 M urea, 50 mM EDTA-Na.

2.5.1. Induction of Cleavage by FMN as Allosteric Cofactor

1. FMN-responsive ribozyme.
2. ATTO680-labeled substrate RNA.
3. 100 mM Tris–HCl, pH 7.5.
4. 100 mM $MgCl_2$.
5. FMN.
6. Stop mix: 7 M urea, 50 mM EDTA-Na.

2.5.2. Cleavage Reaction Under Reducing Conditions

1. FMN-responsive ribozyme.
2. ATTO680-labeled substrate RNA.
3. 100 mM Tris–HCl, pH 7.5.
4. 100 mM $MgCl_2$.
5. Sodium dithionite ($Na_2S_2O_4$).

6. FMN.

7. Stop mix: 7 M urea, 50 mM EDTA-Na.

2.6. Ligation Reaction

2.6.1. Preparation of 2′,3′-Cyclic Phosphate Containing RNA Fragments

1. Reaction buffer: 100 mM Tris–HCl, pH 7.5.

2. 100 mM $MgCl_2$.

3. Ribozyme.

4. ATTO680-labeled substrate.

2.6.2. Cyclic Phosphate Test

1. Calf intestinal alkaline phosphatase (CIP, 10 U/μL, New England Biolabs).

2. Phosphatase buffer (10×, NEB).

3. Oligonucleotide bearing the 2′,3′-cyclic phosphate (see Note 1).

4. Stop mix: 7 M urea, 50 mM EDTA-Na.

2.6.3. Ligation

1. 100 mM Tris–HCl, pH 7.5.

2. 100 mM $MgCl_2$.

3. Ribozyme.

4. ATTO680-labeled 5′-ligation fragment.

5. 3′-Ligation fragment.

6. Stop mix: 7 M urea, 50 mM EDTA-Na.

2.7. Repair Reaction

1. Twin ribozyme.

2. Substrate RNA.

3. ATTO-labeled repair oligonucleotide with 2′,3′-cyclic phosphate.

4. 100 mM Tris–HCl, pH 7.5.

5. 100 mM $MgCl_2$.

6. Stop mix. 7 M urea, 50 mM EDTA-Na.

2.8. Reaction Analysis by PAGE on DNA Sequencer

1. 15 Vol% nonfluorescent acrylamide/bisacrylamide (29:1; Roth) in 1× TBE, 7 M urea.

2. Ammonium persulfate.

3. Running buffer: 0.6× TBE.

4. N,N,N',N'-tetramethylethylenediamine.

5. DNA ladder Gene Ruler low range (Fermentas).

2.9. Determination of Reaction Parameters

1. Software: Chemi Smart:Chemi-Capt 2000 Vers. 12.8. (Chemi Smart) or BaseImagIR Data Collection 4.0 (LICOR 4200L) or NEN Model 4300 DNA Analyzer (LICOR 4300).

3. Methods

Carry out all procedures at room temperature unless otherwise specified.

3.1. In Vitro Transcription

1. The following protocol describes the transcription at 500-μL scale. The given concentrations in brackets are final concentrations. To obtain the given final concentrations take the required volumes of the stock solutions and add water.

2. Mix the dsDNA template (1.5 μM) with the necessary volume of water, the rNTPs (2 mM), and the buffer (1×) and centrifuge (see Note 2).

3. Start the reaction by adding T7 RNA polymerase (0.6 U/μL) (see Note 3).

4. Incubate at 37°C over night.

5. To digest the DNA template, add 2 μL of DNase I and incubate for another 30 min.

6. Precipitate the RNA with ethanol for at least 3 h. Therefore add 300 Vol% of ethanol. For example, if the transcription was done in a volume of 500 μL add 1.5 mL ethanol.

7. Subsequently centrifuge the sample for 15–30 min at 14,000 rpm at 4°C. Discard the supernatant and dry the pellet in vacuo.

8. Purify the RNA by denaturing PAGE. At first dilute 100 mL of 10× TBE buffer with 900 mL water to obtain 1× TBE. Transcriptions in scales up to 500 μL are purified with $100 \times 80 \times 0.75$ mm small gels. Transcription scales larger than 500 μL are purified with $200 \times 150 \times 1.5$ mm gels.

9. For small gels, a final volume of 20 mL gel solution is needed. Therefore, mix the necessary volume of the gel stock solution with 7 M urea in 1× TBE. To start the polymerization, add 200 μL of 10% ammonium persulfate solution and 20 μL TEMED.

   For large gels, a final volume of 100 mL is required. Mix the necessary volume of the gel stock solution with 7 M urea in 1× TBE and add 1,000 μL APS and 100 μL TEMED.

10. Pre-run small gels for at least 30 min, large gels for at least 1 h. To prepare the samples, add 100 Vol% denaturing loading buffer and incubate for at least 5 min at 90°C.

11. Apply 15 μL of the RNA solution to each slot for small gels and 40 μL for large gels, respectively. Use denaturing loading buffer which additionally contains xylencyanol and bromophenol blue as marker in one slot.

12. Run the gel at a constant voltage of 120 V in 1× TBE as running buffer. For large gels use a constant capacity of 20 W.

13. For preparative purification of RNA look at the gels at 254 nm and excise the respective bands. Cut the bands in small pieces and elute the RNA out of the gel.

14. For elution of RNA submerge the gel pieces with an appropriate volume of 0.3 M NaOAc solution (elution buffer), pH 7.5.

15. Constantly shake for at least 2 h at room temperature. Remove the elution buffer and repeat the elution step twice.

16. Pool the eluates and precipitate the RNA (see step 6).

17. For analytical purposes, it may be necessary to stain the gel with ethidium bromide (0.5 µg/mL in 1× TBE). Therefore, soak the gel in ethidium bromide solution and incubate for about 15 min. Analyze the gel at 365 nm with an UV transilluminator.

3.2. RNA Preparation

3.2.1. RNA Synthesis

1. Synthesis of RNA is performed on CPG (see Note 4) using phosphoramidite building blocks. The amino-modified phosphoramidite is incorporated into the growing sequences at the defined position of the substrate.

2. Carry out all synthesis in 1 µmol scale.

3. The reagents and steps in the cycles of the chemical synthesis of RNA according to the phosphoramidite methodology are as follows: Detritylation for 36 s, coupling and activation for 6 min, capping for 48 s, and oxidation for 18 s.

4. Carry out all syntheses in the "trityl-off" mode.

3.2.2. Deprotection

1. Place the column containing the solid support with the synthesized RNA in a vial to remove acetonitrile with a table centrifuge.

2. Divide the dry support into two halves and transfer it into two 1.5-mL screw vials.

3. Add 1 mL of a 1:1 (v/v) mixture of 8 M ethanolic methylamine and concentrated NH_3 (30%) to each vial. Tightly close the screw vials. Oligos up to 20 nt should be kept for 40 min at 65°C; longer oligos are kept for 60 min at 65°C.

4. Cool down to room temperature and keep for 10 min.

5. Collect the supernatant and wash the support 3 times with 200 µL of the deprotection solution.

6. Combine all supernatants and dry in a speed vac (see Notes 5 and 6).

7. Flush the vial with the dry RNA pellet with argon and add 800 µL of the desilylation reagent (see Note 7).

8. Close the vial with parafilm to airtight and keep at 55°C for 90 min.

9. Cool down to room temperature, transfer the reaction solution into a 50-mL falcon tube, and add 200 μL of water to stop the reaction.

10. Add 20 mL of *n*-butanol, mix it vigorously, and keep the mixture at –20°C overnight.

11. Centrifuge for 30 min at 6,000 rpm, decant the *n*-butanol, and dry the pellet in vacuo.

3.2.3. Purification by Denaturing PAGE (see Note 8)

1. Prepare 100 mL of a 15% PAA gel solution by diluting the 20% stock solution of PAA solution with 7 M urea. Prepare a gel of $200 \times 200 \times 1.5$ mm.

2. Induce the polymerization by adding 40 μL TEMED and 1 mL of a 10% APS solution in water.

3. Start a pre-run at 200 V for 30–60 min. Use 1× TBE as buffer.

4. Dissolve the RNA sample in a 1:1 (v/v) mixture of H_2O/loading buffer and denature at 90°C for at least 5 min.

5. Load 40 μL into each slot. Use a mixture of bromophenol blue and xylene cyanol in separate slots as reference.

6. Let the electrophoresis run for 30 min at 200 V. Increase the voltage to 400 V (see Note 9).

7. Irradiate the gel with UV light at 254 nm. Cut off the bands corresponding to the desired RNA, crush it into small pieces, collect in a falcon tube, and shock frost in liquid nitrogen.

8. Add elution buffer to cover all gel pieces. Shake the falcon tube for 90 min and renew the elution buffer for 3–5 times. Keep the last batch of elution buffer overnight at 4°C (see Note 10).

9. Combine all elution solutions and filter through a 0.4-μm aseptic membrane filter.

10. Isolate the RNA by ethanol/acetone precipitation (see Subheading 3.3.2) and dry in vacuo.

3.2.4. Purification by Anion Exchange HPLC (see Note 11)

1. Use the anion exchange column mentioned in Subheading 2 with ÄKTA purifier for the following procedure. The column should be stored in oven at 70°C all time.

2. Wash the column with two column volumes (cv) buffer A, 2 cv buffer b, and then equilibrate with 2 cv buffer A.

3. Make an analytical run with a linear gradient (0%B → 100%B in 6 cv, 2 mL/min) with every RNA. Filter the crude RNA through a 0.2-μm aseptic membrane filter and bring an aliquot up to a volume of 100 μL. Load the column with this sample.

4. Detect the RNA with an UV detector at 254 nm. For preparative scale the linear gradient is changed into a step gradient—depending on the individual separating problem.

5. Load the column with 100 µl of the concentrated crude RNA solution for each HPLC run. Start each run with 1 cv buffer to equilibrate and end with 1 cv buffer B to wash.

6. Collect the fractions containing the desired RNA, combine, desalt by reversed-phase chromatography (see Subheading 3.3.3), and dry the oligo in vacuo.

7. For further desalination solve the RNA in 500 µL water and prepare a gel filtration (see Subheading 3.3.4).

3.3. Desalination

3.3.1. Ethanol Precipitation

1. Solve the RNA in water and add 20 Vol% 50 mM $MgCl_2$, 10 Vol% 3 M NaOAc solution (pH = 5.3), and 250 Vol% ethanol.

2. Mix the solution and keep it overnight at –20°C.

3. Centrifuge the precipitate for 20 min at 4,000 rpm.

4. Decant the supernatant and wash the RNA pellet with 80% ice-cold ethanol.

5. Dry the pellet for a few minutes in vacuo.

3.3.2. Ethanol/Acetone Precipitation (see Note 12)

1. Filter the combined elutions with an aseptic filter membrane (0.2-µm), lyophilize, and solve in 2–4 mL water.

2. Add 5 equivalents of ethanol/acetone 1:1 (v/v) to the solution and mix vigorously.

3. Keep the precipitation at –20°C overnight.

4. Centrifuge the precipitate for 30 min at 6,000 rpm.

5. Decant the supernatant and wash the pellet with 500 µL ice-cold ethanol (80%).

6. Dry the pellet for a few minutes in vacuo.

3.3.3. RP-Chromatography

1. Wash the Sep-Pak cartridge with 10 mL acetonitrile.

2. Equilibrate the column with 10 mL TBK buffer.

3. Solve the RNA in 0.5–1.0 mL water and load the sample on the column.

4. Wash the column with 2 mL TBK buffer and 20 mL water.

5. Elute the RNA by adding a 3:2 (v/v) mixture of acetonitrile/water and collect 1 mL fractions (about 10 fractions).

6. Detect the fractions containing the oligonucleotide sample by UV–VIS spectrometer at 254 nm, combine and dry for a few minutes in vacuo.

3.3.4. Gel Filtration

1. Suspend 1 g Sephadex in 10 mL autoclaved micro-pore water and let it swell for at least 3 h.

2. Use the swollen gel material to pack a column of 4.0×1.5 cm.

3. Equilibrate the column with 20 mL water.

4. Dissolve the oligonucleotide sample in 500 μL water and load the column.

5. Add slowly 10 mL water to elute the oligonucleotide.

6. Collect 0.5 mL fractions and check for the presence of oligonucleotide by UV–VIS spectrometer at 254 nm.

7. Collect fractions containing RNA.

8. Dry the oligo in vacuo.

3.4. Post-synthetic Dye Labeling

3.4.1. Labeling Reaction

1. Lyophilize 10 nmol of amino-modified RNA to dryness and solve into 50 μl 200 mM carbonate buffer.

2. Add the ATTO680 solution to the RNA.

3. Carry out the labeling reaction in the dark for 4 h while stirring.

4. Remove buffer and excess amounts of dye by gel filtration (see Subheading 3.3.4).

5. Isolate the RNA by ethanol precipitation (see Subheading 3.3.1).

3.4.2. Purification of Labeled RNA by RP-HPLC

1. Use the RP column named in Subheading 2 with ÄKTA purifier for the following procedure.

2. Wash the column with 2 cv buffer A, 2 cv buffer B, and finally equilibrated with 2 cv A.

3. Choose a flow rate of 4 mL/min all time.

4. Carry out an analytical run with a linear gradient (0%B 100%B in 6 cv) with every RNA.

5. Filter the crude RNA through a 0.2-μm aseptic membrane filter and dilute an aliquot (1/10) with water up to a volume of 100 μL. Load the column and start the run.

6. Detect the RNA with an UV detector at 254 nm. For preparative scale, change the linear gradient into a step gradient—depending on the individual separating problem.

7. Load 100 μl of the concentrated crude RNA solution for each HPLC run on the column. Start each run with 1 cv buffer A to equilibrate and end with 1 cv buffer B to wash.

8. Collect the fractions containing the desired RNA, combine, desalt by reversed-phase chromatography (see Subheading 3.3.3) and dry the oligo in vacuo.

9. Make a gel filtration for further desalination (see Subheading 3.3.4; Note 13).

3.5. Cleavage Reaction

1. Mix water, Tris–HCl buffer (40 mM), ribozyme (100 nM), and substrate RNA (100 nM) (see Fig. 2 and Note 14).

2. Denature the RNA strands for 2 min at 90°C.

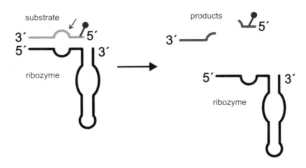

Fig. 2. Cleavage assay. The cleavage site is marked by an *arrow*. The substrate is labeled with the fluorescent dye ATTO680 (indicated by the *red dot*).

3. To allow hybridization of the RNA incubate for 15 min at 37°C.

4. Induce the cleavage reaction by adding $MgCl_2$ (10 mM).

5. Take aliquots of 1 μL at defined time intervals and transfer them in 49 μL of stop mix (see Note 15).

6. Analyze the samples on a DNA sequencer (see Subheading 3.8 for details).

3.5.1. Induction
of Cleavage by FMN
as Allosteric Cofactor

1. The given concentrations in brackets represent the final concentrations for each component in 20 μL reaction volume.

2. Preincubate ribozyme (500 nM) and substrate RNA (25 nM) in separate reaction tubes. Therefore mix both RNA strands separately with the necessary volume buffer (50 mM) and denature for 1 min at 90°C (see Note 16).

3. Subsequently add $MgCl_2$ (10 mM) and FMN (200 μM). Preincubate for another 10 min at 37°C. For control reactions leave out FMN and replace its volume by water.

4. Start the reaction by mixing ribozyme and substrate. Incubate for another 2 h at 37°C.

5. Take aliquots of 1 μL at defined time intervals and transfer them into 49 μL of stop mix (see Note 15).

3.5.2. Cleavage Reactions
Under Reducing Conditions
(see Note 17)

1. Preincubate cleavage reactions in presence of FMN under standard conditions (see Subheading 3.5).

2. In case, FMN should be reduced right from the start, add 2 μL of the reducing agent to ribozyme and substrate.

3. After 2 min, start the reaction by mixing ribozyme and substrate RNA (final volume is 40 μL).

4. In case FMN should be reduced after 2 min, 4 μL of the reducing agent was added to 40 μL of a standard cleavage reaction after 1.5 min (see Note 18).

5. Mix the solution slightly by pipetting. Incubate for another 2 h at 37°C.

6. Take aliquots of 1 μL at defined points of time and transfer them into 49 μL of stop mix (see Note 15).

7. Analyze the samples on a DNA sequencer (see Subheading 3.8 for details).

3.6. Ligation Reaction

3.6.1. Preparation of 2′,3′-Cyclic Phosphate Containing RNA Fragments (see Note 2)

1. Mix the ribozyme (0.4 μM), ATTO-labeled substrate (2 μM), buffer (10 mM), and water. Denature the mixture for 2 min at 95°C.

2. To allow hybridization of both RNA strands incubate for 15 min at 37°C.

3. Start the reaction by adding $MgCl_2$ (10 mM) and incubate for another 3 h at 37°C.

4. Purify the 2′,3′-cyclic phosphate-containing RNA fragment by denaturing PAGE at 10°C.

5. Excise the respective bands, elute the RNA, and precipitate over night with ethanol (for details see Subheading 3.3.1).

3.6.2. Cyclic Phosphate Test (see Note 19)

1. Mix the 2′,3′-phosphate containing RNA fragment (100 nM), phosphatase buffer (1×), and water (see Note 14).

2. Incubate for 5 min at 37°C, add CIP, and keep the mixture at 37°C.

3. Take 1 μL aliquots after 1, 2, 5, and 10 min and transfer them into 49 μL stop mix (see Note 15).

4. Analyze the samples on a DNA sequencer (see Subheading 3.8 for details).

3.6.3. Ligation (see Note 20)

1. Mix water, tris buffer (40 mM), ribozyme (100 nM), and 3′-ligation fragment (100 nM) (see Fig. 3 and Note 14).

2. Denature the RNA strands for 2 min at 90°C.

3. Incubate the mixture at 37°C for 15 min to allow hybridization to proceed.

4. Start the reaction by addition of $MgCl_2$ (10 nM) and the ATTO680-labeled 5′-ligation fragment (100 nM).

5. Take 1 μL aliquots after defined time intervals and transfer into 49 μL stop mix (see Note 15).

6. Analyze the samples on a DNA sequencer (see Subheading 3.8. for details).

3.7. Repair Reaction

1. Mix twin ribozyme (100 nM), substrate RNA (100 nM), and water and denature for 2 min at 95°C (see Fig. 4 and Note 14).

2. Subsequently incubate the mixture at 37°C for 15 min to allow hybridization.

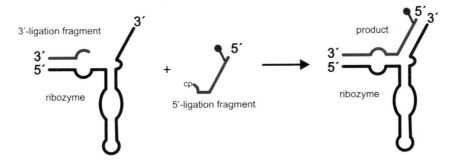

Fig. 3. Ligation assay. The ligation site is marked by an *arrow*. The substrate is labeled with the fluorescent dye ATTO680 (indicated by the *red dot*).

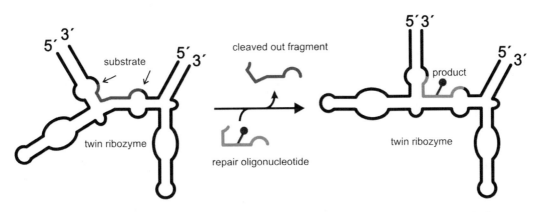

Fig. 4. Twin ribozyme assay. The *arrows* mark the two cleavage sites. The repair oligonucleotide is labeled with the fluorescent dye ATTO680 (indicated by the *red dot*).

3. Add $MgCl_2$ (10 mM) and incubate for another 15 min at 37°C.

4. Start the reaction by adding the repair oligonucleotide (100 nM).

5. Take 1 μL aliquots after 2, 10, 30 min, 1, 2, 4, 6, and 8 h and transfer them into 49 μL stop mix (see Note 15).

6. Analyze the samples on a DNA sequencer (see Subheading 3.8. for details).

3.8. Reaction Analysis by PAGE on DNA Sequencer

1. The detection of ATTO680-labeled oligonucleotides is carried out on a DNA sequencer LICOR 4200L or LICOR 4300. Use 0.6× TBE as running buffer. The fluorescent dye ATTO680 is excited at 680 nm. Bands are detected at 700 nm wavelength.

2. Prepare a gel of 25 cm × 18 cm × 0.25 mm. Therefore, mix 40 mL fluorescent free PAA gel solution with 30 mg APS in 200 μL water and 20 μL TEMED.

3. Allow the gel to polymerize for 1 h.

4. Pre-run the gel for 1 h at 1,500 V.

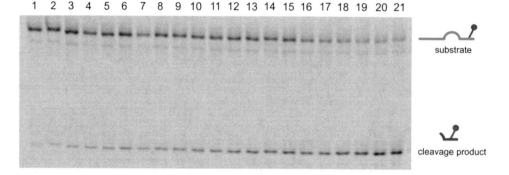

Fig. 5. Gel image of a cleavage reaction. Reaction after (1) 15 s, (2) 30 s, (3) 45 s, (4) 60 s, (5) 75 s, (6) 90 s, (7) 105 s, (8) 120 s, (9) 150 s, (10) 180 s, (11) 210 s, (12) 4 min, (13) 5 min, (14) 6 min, (15) 7 min, (16) 8 min, (17) 10 min, (18) 30 min, (19) 60 min, (20) 90 min, and (21) 120 min.

5. Apply the samples in an amount of 0.4 μL up to 2 μL in each slot of the gel and run the gel at a voltage of 1,500 V, a laser power of 3 mW and a temperature of 55°C (see Fig. 5 and Note 21).

3.9. Determination of Reaction Parameters

3.9.1. Cleavage and Ligation Reaction

1. Determine the fluorescence intensity of individual bands in the gel using the software mentioned in Subheading 2.9.

2. Calculate the concentration of product in relation to the substrate at each time point and plot the product concentration against time.

3. Determine the reaction rate by linear regression of data points in the start phase (linear area) of the reaction.

4. Normalize reaction rates against ribozyme concentration $V/[R]$ (multiple turnover reaction) or substrate concentration $V/[S]$ (single turnover reaction) to determine the observed rate constant k_{obs}.

5. Plot k_{obs} against $k_{obs}/[S]$ (multiple turnover reaction) or $k_{obs}/[R]$ (single turnover reaction) and fit the curve by linear regression according to the Eadie-Hofstee function $k_{obs} = -(k_{obs}/[S])K + k$ or $k_{obs} = -(k_{obs}/[R])K + k$, respectively (see Note 22).

3.9.2. Repair Reaction

1. Determine the fluorescence intensity of individual bands in the gel using the software mentioned in Subheading 2.3.

2. Calculate the percentage of the repair product in relation to the repair oligonucleotide at the desired time (see Fig. 6 and Note 23).

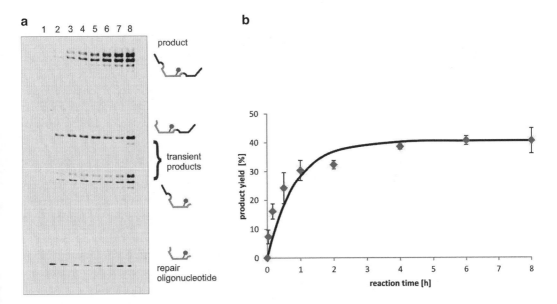

Fig. 6. Gel image (see Note 24) (**a**) and time course (**b**) of a repair reaction. (**a**) Reaction after (1) 2 min, (2) 10 min, (3) 30 min, (4) 1 h, (5) 2 h, (6) 4 h, (7) 6 h, and (8) 8 h.

4. Notes

1. Relevant to the 5′-ligation fragment (ligation assay) and to the repair oligonucleotide (twin ribozyme assay).

2. The final reaction volume is 500 μL.

3. Mix the solution with a pipette tip and centrifuge carefully for a second.

4. Synthesis of RNA is performed on a DNA synthesizer (Gene Assembler Special, Pharmacia). The system is kept under argon during the whole procedure. The solutions of phosphoramidites and activator should be kept over molecular sieve (3 Å).

5. Pay attention to the ammonia!

6. At the end of this procedure, only one 1.5-mL vial containing the RNA should result.

7. Be sure that the DMF is really dry. Water will disturb the desilylation reaction and the TBDMS group will not be cleaved off completely.

8. RNA longer than 20 nt is preferentially purified by a denaturing gel.

9. For RNAs longer than 30 nt, the voltage is recommended to be increased to 600 V and electrophoresis should be carried out in a cool room at 4°C.

10. Check the elution solution for RNA under UV light. If the buffer still contains RNA, continue elution.

11. RNA shorter than 20 nt preferentially is purified by HPLC.

12. This method is used for isolation of RNA that was purified by PAGE.

13. Solve the RNA in 500 μL water.

14. The final reaction volume is 20 μL.

15. Samples can be kept frozen until analysis.

16. In case of the FMN-responsive ribozyme, the activity tests are carried out under single turnover conditions in presence and absence of the cofactor using tubes which are impervious to light.

17. All solutions of the reducing agent have to be prepared freshly. All tubes have to be flushed with argon during preincubation. Also tips need to be rinsed with argon before usage. The entire reaction must be carried out under an argon atmosphere to avoid oxidation as good as possible.

18. Keep in mind that all reaction components are diluted 1:10 according to start conditions.

19. The success of the preparation of the 2′,3′-cyclic phosphate can be observed by alkaline phosphatase digestion. CIP cleaves off free phosphate groups while the enzyme cannot use cyclic phosphates as substrate.

20. For efficient ligation, extension of the 5′-ligation fragment and complementary extension of the ribozyme 3′-terminus are necessary in order to generate an additional helix as schematically shown in Fig. 3. The resulting three-way junction structure corresponds to the right unit of the twin ribozyme shown in Fig. 1. In general the sequence of the additional helix can be freely chosen. However, the stretch of unpaired bases ACCC (comp. Fig. 1c) at the three-way junction is absolutely necessary to ribozyme activity.

21. Add a marker RNA into a separate slot.

22. The rate constant k is obtained from the intercept with the y axis, the equilibrium constant K from the slope of the linear curve. For a multiple turnover reaction, K corresponds to the Michaelis–Menton constant K_M, and k to the turnover constant k_{cat} according to the Michaelis–Menton equation.

23. Determination of kinetic constants is impossible. Due to the complexity of the twin ribozyme reaction, a kinetic model describing the reaction is still missing.

24. Double bands result from the substrate obtained by in vitro transcription and thus having heterogeneous 3′-ends.

References

1. Mulhbacher, J., St-Pierre, P. and Lafontaine, D. A. (2010) Therapeutic applications of ribozymes and riboswitches. *Curr. Opin. Pharmacol.* **10**, 551–556.

2. Liu, J., Cao, Z. and Lu, Y. (2009) Functional nucleic acid sensors. *Chem. Rev.* **109**, 1948–1998.

3. Hampel, A. and Tritz, R. (1989) RNA catalytic properties of the minimum (−)sTRSV sequence. *Biochemistry* **28**, 4929–4933.

4. Feldstein, P. A., Buzayan J. M. and Bruening, G. (1989) Two sequences participating in the autolytic processing of satellite tobacco ringspot virus complementary RNA. *Gene* **82**, 53–61.

5. Haseloff, J. and Gerlach, W. L. (1989) Sequences required for self-catalysed cleavage of the satellite RNA of tobacco ringspot virus. *Gene* **82**, 43–52.

6. Strohbach, D., Novak, N. and Müller, S. (2006) Redox-active riboswitching: Allosteric regulation of ribozyme activity by ligand shape control. *Angew. Chem. Int. Ed.*, **45**, 2127–2129.

7. Strohbach, D., Turku, F., Schuhmann, W. and Müller, S. (2008) Electrochemically induced modulation of the catalytic activity of a reversible redoxsensitive riboswitch. *Electroanalysis* **20**, 935–940.

8. Welz, R., Bossmann, K., Klug, C., Schmidt, C., Fritz, H.-J. and Müller S. (2003) Site-directed alteration of RNA sequence mediated by an engineered twin ribozyme. *Angew. Chem. Int. Ed.* **42**, 2424–2427.

9. Vauléon, S., Ivanov, S. A., Gwiazda, S. and Müller, S. (2005) Site-specific fluorescent and affinity labeling of RNA mediated by an engineered twin ribozyme. *ChemBioChem* **6**, 2158–2162.

10. Drude, I., Vauléon, S. and Müller S. (2007) Twin ribozyme mediated removal of nucleotides from an internal RNA site. *Biochem. Biophys. Res. Commun.* **363**, 24–29.

Chapter 5

Characterization of RNase P RNA Activity

Markus Gößringer, Dominik Helmecke, and Roland K. Hartmann

Abstract

The principle task of the ubiquitous enzyme RNase P is the generation of mature tRNA 5′-ends by removing precursor sequences from tRNA primary transcripts (Trends Genet 19:561–569, 2003; Crit Rev Biochem Mol Biol 41:77–102, 2006; Trends Biochem Sci 31:333–341, 2006). In Bacteria, RNase P is a ribonucleoprotein composed of two essential subunits: a catalytic RNA subunit (P RNA; 350–400 nt) and a single small protein cofactor (P protein; ~14 kDa). In vitro, bacterial P RNA can catalyze tRNA maturation in the absence of the protein cofactor at elevated concentrations of mono- and divalent cations (Cell 35:849–857, 1983). Thus, bacterial P RNA is a trans-acting multiple-turnover ribozyme.

Here we provide protocols for 5′-endonucleolytic ptRNA cleavage by bacterial P RNAs in the absence of any protein cofactor and under single-turnover conditions ($[E] \gg [S]$). Furthermore, we outline a concept that utilizes the bacterial RNase P ribozyme to release RNAs of interest with homogeneous 3′-OH ends from primary transcripts via site-specific cleavage. Also, T7 transcription of mature tRNAs with clustered G residues at the 5′-end may result in 5′-end heterogeneities, which can be avoided by first transcribing the 5′-precursor tRNA (ptRNA) followed by P RNA-catalyzed processing to release the mature tRNA carrying a homogeneous 5′-monophosphate end. Finally, RNase P ribozyme activity can be directly assayed by using total bacterial RNA extracts.

Key words: RNase P, tRNA processing, Ribozyme activity assay, Generation of RNAs with homogeneous 3′- and 5′-ends, In vitro transcription

1. Introduction

The ribonucleoprotein enzyme RNase P universally removes 5′-leader sequences from precursor tRNAs (ptRNAs) to generate the mature tRNA 5′-ends (1–3) as a prerequisite for tRNA participation in protein synthesis. The enzyme harbors a single RNA subunit of common ancestry among all three domains of life (3). The RNA subunit (P RNA, 350–400 nt, gene *rnpB*) of bacterial RNase P (see Fig. 1a) is an efficient catalyst in vitro in the absence of its single protein cofactor (~14 kDa; (4)). The very low RNA-alone

Jörg S. Hartig (ed.), *Ribozymes: Methods and Protocols*, Methods in Molecular Biology, vol. 848,
DOI 10.1007/978-1-61779-545-9_5, © Springer Science+Business Media, LLC 2012

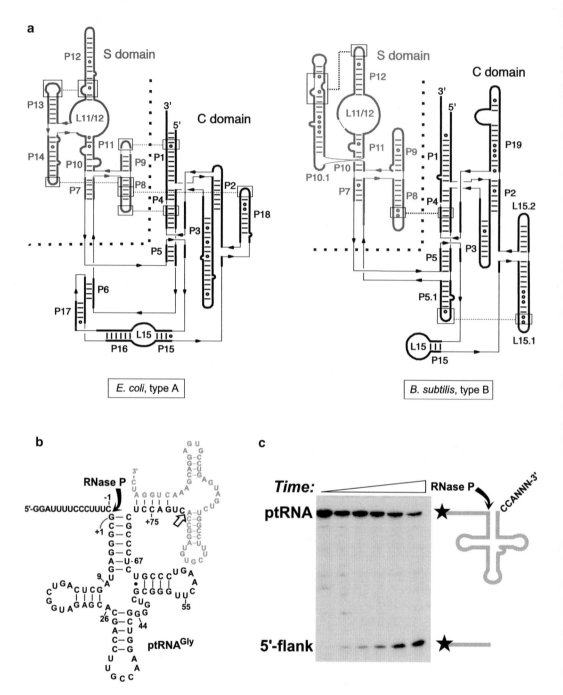

Fig. 1. (a) Secondary structure illustrations of *E. coli* (type A, *left*) and *B. subtilis* (type B, *right*) P RNAs. C domain, catalytic domain (*black*); S domain, specificity domain (*gray*); the two domains are separated by the *thick dotted line*; *P* helical (paired) elements; *L* loop regions. Tertiary interactions are indicated by *boxes* connected with *thin dotted lines*. Type A RNAs usually form three loop–helix interdomain contacts (L8–P4, L9–P1, L18–P8). In type B RNAs, P5.1 interacts with P15.1 and the L9–P1 interaction may also form. Taken from ref. (9). (b) ptRNA^{Gly} primary T7 transcript with a *cis*-hammerhead at the 3′-end (*gray lettering*) designed for the production of ptRNA with homogeneous 3′-ends; the *open arrow* indicates the site of hammerhead self-cleavage. For further details, see ref. 22. (c) Typical phosphorimage of the time course of P RNA-catalyzed cleavage of 5′-^{32}P-end-labeled (*asterisk*) ptRNA^{Gly} (*panel B*) analyzed by 20% denaturing PAGE. The *curved black arrow* in panels b and c indicates the canonical RNase P cleavage site.

activity of archaeal, eukaryal, and organellar P RNAs (5–7) correlates with the presence of multiple protein subunits, 5 in Archaea and 9–10 in Eukarya (8, 9). Recently, also protein-only RNase P enzymes have been discovered in human and land plant mitochondria and chloroplasts (10, 11).

The protein cofactor of bacterial RNase P (encoded by the *rnpA* gene), essential for RNase P function in vivo, affects the structure, function, and kinetics of the holoenzyme under physiological salt conditions (reviewed in ref. (9)). In vitro, the protein subunit is dispensable, but its absence has to be compensated for by increased mono- and particularly divalent cations in order to achieve efficient RNA-alone catalysis (4).

Nature has evolved two architectural subtypes of bacterial P RNAs (see Fig. 1a), the phylogenetically prevailing type A (for ancestral) represented by *E. coli* P RNA, and type B (for *Bacillus*) essentially confined to the low G + C Gram-positive bacteria, the prototype being *B. subtilis* P RNA (9, 12). Bacterial P RNAs are composed of two independent folding domains (13), the specificity (S-) and catalytic (C-) domain. The S-domain interacts with the T-arm module of ptRNAs which is crucial for substrate affinity and positioning (14, 15). The C-domain includes all structural elements forming the active site and provides the binding interface for the protein cofactor (16).

1.1. Single-Turnover Cleavage by E. coli P RNA

Single-turnover conditions have been adapted to most mechanistic studies in recent years, because dissociation of the tRNA product from the catalytic RNA usually limits the rate of the RNA-alone reaction under multiple-turnover conditions (mto; $(S)>>(E)$). By contrast, single-turnover conditions (sto; $(E)>>(S)$) permit the experimenter to analyze steps preceding product release, such as the chemical step, because an enzyme molecule traverses the catalytic cycle at most once under these conditions. As a consequence, the product release step has no effect on the kinetics. All ribozyme-product (mature tRNA) complexes existing under equilibrium conditions are then disrupted in the course of denaturing (8 M urea) PAGE analysis.

1.2. P RNA Processing to Generate RNAs with Homogeneous 3′-OH and 5′-Phosphate Termini

While T7 RNA polymerase usually initiates transcription at a defined position, it tends to append one or occasionally even a few more non-templated nucleotides to the product 3′-terminus (17). Also, 5′-end heterogeneity may become a problem when the template encodes unusual 5′-terminal sequences, such as 5′-CACUGU, 5′-CAGAGA, or 5′-GAAAAA (18), or when transcripts are initiated with multiple guanosines (19). For example, in the case of mature tRNA transcripts starting with 5′-GGGGG, 75% had canonical 5′-ends, relative to >99% for 5′-GCGGA, 87% for 5′-GGGCC, 97% for 5′-GGGAG, and only 66% for 5′-GGGGC (19). Thus, templates for T7 transcription employing the commonly used T7 class III promoter may give rise to 5′-heterogeneous transcripts if they encode more than two consecutive G residues at the 5′-end.

Since P RNA can be forced to cleave site specifically and because the enzyme generates 3'-OH and 5'-phosphate termini, representing the appropriate functional end groups for most applications, processing by P RNA is a method of choice to produce RNAs of interest with homogeneous 3'-ends; a more specialized application is to produce a 5'-mature tRNA by P RNA processing in cases where direct T7 transcription of 5'-mature tRNA is expected to yield 5'-heterogeneous products owing to problematic sequences at the 5'-end of the tRNA moiety (see above); also, such tRNAs derived from P RNA processing carry a 5'-monophosphate (as present in vivo), while direct production of 5'-mature tRNAs with a 5'-mono-phosphate by T7 transcription requires to include an excess of usually GMP over GTP in the transcription reaction. Figure 2 illustrates the approach to produce (i) an RNA of interest with homogeneous 3'-OH termini, e.g., for subsequent 3'-end labeling with $(5'-^{32}P)$-pCp or (ii) 5'-mature tRNAs with homogeneous 5'-monophosphate termini. The approach may also become the method of choice to generate smaller RNA oligonucleotides (~10–30 nt), released by P RNA cleavage as the 5'-flank of the ptRNA, (i) because direct T7 transcription of short RNAs may be inefficient and (ii) because the T7 primary transcript can, in a first step, be efficiently separated from smaller premature transcription termination products generated during T7 transcription; the latter may otherwise impair identification and separation of short T7 transcripts from side products.

1.3. Assaying Total Bacterial RNA Extracts for P RNA Activity

This approach has been applied to the search for RNase P RNA in *Aquifex aeolicus*, using total cellular RNA pools from several thermophilic bacteria of the genus *Thermus* as positive controls (20).

2. Materials

Prepare all solutions using double-distilled RNase-free water (ddH$_2$O) (exception: deionized H$_2$O to dilute 5× TBE to 1× TBE electrophoresis buffer) and analytical grade reagents. All stock solutions except for electrophoresis buffers were subjected to sterile filtration (Filtropur S 0.2 µM; Sarstedt) and stored at –20°C if not stated otherwise.

2.1. Single-Turnover Cleavage by E. coli P RNA

1. 0.25 (final assay conc.: 10 nM) to 125 (final assay conc.: 5 µM) pmol *E. coli* P RNA per reaction; 2.5 pmol *E. coli* P RNA in the example of Subheading 3.1.
2. ptRNA^{Gly} (5'-flank: 5'-GGAUUUUCCCUUUC/3'-end: CCAGUC-3'; Fig. 1b), 5'-^{32}P-end labeled (e.g. 1.3×10^6 Cerenkov cpm/µL/3 pmol; total volume: 10 µL).
3. 5× RNase P buffer A: 250 mM MES (2-(N-morpholino) ethanesulfonic acid) pH 6.0, 0.5 mM EDTA (disodium salt), 0.5 M NH$_4$OAc.

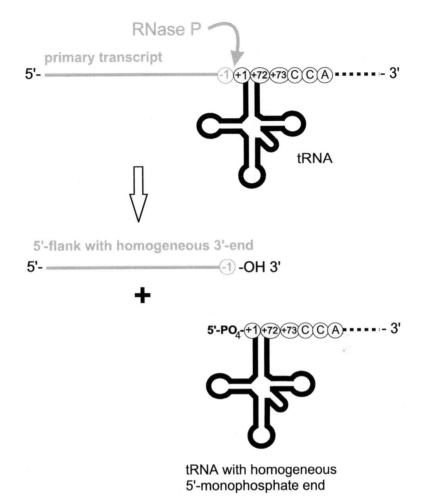

Fig. 2. P RNA processing to generate RNAs with homogeneous 3'-OH and 5'-monophosphate termini. The primary transcript comprises the 5'-RNA fragment (*gray line*) and the tRNA moiety positioned immediately downstream (*black line*). P RNA cleavage (*gray arrow*) of the primary transcript generates a 5'-RNA fragment with a homogeneous 3'-OH terminus at nucleotide position −1 and a tRNA molecule with a homogeneous 5'-monophosphate end at nucleotide position +1. For efficient and site-specific cleavage by P RNA, the first acceptor stem base pair should be G(+1)–C(+72); base pairing between N−1 and the discriminator nucleotide (N+73) should be avoided; nucleotide −1 should be preferably U, C, or A, and a C should be avoided at +73 when using *E. coli* P RNA in order to prevent aberrant cleavage; the discriminator nucleotide +73 should be followed by the sequence CCA; a 3'-extension of up to 6 nt beyond CCA does not interfere with P RNA cleavage; however, a second CCA trinucleotide should be avoided within the extension.

4. 0.5 M Mg(OAc)$_2$.

5. 5× TBE buffer: 445 mM Tris base, 445 mM boric acid, 10 mM EDTA (disodium salt), pH 8.3.

6. 2× denaturing loading buffer: 2× TBE buffer, 2.6 M urea, 66% deionized formamide, 0.02% (w/v) each bromophenol blue (BPB), and xylene cyanol blue (XCB).

2.2. P RNA Processing to Generate RNAs with Homogeneous 3'-OH and 5'-Phosphate Termini

1. 10 pmol *E. coli* P RNA.
2. 20 pmol RNA-tRNA primary transcript.
3. 5× RNase P cleavage buffer B: 250 mM Tris–HCl, pH 7.5, 0.5 mM EDTA (disodium salt), 0.5 M Mg(OAc)$_2$, 0.5 M NH$_4$OAc, 25% PEG 6000.

2.3. Preparation of Total Bacterial RNA Extracts for Assaying P RNA Activity

1. Total RNA extraction buffer (TREB): 20 mM NaOAc, 1 mM EDTA (disodium salt), 0.5% SDS, pH 5.5.
2. Phenol saturated with TREB buffer.

2.4. Assaying Total Bacterial RNA Extracts for P RNA Activity

1. 2.5× RNase P cleavage buffer C: 125 mM HEPES-NaOH, pH 7.0, 0.25 M Mg(OAc)$_2$, 0.25 M NH$_4$OAc.
2. Total cellular RNA from bacteria in 1× TE buffer (10 mM Tris–HCl pH 7.4, 1 mM EDTA), in this example mainly from strains of the genus *Thermus*.
3. ptRNA^{Gly} (see item 2 of Subheading 2.1 and Fig. 1b) 5'-^{32}P-end labeled (approx. 1.5×10^6 Cerenkov cpm/μL/3 pmol).

3. Methods

3.1. Single-Turnover Cleavage by E. coli P RNA

1. Prepare P RNA mix:

	Final concentration	MS
E. coli P RNA, 2.5 μM	125 nM	1 μL
5× RNase P buffer A	1×	4 μL
0.5 M Mg(OAc)$_2$	100 mM	4 μL
RNase-free ddH$_2$O		11 μL

Incubate at 55°C for 5 min, followed by 55 min at 37°C.

2. Prepare substrate mix:

	Final concentration	MS
ptRNA^{Gly}, ~20,000 Cerenkov cpm (~45 fmol)	–	1 μL
5× RNase P buffer A	1×	1.6 μL
0.5 M Mg(OAc)$_2$	100 mM	1.6 μL
RNase-free ddH$_2$O		3.8 μL

Incubate at 55°C for 5 min, followed by 25 min at 37°C.

3. Combine 16 μL of P RNA mix and 4 μL of substrate mix (final assay concentrations in this example: 100 nM P RNA and trace

amounts of ptRNA^{Gly}, usually indicated as <1 nM), keep at 37°C in a PCR machine (Biometra T Gradient); also incubate the remaining substrate mix at 37°C.

4. Withdraw 4 μL aliquots at defined time points (in this example: 10, 30 s, 1, 15 min); mix aliquots immediately with 6 μL ice-cold 2× denaturing loading buffer, keep on ice until gel loading. As control: at the last kinetic time point, also withdraw 1 μL from the original substrate mix kept at 37°C, mix with 3 μL ddH$_2$O and add 6 μL 2× denaturing loading buffer.

5. Prepare a 20% polyacrylamide (PAA) gel (thickness: 1 mm; width: 30 cm; height: 17 cm) containing 8 M urea in 1× TBE buffer.

6. Rinse the gel pockets with a syringe immediately before gel loading; load the samples and start electrophoresis in 1× TBE buffer at 20 mA until BPB has migrated to approximately 5 cm above the bottom of the gel.

7. Remove the glass plates, wrap up the gel in kitchen wrapping film, and expose an image plate to the gel overnight.

8. Scan the image plate with a phosphorImager (see example in Fig. 1c); for quantitation of cleavage extent at individual time points, encircle each band, either by a rectangle or ellipsoid; quantify the image quants therein (see Note 3).

9. For data evaluation and determination of kinetic parameters, see Notes 3–5.

3.2. P RNA Processing to Generate RNAs with Homogeneous 3′-OH and 5′-Phosphate Termini

1. Preincubate 10 pmol of *E. coli* P RNA for 5 min at 55°C and 55 min at 37°C in a volume of 15 μL 1× RNase P cleavage buffer B.

2. Combine 20 pmol of an RNA-tRNA primary transcript of the type shown in Fig. 2 (in 15 μL 1× RNase P cleavage buffer B) with the P RNA preincubation mixture, followed by incubation for 1.5 h at 37°C.

3. Run a 5–20% PAA/8 M urea gel (depending on the size of the RNA of interest released by P RNA cleavage) and stain the gel with SYBR® Gold (see Note 10).

4. Prepare multiple assays of the type above or scale up for preparative denaturing PAGE followed by elution of the RNA of interest released by P RNA cleavage. For estimation of required RNA amounts, see Note 10.

3.3. Preparation of Total Bacterial RNA Extracts for Assaying P RNA Activity

1. Resuspend ~0.4 g of a frozen bacterial cell pellet in 4.8 mL TREB buffer in a 15-mL Falcon tube (Sarstedt).

2. Add 5 mL of 60°C hot phenol saturated with TREB, incubate for 15 min at 60°C (or 5 min at 65°C), mix every 1–2 min (some protocols advise continuous vigorous shaking, although this is not necessary according to our experience).

3. Centrifuge for 15 min at 8,200 × g (higher g numbers may lead to disruption of the Falcon tube) in a table centrifuge (Eppendorf 5810 R) at room temperature (RT) to separate the phases; transfer the upper aqueous phase to a new Falcon tube and repeat step 2.

4. Extract the aqueous phase with 5 mL CHCl$_3$, centrifuge as in step 3; repeat the CHCl$_3$ extraction.

5. Add 12.5 mL EtOH (2.5-fold volume) to the aqueous phase (5 mL), mix, leave at –20°C for 30 min, then centrifuge in the Falcon tube at 4°C for 1 h at 8,200 × g. *Alternatively, one may distribute the ethanol precipitation to several 1.5-mL Eppendorf tubes, permitting the experimenter to increase the centrifugal force to 11,000 × g (30 min, 4°C).*

6. Cautiously discard the ethanolic supernatant and air-dry the pellet before redissolving the RNA in a total volume of 60–100 μL RNase-free ddH$_2$O.

7. Measure concentration by UV spectroscopy, shock-freeze in liquid nitrogen, and store the total RNA at –80°C.

3.4. Assaying Total Bacterial RNA Extracts for P RNA Activity

1. Prepare total cellular RNA mix:

		Final amount/conc.	MS
Total cellular RNA		5–11 μg	X μL
2.5× RNase P buffer C	1×		6 μL
RNase-free ddH$_2$O			to 15 μL

Incubate at 55°C for 15 min

2. Prepare substrate mix:

		Final amount/conc.	MS
ptRNA^{Gly} (1:10 diluted) 150,000 Cerenkov cpm/μL		0.3 pmol	1 μL
2.5× RNase P buffer C	1×		4 μL
RNase-free ddH$_2$O			5 μL

Incubate at 55°C for 5 min.

3. Add 7 μL of substrate mix to the total RNA mix (final volume 22 μL); incubate at 55°C and withdraw 6-μL aliquots at defined time points (e.g., 0.5, 5, and 35 min); mix aliquots immediately with 6 μL ice-cold 2× denaturing loading buffer (see Subheading 2.1), keep on ice until gel loading. The assay temperature of 55°C was chosen because total RNAs were isolated from thermophilic bacteria harboring thermostable P RNAs.

4. For denaturing PAGE and further steps, see Subheading 3.1 (steps 5–8).

Examples of such activity assays are shown in Fig. 3.

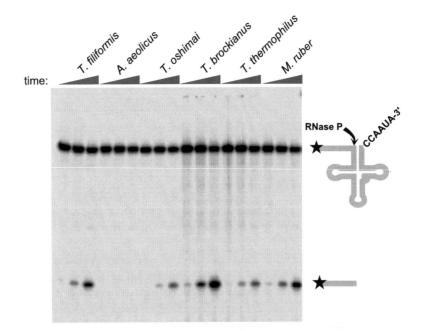

Fig. 3. Processing of 5′-³²P-end-labeled *T. thermophilus* ptRNA^Gly (see Fig. 1b) by P RNA-alone activity in total cellular RNA pools from several bacteria of the genus *Thermus* and one *Meiothermus* (*M. ruber*) strain in comparison with total cellular RNA from *Aquifex aeolicus*. After 15-min preincubation of total RNAs in RNase P buffer C at 55°, trace amounts of 5′-³²P-labeled substrate were added followed by further incubation at 55° for 0.5, 5, and 35 min, respectively. Samples were analyzed by 17% denaturing PAGE. Adapted from ref. (20).

4. Notes

4.1. Single-Turnover Cleavage by E. coli P RNA

1. *E. coli* P RNA was transcribed with T7 RNA polymerase from plasmid pDW98 (21) linearized with BsaA I, followed by phenol and chloroform extractions, preparative 5% denaturing PAGE, UV shadowing, gel elution in 1 M NaOAc (pH 5.0), and ethanol precipitation. The RNA was redissolved in ddH$_2$O, the concentration was determined by UV spectroscopy (260 nm), and the RNA solution was shock-frozen in liquid nitrogen for storage at −80°C.

2. The ptRNA^Gly substrate was transcribed with T7 RNA polymerase from plasmid pSBpt3′hh linearized with BamH I (22). The purification after transcription was as for P RNA, except that 8% denaturing PAGE was used note that three major RNA bands will appear on a denaturing PAA gel: the primary transcript (143 nt) and the two products resulting from hammerhead *cis*-cleavage, i.e., ptRNA^Gly (93 nt) and the hammerhead moiety (50 nt).

3. We use a Bio-Imaging Analyzer FLA 3000-2R (Fujifilm) and the analysis software PCBAS/AIDA (Raytest) for quantification of the 5′-^{32}P-labeled ptRNA substrate and the 5′-cleavage product.

4. For kinetic analyses, determine the fraction of cleaved ptRNA ($f_{cleaved}$) at each time point by dividing the image quants for the released 5′-flank by the sum of the image quants for the 5′-flank and the uncleaved ptRNA. Pseudo-first-order rate constants of cleavage (k_{obs}) are calculated by nonlinear regression analysis (e.g., program Grafit, Erithacus Software), fitting the data to the equation for a single exponential: $f_{cleaved} = f_{endpoint}(1 - e^{k_{obs} \times t})$, where $f_{cleaved}$ = fraction of ptRNA cleaved at a defined time point, t = time, $f_{endpoint}$ = maximum cleavable fraction of ptRNA. Try to design the kinetic experiment such that the time points homogeneously cover the entire range of substrate conversion; if substrate conversion at the first time point has already exceeded 50%, data fitting will become less accurate, although this is hardly avoidable in the case of fast kinetics (k_{obs} approximately >5 min^{-1}). The earliest time point that can be handled is 5 s. If the limits of manual kinetics are reached, and rapid mixing techniques such as stopped-flow kinetics are not considered, the following strategies can be pursued to make manual kinetics still feasible: (i) decrease the pH from 6.0 to 5.5, which will slow down the rate constant for the chemical step by three to fourfold (23); (ii) decrease the Mg^{2+} concentration down to 20 mM, while simultaneously increasing the NH$_4$OAc concentration (from 0.1 to 1 M; (24)); high concentrations of monovalent cations enhance ribozyme-substrate binding and improve RNA folding through counteracting electrostatic repulsion effects within and between polyanionic RNA molecules (21). For example, the maximum rate constant for sto cleavage of ptRNA^{Gly} (<1 nM) by E. coli P RNA (5 μM) reduced to ~1 min^{-1} at 20 mM Mg(OAc)$_2$, 1 M NH$_4$OAc, and pH 5.5 (24), thus being in a range that is easy to handle by manual kinetics.

5. To determine the kinetic parameters $K_{m(sto)}$ and k_{react} (representing the single-turnover K_m and V_{max} parameters), we usually perform single-turnover reactions with constant trace amounts of substrate and increasing excess concentrations of P RNA at a pH < 6.5 to ensure that steps associated with cleavage chemistry limit the reaction. Such plots (k_{obs} over P RNA concentration) are then fitted to the equation: $k_{obs} = k_{react}[\text{P RNA}] / (K_{m(sto)} + [\text{P RNA}])$ (22, 25).

4.2. P RNA Processing to Generate RNAs with Homogeneous 3'-OH and 5'-Phosphate Termini

6. A ribozyme:substrate concentration ratio of 1:2 is considered to be a reasonable starting point, representing a compromise between cleavage efficiency and saving of ribozyme material. 20 pmol of an RNA-tRNA primary transcript is a lower limit amount for the preparation of an RNA of interest. When larger amounts of primary transcripts are subjected to P RNA processing, the ratio of P RNA:substrate RNA may be reduced to save P RNA material.

7. Note that the cleavage rate by *E. coli* P RNA under multiple-turnover conditions will not exceed 1 min^{-1} at 37°C under the conditions specified above, because product release limits the reaction (25). However, when processing assays are conducted at 55°C with the thermostable P RNA from *Thermus thermophilus*, cleavage rates of up to 25 min^{-1} have been achieved (25). Cleavage by *T. thermophilus* P RNA can be performed in the same buffer as used for *E. coli* P RNA (see above). In the case of higher assay temperatures, it is advisable to design primary transcripts harboring a thermostable tRNA moiety, such as the tRNA^{Gly} of *T. thermophilus* (see Fig. 1b), to minimize partial tRNA melting during processing at elevated temperature.

8. The use of a thermostable P RNA, such as the one from *T. thermophilus*, may also be considered if the primary transcript forms stable structures that impair access of P RNA to the cleavage site. Cleavage can be performed at a typical incubation temperature of 55°C (above 70–75°C partial melting of the thermostable P RNA and tRNA moiety will become significant (25, 26), apart from the temperature-dependent affinity loss).

9. Preincubation of *T. thermophilus* P RNA for 20 min at 55°C in RNase P cleavage buffer is essential for ribozyme activation if the cleavage reaction is to be performed at temperatures below 55°C, but can be omitted for cleavage assays conducted at 55–75°C.

10. For preparative denaturing gel electrophoresis and elution of the RNA of interest, one may use UV shadowing (27) if sufficient RNA amounts are available. An alternative is staining with SYBR® Gold (Invitrogen; note that degradation of the dye after repeated freeze–thaw cycles has been reported); with SYBR® Gold we succeeded in detecting 5–10 ng of a 14-nt long RNA 5'-flank (5'-GUU CGG UCA CGA CU) using a gel documentation system (Biostep DH-50) with EtBr filter (575 nm), thus being tenfold more sensitive than EtBr and SYBR Safe® (Invitrogen) or ~100-fold more sensitive than Roti®-Safe (Roth) in our hands.

References

1. Hartmann E, Hartmann RK (2003) The enigma of ribonuclease P evolution. Trends Genet 19: 561–569

2. Walker SC, Engelke DR (2006) Ribonuclease P: the evolution of an ancient RNA enzyme. Crit Rev Biochem Mol Biol 41: 77–102

3. Evans D, Marquez SM, Pace NR (2006) RNase P: interface of the RNA and protein worlds. Trends Biochem Sci 31: 333–341

4. Guerrier-Takada C, Gardiner K, Marsh T, Pace NR, Altman, S (1983) The RNA moiety of ribonuclease P is the catalytic subunit of the enzyme. Cell 35: 849–857

5. Pannucci JA, Haas ES, Hall TA, Harris JK, Brown, JW (1999). RNase P RNAs from some Archaea are catalytically active. Proc Natl Acad Sci USA 96: 7803–7808

6. Kikovska E, Svard SG, Kirsebom, LA (2007) Eukaryotic RNase P RNA mediates cleavage in the absence of protein. Proc Natl Acad Sci USA 104: 2062–2067

7. Li D, Willkomm DK, Schön A, Hartmann RK (2007) RNase P of the *Cyanophora paradoxa* cyanelle: a plastid ribozyme. Biochimie 89: 1528–1538

8. Cho I-M, Lai LB, Susanti D, Mukhopadhyay B, Gopalan V (2010) Proc Natl Acad Sci USA 107: 14573–14578

9. Hartmann RK, Gößringer M, Späth B, Fischer S, Marchfelder A (2009) The making of tRNAs and more - RNase P and tRNase Z. Prog Mol Biol Transl Sci 85: 319–368

10. Holzmann J, Frank P, Löffler E, Bennett KL, Gerner C, Rossmanith W (2008) RNase P without RNA: identification and functional reconstitution of the human mitochondrial tRNA processing enzyme. Cell 135: 462–474

11. Gobert A, Gutmann B, Taschner A, Gößringer M, Holzmann J, Hartmann RK, Rossmanith W, Giegé P (2010) A single *Arabidopsis* organellar protein has RNase P activity. Nat Struct Mol Biol 17: 740–744

12. Haas ES, Banta AB, Harris JK, Pace NR, Brown JW (1996). Structure and evolution of ribonuclease P RNA in Gram-positive bacteria. Nucleic Acids Res 24: 4775–4782

13. Loria A, Pan T (1996) Domain structure of the ribozyme from eubacterial ribonuclease P. RNA 2: 551–563

14. Kirsebom LA, Trobro S (2009) RNase P RNA-mediated cleavage. IUBMB Life 61:189–200

15. Lai LB, Vioque A, Kirsebom LA, Gopalan V (2010) Unexpected diversity of RNase P, an ancient tRNA processing enzyme: challenges and prospects. FEBS Lett 584:287–296

16. Reiter NJ, Osterman A, Torres-Larios A, Swinger KK, Pan T, Mondragón A (2010) Structure of a bacterial ribonuclease P holoenzyme in complex with tRNA. Nature 468: 784–789

17. Milligan JF, Uhlenbeck OC (1989) Synthesis of small RNAs using T7 RNA polymerase. Methods Enzymol 180: 51–63

18. Helm M, Brulé H, Giegé R, Florentz C (1999) More mistakes by T7 RNA polymerase at the 5′ ends of in vitro-transcribed RNAs. RNA 5: 618–621

19. Pleiss A, Derrick ML, Uhlenbeck OC (1998) T7 RNA polymerase produces 5′ end heterogeneity during in vitro transcription from certain templates. RNA 4: 1313–1317

20. Willkomm DK, Feltens R, Hartmann RK (2002) tRNA maturation in *Aquifex aeolicus*. Biochimie 84: 713–722

21. Smith D, Burgin A, Haas ES, Pace NR (1992) Influence of metal ions on the ribonuclease P reaction. J Biol Chem 267:2429–2436

22. Busch S, Kirsebom LA, Notbohm H, Hartmann RK (2000) Differential role of the intermolecular base pairs G292-C75 and G293-C74 in the reaction catalyzed by *Escherichia coli* RNase P RNA. J Mol Biol 299: 941–951

23. Warnecke JM, Fürste JP, Hardt WD, Erdmann VA, Hartmann RK (1996) Ribonuclease P (RNase P) RNA is converted to a Cd(2+)-ribozyme by a single Rp-phosphorothioate modification in the precursor tRNA at the RNase P cleavage site. Proc Natl Acad Sci USA 93: 8924–8928

24. Cuzic-Feltens S, Weber MH, Hartmann RK (2009) Investigation of catalysis by bacterial RNase P via LNA and other modifications at the scissile phosphodiester. Nucleic Acids Res 37: 7638–7653

25. Hardt WD, Schlegl J, Erdmann VA, Hartmann RK (1995) Kinetics and thermodynamics of the RNase P RNA cleavage reaction: Analysis of tRNA 3′-end variants. J Mol Biol 247: 161–172

26. Hartmann RK, Erdmann VA (1991) Analysis of the gene encoding the RNA subunit of ribonuclease P from *T. thermophilus* HB8. Nucleic Acids Res. 19: 5957–5964

27. Frilander MJ, Turunen JJ (2005) RNA ligation using T4 DNA ligase. In: Handbook of RNA Biochemistry (eds. R. K. Hartmann, A. Bindereif, A. Schön, E. Westhof), WILEY-VCH, Weinheim, Germany, pp. 36–52

Chapter 6

Group I Intron Ribozymes

Henrik Nielsen

Abstract

Group I intron ribozymes constitute one of the main classes of ribozymes and have been a particularly important model in the discovery of key concepts in RNA biology as well as in the development of new methods. Compared to other ribozyme classes, group I intron ribozymes display considerable variation both in their structure and the reactions they catalyze. The best described pathway is the splicing pathway that results in a spliced out intron and ligated exons. This is paralleled by the circularization pathway that leads to full-length circular intron and un-ligated exons. In addition, the intronic products of these pathways have the potential to integrate into targets and to form various types of circular RNA molecules. Thus, group I intron ribozymes and associated elements found within group I introns is a rich source of biological phenomena. This chapter provides a strategy and protocols for initial characterization of new group I intron ribozymes.

Key words: Group I intron, Splicing, Hydrolysis, Circular RNA

1. Introduction

Introns are genetic elements interrupting the parts of the gene that encode a product. Group I introns constitute a major class of introns that catalyze their own excision from the flanking exons at the level of the primary transcript. Thus, a group I intron harbors a ribozyme termed a "group I intron ribozyme" that catalyze the splicing reaction. The ribozyme part of the intron consists of a core of 250–500 nt and peripheral elements that mostly serve to stabilize the core. In addition, group I introns can harbor protein coding sequences, repetitive elements, and other sequences. The core of group I intron ribozymes has a characteristic architecture consisting of three helical stacks that all contribute to the formation of the active site (see Fig. 1a). In a useful, although not precise, way of describing this, one helical stack (P4–P6) is referred to as the scaffolding domain and another (P3–P9) as the catalytic domain

Jörg S. Hartig (ed.), *Ribozymes: Methods and Protocols*, Methods in Molecular Biology, vol. 848,
DOI 10.1007/978-1-61779-545-9_6, © Springer Science+Business Media, LLC 2012

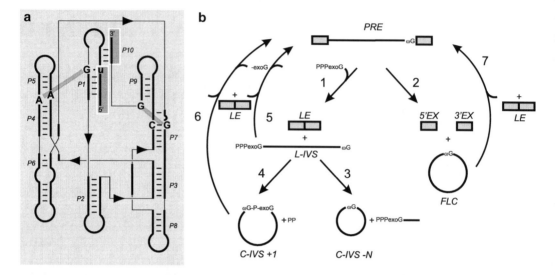

Fig. 1. (a) Helix diagram of the consensus structure of group I intron ribozymes with an indication of conserved nucleotides. (b) Processing of group I introns. See the main text for explanation.

because this domain harbors the binding site for a guanosine involved in the splicing reaction (see below). A substrate domain (P1–P2) that contains the 5′ splice site (5′ SS) and comprises intron as well as exon sequences is docked into a cleft formed by the two other domains. In contrast to the well-conserved architecture of group I intron ribozymes, very little is conserved at the sequence level. The last nucleotide in the upstream exon is almost always a U, and the last nucleotide of the intron is predominantly a G. The P7 stem is well conserved and can be found in sequences using covariation constraints. It has an invariant G-C pair that is directly involved in binding of the guanosine cofactor. The U found as the last nucleotide of the upstream exon is base paired to a G within the intron and the resulting G-U pair is recognized by a structure that includes two conserved A residues.

The best known reaction catalyzed by group I intron ribozymes is the splicing reaction that occurs by two coupled transesterifications (see Fig. 1b, reaction #1). In the first step, the exogenous guanosine cofactor makes a nucleophilic attack at the 5′ SS. The bond between the 5′ exon and the intron is cleaved and the guanosine cofactor becomes linked to the 5′ end of the intron. After a conformational change within the ribozyme structure, the second step consists of an attack by the 3′ terminal nucleotide of the 5′ exon on the 3′ SS. This leads to the release of the intron as a linear molecule and ligation of the exons. Many group I introns are self-splicing in vitro but others require protein factors derived from intronic genes or from the host. The splicing process is reversible (Fig. 1b, reaction #5). Alternatively, the spliced out intron can undergo circularization by attack of the 3′ OH of the intron terminal G (the ωG) at an internal phosphodiester bond to form truncated circles and a short RNA oligo derived from the 5′ end of the intron (Fig. 1b, reaction #3). Recently, an intron was shown to circularize by attack at the triphosphate carried by the guanosine cofactor to form full-length circles including the guanosine and pyrophosphate (Fig. 1b, reaction #4). These circular RNAs can potentially reopen by hydrolysis and reverse splice into ligated exons to restore the unspliced precursor (Fig. 1b, reaction #6). In addition to the splicing pathway, many group I introns can undergo processing following the circularization pathway (Fig. 1b, reaction #2). Here, hydrolysis at the 3′ SS is followed by attack by the ωG at the 5′ SS leading to formation of full-length circular intron RNA (FLC). This pathway does not lead to ligation of the exons and thus appears to benefit the intron at the expense of the host. The full-length circular RNA can integrate into linear targets by an unknown mechanism (Fig. 1b, reaction #7; our unpublished results).

Group I introns have been studied intensively over several decades and the basic aspects of the splicing pathway have been reviewed several times in monographs of RNA biology (1–3). Structural aspects have been the emphasis of many papers and three group I intron ribozymes have been analyzed by X-ray crystallography (reviewed in ref. (4)). The natural history of group I introns with emphasis on distribution and phylogeny was reviewed by Haugen et al. (5) and a recent review on the broader biological aspects of group I introns provided by Nielsen and Johansen (6). Surprisingly, studies are continuously turning up new and interesting aspects of group I intron ribozymes. In 2008, Beckert et al. (7) showed that a group I intron ribozyme most likely had given rise to a new class of ribozymes, the GIR1 branching ribozyme (8) by a perturbation of the active site. In 2009, Vicens and Cech described a 3′, 5′ ligase activity in a group I intron ribozyme leading to the formation of a new type of circular RNA (9), and in 2010, Lee et al. found the first example of an allosteric group I intron ribozyme, that coupled binding of a second messenger to expression of a putative virulence gene (10). In addition to these new findings, group I intron ribozymes are studied in relation to their possible role in the ancient RNA World (e.g., (11)) and as molecular tools, e.g., in repair of cellular mRNAs by trans-splicing (12) (see Chapter 25).

The aim of this chapter is to provide a guide to how new examples of group I intron ribozymes are found as well as protocols on the initial characterization of the reactions that they may catalyze. After the initial discovery of group I intron ribozymes as one of the first two examples of catalytic RNA (together with RNase P RNA) (13), group I intron ribozymes became the testing ground for new methods in RNA biology for many years. The protocols that were developed for the more detailed aspects of these ribozymes are not included in this chapter but references can be found by consulting the reviews mentioned above (in particular (1–3)).

1.1. Identification of New Group I introns

Group I introns are found in genes encoding rRNA, mRNA, and tRNA in the genomes of bacteriophages, bacteria, mitochondria, chloroplasts, the nuclei of eukaryotic microorganisms, and some eukaryotic viruses. Occurrences among metazoans are restricted to sea anemones and sea corals. Large collections of group I introns are found in the Rfam database (http://rfam.sanger.ac.uk) (14), GISSD (http://www.rna.whu.edu.ch/gissd) (15), and CRW (http://ww.rna.ccbb.utexas.edu) (16). Most group I introns are identified as inserts at the level of the gene. A first clue to identifying an insert as a group I intron comes from inspection of the site of insertion. First, group I introns are preferentially found in highly conserved/functionally important sequences of the host gene. Second, the insertion site mostly has a T as the nucleotide immediately preceding the insertion site, and a G as the last nucleotide of the insert. Next, the size of the insert should be assessed. The core of group I introns is 250–500 nt. If the size is larger than this, it may be relevant to analyze for open reading frames expressed from within the intron (in rare cases the ORF may extend into one of the flanking exons). The most important step is to model the sequence of the putative group I intron ribozyme into its characteristic structure. Unfortunately, there are no bioinformatics tools that allow for correct folding of the entire core of the ribozyme, but a combination of alignment with related sequences (to identify covariations), local folding of helices using, e.g., RNAfold (http://rna.tbi.univie.ac.at/cgi-bin/RNAfold.cgi), and manual modeling is feasible. The most important step is to localize the highly conserved P7 architecture. Once this is in place, most of the modeling is straightforward. In folding of a group I intron, it is important to know that 13 subgroups have been described based on differences in P7 and peripheral elements (17, 18).

As an alternative, group I intron ribozymes can be experimentally identified by taking advantage of the uniqueness of the first step of splicing. If an RNA extract including unspliced group I introns is incubated in the presence of [^{32}P]GTP, the intron will "self-label" by addition of the GTP at its 5' end (19). In principle, both [α-^{32}P]GTP and [γ-^{32}P]GTP can be used, but only [γ-^{32}P] GTP ensures incorporation by other means than polymerization. On the other hand, circular RNA that include the guanosine cofactor is not detected if [γ-^{32}P]GTP is used as the label because circularization occurs by attack at the pyrophosphate. After G-labeling, the intron can be sequenced directly by the chemical sequencing method or used as a hybridization probe for cloning of the intron. As an alternative, biotin-GTP can be used in the labeling reaction followed by purification on streptavidin beads followed by cloning or sequence characterization. Obviously, these tricks only function if the intron is self-splicing or if a protein extract that promote splicing can be added.

1.2. Strategy and Protocols for Characterization of the Reactions Catalyzed by a Group I Intron Ribozyme

The group I intron is synthesized as an in vitro transcript using any of the known phage polymerases (T7 RNA polymerase is used here as the example). The first step is to design a collection of oligonucleotides used to make and analyze the transcripts (see Fig. 2). This is not a trivial matter because synthesis of a ribozyme out of its natural context and using a polymerase that has a much higher polymerization rate than the natural polymerase may induce misfolding of the RNA and lead to incorrect conclusions about the properties of the ribozyme. Oligos A and B (or B′) are used in a PCR reaction to synthesize the precursor transcript. The sequence encoding the T7 RNA polymerase promoter is appended to the 5′ end of oligo A. The choice of these oligos is important for several reasons. First, co-folding of the intron with exon sequences is important for the activity of the ribozyme. The 5′ SS is presented in P1 that is made from exon and intron sequences but other parts of the exons may interfere with the proper folding of the intron. To some extent this can be predicted by folding algorithms but mostly it is a matter of trial-and-error. Second, the size of the 5′ exon and 3′ exon should be chosen such that these as well as ligated exons can be easily identified in a gel analysis. Oligo C is used in combination with A to study the second step of the circularization pathway. This oligo has a C complementary to the G at the last position of the intron at the 5′ end. Oligo D is used in combination with B to study the first step (hydrolysis) of the circularization pathway. This oligo is designed to exclude the 5′ SS from the transcript and has a T7 promoter sequence appended to its 5′ end. Oligos E and F are used to make RT-PCR of processing reactions in order to sequence the circularization junctions. Oligo F can also be used to make primer extension analysis of the 5′ end of the spliced out intron and thus to discriminate between linear intron originating from splicing (with an additional G at the 5′ end), reopening of FLC (no G added), and reopening of truncated circular RNA.

The protocols provided in this chapter starts with preparation of in vitro transcripts. Then follows a protocol for folding of the

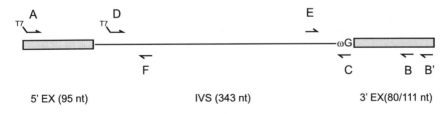

Fig. 2. Strategy for oligonucleotides used in synthesis of transcripts and their subsequent analysis in a first characterization of a group I intron. The intron depicted in the figure is a deletion variant (GIR2) of the Dir.S956-1 (the twin-ribozyme intron from the myxomycete *Didymium iridis*) also used in Fig. 3.

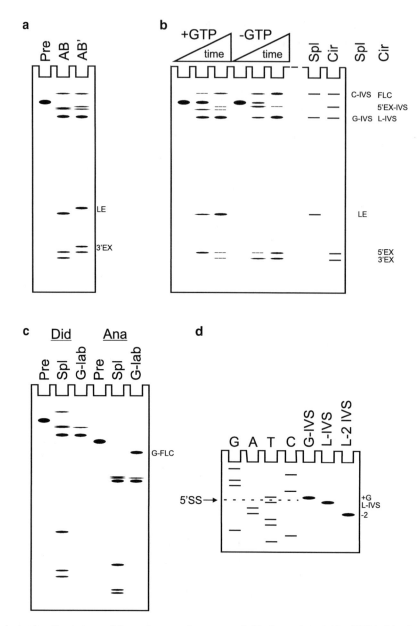

Fig. 3. Gel electrophoretic analyses of the various reactions suggested in the main text. The "AB" in (**a**) refers to the oligo nomenclature introduced in Fig. 2. The part of (**b**) to the *right* is an interpretation of the gel run divided into products of the splicing ("Spl") and circularization ("Cir") pathways, respectively. "Did" in (**c**) is the result obtained with a derivative of the Dir.S956-1 ribosomal RNA intron from *D. iridis* and "Ana" the result with the tRNA[Leu] intron from the cyanobacterium *Anabaena sp.* PCC7120. Three different forms of L-IVS are analyzed by primer extension in (**d**). The figure is explained in Subheading 3.11.

RNA and suggestions for incubation conditions for the various types of reactions. Finally, a few protocols are provided for analysis of the processing products at the sequence level.

The templates for transcription are made by PCR using a DNA polymerase with proofreading activity (e.g., Pfu polymerase) to avoid transcripts with errors that inhibit the reactions. After template purification, unlabeled or radioactively labeled transcripts are made in a standard in vitro transcription reaction. If some of the transcripts are too short (mostly because of premature termination), the transcripts are purified on a denaturing gel. The RNA folds during transcription and in some cases, the co-transcriptional folding combined with the folding that occurs upon incubation in activity buffer has proven to be favorable. In other cases, the transcripts are denatured and refolded. This provides a more well-defined starting point, e.g., for kinetic cleavage analysis. The optimal conditions for ribozyme activity have proven to vary considerably for different group I intron ribozymes. Thus, a panel of buffers is tested for initial analysis and should be followed by careful optimization of the reaction conditions. Some ribozymes show very little in vitro activity, possibly reflecting a requirement for protein factors for stabilization of the active conformation of the ribozyme. Preparation of such proteins is beyond the scope of this chapter. The processing pattern of the ribozyme can often be deduced from running different types of processing reactions next to one another on a gel. Some of the logic behind this is presented in Fig. 3. Verification of the interpretation is done by analysis of individual processing products, e.g., by primer extension or RT-PCR.

2. Materials

2.1. In Vitro Transcription

1. Template DNA at 1 μg/μL of a 3 kb linearized plasmid or 0.2 μg/μL of a 600 bp PCR-product. This will result in a final concentration of T7 promoter in the transcription of ~20 nM.

2. 10× polymerase buffer (for making a standard unlabeled transcript): 100 mM NaCl, 80 mM $MgCl_2$, 20 mM spermidine, 800 mM Tris–HCl, pH 8.0.

3. 10× polymerase buffer ("low $MgCl_2$" for making precursor transcripts): 100 mM NaCl, 20 mM $MgCl_2$, 20 mM spermidine, 800 mM Tris–HCl, pH 8.0 (see Note 1).

4. 100 mM DTT.

5. 10× rNTP mix (for making a standard unlabeled transcript): 10 mM of each rNTP.

6. 10× rNTP mix ("low UTP" for radiolabeled transcripts): 1 mM UTP, 10 mM of each of ATP, CTP, and GTP.

7. T7 RNA polymerase (20 U/μL).

8. [α-^{32}P]UTP (3,000 Ci/mmol; 10 mCi/mL) (this corresponds to ~3 μM in UTP) (see Notes 2 and 3).

2.2. Gel Purification of Transcripts	1. Denaturing polyacrylamide gel (5% polyacrylamide/bisacrylamide [19:1], 7.0 M urea, 1× TBE).	
	2. Elution buffer: 1 mM EDTA, 0.25 M sodium acetate, pH 6.0.	
2.3. Folding (Optional)	1. In vitro transcribed RNA (see Subheadings 2.1 and 2.2).	
	2. 5× renaturation buffer (1×: 20 mM Tris–HCl, pH 7.8, 140 mM KCl).	
	3. 1 M MgCl$_2$.	
2.4. Splicing	1. Labeled precursor transcript AB or AB′ (see Subheadings 2.1 and 2.2).	
	2. 5× reaction buffer. See Table 1 for composition of 1× buffers.	
	3. 2 mM GTP.	
2.5. G Labeling	1. Unlabeled precursor transcript AB or AB′ (see Subheadings 2.1 and 2.2).	
	2. 5× reaction buffer: See Table 1 for composition of 1× buffers.	
	3. 2 mM GTP.	
	4. [α-^{32}P]GTP or [γ-^{32}P]GTP (3,000 Ci/mmol; 10 mCi/mL).	
2.6. Hydrolysis	1. Labeled precursor transcript AB or AB′ (see Subheadings 2.1 and 2.2).	
	2. 5× reaction buffer. See Table 1 for composition of 1× buffers.	
2.7. Circularization	1. Labeled precursor transcript AC (see Subheadings 2.1 and 2.2)	
	2. 5× reaction buffer. See Table 1 for composition of 1× buffers.	

Table 1
Suggestions for 1× reaction buffers (modified from ref. (20))

	Monovalent salt	Divalent salt	Buffer	Additive	T (°C)
I	25 mM NaCl	15 mM Mg$_2$Cl	25 mM HEPES (pH 7.5)	–	32
II	25 mM NaCl	15 mM Mg$_2$Cl	25 mM HEPES (pH 7.5)	–	42
III	1.0 M NaCl	15 mM Mg$_2$Cl	25 mM HEPES (pH 7.5)	–	32
IV	25 mM NaCl	200 mM Mg$_2$Cl	25 mM HEPES (pH 7.5)	–	32
V	1.0 M NH$_4$OAc	100 mM Mg$_2$Cl	25 mM HEPES (pH 7.5)	–	37
VI	0.5 M KCl	15 mM Mg$_2$Cl	40 mM Tris–HCl (pH 7.5)	5 mM DTT 2 mM spermidine	45

2.8. Gel Electrophoretic Analysis of Processing Reactions

1. UBB: 1× TBE (TBE is the electrophoresis buffer used in UPAGE), 7.0 M urea, 1 mg/mL bromophenol blue, 1 mg/mL xylene cyanol.

2. 5% Denaturing polyacrylamide gel: 5% polyacrylamide/bisacrylamide [19:1], 7.0 M urea, 1× TBE.

3. 4% Denaturing polyacrylamide gel (4% polyacrylamide/bisacrylamide [19:1], 7.0 M urea, 0.4× TBE) (see Note 4).

4. 1× TBE: 100 mM Tris-base, 83 mM boric acid, 1.0 mM EDTA.

2.9. Primer Extension Analysis of Linear Intron

1. RNA. An aliquot of a processing reaction or a gel purified product can be used.

2. Oligo F (10 pmol/μL).

3. 5× PNK (forward) buffer : 50 mM $MgCl_2$, 25 mM DTT, 0.05 mM spermidine-HCl, 0.05 mM EDTA, 250 mM Tris–HCl, pH 7.6.

4. [γ-^{32}P]ATP (3,000 Ci/mmol; 10 mCi/mL).

5. T4 Polynucleotide kinase (PNK).

6. 5× RT buffer: 250 mM KCl, 20 mM $MgCl_2$, 50 mM DTT, 250 mM Tris–HCl, pH 8.3 (at 25°C) (normally provided with the enzyme).

7. dNTP (2 mM each of dATP, dCTP, dGTP, dTTP).

8. Reverse transcriptase (RT), e.g., Revertaid from Fermentas 200 U/μL.

9. UBB: 1× TBE (TBE is the electrophoresis buffer used in UPAGE), 7.0 M urea, 1 mg/mL bromophenol blue, 1 mg/mL xylene cyanol.

10. Sequencing ladder of PCR product AB made with oligo F.

2.10. RT-PCR and Sequencing Analysis of Ligated Exons and Circularization Sites

1. RNA. An aliquot of a processing reaction or a gel purified product can be used.

2. Oligos A, B, E, and F (each 10 pmol/μL).

3. (Optional) Radioactively labeled oligos A, B, E, and F (each 50 pmol/μL) (see Subheadings 2.9 and 3.9 for labeling).

4. 5× RT buffer: 250 mM KCl, 20 mM $MgCl_2$, 50 mM DTT, 250 mM Tris–HCl, pH 8.3 (at 25°C) (normally provided with the enzyme).

5. dNTP (2 mM each of dATP, dCTP, dGTP, dTTP).

6. Reverse transcriptase (RT), e.g., Revertaid from Fermentas 200 U/μL.

7. Standard PCR reagents and utensils.

8. DNA sequencing reagents.

3. Methods

3.1. In Vitro Transcription

1. Set up the transcription reaction by adding at room temperature the components in a siliconized or Teflon-coated tube in the following order (see Note 5):
   - 5 μL of 5× transcription buffer
   - 4 μL of 10× rNTP mix ("low UTP" for radioactive transcripts)
   - (Optional) 1–5 μL of 3,000 Ci/mmol, 10 mCi/mL [α-^{32}P]UTP
   - 2.5 μL of 100 mM DTT
   - 11.5 μL DEPC-treated dH$_2$O (less if radiolabel was added in previous step)
   - 1 μL of template DNA (linearized plasmid or PCR-product)
   - 1 μL 10 U of the appropriate (in this case T7) RNA polymerase

2. Incubate for 30–60 min at 37°C.

3. Phenol extract and ethanol precipitate the transcripts (see Note 6).

3.2. Gel Purification of Transcripts

3.2.1. Unlabeled RNA

1. The RNA is localized in the gel by UV shadowing. Wrap the gel in plastic wrap and place it over a sheet of Xerox paper or a Thin-Layer Chromatography plate. Inspect the gel under UV-light (e.g., a UV$_{254}$ handheld lamp). Due to absorption of the UV light by the RNA, the RNA band will appear as a shadow on the Xerox paper or TLC-plate. Cut out a gel slice (as small as possible) using a disposable scalpel and place it in a tube (see Note 7).

2. Add approximately 400 μL of elution buffer (0.25 M sodium acetate (pH 6.0), 1 mM EDTA) and 200 μL of buffer saturated phenol. Wrap the tube in parafilm to avoid leakage.

3. Elute the RNA at room temperature with continuous shaking for a few hours or overnight in the cold room with or without shaking.

4. Transfer the liquid to a second tube, separate the phases and ethanol precipitate the RNA from the aqueous phase. The recovery is usually well in excess of 50%. The gel slice can be subjected to a second round of elution to increase the recovery (see Note 8).

3.2.2. Labeled RNA

1. The RNA is localized in the gel by autoradiography. Proceed as described in Subheading 3.2.1.

3.3. Folding (Optional)

1. Heat denature the RNA at 90°C for 1 min in 20 mM Tris–HCl, pH 7.8, 140 mM KCl.

2. Transfer to 60°C and leave for 15 min.

3. Cool slowly to 30°C over a 15-min period.

4. Add MgCl$_2$ to a final concentration of 2.5 mM and leave at 30°C for 15 min.

5. Transfer to 0°C (see Note 9).

3.4. Splicing (see Notes 10 and 11)

1. Combine in a siliconized or Teflon-coated tube:

Radiolabeled RNA (AB or AB′)	x μL
5× reaction buffer	4 μL
2 mM GTP	2 μL
dH$_2$O to make the final volume	20 μL

2. Incubate at the desired temperature (see Table 1) for 30 min to 24 h.

3. Terminate the reaction by ethanol precipitation if the reaction is to be used for, e.g., primer extension or RT-PCR and by addition of 1 vol. of UBB for gel electrophoretic analysis.

3.5. G Labeling

1. Combine in a siliconized or Teflon-coated tube:

Unlabeled RNA (AB or AB′; final 10–50 nM)	x μL
5× reaction buffer	4 μL
2 mM GTP (see Note 12)	2 μL
[α-^{32}P]GTP or [γ-^{32}P]GTP	2 μL
dH$_2$O to make the final volume	20 μL

2. Incubate at the desired temperature (see Table 1) for 30 min to 24 h.

3. Terminate the reaction by ethanol precipitation if the reaction is to be used for, e.g., primer extension or RT-PCR and by addition of 1 vol. of UBB for gel electrophoretic analysis.

3.6. Hydrolysis

1. Combine in a siliconized or Teflon-coated tube:

Radiolabeled RNA (AB or AB′)	x μL
5× reaction buffer	4 μL
dH$_2$O to make the final volume	20 μL

2. Incubate at the desired temperature (see Table 1) for 30 min to 24 h.

3. Terminate the reaction by ethanol precipitation if the reaction is to be used for, e.g., primer extension or RT-PCR and by addition of 1 vol. of UBB for gel electrophoretic analysis.

3.7. Circularization

1. Combine in a siliconized or Teflon-coated tube:

Radiolabeled RNA (AC)	x µL
5× reaction buffer	4 µL
dH$_2$O to make the final volume	20 µL

2. Incubate at the desired temperature (see Table 1) for 30 min to 24 h.

3. Terminate the reaction by ethanol precipitation if the reaction is to be used for, e.g., primer extension or RT-PCR and by addition of 1 vol. of UBB for gel electrophoretic analysis.

3.8. Gel Electrophoretic Analysis of Processing Reactions

1. Run the samples from processing reactions according to the outline in Fig. 3.

3.9. Primer Extension Analysis of Linear Intron (see Note 13)

1. Label the oligo by mixing:

Oligonucleotide (10 pmol/µL)	1 µL
5× PNK forward reaction buffer	4 µL
[γ-^{32}P]ATP	5 µL
T4 PNK	1 µL
dH$_2$O to make the final volume	20 µL

2. Incubate for 15 min at 37°C.

3. Inactivate the enzyme by incubation for 10 min at 68°C (see Note 14).

4. Set up the primer extension reaction by mixing in a siliconized (or Teflon-coated) tube:

RNA	3.8 µL
5× RT buffer	1.2 µL
Oligo (0.5 pmol/µL) radioactively labeled	1.0 µL
dH$_2$O to to make the final volume	6.0 µL

5. Heat to 90°C for 30 s and transfer to the appropriate annealing temperature (mostly 42°C) for 5 min. Spin briefly to collect.

6. Add

dNTP mix	1.0 µL
5× RT buffer	0.8 µL
dH$_2$O/reverse transcriptase (0.1 µL/reaction)	2.2 µL

7. Incubate at 42°C for 15 min.

8. Add 1 vol. of UBB (10 µL) to terminate the reaction.

3.10. RT-PCR and Sequencing Analysis of Ligated Exons and Circularization Sites

1. Make a first strand cDNA as described in Subheading 3.9, steps 1–7. For analysis of ligated exons, use oligo B for the RT step. For analysis of circular RNA, use oligo F for the RT step.

2. Use 2 µL of the RT reaction for the PCR step. For ligated exons, use oligos A and B. For circular RNA use oligos E and F.

3. Purify and sequence the PCR products using the same primers as used for PCR as sequencing primers.

3.11. Interpretation of the Processing Reactions (see Fig. 3)

The gel electrophoretic analysis of the processing reactions can be quite complex depending on the pathways that are active (see Fig. 1). A relatively simple approach to the analysis is outlined in Fig. 3. First, in Fig. 3a, two length variants (AB and AB′) that differ in the length of the 3′EX are run in parallel. This results in identification of ligated exon (LE) that is the hallmark of the splicing reaction because of the predicted shift in size of this product. Similarly, the 3′EX resulting from the first step of circularization can be identified. Other products that show a shift in size (IVS-3′EX) may or may not be resolved on the gel. Next, in Fig. 3b, precursor transcripts are incubated in parallel with and without GTP. This experiment is done as a time-course experiment because the disappearance and appearance of bands as a function of time can aid in their identification. In the present case, the incubation with GTP results in a mixed reaction with products of both the splicing and circularization pathways, respectively. However, only the circularization pathway is active in the absence of GTP. Here, the products of the first step (5′EX-IVS and 3′EX) and second step (5′EX and FLC) as well as reopened circular RNA (L-IVS) are seen to accumulate while the precursor is used up. With this interpretation at hand, the processing pattern obtained in the presence of GTP can be understood by subtraction of the above mentioned products from the band pattern. The splicing products that are seen to accumulate are then the spliced out intron and ligated exons (G-IVS and LE), and the truncated circular RNA originating from circularization of the spliced out intron (C-IVS). The 5′ end of the intron that is cleaved of in this circularization reaction is very short and usually not seen in this type of gel analysis, but it can easily be detected if G-labeled intron RNA is incubated at circularization conditions and subsequently analyzed on a high percentage polyacrylamide gel. The products of the first step of splicing (5'EX and G-IVS-3'EX) are sometimes seen transiently in the analysis. Figure 3c shows a G-labeling experiment in which the G-labeling reaction (unlabeled precursor RNA reacted in the presence of [γ-^{32}P]GTP) is run in parallel with a body-labeled splicing reaction. Two different types of introns are analyzed. In the first ("Did"), the 3' product of the first step of splicing and the intron are labeled, as expected. In the second ("Ana"), an additional band ("G-FLC") appears and can be shown by other types of analysis to be the result of attack of the 3' end of the intron on the pyrophosphate of the guanosine cofactor added to the 5' end of the intron, as described previously. In case of difficulties in assigning a band on the gel to a particular processing product, it is possible to excise the band from the gel and analyze it by re-incubation at reaction conditions or by sequence analysis. Finally, in Fig. 3d, a primer extension (px) analysis using primer F and linear intron isolated from three different reactions is shown. RNA from the splicing reaction (G-IVS) gives a px product one nucleotide longer than the intron 5' end. The px product from reopening of circular RNA in the circularization pathway (L-IVS) corresponds exactly to the intron 5' end, and RNA from reopened circular RNA in the splicing pathway (L-2 IVS) is shorter than the 5' end by two nucleotides in this case. The latter can vary considerably and often, several shorter products are made.

4. Notes

1. The transcripts may react during in vitro transcription. In order to obtain a precursor transcript for use in the various types of reactions, the in vitro transcription can be performed at 2 mM MgCl$_2$ that supports transcription but in many cases is insufficient for splicing and other reactions. The yield of transcripts is less at low MgCl$_2$ and it may be necessary to scale up the reaction. As an alternative, precursor transcripts can be gel purified from a standard in vitro transcription reaction as described in Subheading 3.2.

2. ^{32}P is a high energy β-emitter. Avoid exposure to the radiation and radioactive contamination. Wear disposable gloves when handling radioactive solutions. Check your gloves and pipettes frequently for radioactive contamination. Use protective laboratory equipment (protective eyeglasses, Plexiglas shields) to minimize exposure to radiation. Dispose of radioactive waste in accordance with the rules and regulations established at your institution.

3. Any of the four rNTPs can be used as label. The main concern is to avoid using a nucleotide that is prevalent in the first 10–12 nt of the transcript and this criteria will in many cases argue against GTP because G's are required at +1 and +2 and preferred at +3 positions.

4. The gel electrophoretic mobility of circular RNA depends on the gel% and the buffer strength. Thus, the circular reaction products are better separated from the linear counterparts on a 4% UPAG, 0.4× TBE, than on the standard 5% UPAG, 1× TBE. In fact, the mobility shift observed when comparing gel runs on the two types of gels serves as a first indication of the presence of circular RNA (or other isoforms) in the reaction.

5. The order of assembling the reaction is to avoid spermidine precipitation of the template DNA, especially at low temperatures.

6. Unincorporated radioactive nucleotides can be removed from the transcription reaction by gel filtration on a Sephacryl S-200 spin column (GE Healthcare) or similar. This will also desalt the sample.

7. At this stage, many protocols recommend crushing of the gel with a pipette tip to facilitate the elution. However, polyacrylamide fragments will make subsequent pipetting steps difficult, and we prefer to compromise on the recovery and leave the gel slice intact.

8. An alternative to elution by diffusion is electroelution. Here, the gel slice is placed in a dialysis bag containing electrophoresis buffer and placed in the electrophoresis chamber perpendicular to the electrical field. After 1 h of electrophoresis at 10 V/cm, the field is reversed for a few minutes, and the RNA can be recovered from the buffer in the dialysis bag. Several companies have made specialized electrophoresis units for electroelution of nucleic acids (e.g., BioRad).

9. The folding of the RNA can be assessed by running it on a non-denaturing (native) polyacrylamide gel (6% polyacrylamide/bisacrylamide [19:1], 15 mM $MgCl_2$, 25 mM NaCl, 0.5× TBE, run at 4°C).

10. This and the following processing protocols define a starting point for the experiments. The RNA can be used directly from the transcription reaction with the co-transcriptional folding preserved or it can be denatured and folded under more well-defined conditions as described in Subheading 3.3 (required for kinetic analyses). The amount of RNA used in a reaction depends on the purpose and 10–50 nM can serve as a starting point. The reaction buffers provided in Table 1 are buffers found in the literature for characterization of different group I introns or other ribozyme catalyzed reactions and can be used

for inspiration. Likewise, the optimal temperature and length of incubation has to be determined empirically.

11. The various processing steps can be isolated for kinetic analyses. For this the starting point has to be well defined. The precursor RNA should be gel purified and the RNA renatured at conditions that are nonpermissive for the reaction. Splicing is analyzed by precursor removal in a splicing reaction provided that hydrolysis at the 3′ SS (first step of the circularization pathway) can be neglected. Hydrolysis at the 3′ SS is analyzed in a similar reaction omitting GTP. This step can in principle also be analyzed using RNA from a hydrolysis construct (DB). However, this type of constructs frequently uncovers a very fast reaction that is suppressed in the full-length construct and possibly irrelevant in understanding the biology of the ribozyme. The second step of the circularization pathway is analyzed in construct AC and reopening of C-IVS is analyzed by incubation of gel purified and renatured RNA, although this is not optimal.

12. Cold GTP can be omitted in order to increase the specific activity of the radiolabel. This is required for G labeling of whole cell RNA in search for new group I introns.

13. This is a basic scheme that should be adapted to the experiment in question. In experiments involving several samples, two master mixes should be prepared—one containing the oligo-nucleotide/5× buffer/dH$_2$O and the other containing dNTP mix/5× buffer/dH$_2$O and enzyme. The reaction can be scaled as required. The ratio between RNA and oligo should be considered in case of problems. The enzyme can be any reverse transcriptase. We mostly use M-MuLV RT, but AMV RT is a good alternative with structured RNAs.

14. The oligonucleotide can be purified by gel filtration on a spin column (e.g., G25 Sephadex) or by ethanol precipitation.

References

1. T.R. Cech and B.L. Golden (1999) Building a Catalytic Active Site Using Only RNA. In R.F. Gesteland, T.R. Cech and J.F. Atkins (eds.), *The RNA World*. Cold Spring Harbor Laboratory Press, Cold Spring Harbor; New York, pp. 321–349.

2. Hougland, J.L., Piccirilli, J.A., Forconi, M., Lee, J. and Herschlag, D. (2006) How the Group I Intron Works: A Case Study of RNA Structure and Function. In Gesteland, R.F., Cech, T.R. and Atkins, J.F. (eds.), *The RNA World*. CSHL Press, pp. 133–205.

3. Golden, B.L. (2008) Group I Introns: Biochemical and Crystallographic Characterization of the Active Site Structure. In Lilley, D.M. and Eckstein,F. (eds.), *Ribozymes and RNA Catalysis*. RSC Publishing, pp. 178–200.

4. Vicens, Q. and Cech, T.R. (2006) Atomic level architecture of group I introns revealed. *Trends Biochem. Sci.*, **31**, 41–51.

5. Haugen, P., Simon, D.M. and Bhattacharya, D. (2005) The natural history of group I introns. *Trends Genet.*, **21**, 111–119.

6. Nielsen, H. and Johansen, S.D. (2009) Group I introns: Moving in new directions. *RNA. Biol.*, **6**, 375–383.

7. Beckert, B., Nielsen, H., Einvik, C., Johansen, S.D., Westhof, E. and Masquida, B. (2008) Molecular modelling of the GIR1 branching ribozyme gives new insight into evolution of structurally related ribozymes. *EMBO J.*, **27**, 667–678.

8. Nielsen, H., Westhof, E. and Johansen, S. (2005) An mRNA is capped by a 2′, 5′ lariat catalyzed by a group I-like ribozyme. *Science*, **309**, 1584–1587.

9. Vicens, Q. and Cech, T.R. (2009) A natural ribozyme with 3′,5′ RNA ligase activity. *Nat. Chem. Biol.*, **5**, 97–99.

10. Lee, E.R., Baker, J.L., Weinberg, Z., Sudarsan, N. and Breaker, R.R. (2010) An allosteric self-splicing ribozyme triggered by a bacterial second messenger. *Science*, **329**, 845–848.

11. Vaidya, N. and Lehman, N. (2009) One RNA plays three roles to provide catalytic activity to a group I intron lacking an endogenous internal guide sequence. *Nucleic Acids Res.*, **37**, 3981–3989.

12. Fiskaa, T. and Birgisdottir, A.B. (2010) RNA reprogramming and repair based on trans-splicing group I ribozymes. *N. Biotechnol.*, **27**, 194–203.

13. Kruger, K., Grabowski, P.J., Zaug, A.J., Sands, J., Gottschling, D.E. and Cech, T.R. (1982) Self-splicing RNA: autoexcision and autocyclization of the ribosomal RNA intervening sequence of Tetrahymena. *Cell*, **31**, 147–157.

14. Gardner, P.P., Daub, J., Tate, J.G., Nawrocki, E.P., Kolbe, D.L., Lindgreen, S., Wilkinson, A.C., Finn, R.D., Griffiths-Jones, S., Eddy, S.R. *et al.* (2009) Rfam: updates to the RNA families database. *Nucleic Acids Res.*, **37**, D136–D140.

15. Zhou, Y., Lu, C., Wu, Q.J., Wang, Y., Sun, Z.T., Deng, J.C. and Zhang, Y. (2008) GISSD: Group I Intron Sequence and Structure Database. *Nucleic Acids Res.*, **36**, D31-D37.

16. Cannone, J.J., Subramanian, S., Schnare, M.N., Collett, J.R., D'Souza, L.M., Du, Y., Feng, B., Lin, N., Madabusi, L.V., Muller, K.M. *et al.* (2002) The comparative RNA web (CRW) site: an online database of comparative sequence and structure information for ribosomal, intron, and other RNAs. *BMC. Bioinformatics.*, **3**, 2.

17. Michel, F. and Westhof, E. (1990) Modelling of the three-dimensional architecture of group I catalytic introns based on comparative sequence analysis. *J. Mol Biol*, **216**, 585–610.

18. Suh, S.O., Jones, K.G. and Blackwell, M. (1999) A Group I intron in the nuclear small subunit rRNA gene of Cryptendoxyla hypophloia, an ascomycetous fungus: evidence for a new major class of Group I introns. *J. Mol. Evol.*, **48**, 493–500.

19. Reinhold-Hurek, B. and Shub, D.A. (1993) Experimental approaches for detecting self-splicing group I introns. *Methods Enzymol.*, **224**, 491–502.

20. Vicens, Q., Paukstelis, P.J., Westhof, E., Lambowitz, A.M. and Cech, T.R. (2008) Toward predicting self-splicing and protein-facilitated splicing of group I introns. *RNA.*, **14**, 2013–2029.

Chapter 7

Kinetic Characterization of Group II Intron Folding and Splicing

Olga Fedorova

Abstract

Group II introns are large self-splicing ribozymes found in bacterial genomes, in organelles of plants and fungi, and even in some animal organisms. Many organellar group II introns interrupt important housekeeping genes; therefore, their splicing is critical for the survival of the host organism. Group II introns are versatile catalytic RNAs: they facilitate their own excision from a pre-mRNA, they promote ligation of exons to form a translation-competent mature mRNA; they can act like mobile genomic elements and insert themselves into RNA and DNA targets with remarkable precision, which makes them attractive tools for genetic engineering. The first step in characterization of any group II intron is the evaluation of its catalytic activity and its ability to properly fold into the native functionally active structure. This chapter describes kinetic assays used to characterize folding and catalytic properties of group II intron-derived ribozymes.

Key words: Ribozyme, Group II intron, Folding, Kinetics

1. Introduction

Group II introns are large natural ribozymes capable of facilitating multiple biochemical reactions (1). Remarkably, they self-splice by a mechanism resembling spliceosomal splicing of an mRNA (see Fig. 1), suggesting that group II introns and the spliceosome may have a common ancestor (1, 2). In addition, some group II introns are mobile genomic elements, which efficiently and accurately insert themselves into RNA and double-stranded DNA targets (2, 3). The latter quality makes them attractive for potential application in gene therapy.

Group II intron RNAs form a conserved secondary structure consisting of six domains (2, 4, 5) (see Fig. 1a). Only Domains 1 and 5 are indispensable for the catalytic function (1, 2, 5). Most phylogenetically conserved Domain 5 is the catalytic center of the ribozyme. Domain 1 (D1) serves as a scaffold for docking of other domains and

Jörg S. Hartig (ed.), *Ribozymes: Methods and Protocols*, Methods in Molecular Biology, vol. 848,
DOI 10.1007/978-1-61779-545-9_7, © Springer Science+Business Media, LLC 2012

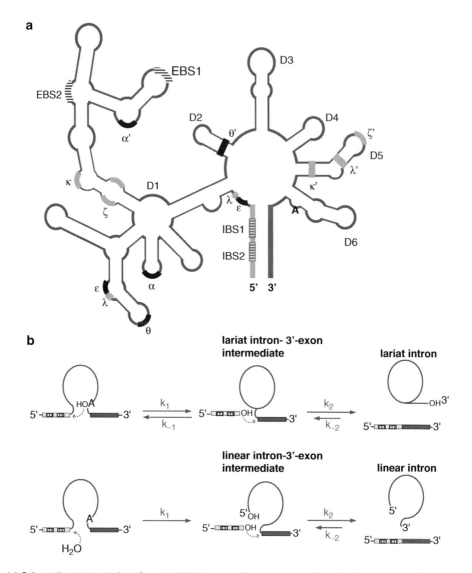

Fig. 1. (**a**) Schematic representation of a group II intron secondary structure. *Gray lines* indicate main long-range tertiary interactions between catalytically critical elements in D5 and D1; other tertiary interactions are shown in *black. Patterned lines* show exonic substrate recognition sequences in D1 (EBS1 and EBS2) and their counterparts IBS1 and IBS2 in the 5′-exon. (**b**) Schematic of two major pathways of the group II intron splicing: branching (*top*) and hydrolysis (*bottom*).

recognition of exonic substrates (5). It is an independently folding unit; folding of D1 is the rate-limiting step for folding of the entire intron (6). Domain 3 (D3) functions as an allosteric catalytic effector increasing the rate of any reaction catalyzed by a group II ribozyme (5). Domain 6 (D6) provides a branch-point adenosine, which serves as a nucleophile in the branching pathway of splicing (1, 2, 5). Domain 2 (D2) mediates conformational rearrangements between two steps of splicing, and Domain 4 (D4) may contain an ORF encoding a maturase protein, which is required for folding, splicing, and mobility functions of some group II introns (2, 5).

Group II introns splice via two sequential transesterifications; both in vitro and in vivo the first step can occur via two parallel pathways: branching and hydrolysis (1, 2) (see Fig. 1b). Notably, different monovalent ions facilitate different pathways of the first step of splicing: ammonium ions, especially ammonium sulfate, tend to promote the branching pathway, resulting in the formation of a lariat intron (see Figs. 1b and 3a) (1, 2). Potassium ions, on the other hand, promote hydrolytic pathway, resulting in a release of the linear form of the intron (see Figs. 1b and 3a) (1, 2).

One of the most striking features of group II introns is their modularity. These ribozymes can be disassembled into two or more building blocks and then put back together to restore catalytic activity (see Fig. 2) (7). The most commonly used trans-constructs were designed to test the first step of splicing via branching or hydrolysis. These constructs consist of two pieces: the first piece

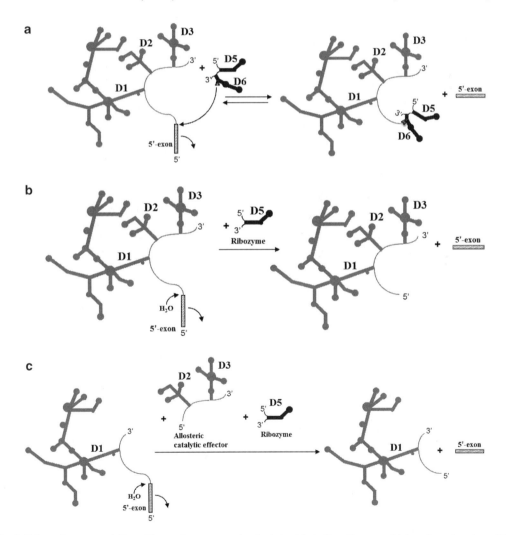

Fig. 2. Schematic representation of trans-ribozyme constructs derived from the ai5γ group II intron: two-piece branching system (**a**), two-piece hydrolysis system (**b**) and three-piece hydrolysis system (**c**).

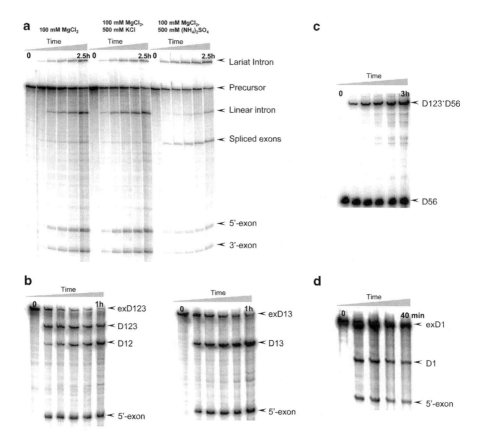

Fig. 3. Analysis of reaction products for different ai5γ ribozyme constructs on a 5% denaturing polyacrylamide gel. Reaction products are listed on the *right* side of each gel. (**a**) Splicing of the ai5γ intron under different ionic conditions. Reaction time points are 0, 5, 15, 30 min, 1, 2.5 h. (**b**) Hydrolysis reactions in the two-piece hydrolysis systems consisting of D5 and exD123 (*left*) or D13 (*right*) RNAs. Time points are 0, 10, 20, 30, 40 min, and 1 h. (**c**) Branching reaction in the two-piece system consisting of D56 and exD123 RNAs. Time points are 0, 10, 30 min, 1, 2, 3 h. (**d**) Hydrolysis reaction in the three-piece construct consisting of D5, exD1, and D23 RNAs. Time points are 0, 10, 20, 30, and 40 min.

contains 5′-exon and domains 1, 2, and 3 (exD123 RNA) and the second piece contains either domains 5 and 6 (D56 RNA, trans-branching system) (8–11) (see Fig. 2a) or domain 5 alone (D5 RNA, trans-hydrolysis system) (see Fig. 2b). In this context exD123 RNA is treated as a substrate and D56 or D5 RNA as a ribozyme. These two-piece constructs are frequently used to determine the effect of base mutations or single atom substitutions on the ribozyme catalytic activity. Kinetic analyses are carried out either under single turnover conditions (excess of ribozyme D5 or D56 over the substrate (exD123)) or under multiple turnover conditions (excess of the substrate over the ribozyme). These ribozyme constructs can be divided even further, into three-piece systems, where exD1 RNA is a substrate and separately transcribed D5 (D56) and D23 RNAs are a ribozyme and an allosteric catalytic effector, respectively (12) (see Fig. 2c). Notably, reaction rates of hydrolysis, and especially branching, are much lower in three-piece systems compared to two-piece systems.

Folding of group II introns has been extensively studied using the ai5γ intron as a model system (6). Folding of this RNA into the native structure requires either high monovalent and divalent salt concentration or help of protein cofactors, for example, yeast proteins Mss 116 or Ded1 or *Neurospora crassa* protein CYT 19 (6). Notably, ribozyme constructs used in folding studies contain either no exons or very short exons, because long exons have been shown to promote misfolding (13). Multiple studies suggest that the ai5γ intron folds slowly, but directly to the native state via formation of on-pathway folding intermediate(s) in D1 (14–16). The D135 (or D1356) ribozyme is most commonly used for folding studies. It contains full-length intronic domains 1, 3, and 5 (or 5 and 6) and domains 2 and 4 shortened to 5 and 6 bp stem-loops, respectively. It is an active ribozyme, which cleaves an oligonucleotide substrate with multiple turnover (14) (see Fig. 4).

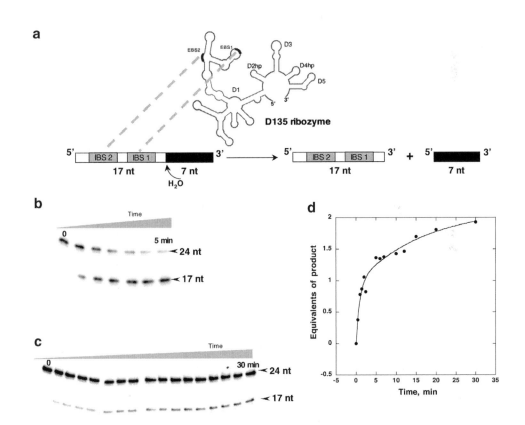

Fig. 4. (a) Schematic of the RNA oligonucleotide hydrolysis by the D135 ribozyme. *Dashed lines* indicate base-pairing between EBS1 and EBS2 sequences of the ribozyme and IBS1 and IBS2 sequences of the substrate, respectively. (b) Products of the RNA oligo substrate hydrolysis by the D135 ribozyme under single turnover conditions (100 nM ribozyme and 5 nM 5′-end labeled substrate) were analyzed on a 20% denaturing polyacrylamide gel. Time points are 0, 30 s, 1, 1.5, 2, 3, 5 min. (c) Analysis of the RNA oligo substrate cleavage by the D135 ribozyme under multiple turnover conditions (20 nM ribozyme and 200 nM 5′-end labeled substrate). Time points are listed in Subheading 3.4.2, step D. Band intensities were quantified using Storm 820 Phospholmager (GE Healthcare). Fraction of hydrolyzed product at each time point was calculated, converted into equivalents of product per one equivalent of ribozyme, plotted vs. time, and fit to a double-exponential equation (d).

Folding of this ribozyme can be monitored either directly, using native gel electrophoresis or single-molecule approach, or indirectly, using the oligo substrate cleavage as a measure of the correctly folded ribozyme.

Based on the phylogenetic comparison, group II introns were initially subdivided into three main classes: IIA, IIB, and IIC (17). This classification was subsequently expanded to include bacterial classes D, E, and F (17–19). However, so far biochemical studies of group II introns have been mostly limited to the yeast intron ai5γ, which belongs to the IIB class. Recent studies suggest that group II introns from different classes may vary in their catalytic properties. They greatly differ in splicing efficiency and may even generate different reaction products (20). Therefore biochemical characterization of folding and splicing of group II introns from other classes is essential for better understanding of folding pathways of large RNAs, as well as mechanisms of group II intron splicing, spliceosomal splicing, and intron mobility. This chapter summarizes essential protocols, which were developed over the years in the laboratory of Prof. A.M. Pyle at Yale University to study folding and catalytic function of other group II introns and possibly other large ribozymes.

2. Materials

2.1. Preparation of RNA Molecules

1. RNase-free water. All solutions must be prepared using RNase-free water. We use deionized water purified by Milli Q water purification system (Millipore). All solutions are passed through Nalgene filter units for further sterilization. Alternatively, distilled water autoclaved in the presence of diethylpyrocarbonate may be used as well, although it is less desirable due to potential introduction of contaminants. It is important that all experiments are carried out using RNase-free plasticware and glassware.

2. DNA templates for transcription (linearized plasmids or PCR-generated templates) at 1 μg/μL in TE buffer (see below). Stored at –20°C.

3. T7 RNA polymerase. We use in-house expressed and purified T7 RNA polymerase (1.6–2 μg/μL stock). Alternatively, T7 RNA polymerase may be purchased from Epicentre Biotechnologies or Ambion (now owned by Applied Biosystems).

4. Protector RNase inhibitor (Roche).

5. 30 mM Solutions of ATP, GTP, CTP, and UTP in 10 mM Tris–HCl, final pH 7.0, stored at –20°C. We prepare our NTP solutions from powder, which is available from Amersham Biosciences (ATP, GTP, and CTP) or Sigma-Aldrich (UTP). However, premade solutions may be used as well.

Adjusting pH of NTP solutions to 7.0 makes them more stable and allows several freeze–thaw cycles.

6. 2 mM UTP solution in 10 mM Tris–HCl, final pH 7.0, stored at −20°C.

7. 1 M Tris–HCl stock solution (pH 7.5 at 25°C) for making transcription buffer and TE buffer.

8. 1 M MOPS-KOH, pH 6.0, for making RNA elution and storage buffers.

9. 1 M DTT, stored at −20°C.

10. 0.5 M EDTA-NaOH (pH 8.0), stored at 4°C.

11. TE buffer for DNA template storage: 10 mM Tris–HCl, pH 7.5, 1 mM EDTA.

12. 10× Transcription buffer: 400 mM Tris–HCl, pH 7.5, 150 mM $MgCl_2$, 20 mM Spermidine, 50 mM DTT. Store at −20°C in 1 mL aliquots.

13. RNA storage buffer ($M_{10}E_1$): 10 mM MOPS, pH 6.0, 10 mM EDTA.

14. RNA elution buffer: 10 mM MOPS, pH 6.0, 1 mM EDTA, 400 mM NaCl.

15. 5 M NaCl for RNA precipitation.

16. Calf Intestine alkaline phosphatase (CIP, 1 U/μL) (Roche).

17. T4 Polynucleotide kinase (PNK), (10,000 U/mL) (NEB).

18. Electrophoresis materials:

    (a) 40% Acrylamide (29:1 acrylamide/bisacrylamide ratio, Fisher). Store at 4°C.

    (b) TEMED (N,N,N′,N′-tetramethyl ethylenediamine, Sigma-Aldrich).

    (c) 10% APS (ammonium persulfate) solution. Store at 4°C no longer than 3 weeks.

    (d) Urea (J.T. Baker).

    (e) 10× TBE buffer for electrophoresis: 0.89 M Tris base, 0.89 M Boric acid, 20 mM EDTA. Autoclave and store at room temperature.

    (f) Denaturing gel-loading buffer (2×): 83 mM Tris–HCl, pH 7.5, 8 M urea, 1.7 mM EDTA, 17% sucrose, 0.05% each of bromophenol blue and xylene cyanol. Heat at 65°C to dissolve urea, aliquot hot buffer into eppendorf tubes (1-mL/tube, store at room temperature). Heat up at 65°C before adding to a sample (urea will precipitate over time, and heating is needed to redissolve it).

    (g) Denaturing gel solutions: 5 or 20% acrylamide, 7 M Urea, 1× TBE. Filter through Nalgene filter unit and store at 4°C no longer than 2 months (see Note 1).

19. Radioactive materials.

    (a) (α-^{32}P) UTP, 3,000 Ci/mmol, 10 μCi/μL (Perkin Elmer) (see Note 2).

    (b) (γ-^{32}P) ATP, 6,000 Ci/mmol, 150 μCi/μL (Perkin Elmer).

2.2. Kinetic Analysis of Group II Intron Splicing In Vitro

1. 1 M MOPS, MES, and HEPES stock solutions at different pH for kinetic studies. The most commonly used buffer is 40 or 80 mM MOPS, pH 7.5. At this pH the ai5γ and other introns are at maximum catalytic activity, while the reaction rate is still limited by the chemical step. Group II introns are sensitive to monovalent ions present in the reaction mixture. For example, the catalytic activity of ai5γ ribozymes is supported by ammonium and potassium ions, but inhibited by sodium ions. When adjusting buffers to a proper pH, it is important to use monovalent ions that do not inhibit catalysis. When preparing MOPS stock solutions, it is important to keep in mind that its pH slightly decreases upon dilution. Thus, the pH of the stock solution should be adjusted so that upon dilution to the final concentration the reaction buffer would have a correct pH.

2. 1 M MgCl$_2$ stock solution. It is important to use the highest purity magnesium chloride, because traces of transition metal ions will induce cleavage of the RNA. Magnesium chloride is hygroscopic; therefore, weighing out the calculated amount of the powder may result in getting lower concentration of MgCl$_2$ solution. It is important to verify the concentration of the stock solution using density vs. concentration plots (*Handbook of Chemistry and Physics*, CRC Press). Since catalytic activity of group II ribozymes is magnesium-dependent, it is critical to use accurate concentration of MgCl$_2$ in all experiments. Therefore, after the stock solution of MgCl$_2$ is prepared and its concentration is verified, it is recommended to aliquot it into eppendorf tubes and store at $-20°$C.

3. Solutions of monovalent ions for kinetic experiments. Prepare 2 M stock solutions of KCl and (NH$_4$)$_2$SO$_4$ (higher concentrations are not recommended as the salts will precipitate over time) and 4 M solution of NH$_4$Cl.

4. Formamide quench buffer: 14.5% sucrose, 0.7× TBE, 72% formamide, 0.05 M EDTA, 0.1% bromophenol blue, 0.1% xylene cyanol.

5. 5% Denaturing gel solution (see Subheading 2.1).

6. 20% Denaturing gel solution (see Subheading 2.1).

2.3. Kinetic Analysis of the Group II Intron Folding

1. For buffer and salt solutions, see Subheading 2.2.

2. Native gel-running buffer stock (Tris-HEPES-EDTA (THE)) (20×): 680 mM Tris, 1,320 mM HEPES, 2 mM EDTA. Store

at 4°C. Use to prepare 1× native gel running buffer and native gel solutions.

3. 1× Native gel running buffer: 34 mM Tris, 66 mM HEPES, 0.1 mM EDTA, 3 mM MgCl$_2$. Store at 4°C.

4. 6% Native gel solution: 6% acrylamide, 1× THE, 3 mM MgCl$_2$. Store at 4°C.

5. Native gel loading dye (4×): 80% (v/v) glycerol, 0.05% each bromophenol blue and xylene cyanol. Store at 4°C.

3. Methods

3.1. Preparation of a Body-Labeled Intron RNA Transcript

1. Thaw on ice, vortex, and briefly spin down 10× transcription buffer, 1 M DTT solution, 30 mM ATP, GTP, and CTP stocks, 2 mM UTP stock and a DNA template solution.

2. Thaw (α-^{32}P) UTP at room temperature using proper shielding.

3. In a 50 μL of total volume, mix 4–5 μg of the DNA template, 5 μL of the 10× transcription buffer, 0.5 mL of 1 M DTT, 1.6 μL of each NTP stock, 5–8 μL of (α-^{32}P) UTP, 1 μL of Protector RNase Inhibitor, 3 μL of T7 RNA polymerase, and RNase-free water.

4. Incubate the reaction mixture at 37°C for 1.5 h. Since some group II introns can splice co-transcriptionally, longer transcription times are not recommended.

5. After transcription, add 50 μL of the denaturing urea-loading buffer and purify the transcript on a 5% denaturing gel.

6. Visualize the band corresponding to the transcribed RNA using PhosphoImager (GE Healthcare) and excise it from the gel. Place the acrylamide piece in a 1.7-mL Eppendorf tube, crush it with a sterile pipette tip, then add 400 μL of elution buffer, freeze the sample on dry ice, and incubate on a shaking platform at 4°C for 3–4 h.

7. Filter the RNA solution from acrylamide and precipitate the RNA by addition of 1.5 μL of glycogen (Roche) and 1 mL of ethanol followed by incubation at –80°C for at least 1 h. Addition of glycogen is recommended for precipitation of small amounts of RNA, because it facilitates precipitation and helps visualize the pellet.

8. Centrifuge the sample and dispose of the supernatant in accordance with proper procedures. Redissolve RNA pellet in 50–60 μL of the RNA storage buffer and measure the RNA concentration either by UV absorbance (see Note 3) or by using liquid scintillation counter (see Note 4).

3.2. Preparation of a Cold (Unlabeled) RNA Transcript

1. Thaw on ice, vortex, and briefly spin down 10× transcription buffer, 1 M DTT solution, 30 mM ATP, GTP, UTP, and CTP stocks, and a DNA template solution.

2. In a 1 mL of a total volume, combine 20 μg of the DNA template, 100 μL of the 10× transcription buffer, 10 μL of 1 M DTT, 32 μL of each 30 mM NTP stock, 5 μL of Protector RNase inhibitor, RNase-free water, and 30–50 μL of T7 RNA polymerase.

3. Incubate the resulting mixture at 37°C for 4–5 h. After transcription, RNA can be either purified on a denaturing polyacrylamide gel (see steps below), or natively purified (without denaturation; see Note 5).

4. Prior to gel-purification, centrifuge the transcription in order to precipitate inorganic pyrophosphate (see Note 6). Transfer the supernatant into a 14-mL culture tube, add 40 μL of 0.5 M EDTA, 50 μL of 5 M NaCl, and 10 mL of ethanol, and incubate at −80°C for 1 h or overnight for precipitation.

5. Centrifuge ethanol-precipitated RNA, redissolve the pellet in water, mix with an equal volume of the urea-loading buffer, and purify on a denaturing polyacrylamide gel. Visualize the band corresponding to the RNA transcript by UV shadowing and elute as above. Use larger (14 mL) tube for elution and subsequent precipitation instead of an Eppendorf tube.

3.3. Preparation of the 5′-End Labeled RNA Substrates

If the RNA is obtained by in vitro transcription, it has to be dephosphorylated prior to 5′-end labeling. In order to dephosphorylate the transcribed RNA, follow the protocol below.

1. Mix 100 pmol of RNA in 43 μL of ddH_2O with 5 μL of 10× buffer for CIP and 2 μL of CIP (Roche).

2. Incubate the reaction mixture at 37°C for 1 h, then phenol-extract and ethanol-precipitate the RNA.

3. Centrifuge the precipitated RNA, redissolve the pellet in the RNA storage buffer, and store at −20°C.

Chemically synthesized RNAs do not require this step. The following protocol is used for 5′-end labeling of chemically synthesized RNA oligos or dephosphorylated RNA transcripts.

1. Combine 30 pmol of RNA with 1 μL of 10× PNK buffer, 1 μL of $(\gamma\text{-}^{32}P)$ ATP, and 1 μL of T4 PNK in 10 μL of a total volume.

2. The reaction mixture is incubated for 1 h at 37°C.

3. Add 10 μL of urea-loading buffer and purify the labeled RNA on a denaturing polyacrylamide gel as described above (see Note 7).

**3.4. Kinetic Analysis
of the Group II Intron
Splicing**

*3.4.1. Analysis of the
Group II Intron Self-
Splicing*

Below is a general protocol for kinetic analysis of a group II intron *cis*-splicing. Optimal reaction conditions may vary for different introns. For example, yeast intron ai5γ requires high concentration of $MgCl_2$ (100 mM) for optimal reactivity (21, 22), while brown algae intron Pl. LSU/2 is active at 10 mM $MgCl_2$ (23) and recently discovered Lt SSU/1 group II intron from *Leptographium truncatum* is active at magnesium ion concentrations as low as 6 mM (24). In addition, optimal reaction temperature for introns with high AU content is generally 42–45°C (21, 22), while introns with high GC content (like intron AV from *Azotobacter vinelandii*) require higher temperatures (50–55°C) for optimal catalytic activity, and they tend to misfold at lower temperatures (25). In general, when testing new group II introns for splicing activity, it is important to test a variety of temperatures and ionic conditions. The following conditions may be recommended as a start. Temperatures: 25, 37, 42, 50, 55, 60°C (test the last two temperatures only if the intron is GC-rich). Ionic conditions: (a) high salt—100 mM $MgCl_2$; 100 mM $MgCl_2$, 0.5 M KCl; 100 mM $MgCl_2$, 0.5 M NH4Cl; 100 mM $MgCl_2$, 0.5 M $(NH_4)_2SO_4$. If the level of reactivity under high salt conditions and all temperatures listed above is still low, the concentration of the monovalent ion used can be raised to 1 M. (b) low salt—8 or 10 mM $MgCl_2$; 8 or 10 mm $MgCl_2$, 50 or 100 mM KCl, 8 or 10 mM $MgCl_2$, 50 or 100 mM NH_4Cl; 8 or 10 mm $MgCl_2$, 50 or 100 mM $(NH_4)_2SO_4$; 8 or 10 mm $MgCl_2$, 5 nM protamine. Higher concentration of monovalent ions in combination with low concentration of magnesium ions was found to be optimal for Pl. LSU/2 intron (23). However, this choice of conditions may also result in monovalent ions outcompeting magnesium for Mg-specific binding sites, which would inhibit the reaction. In case low salt conditions result in very little reactivity, it is recommended to test other magnesium ion concentrations between 10 and 100 mM.

1. Mix body-labeled RNA transcript containing the intron and flanking exons (2–10 nM final concentration) with ddH_2O and reaction buffer (40–80 mM (final concentration) MOPS-KOH, pH 7.5 is the most commonly used buffer (see also Note 8)) in an Eppendorf tube, heat at 95°C for 1 min (see Note 9), and cool down at room temperature for 1–2 min (see Note 10).

2. Simultaneously add monovalent and divalent salts to start the reaction and incubate the resulting mixture at a reaction temperature of choice (see options listed above).

3. Withdraw 0.5–1 μL aliquots from the reaction mixture at chosen time points, mix with 3 μL of the formamide quench buffer, and immediately chill on ice.

4. Analyze splicing products on a 4 or 5% polyacrylamide gel (see Fig. 3a). If some of the products are much shorter than others (e.g., if the construct contains a full-length intron and short (20–30 nt) exons), use a longer gel or a stacking 4–20% gel to resolve all products.

For the very first experiment designed to test a wide range of reaction conditions, it is recommended to use only 4–5 time points: 0, 15, 30 min, 1, 3 h. This experiment determines whether the ribozyme is fast-reacting or slow-reacting under the chosen conditions, and allows one to choose an appropriate range of time points for subsequent time courses in order to determine kinetic parameters of the reaction. If the ribozyme is very fast-reacting and time points earlier than 30 s need to be taken, it is recommended to use QuenchFlow (KinTek) for more accurate measurements.

Data analysis

1. Dry the gels and expose them to a PhosphoImager screen (GE Healthcare or Fuji) for a few hours or overnight, depending on the amount of radioactive material in the samples. For example, if each aliquot contains 10,000 CPM (counts per minute) or more of radioactive material, 1–2 h of exposure may be sufficient, but samples containing 1,000 CPM or less are better exposed overnight.

2. Scan the screens, quantify band intensities corresponding to each product using ImageQuant (if Storm PhosphoImager from GE Healthcare is used) or other appropriate software, and adjust the resulting values for background.

3. Calculate fractions of each product, and adjust fraction of every product (except for precursor) for a difference in number of uridine residues (or other nucleotides, depending on the radioactive NTP used for body-labeling) between this product and the precursor using the following formula:

$$F_{pro} = I_{pro} \, / \, I_{tot} * NU_{pre} \, / \, NU_{pro},$$

where F_{pro} is the fraction of a splicing product (lariat intron, linear intron, spliced exons, etc.). I_{pro} is the band intensity corresponding to that product, I_{tot} is a sum of band intensities for all bands in the lane (precursor and all products), NU_{pre} is the number of all uridines in the precursor, and NU_{pro} is the number of all uridines in the respective product. If fractions of splicing products are calculated without this adjustment, then the reaction rates for the formation of shorter products will be significantly underestimated.

4. Plot the calculated fractions of precursor and products vs. time in order to obtain respective rate constants.

Various ribozyme constructs may exhibit slightly different kinetic behavior. In the simplest case all RNA precursor behaves like one homogeneous population, and all molecules react at the same rate. This behavior has been reported for the ai5γ intron constructs with short exons (13). If all precursor molecules react at the same rate, data are fit to a single exponential equation:

$$F_{pro} = F_{end} * (1 - e^{-kt}),$$

where F_{pro} is fraction of a respective splicing product, F_{end} is fraction of this product at the reaction endpoint, and k is the first order rate constant.

If two populations of molecules reacting with different rates are observed, the results are fit to a double exponential equation:

$$F_{pro} = F_{fast} * (1 - e^{-kfastt}) + F_{slow} * (1 - e^{-kslowt}),$$

where F_{pro} is the total fraction of the respective splicing product, F_{fast} is the fraction reacting fast, $kfast$ is the rate constant for the fast reaction, F_{slow} is the fraction reacting slowly, and $kslow$ is the rate constant for the slow reaction.

This type of behavior is characteristic of the ai5γ intron constructs with long (~300 nt) 5'- and 3'-exons (22). Slow reaction rates for one of the populations are likely caused by RNA misfolding. Extreme case of this behavior was observed for a GC-rich AV intron at low temperatures: the misfolded population of the precursor simply remained unreacted (25). This type of data is fit using the following equation:

$$F_{pre} = F_{u} + F_{r} * e^{-kt},$$

where F_{pre} is total fraction of the precursor at a given time point, F_{u} is the fraction of the unreactive species, F_{r} is the fraction of the active species, and k is the first order rate constant.

3.4.2. Kinetic Analysis of trans-Splicing Constructs Derived from Group II Introns

A. Two-piece hydrolysis system (see Fig. 2b). Most commonly used reaction conditions are 40 mM MOPS-KOH, pH 7.5, 100 mM MgCl$_2$, 500 mM KCl. D5 is frequently used at saturation to test the effect of various exD123 and D5 mutants on the catalytic activity of the ribozyme. In order to obtain an apparent Km value for D5 under single turnover conditions, D5 concentrations are typically varied from 20 nM to 5 μM (see Note 11).

1. Denature body-labeled exD123 (2 nM final concentration) and unlabeled D5 (3 μM final concentration, if used at saturation) RNAs at 95°C for 1 min in separate Eppendorf tubes. Each tube should contain half of the total amount of MOPS buffer (40 mM final concentration, pH 7.5) and ddH$_2$O.

2. Cool the two RNAs at room temperature for 1 min and combine them in one tube, then add KCl to the 500 mM final concentration.

3. Initiate the reaction by adding $MgCl_2$ (100 mM final concentration). Total reaction volume is usually 20–40 μL. Withdraw "0" time point (1 μL) immediately after $MgCl_2$ addition, mix with 3 μL of formamide quench buffer, and chill on ice. Incubate the reaction mixture at 45°C.

4. Take aliquots corresponding to different time points (1 μL), mix with 3 μL of formamide quench buffer, and chill on ice. Recommended time points for the first experiment are 0, 15, 30, 45 min, 1, 2 h.

5. Analyze samples on a 5% denaturing polyacrylamide gel (see Fig. 3b). For best resolution, conduct the electrophoresis until xylene cyanol dye (XC) reaches the bottom of the 10" (25 cm)-long gel.

6. Dry the gel and analyze the results on a PhosphoImager as described above.

   Notably, in addition to D123 and 5′-exon, a secondary cleavage product between Domains 2 and 3 is generally observed (see Fig. 3b, left). Interestingly, this product does not form if D2 is shortened to a 5-bp stem-loop structure (see Fig. 3b, right).

B. Two-piece branching system (see Fig. 2a). Reaction is set up and carried out essentially as described above, with the following exceptions.

1. D56 is used instead of D5.

2. Since KCl promotes hydrolysis, in the branching reaction ammonium sulfate or ammonium chloride is used instead at the same concentration (500 mM) (see Note 12).

3. The main difference between branching and hydrolysis reaction is that in the latter D5 acts as a true ribozyme. It only stimulates the reaction but remains unaffected by it. By contrast, as a result of the trans-branching reaction, D56 becomes covalently attached to the D123 RNA. Thus, branching reaction can be monitored by either using body-labeled exD123 at 2 nM and unlabeled D56 in excess (at 3 μM) or using 5′-end labeled D56 at 2 nM and unlabeled exD123 in excess (at 3 μM). If reaction is monitored using body-labeled exD123, the samples are analyzed on a 5% polyacrylamide denaturing gel as described above (see Fig. 3c). If, however, D56 is labeled, reaction products (only D56 and the D56·D123 branched product will be visualized) are analyzed on a stacking 5–15% (or 5–20%) stacking gel because of the significant difference in size between them.

C. Multi-piece systems. A three-piece construct derived from the ai5γ intron consists of the exD1 RNA substrate (2 nM), D5 ribozyme (3 μM), and D23 allosteric effector (5 μM) (see Fig. 2c). In the effector molecule, D2 can be reduced to a short stem-loop as described above (Hp2D3 RNA) without any significant loss of activity (12). Reaction is carried out in 40 mM MOPS-KOH, pH 7.5, 100 mM $MgCl_2$, 1 M KCl according to the protocol below

1. Denature exD1 (body-labeled), D5 (unlabeled), and D23 (Hp2D3) (unlabeled) RNAs at 95°C for 1 min in three separate Eppendorf tubes. Each tube should contain a third of the total amount of MOPS buffer (final concentration 40 mM) and ddH_2O.

2. Cool the three RNAs at room temperature for 1 min, and combine them together. Add KCl to the final concentration of 1 M.

3. Initiate the reaction by adding $MgCl_2$ (final concentration 100 mM). Withdraw "0" time point (1 μL) immediately after $MgCl_2$ addition, mix with the formamide quench buffer (3 μL), and chill on ice. Incubate the reaction mixture at 45°C.

4. Withdraw aliquots corresponding to different time points (1 μL), mix with 3 μL of formamide quench buffer, and chill on ice. Recommended time points for the first experiment are 0, 10, 20, 30 min, 1, 2, 3 h.

5. Analyze samples on a 5% denaturing polyacrylamide gel (see Fig. 3d) as described above for the two-piece hydrolysis construct.

D. Multiple-turnover ribozyme reactions—active site titration and evaluation of the folding rate. The D135 and D1356 ribozymes derived from the ai5γ group II intron are good model systems for studying the group II intron folding. The fast and efficient reaction of these ribozymes with the 24-mer RNA oligo substrate (see Fig. 4a, b) is often used to both evaluate the quality of the ribozyme and determine its folding rate. Active site titration is generally used to determine the percentage of active homogeneously folded population in a ribozyme sample (26). This is a protocol of active site titration of D135 or D1356 ribozyme under optimal conditions for catalysis, which are most frequently used for folding studies (see Note 13).

1. Separately denature the ribozyme (20 nM final concentration) and the 5′-end labeled 24-mer oligo substrate (200 nM final concentration) in 80 mM MOPS, pH 7.5, at 95°C for 1 min and cool at room temperature for 1 min.

2. Add $MgCl_2$ and KCl (final concentrations are 100 mM and 0.5 M, respectively) to each ribozyme and substrate;

separately incubate the samples at 42°C for 10 min. The volume of each ribozyme and substrate sample after addition of the salts is 20 μL, thus, the final reaction volume is 40 μL.

3. Initiate the reaction by combining ribozyme and substrate together. Withdraw "0" time point (1 μL) immediately after mixing the ribozyme and the substrate together, mix with 3 μL of the formamide quench buffer, and chill on ice.

4. Incubate reaction mixture at 42°C, withdraw aliquots at respective time points and treat as above. Suggested time points are 0, 0.5, 1, 1.5, 2, 2.5, 3, 3.5, 4, 4.5, 5, 6, 7, 10, 12, 15, 20, and 30 min.

5. Analyze the samples on a 20% polyacrylamide denaturing gel. Dry the gel and quantify the results using PhosphoImager as above (see Fig. 4c). Calculate fraction of the cleavage product at each time point and convert the data into equivalents of product per equivalent of ribozyme by multiplying the resulting value by the substrate/ribozyme concentration ratio. Plot the result vs. time and fit to a double exponential equation (see Fig. 4d).

In this plot the first phase (burst) represents the first turnover, which is not limited by the rate of product release. If the ribozyme sample represents a homogeneously folded population, the burst magnitude is close to one equivalent of product per one equivalent of ribozyme (the burst is 0.94 in the example shown in Fig. 4d).

A variation of this experiment can be used to indirectly measure the ribozyme folding rate (7). In this setup, the reaction is carried out in 80 mM HEPES, pH 8.0, so that the chemical step is fast and not rate-limiting. Ribozyme and substrate are separately denatured in the above buffer at 95°C for 1 min, cooled down for 1 min at room temperature, and the reaction is initiated by mixing the two RNAs together with simultaneous addition of salts ($MgCl_2$ and KCl). Note that in this setup the ribozyme pre-folding step is omitted so that it will react as it folds, and thus the reaction rate will reflect the folding rate. If the folding rate is measured under the optimal conditions for catalysis, the same salt concentrations and temperature as listed above for active site titration are used.

3.5. Kinetic Analysis of Global Compaction of ai5γ Ribozyme Constructs Using Native Gel Electrophoresis

1. Denature body-labeled RNA (D1, D135, D1356 or full-length splicing construct with short exons (SE), 5 nM) at 95°C in 80 mM MOPS, pH 6.0 (final concentration) and cool at room temperature for 2 min.

2. Initiate folding by simultaneous addition of salts ($MgCl_2$ and KCl), withdraw "0" time point (5 μL), immediately mix with prechilled native gel loading buffer (2 μL), and put on ice. In order to measure rates of global compaction as a function of magnesium ion concentration, use the following final concentrations

of $MgCl_2$: 3, 5, 10, 20, 30, 50, 100, and 200 mM. KCl is used at 100 mM final concentration at lower (3–10) magnesium ion concentrations. At higher concentrations of magnesium ions, it is also useful to compare the compaction rates in the presence of 100 mM KCl and 500 mM KCl.

3. Incubate RNA samples either at 30°C (physiological temperature of yeast growth) or at 42°C (optimal temperature for in vitro catalysis) in a final volume of 100 μL. Withdraw aliquots at chosen time points (5 μL), immediately mix with prechilled native gel loading buffer (2 μL) and put on ice (see Note 14). At 42°C and higher than 10 mM concentrations of $MgCl_2$ compaction generally occurs much faster than at 30°C (15); therefore, recommended time points are 15, 30, 45 s, 1, 3, 5, 7, 10, 20, 30 min. At 30°C and magnesium ion concentrations 3–10 mM, recommended time points are 0, 20, 40 min, 1, 2, 3, 4, 5, 6, 11, 24 h. At 30°C and magnesium ion concentrations 20–200 mM, recommended time points are 0, 5, 10, 20, 30, 40 min, 1, 2, 3, 4, 5, 6, 11, 24 h (see Note 15).

4. Analyze the samples on a 6% native polyacrylamide gel, containing 1× THE buffer and 3 mM $MgCl_2$ (27) (see Fig. 5a). Conduct the electrophoresis at 4°C, run gels (23.5 cm long × 26 cm

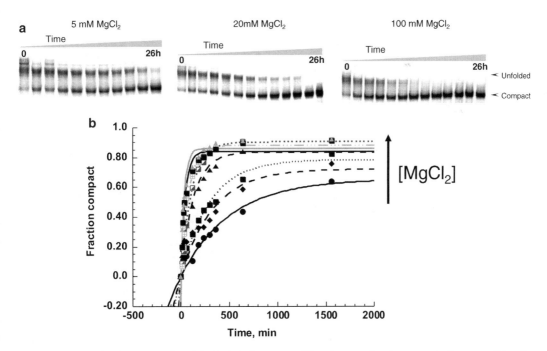

Fig. 5. (a) Analysis of global compaction of the D1356 ribozyme at different magnesium ion concentrations on a 6% native polyacrylamide gel. Band intensities were quantified as above, fraction of compact species was calculated, plotted vs. time, and fit to a single-exponential equation (b) (15).

wide × 0.2 mm thick) at 12 W for 5–6 h, dry and analyze on a PhosphoImager as described above.

5. Calculate the fraction of compact species, plotted vs. time, and fit to a single-exponential or a double-exponential equation as described above (see Fig. 5b).

4. Notes

1. When stored over prolonged periods of time, acrylamide solutions form aromatic compounds, which absorb UV light at 254 nm. Gels prepared from these solutions will be impossible to use for UV shadowing of nucleic acids.

2. Since the sequence of the ai5γ group II intron is U-rich, we generally use α-UTP for preparation of body-labeled transcripts. Depending on the nucleotide content of a particular intron sequence, other radiolabeled nucleotides may be more useful for body labeling. For example, if the intron sequence is GC-rich, one may consider using α-GTP or α-CTP for body-labeling.

3. The yield of RNA from the transcription scale used for body-labeling is generally not very high. Diluting the gel-purified RNA even 100 times for measuring the absorbance in regular spectrophotometer cuvettes will not produce detectable signal. Thus, one can either measure undiluted samples in regular 100-μL cuvettes (not recommended because the cuvette will likely be highly contaminated after such measurement) or measure UV absorbance of such samples on a NanoDrop Spectrophotometer (Thermo Scientific), which requires only 1.5–2 μL of the undiluted sample.

4. In order to determine the RNA concentration by using liquid scintillation counter, several samples (undiluted RNA and 2–3 dilutions, e.g., 1:10, 1:50, and 1:100; 1 μL each) are spotted onto pieces of Whatman paper, dried and placed into respective scintillation vials. Scintillation liquid is then added (volume depends on the instrument used) and the CPM values of the samples are measured on a liquid scintillation counter. CPM for the undiluted sample is then calculated based on dilution factors, and the average CPM value for the undiluted sample from all measurements is determined. Specific activity of the radioactive material is calculated by using the table provided by the manufacturer. CPM values are then converted to millicurie, and RNA concentration is calculated taking into account the ratio of hot and cold UTP and number of uridine residues in the transcript.

5. In some cases, it is beneficial not to denature RNA after transcription. Some RNAs may not be able to correctly fold into their native structure after denaturation. In order to purify these RNAs after transcription, Turbo DNAse (Ambion) (2 U/20 μL of the transcription mixture) is added to cleave the DNA template and the mixture is incubated at 37°C for 30 min. Then Proteinase K (50 μL of 30 mg/mL solution per 1 mL of the transcription mixture) is added, and reaction mixture is incubated at 37°C for 30 min. RNA (it must be substantially larger than proteinase K (28.5 kDa) for this protocol to work) is then isolated using Amicon Ultra centrifugal devices with the appropriate cut-off (usually 50–100 kDa).

6. Some transcription protocols recommend adding pyrophosphatase to the transcription mixture in order to hydrolyze inorganic pyrophosphate, but in our hands this makes no difference with respect to the transcription yield.

7. In order to avoid accumulation of unreacted radioactive nucleotide in the bottom chamber of the electrophoresis apparatus, labeled samples may be passed through Microspin G-25 columns (GE Healthcare) prior to loading onto the gel. Used columns must be disposed of in tightly capped 14-mL tubes to prevent possible radioactive contamination.

8. Buffers with lower pH are used in order to obtain pH-rate profile and determine whether the rate of catalytic reaction is limited by the chemical step (and not folding and/or conformational rearrangements). This experiment would necessitate the use of buffers other than MOPS (potassium acetate, MES, HEPES) in order to cover a wide range of pH values.

9. Harsh denaturation conditions may cause some RNAs to degrade or otherwise lose activity. Therefore, in addition to the 95°C denaturation step, it is recommended to test other possibilities: milder denaturation at 65°C and no denaturation and refolding (natively purified ribozyme, see also Note 5).

10. At high temperatures Mg^{2+} ions more actively facilitate hydrolysis of the phosphodiester linkage, resulting in RNA degradation. Therefore, it is important to let the sample cool down prior to adding magnesium salts.

11. In order to maintain the pseudo-first order kinetic regime, the lowest concentration of D5 should be ~10 times higher than the concentration of the exD123 substrate.

12. Although ammonium sulfate actively promotes branching in cis-splicing constructs, it has been reported to cause long irreproducible lags in the trans-branching reaction (11). Therefore the use of ammonium chloride is recommended in this case.

13. This experiment can also be conducted under other conditions if the rate of the first turnover is still much higher (at least ten-fold) than the rate of subsequent turnovers.

14. In order to determine whether samples can be stored on ice before loading them onto a native gel, a control experiment was conducted: samples were divided in half; the first half was analyzed on a native gel immediately after adding the chilled loading buffer and the second half after overnight storage at 4°C. Electrophoresis analysis of both series produced identical results, suggesting that prolonged storage at 4°C does not result in any additional compaction or unfolding of the samples (15).

15. These time points are recommended for a time course conducted at pH 6.0. If compaction is studied at pH 7.0 and higher, it will be faster.

Acknowledgments

I would like to thank Prof. A.M. Pyle for guidance and support and members of the Pyle lab for helpful discussions. I would like to apologize to all our colleagues whose work could not be extensively cited in this chapter. O.F. is a Research Specialist at Howard Hughes Medical Institute.

References

1. Lehmann, K., and Schmidt, U. (2003) Group II introns: structure and catalytic versatility of large natural ribozymes. *Crit. Rev. Biochem. Mol. Biol.* **38**: 249–303.

2. Pyle, A.M., Lambowitz, A.M. (2006) Group II introns: ribozymes that splice RNA and invade DNA. in *The RNA world*, R.F. Gesteland, Cech T.R., Atkins J.F. (ed) Cold Spring Harbor Laboratory Press: Cold Spring Harbor, New York, 469–506.

3. Lambowitz, A.M., Zimmerly, S. (2004) Mobile group II introns. *Annu Rev Genet.* **38**: 1–35.

4. Michel, F., K. Umesono, and H. Ozeki (1989) Comparative and functional anatomy of group II catalytic introns-a review. *Gene* **82**: 5–30.

5. Qin, P.Z. and A.M. Pyle (1998) The architectural organization and mechanistic function of group II intron structural elements. *Curr. Opin. Struct.Biol.* **8**: 301–308.

6. Pyle, A.M., Fedorova, O., Waldsich, C. (2007) Folding of group II introns: a model system for large, multidomain RNAs? *Trends Biochem. Sci.* **32**: 138–145.

7. Fedorova, O., Su, L.J., Pyle, A.M. (2002) Group II introns: highly specific endonucleases with modular structures and diverse catalytic functions. *Methods* **28**: 323–335.

8. Jarrell, K.A., R.C. Dietrich, and P.S. Perlman (1988) Group II Intron Domain 5 facilitates a *trans*-splicing reaction. *Mol. Cell. Biol.* **8**: 2361–2366.

9. Franzen, J.S., M. Zhang, and C.L. Peebles (1993) Kinetic analysis of the 5'-splice junction hydrolysis of a Group II intron promoted by Domain 5. *Nucleic Acids Res.* **21**: 627–634.

10. Pyle, A.M. and J.B. Green (1994) Building a kinetic framework for group II intron ribozyme activity: quantitation of interdomain binding and reaction rate. *Biochemistry* **33**: 2716–2725.

11. Chin, K. and A.M. Pyle, (1995) Branch-point attack in group II introns is a highly reversible transesterification, providing a possible proofreading mechanism for 5'-splice site selection. *RNA* **1**: 391–406.

12. Fedorova, O., Mitros T., Pyle A.M. (2003) Domains 2 and 3 interact to form critical

elements of the group II intron active site. *J. Mol. Biol.* **330**: 197–209.

13. Zingler, N., Solem, A., Pyle, A.M. (2010) Dual roles for the Mss116 cofactor during splicing of the ai5{gamma} group II intron. *Nucleic Acids Res.* **38**: 6602–6609.

14. Su, L.J., Waldsich, C., and Pyle, A.M. (2005) An obligate intermediate along the slow folding pathway of a group II intron ribozyme. *Nucleic Acids Res.* **33**: 6674–6687.

15. Fedorova, O., Waldsich, C., Pyle A.M. (2007) Group II Intron Folding under Near-physiological Conditions: Collapsing to the Near-native State. *J. Mol. Biol.* **366**: 1099–1114.

16. Steiner, M., Karunatilaka, K.S., Sigel, R.K., Rueda, D. (2008) Single-molecule studies of group II intron ribozymes. *Proc Natl Acad Sci U S A* **105**: 13853–13858.

17. Toor, N., G. Hausner, and S. Zimmerly (2001) Coevolution of group II intron RNA structures with their intron-encoded reverse transcriptases. *RNA* 7: 1142–52.

18. Toro, N., Molina-Sánchez,M.D., Fernández-López, M.(2002) Identification and characterization of bacterial class E group II introns. *Gene* **299**: 245–250.

19. Simon, D.M., Clarke, N.A., McNeil, B.A., Johnson, I., Pantuso, D., Dai, L., Chai, D., Zimmerly, S. (2008) Group II introns in eubacteria and archaea: ORF-less introns and new varieties. *RNA* **14**: 1704–1713.

20. Toor, N., Robart, A.R., Christianson, J., Zimmerly, S. (2006) Self-splicing of a group IIC intron: 5' exon recognition and alternative 5' splicing events implicate the stem-loop motif

of a transcriptional terminator. *Nucleic Acids Res.* **34**: 6461–6471.

21. Jarrell, K.A., Peebles, C.L., Dietrich, R.C., Romiti, S.L., Perlman, P.S. (1988) Group II intron self-splicing: Alternative reaction conditions yield novel products. *J. Biol. Chem.* **263**: 3432–3439.

22. Daniels, D., W.J. Michels, and A.M. Pyle (1996) Two competing pathways for self-splicing by group II introns; a quantitative analysis of in-vitro reaction rates and products. *J. Mol. Biol.* **256**: 31–49.

23. Costa, M., Fontaine, J.M., Goer, S.L, Michel, F. (1997) A group II self-splicing intron from the Brown Alga Pylaiella littoralis is active at unusually low magnesium concentrations and forms populations of molecules with a uniform conformation. *J. Mol. Biol.* **274**: 353–364.

24. Mullineux, S.T., Costa, M., Bassi, G.S., Michel, F., Hausner, G. (2010) A group II intron encodes a functional LAGLIDADG homing endonuclease and self-splices under moderate temperature and ionic conditions. *RNA*, 16: 1818–1831.

25. Adamidi, C., Fedorova, O., Pyle, A.M. (2003) A group II intron inserted into a bacterial heat-shock operon shows autocatalytic activity and unusual thermostability. *Biochemistry* **423409–18**: 3409–3418.

26. Swisher, J., Su, L.J., Brenowitz, M., Anderson, V.E., Pyle, A.M. (2002) Productive Folding to the Native State by a Group II Intron Ribozyme. *J. Mol. Biol.* **315**: 297–310.

27. Rangan, P., Masquida, B., Westhof, E. & Woodson, S.A. (2003). Assembly of core helices and rapid tertiary folding of a small bacterial group I ribozyme. *Proc. Natl. Acad. Sci. USA* **100**: 1574–1579.

Chapter 8

Mechanism and Distribution of *glmS* Ribozymes

Phillip J. McCown, Wade C. Winkler, and Ronald R. Breaker

Abstract

Among the nine classes of ribozymes that have been experimentally validated to date is the metabolite-responsive self-cleaving ribozyme called *glmS*. This RNA is almost exclusively located in the 5′-untranslated region of bacterial mRNAs that code for the production of GlmS proteins, which catalyze the synthesis of the aminosugar glucosamine-6-phosphate (GlcN6P). Each *glmS* ribozyme forms a conserved catalytic core that selectively binds GlcN6P and uses this metabolite as a cofactor to promote ribozyme self-cleavage. Metabolite-induced self-cleavage results in down-regulation of *glmS* gene expression, and thus the ribozyme functions as a key riboswitch component to permit feedback regulation of GlcN6P levels. Representatives of *glmS* ribozymes also serve as excellent experimental models to elucidate how RNAs fold to recognize small molecule ligands and promote chemical transformations.

Key words: Glucosamine-6-phosphate, Metabolite sensing, Riboswitch, RNA catalysis, Self-cleaving

1. Introduction

1.1. Regulation of Bacterial Gene Expression by Riboswitches

The typical bacterium encodes for at least 100 different transcription factors, which individually exert regulatory control over the efficiency of transcription initiation. However, each of the stages of gene expression after transcription initiation is also subjected to regulatory control in bacteria. These post-initiation processes include transcription elongation, transcription termination, translation, and mRNA stability. A diverse array of cis- and trans-acting RNA elements has been discovered that can regulate these post-initiation processes. For example, an increasingly large collection of metabolite-sensing, cis-acting regulatory RNAs called ribo-switches has been identified (1, 2). Riboswitches are mostly located in 5′-leader regions of mRNAs, although some are located within intercistronic portions of polycistronic transcripts. These RNA

Jörg S. Hartig (ed.), *Ribozymes: Methods and Protocols*, Methods in Molecular Biology, vol. 848,
DOI 10.1007/978-1-61779-545-9_8, © Springer Science+Business Media, LLC 2012

elements control gene expression in response to direct association with the appropriate metabolite ligand. Over 20 classes have been discovered, each individually responding to a different metabolite ligand.

A variety of metabolite ligands are sensed by riboswitches (1), including cofactors, amino acids, nucleobase-containing metabolites, a second messenger signaling molecule, and an amino sugar precursor of cell wall synthesis. Sensing of the appropriate metabolite is achieved through a ligand-responsive domain (the aptamer), which controls gene expression by altering the configuration of regulatory sequences that interface between the aptamer and the downstream gene (the expression platform). Riboswitch aptamers associate with their target ligands with high affinity and with exquisite specificity. Structural models of ligand-bound riboswitch aptamers (3–5) have revealed that the RNA-bound ligands are stabilized by a combination of hydrogen-bonding interactions, electrostatic interactions, and base-stacking energies. Typically, association of the appropriate metabolite ligand is integrated into a feedback inhibition regulatory response, although many riboswitch examples have also been discovered for other types of regulatory circuitry topology, such as substrate-mediated activation of catabolism genes (6).

The majority of bacterial riboswitches couple association of the ligand(s) to gene expression by controlling ribosomal access to a downstream translation initiation site or by controlling formation of a transcription termination site (transcription attenuation). One particular riboswitch class, which responds to thiamin pyrophosphate (TPP), has been demonstrated to affect mRNA processing and stability in plants and lower eukaryotes (7). However, one riboswitch class has been discovered to regulate gene expression through a mechanism other than classical transcription attenuation or ligand-mediated sequestration of the ribosome-binding site. Members of this particular riboswitch class, generally called "*glmS* ribozymes" respond to the cell wall precursor, glucosamine-6-phosphate (GlcN6P) and regulate expression of the glucosamine-6-phosphate synthetase gene (*glmS*). Instead of transcription attenuation or translational regulation, *glmS* ribozymes control mRNA stability in bacteria in a ligand-responsive manner, as reviewed herein.

1.2. Distribution of glmS Ribozymes

The known phylogenetic distribution of *glmS* ribozymes has greatly increased since the initial identification of this riboswitch class in organisms from the Firmicutes and Fusobacterium phyla (8, 9). Representatives have subsequently been identified within the Actinobacteria, Deinococcus-Thermus, Tenericutes, Chloroflexi, Synergistetes, and Dictyoglomi phyla, as well as *Thermobaculum terrenum*, which has yet to be assigned to a phylum (10, 11). Despite this expansion in the phyla containing *glmS* ribozymes, most examples still reside in Firmicutes.

A total of 463 *glmS* ribozyme representatives have been identified to date (11). Almost without exception, each organism carries a single representative that resides immediately upstream of the open reading frame coding for the GlmS protein. There are only two known instances of organisms that carry more than one *glmS* ribozyme. The Firmicute *Dethiobacter alkaliphilus* carries one representative located upstream of the *glmS* ORF and a second representative located upstream of the *glmM* ORF. GlmM is an enzyme (phosphoglucomutase) that converts GlcN6P into glucosamine-1-phosphate. The Fusobacterium *Sebaldella termitidis* also carries a *glmS* ribozyme in the normal location and has a second representative located upstream of the *glnA* ORF. GlnA is an enzyme (glutamine synthase) that converts glutamate into glutamine (the immediate precursor of GlcN6P). Thus, both additional riboswitches appear to control the expression of genes that are immediately adjacent to GlcN6P in its synthesis and utilization pathways.

It has been suggested (12, 13) that *glmS* ribozymes may be useful targets for the design of novel antibacterial compounds because they control the expression of the *glmS* gene whose protein product is essential for cell survival (14), Indeed, the *glmS* ribozyme is also a potentially attractive drug target due to its seemingly immutable catalytic core structure, and the abundance of representatives in pathogenic Firmicutes (11, 15). A number of GlcN6P analogs have been prepared and shown to trigger ribozyme function in biochemical assays (12, 16). A structural rationale for function of each analog is evident on examining the atomic-resolution structures of *glmS* ribozymes (e.g., (17)). It appears possible to create additional GlcN6P analogs or perhaps even compounds with completely different chemical scaffolds that could trigger *glmS* ribozyme action in cells. Such compounds would trick bacteria into repressing *glmS* gene expression despite the fact that cells are starved for the natural metabolite.

1.3. The Structural Features of glmS Ribozymes

Nearly all *glmS* ribozyme representatives conform to a complex consensus sequence and secondary structure model that can be divided into catalytic core and core support sub-domains. Each ribozyme carries a well-conserved ligand-binding and catalytic core, but with greater variability in sequence and structure outside of this active site. This same pattern of sequence and structural conservation holds true for a mutant *glmS* ribozyme population that was subjected to *in vitro* selection (18), suggesting that there are very few if any variants close in sequence space that retain GlcN6P-mediated catalytic activity. Most likely, the strict conservation of the catalytic core is due to the dual necessity to selectively bind GlcN6P and to position this ligand to promote efficient RNA transesterification.

A 3-D structure (Fig. 1) comprised of several roughly coaxial RNA helices and three pseudoknots was identified upon analysis by X-ray crystallography of a *glmS* ribozyme representative from *Thermoanaerobacter tengcongensis* (19) and another from *Bacillus anthracis* (20). Specifically, the P2.1 helix centrally located between the P4 helix on one side and the P1, P2.2, P2, P3, and P3.1 helices. In every *glmS* ribozyme that possesses a P4 helix, it serves as a support for the catalytic core and binding pocket by employing its GNRA tetraloop as a docking component to form a tertiary interaction with the P1 helix (19, 20). In almost every instance when the P4 helix is removed, catalysis is retained (albeit with a substantially reduced rate constant) when high concentrations of cations are present to aid in stable structural formation (21).

The catalytic core of the ribozyme is largely contained within the P2.1 and P2.2 segments of the ribozyme (Fig. 1; (8, 11, 19, 20)). The nucleotides within these regions that have been determined to be responsible for direct binding to GlcN6P are A(-1), G1, C2, A50, U51, and G65 (see numbering of the consensus model in Fig. 1) (19, 20). These contacts are located within the most highly conserved part of the ribozyme, and therefore efforts to design analogs that trigger ribozyme function will likely need to consider this chemical landscape. Also, some parts of the ligand may not be recognized by the ribozyme. For example, there are no direct contacts to the hydroxyl at the 4 position of GlcN6P, which could be exploited in future analog constructions. In contrast, G40 is strictly conserved and is essential for catalysis, but does not participate in ligand binding (17, 19, 20).

Some of the highly conserved nucleotides make contacts with others to form the rigid structure of the active site, while others establish key contacts between the ribozyme and GlcN6P (19, 20). For example, the ribose of nucleotide A(-1) carries the 2'-oxygen nucleophile for the phosphoester transfer reaction that results in RNA cleavage. The 2'-endo ribose pucker of this nucleotide assists in forming a ribose-phosphate backbone conformation that adopts the near in-line geometry necessary for catalysis to occur. The guanine nucleobase of G1, which stacks on top of the sugar ring of GlcN6P, uses its N1 position to form a hydrogen bond to the phosphate portion of the ligand. This may be the source of the ribozyme's ability to discriminate to some extent against the sulfate analog of GlcN6P (8). Most intriguingly, the amine at the 2 position of the ligand (or hydroxyl of other inactive analogs such as Glc6P) forms a hydrogen bond with the 5' oxygen of G1. This oxygen is the leaving group during RNA chain scission, and therefore neutralization of developing negative charge on this leaving group oxygen may be a major contributor to the catalytic rate enhancement generated by the ribozyme. Interestingly, the epimeric configuration of the amino group of GlcN6P has little effect on the activity and the structure of the ribozyme (12, 17).

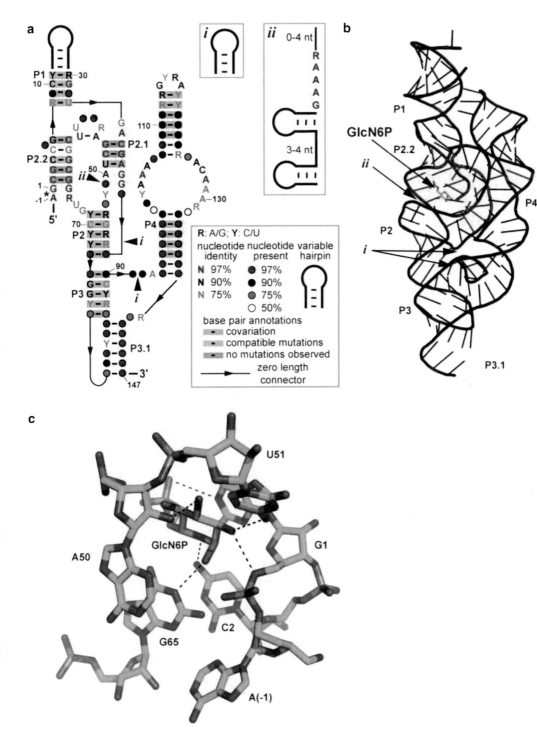

Fig. 1. Key sequence and structural features of *glmS* ribozymes, its relation to an atomic-resolution structural model, and RNA contacts to GlcN6P. (a) Consensus sequence and secondary structure model for *glmS* ribozymes. The *asterisk* designates the site of self-cleavage. Optional hairpins (*i*) or optional hairpins and conserved nucleotides (*ii*) are present in some *glmS* ribozymes. When provided, numbered nucleotides correspond to positions from a previously reported *glmS* ribozyme structure (19) as depicted in (**b**). (**b**) Atomic-resolution structural model of a *glmS* ribozyme from *T. tengcongensis* ((28); Protein Databank 3B4C). Nucleotides that are shaded in *red* are conserved in at least 97% of *glmS* ribozymes. *Arrows* depict the locations of optional sub-structures (*i, ii*) when present. (**c**) Model of the nucleotides forming the coenzyme-binding active site of the *T. tengcongensis glmS* ribozyme ((19); Protein Databank 3B4C). Nucleotides are labeled and numbered according to the consensus model in (**a**). (**a, b**) Are adapted from a previous publication (11).

A number of other contacts help bind and position the ligand in the active site. For example, the N4 of C2 forms a hydrogen bond with the oxygen at the 5 position of GlcN6P, and another hydrogen bond is formed between the 2' oxygen of A50 and the oxygen at the 3 position of the ligand. The contact that C2 makes to the ligand may explain why it is necessary for the ligand to be cyclic for robust activation of the ribozyme (12, 22). However, the contact that A50 makes to GlcN6P does not appear to discriminate between an axial or equatorial position for the hydroxyl group at the 3 position. The O4 of U51 forms a hydrogen bond with the amine of GlcN6P and the N1 of G65 forms a hydrogen bond to the hydroxyl of the anomeric carbon of GlcN6P (20). The N1 of G65 exclusively binds to the α-conformation of anomeric carbon in GlcN6P (12, 13).

Despite the extensive conservation of the catalytic cores of *glmS* ribozymes, there are rare exceptions that tolerate mutations. In every Deinococcales species sequenced to date, the nucleotide that corresponds to U51 is a guanosine (11). Within all Thermales species thus far sequenced, as well as within *Truepera radiovictrix*, there is an additional structural sub-domain that resides just before A50 (Fig. 1). This sub-domain consists of two variable-sequence hairpins separated by a short linker sequence, and a conserved sequence GAAAR that immediately follows the second hairpin. The loop of the second hairpin is almost always a tetraloop and often consists of structured loops such as GNRA or UNCG. Though this module is positioned very close to the binding pocket, it does not appear to affect substrate specificity in biochemical assays (11).

Far more sequence variation is observed outside of the *glmS* ligand-binding catalytic core, although the vast majority of the variation occurs with retention of the P3 and P4 structures. Some organisms within the Deinococcus-Thermus phylum carry an additional variable-sequence hairpin between the P3 and P4 helices (11). Previously, it was proposed that novel core support structures may take the place of the P4 helical stack (21). Indeed, *Truepera radiovictrix* carries a *glmS* ribozyme that appears to dispense with the entire P4 helix (Fig. 2). This ribozyme functions efficiently without the need for high cation concentrations that were required for other *glmS* ribozymes without the P4 helix (21). Perhaps other *glmS* ribozymes with similar structural changes that retain robust catalytic activity can be found as more genomic sequences become available.

1.4. A Possible Catalytic Mechanism for glmS Ribozymes

Both biochemical and structural data provide some clues regarding the mechanism of RNA transesterification used by *glmS* ribozymes. The rate-limiting step for catalysis is the presence of an appropriately formed catalytic core, although the rate of formation of the core is dependent upon specific supporting divalent cations

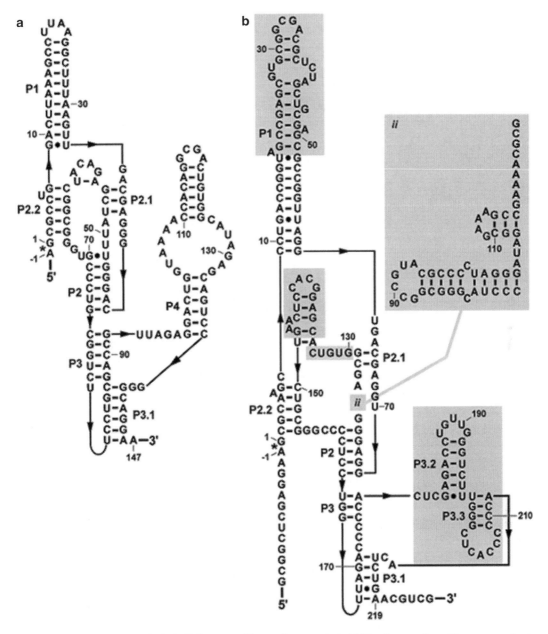

Fig. 2. Comparison of a canonical *glmS* ribozyme with an extreme variant. (**a**) The ribozyme from *Thermoanaerobacter tengcongensis* corresponds well to the consensus *glmS* ribozyme architecture typical of the vast majority of representatives. The *asterisk* identifies the site of self-cleavage and arrows denote zero-length connections. (**b**) A variant *glmS* ribozyme present in *Truepera radiovictrix*. Shaded regions identify sequences and structures that are substantially different from the typical *glmS* ribozyme consensus architecture.

(21, 23, 24). However, the activity of the catalytic core was determined to be independent of divalent metal ions, which only serve to assist in structure formation rather than having a direct involvement in bond making or breaking events during RNA cleavage (19–21, 24).

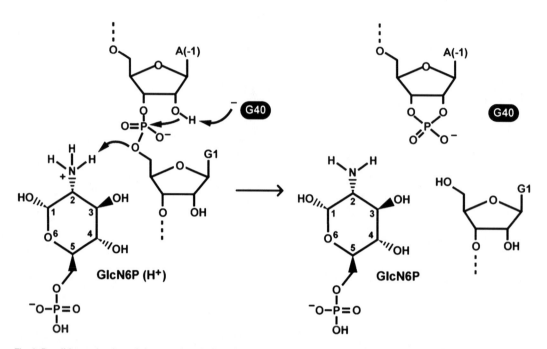

Fig. 3. Possible mechanism of cleavage for *glmS* ribozymes. *Left*: the reaction is initiated by deprotonation of the 2′-hydroxyl group of nucleotide A(-1). This creates a strong nucleophile for attack on the adjoining phosphorus center. The reaction is completed by protonation of the 5′-oxygen leaving group by the amine of GlcN6P. *Right*: products of cofactor-mediated RNA chain cleavage. *Dashed lines* indicate continuation of the RNA chain.

Although GlcN6P is the preferred ligand for *glmS* ribozymes (8, 12), surprisingly a variety of compounds that carry an amine in a beta position relative to a hydroxyl group (e.g., glucosamine, serinol, ethanolamine, etc.) also can support RNA catalysis, albeit with vastly reduced binding affinities (25). Even Tris buffer at concentrations typically used for biological assays (50 mM) slowly triggers ribozyme self-cleavage (21, 25). In contrast to the activities of compounds remotely similar to GlcN6P, glucose-6-phosphate (Glc6P), which is a very close structural analog of GlcN6P, does not support catalysis (8). Rather, Glc6P functions as a weak competitive inhibitor of ribozyme function (25), which was also shown in an *in vivo* system (26). Moreover, there are no major structural changes brought about by ligand binding to the RNA (8, 19, 27, 28), as would be likely if an allosteric mechanism for ribozyme activation were involved. Finally, both biochemical probing (29) and structural analyses (19, 20) reveal that the ligand resides close to the ribozyme cleavage site as described above. These results strongly implicate GlcN6P as a cofactor for ribozyme catalysis, most likely involving the general acid–base properties of the amine functional group (25).

A possible mechanism for RNA cleavage has emerged (Fig. 3) that accounts for many of the structural and functional characteristics observed. A deprotonated N1 of G40 could serve as a general base to deprotonate the 2′-hydroxyl group of A(-1). Recall that this 2′ oxygen is the nucleophile that initiates the cleavage reaction, and is prepositioned for in-line attack. To further accelerate RNA chain cleavage, a proton from the amine group of GlcN6P is donated to the 5′-oxyanion-leaving group of G1. The cleavage mechanism includes features that are similar to aspects of the mechanism employed by RNase A (13, 30).

Several observations are consistent with this proposed mechanism for *glmS* ribozymes. For example, changing the G40 nucleotide to any other nucleotide abolishes catalysis despite the fact that there is no expected adverse effect on the active site structure and ligand-binding determinants (17, 31). This finding suggests that the properties of the guanosine residue are important only for catalysis, and molecular dynamics models of the catalytic core support a general base role for G40 (32, 33). However, some observations remain puzzling and therefore suggest the detailed mechanism for *glmS* ribozyme function remains incomplete. Particularly notable is the fact that G40 mutation renders the ribozyme completely devoid of self-cleavage activity. This finding is perplexing since all the catalytic machinery within the ribozyme is ready for catalysis, and mutating G40 to adenosine does not drastically alter the binding site of the *glmS* ribozyme (13). Positioning the cleavage site phosphoester linkage for in-line nucleophilic attack and protonation of the leaving group by GlcN6P should yield measurable rate enhancements even if nucleophile activation does not occur (34, 35). Furthermore, the ribozyme appears to favor binding of GlcN6P in its uncharged amino form (20), which yields a bound cofactor that is not well suited for protonation of the 5′-oxygen leaving group. These findings have led to the suggestion that G40 and the GlcN6P cofactor may be chemically coupled wherein each tunes the other's chemical properties to promote RNA cleavage (30). Regardless, rate constants exhibited by some *glmS* ribozymes are in excess of $10 \ \text{min}^{-1}$ (36), and such rate enhancements require multiple catalytic strategies to be employed in concert (35).

1.5. Regulation of mRNA Stability by glmS Ribozymes

Despite extensive data demonstrating that binding of GlcN6P to the *glmS* ribozyme stimulates self-cleavage, the basis for intracellular control of gene expression was not immediately obvious upon initial discovery of the ribozyme. The obvious expectation was that self-cleavage is directly coupled to a change in mRNA stability. However, studies on mRNA degradation pathways in *Escherichia coli* present a conundrum for self-cleaving ribozymes like *glmS*, where the precursor transcript would be predicted to be similar in stability to the main product generated by ribozyme self-cleavage.

In aggregate, investigations of mRNA degradation in *E. coli* have demonstrated that molecular features at the 5′ terminus can be critical determinants for mRNA stability (37). Specifically, the degree of phosphorylation at the 5′ terminus can exert an important influence on an initial, rate-limiting step in degradation. For example, mRNA transcripts with a 5'-triphosphate group exhibit a much longer intracellular half-life as compared to transcripts with a 5' monophosphate, due to lowered affinity for the global RNase enzyme, RNase E. However, transcripts with a 5'-hydroxyl group also exhibit a long intracellular half-life, almost identical to triphosphorylated mRNAs (38, 39). Therefore, an mRNA containing a ribozyme at its 5' terminus, such as *glmS*, would be predicted to be stable both for the full-length mRNA and after ribozyme self-cleavage, since the corresponding transcripts contain a 5' terminal triphosphate and hydroxyl, respectively.

A potential solution to this problem was suggested by studies of mRNA degradation by *Bacillus subtilis*, which differs from *E. coli* in that it lacks the global RNase E enzyme (39, 40). Instead, *B. subtilis* encodes for at least one different global RNase enzyme, RNase J, which also exhibits 5'-end preferences for a subset of its RNA substrates and has a widespread distribution in bacteria that often contain a *glmS* ribozyme (40). In *B. subtilis*, the 3' cleavage product of the *glmS* ribozyme self-cleavage reaction, which includes the *glmS* coding region, is degraded in an RNase J-dependent manner (41). If, however, the same transcript contains a 5' triphosphate instead of a 5'-hydroxyl group, it is no longer rapidly degraded in an RNase J-dependent manner and the resulting transcripts accumulate within cells.

For *B. subtilis*, the presence of a 5' hydroxyl, such as imparted on downstream transcripts by self-cleaving ribozymes, can be a signal for RNase J-mediated degradation. As further proof for this model, heterologous addition of the *B. subtilis* RNase J enzyme to the *E. coli* cytosol led to rapid degradation of 5'-hydroxyl-containing transcripts, which otherwise accumulated in the absence of RNase J expression (41). Therefore, ribozyme-mediated control of mRNA stability is likely to be dependent on the appropriate expression of RNase enzymes.

1.6. Methods for glmS Ribozyme Cleavage Assays

Various *glmS* ribozymes can operate under a wide range of conditions *in vitro*, though there are a few characteristics and limitations that must be considered. A few key considerations are noted herein. For example, representative *glmS* ribozymes examined in the greatest detail function optimally within a pH range of 6.8–8.5 (25). Low millimolar concentrations of divalent magnesium and high millimolar concentrations of monovalent cations support robust activity, although Mg^{2+} can be substituted for by a variety of other divalent cations (16, 21). The apparent K_D for the GlcN6P cofactor is approximately 200 µM, and therefore concentrations above this value will yield optimal cleavage rates (8, 25). Other molecules that mimic the amine and adjacent hydroxyl group arrangement of GlcN6P will trigger ribozyme cleavage, albeit with reduced apparent K_D values (25).

Particularly noteworthy is the fact that the commonly-used buffer Tris (2-amino-2-hydroxymethyl-propane-1,3-diol) will also trigger ribozyme activity due to its chemical similarity to the natural cofactor GlcN6P (21, 25). The use of other buffers such as HEPES (4-(2-hydroxyethyl)-1-piperazineethanesulfonic acid) is recommended to prevent unwanted ribozyme action (11, 25). Some *glmS* ribozymes may require elevated temperatures to function efficiently, such as the *Thermus thermophilus glmS* ribozyme (11).

When constructing *glmS* ribozymes, it may be useful to retain a sufficient amount of RNA in the portion 5′ of the cleavage site to differentiate the 3′ fragment from unprocessed RNA. Inclusion of too many flanking nucleotides could interfere with efficient folding of the catalytic domain, particularly if precursor RNAs are purified by denaturing polyacrylamide gel electrophoresis (PAGE) before initiating a cleavage assay. If ribozyme refolding is a problem, adding GlcN6P to a transcription reaction when preparing precursor RNAs can be attempted to conduct ribozyme cleavage assays during transcription.

2. Materials

2.1. Preparation of a DNA Template

1. PCR Buffer (10×). 15 mM $MgCl_2$, 500 mM KCl, 100 mM Tris–HCl (pH 8.3 at 23°C), and 0.1% gelatin in deionized water (dH_2O).

2. Deoxyribonucleoside 5′-triphosphate Mix (10×). 2 mM of each of the following in deionized water: dATP, dGTP, dCTP, and dTTP.

3. Appropriate primers at 10 µM in dH_2O.

4. Trace amount of the appropriate template DNAs. This template input could come from natural sources such as bacterial genomic DNA, or from synthetic DNA templates.

5. Thermostable DNA Polymerase (e.g., *Taq*) (New England BioLabs).

6. Optional: 100% dimethylsulfoxide (DMSO). Typically used for DNA templates that are above 60% G-C content.

7. PCR Cleanup Kit (Qiagen).

8. Gel Extraction Kit (Qiagen).

9. GPG/Low-melt agarose (Applied Biosystems).

10. Gel electrophoresis rig.

11. Gel loading buffer (2×). 8 M urea, 20% (w/v) sucrose, 0.1% (w/v) SDS, 0.05% (w/v) bromophenol blue sodium salt,

0.05% (w/v) xylene cyanol, 90 mM Tris–HCl, 90 mM borate, and 10 mM ethylenediaminetetraacetic acid (EDTA) (pH 8.0 at 23°C).

12. Tris-Borate-EDTA gel-running buffer (1×). 90 mM Tris, 90 mM borate, and 10 mM EDTA (pH 8.0 at 23°C).

2.2. RNA Transcription of glmS Ribozymes

1. Transcription buffer (5×). 120 mM $MgCl_2$, 20 mM spermidine, 400 mM HEPES-KOH (pH 7.5 at 23°C), 200 mM dithiothreitol (DTT).

2. Nucleoside 5′-triphosphate (NTP) mixture (6×). 20 mM in dH_2O each of ATP, GTP, CTP, and UTP.

3. 10 µCi of a relevant α-[^{32}P] NTP, usually the most abundant nucleotide in the RNA.

4. Bacteriophage T7 RNA polymerase.

5. DNA Template (typically 50 pmol of PCR product).

6. Reagents and apparatus for PAGE.

7. Denaturing 6% polyacrylamide gel (0.75 mm thickness).

8. Tris-Borate-EDTA gel-running buffer (1×; see above).

9. Gel-loading buffer (2×; see above).

10. Hand-held shortwave (254 nm) UV light.

11. Fluor-coated thin layer chromatography (TLC) plate (Applied Biosystems).

12. Crush-soak buffer. 200 mM NaCl, 10 mM HEPES-KOH (pH 7.5 at 23°C), and 1 mM EDTA (pH 8.0 at 23°C).

13. Ethanol (70 and 100%) chilled to −20°C.

14. 3 M sodium acetate (pH 5.0 at 23°C).

15. X-ray film (BioMax MR; Kodak).

2.3. RNA Transcription with Ribozyme Cleavage Assay of glmS Ribozymes

1. Same materials as noted above for typical transcription assays, except that a 2-mM stock solution of GlcN6P (or any other test ligands) is prepared.

2. PhosphorImager cassette and PhosphorImager with ImageQuant Software (GE Healthcare).

3. Blotting paper 703 (VWR International).

4. Gel dryer (e.g., Bio-Rad model 583).

2.4. glmS Ribozyme Cleavage Assay

1. PAGE apparatus and materials as noted above.

2. Radiolabeled *glmS* ribozyme (no more than 10 nM final concentration).

3. Ribozyme reaction buffer (2×). 50 mM HEPES-KOH (pH 7.5 at 23°C), 50 mM $MgCl_2$, and 200 mM KCl.

3. Methods

In general, maintaining clean, RNase-free laboratory settings and materials is essential. Also, safety protocols for handling radioactive samples must be observed.

3.1. Preparation of a DNA Template

Normal PCR protocols can be followed using either genomic DNA or synthetic DNA as the input source. Template product integrity should be confirmed by DNA sequencing.

1. Design reciprocal primers to generate desired construct for transcription. If performing PCR from genomic DNA, one set of primers is sufficient. If building a construct from synthetic DNA, performing sequential PCR reactions may be necessary due to the limits on the lengths of synthetic DNAs. The initial 5′ primer of the entire sequence should have the T7 RNA polymerase recognition sequence of 5′-TAATACGACTCACTATA, and the first two transcribed RNA nucleotides should be GG to maximize transcription yields.

2. Combine the following into a microfuge tube to perform a PCR to amplify the region of interest.
   (a) 10 μL of 10× PCR buffer.
   (b) 10 μL of 10× dNTP mix.
   (c) 5–100 ng of genomic DNA template (not applicable if using only synthetic DNAs).
   (d) 5 μL of 10 μM 5′ primer.
   (e) 5 μL of 10 μM 3′ primer.
   (f) *Taq* DNA polymerase to 0.05 U μL^{-1} final concentration.
   (g) 5 μL DMSO (required if final construct is above 60% G-C content).
   (h) dH$_2$O to 100 μL.

3. Using standard thermocycler conditions, conduct 20–25 cycles with appropriate temperatures for the primers.

4. Verify PCR product yield by using agarose gel electrophoresis.

5. Use PCR Cleanup Kit if PCR produced a single desired product. If more than one product is generated, use Gel Extraction Kit after extracting the relevant band.

6. Sequence the resulting PCR products to confirm integrity.

3.2. RNA Transcription of glmS Ribozymes

1. Combine the following in a 1.5-mL microfuge tube:
   (a) Approximately 50 pmol DNA template.
   (b) 6 μL of 5× transcription buffer.
   (c) 6 μL of 5× NTPs.

(d) 10 μCi of α-[^{32}P]NTP whose identity is chosen to maximize radiolabeling. Typically, this will be UTP, though for G-C-rich constructs, using GTP is preferred.

(e) 25 U μL^{-1} T7 RNA polymerase final concentration.

(f) dH$_2$O to 30 μL.

2. Incubate for 0.5–1 h at 37°C.

3. Add 30 μL of 2× loading dye to the reaction to quench transcription and load sample into a 6% denaturing polyacrylamide gel for purification of the desired RNA products.

4. Run the gel until the product of interest has traversed approximately halfway through the gel, using the bromophenol blue and xylene cyanol markers to estimate the location of the product.

5. Separate the glass plates of the PAGE rig and place gel into plastic wrap.

6. Use either UV-shadowing or X-ray film to determine the location of the product in the gel.

7. Excise the product with a razor blade and transfer gel slice to a fresh microfuge tube using razor blade. Avoid contact between the gel slice with any surfaces that may have contaminating nucleases.

8. Add 300 μL of crush-soak buffer (no Tris) to the gel and either incubate at room-temperature for 0.5 h or at 4°C overnight.

9. Transfer the crush-soak buffer and RNA product to a new tube and add 30 μL of 3 M-sodium acetate (pH 5.0 at 23°C) and 1 mL of cold 100% ethanol. Store at −20°C for at least 20 min.

10. Centrifuge for 20 min at 14,000×g at 4°C.

11. Decant supernatant and add 70% cold ethanol to the tube containing the RNA pellet. Repeat centrifugation step.

12. Decant supernatant and dry pellet. Resuspend in 50 μL of dH$_2$O and quantify RNA.

3.3. RNA Transcription with Ribozyme Cleavage Assay of glmS Ribozymes

1. Make two reactions and follow steps 1–4 from the transcription reaction listed above, except add 3 μL of 2 mM GlcN6P to one reaction.

2. Place gel onto one piece of plastic wrap on one side and one piece of blotting paper.

3. Place gel into gel dryer for at least 1 h and place into a PhosphorImager cassette overnight.

4. Collect image from the PhosphorImager cassette and quantify the data using ImageQuant.

3.4. glmS Ribozyme
Cleavage Assay

1. Add the following into two 1.5-mL microfuge tubes:

   (a) 10 μL of 2× reaction buffer (no Tris).

   (b) Add *glmS* ribozyme RNA solution, to yield no more than a 10-nM final concentration.

2. Add the following into a 1.5-mL microfuge tube:

   (a) 2 μL of GlcN6P stock solution.

   (b) Fill with dH$_2$O to 20 μL final volume for the combined total in steps 1 and 2.

3. In another 1.5-mL microfuge tube, only add dH$_2$O to fill a reaction to 20 μL from step 1. There should be four microfuge tubes at this point.

4. Incubate all four tubes at the desired temperature (37°C is default) for 5 min.

5. Add the contents from tube #2 into a tube from #1. Add the contents from tube #3 into the other tube from #1.

6. Incubate at the desired temperature for 5 min, or take aliquots at various time points as desired. Reactions should be terminated by addition to an equal volume of 2× gel loading buffer.

7. Perform PAGE to separate ribozyme reaction products.

8. Image the gel using a PhosphorImager and quantify the resulting gel bands.

4. Conclusions

In total, the discovery and characterization of the *glmS* ribozyme has provided an important model system for understanding the basis of gene regulation by synthetic and natural, signal-responsive ribozymes. Moreover, the widespread distribution of both the *glmS* ribozyme and the RNase enzymes that it depends upon provide a mechanistic precedence for how other possible self-cleaving ribozymes could sense and respond to cellular metabolites. Perhaps other novel ribozymes with allosteric- or cofactor-based responses to metabolites still await discovery.

Acknowledgments

We are thankful to the members of the Breaker lab for helpful comments and advice. P.M. is supported by the NIH Training Grant T32GM007499. This work is supported in the Breaker lab by the NIH Grant PO1 GM022778-34 and by the Howard Hughes Medical Institute. R.R.B. is a Howard Hughes Medical Institute

Investigator. Research on the *glmS* ribozyme in the Winkler lab was supported by the University of Texas Southwestern Medical Center Endowed Scholars Fund.

References

1. Mandal M, Breaker RR. 2004. Gene regulation by riboswitches. *Nat. Rev. Mol. Cell Biol.* **5**:451–463.

2. Breaker RR. 2010. Riboswitches and the RNA World. *Cold Spring Harb. Perspect. Biol.* doi: 10.1101/cshperspect.a003566.

3. Montange RK. Batey RT. 2008. Riboswitches: emerging theme in RNA structure and function. *Annu. Rev. Biophys.* **37**:117–133.

4. Serganov A, Patel DJ. 2009. Amino acid recognition and gene regulation by riboswitches. *Biochim. Biophys. Acta.* **1789**:592–611.

5. Zhang J, Lau MW, Ferré-D'Amaré AR. 2010. Ribozymes and riboswitches: modulation of RNA function by small molecules. *Biochemistry* **49**:9123–9131.

6. Dambach MD, Winkler WC. 2009. Expanding roles for metabolite-sensing regulatory RNAs. *Curr. Opin. Microbiol.* **12**:161–169.

7. Wachter A. 2010. Riboswitch-mediated control of gene expression in eukaryotes. *RNA Biol.* **7**:67–76.

8. Winkler WC, Nahvi A, Roth A, Collins JA, Breaker RR. 2004. Control of gene expression by a natural metabolite-responsive ribozyme. *Nature* **428**:281–286.

9. Barrick JE, Corbino KA, Winkler WC, Nahvi A, Mandal M, Collins J, Lee M, Roth A, Sudarsan N, Jona I et al. 2004. New RNA motifs suggest an expanded scope for riboswitches in bacterial genetic control. *Proc. Natl. Acad. Sci. USA* **101**:6421–6426.

10. Barrick JE, Breaker RR. 2007. The distributions, mechanisms, and structures of metabolite-binding riboswitches. *Genome Biol.* **8**:R239.

11. McCown PJ, Roth A, Breaker RR. 2011. An expanded collection and refined consensus model of *glmS* ribozymes. *RNA* **17**:728–736.

12. Lim J, Grove BC, Roth A, Breaker RR. 2006. Characteristics of ligand recognition by a *glmS* self-cleaving ribozyme. *Angew. Chem. Int. Ed.* **45**:6689–6693.

13. Ferré-D'Amaré AR. 2010. The *glmS* ribozyme: use of a small molecule coenzyme by a gene-regulatory RNA. *Q. Rev. Biophys.* **43**:423–447.

14. Milewski S. 2002. Glucosamine-6-phosphate synthase – the multi-facets enzyme. *Biochim. Biophys. Acta* **1597**:173–192.

15. Blount KF, Breaker RR. 2006. Riboswitches as antibacterial drug targets. *Nat. Biotechnol.* **24**:1558–1564.

16. Lünse CE, Schmidt MS, Wittmann V, Mayer G. 2011. Carba-sugars activate the *glmS*-riboswitch of *Staphylococcus aureus*. *ACS Chem. Biol.* DOI: 10.1021/cb200016d.

17. Cochrane JC, Lipchock SV, Smith KD, Strobel SA. 2009. Structural and chemical basis for glucosamine 6-phosphate binding and activation of the *glmS* ribozyme. *Biochemistry* **48**:3239–3246.

18. Link KH, Guo L, Breaker RR. 2006. Examination of the structural and functional versatility of *glmS* ribozymes by using *in vitro* selection. *Nucleic Acids Res.* **34**:4968–4975.

19. Klein DJ, Ferré-D'Amaré AR. 2006. Structural basis of *glmS* ribozyme activation by glucosamine-6-phosphate activation. *Science.* **313**:1752–1756.

20. Cochrane JC, Lipchock SV, Strobel SA. 2007. Structural investigation of the *glmS* ribozyme bound to its catalytic cofactor. *Chem. Biol.* **14**:97–105.

21. Roth A, Nahvi A, Lee M, Jona I, Breaker RR. 2006. Characteristics of the *glmS* ribozyme suggest only structural roles for divalent metal ions. *RNA* **12**:607–619.

22. Blount KF, Puskarz I, Penchovsky R, Breaker RR. 2006. Development and application of a high-throughput assay for *glmS* riboswitch activators. *RNA Biol.* **3**:77–81.

23. Brooks KM, Hampel KJ. 2009. A rate-limiting conformational step in the catalytic pathway of the *glmS* ribozyme. *Biochemistry* **48**: 5669–5678.

24. Klawuhn K, Jansen JA, Souchek J, Soukup GA, Soukup JK. 2010. Analysis of metal ion dependence in *glmS* ribozyme self-cleavage and coenzyme binding. *Chem. Biochem.* **11**:2567–2571.

25. McCarthy TJ, Plog MA, Floy SA, Jansen JA, Soukup JK, Soukup GA. 2005. Ligand requirements for *glmS* ribozyme self-cleavage. *Chem. Biol.* **12**:1221–1226.

26. Watson PY, Fedor MJ. 2011. The *glmS* riboswitch integrates signals from activating and inhibitory metabolites *in vivo*. *Nat. Struct. Mol. Biol.* **18**:359–363.

27. Hampel KJ, Tinsley MM. 2006. Evidence for preorganization of the *glmS* ribozyme ligand binding pocket. *Biochemistry* **45**:7861–7871.

28. Klein DJ, Wilkinson SR, Been MD, Ferré-D'Amaré AR. 2007. Requirement of helix P2.2 and nucleotide G1 for positioning of the cleavage site and cofactor of the *glmS* ribozyme. *J. Mol. Biol.* **373**:178–189.

29. Jansen JA, McCarthy TJ, Soukup GA, Soukup JK. 2006. Backbone and nucleobase contacts to glucosamine-6-phosphate in the *glmS* ribozyme. *Nat. Struct. Mol. Biol.* **13**:517–523.

30. Ferré-D'Amaré AR, Scott WG. 2010. Small self-cleaving ribozymes. *Cold Spring Harbor Perspect. Biol.* **2**:a003574.

31. Klein DJ, Been MD, Ferré-D'Amaré AR. 2007a. Essential role of an active-site guanine in *glmS* ribozyme catalysis. *J. Am. Chem. Soc.* **129**:14858–14859.

32. Banáš P, Walter NG, Šponer J, Otyepka M. 2010. Protonation states of the key active site residues and structural dynamics of *glmS* riboswitch as revealed by molecular dynamics. *J. Phys. Chem.* **114**:8701–8712.

33. Xin Y, Hamelberg, D. 2010. Deciphering the role of glucosamine-6-phosphate in the riboswitch action of *glmS* ribozyme. *RNA* **16**:2455–2463.

34. Breaker RR, Emilsson GM, Lazarev D, Nakamura S, Puskarz IJ, Roth A, Sudarsan N. 2003. A common speed limit for RNA-cleaving ribozymes and deoxyribozymes. *RNA* **9**:949–957.

35. Emilsson GM, Nakamura S, Roth A, Breaker RR. 2003. Ribozyme speed limits. *RNA* **9**:907–918.

36. Wilkinson SR, Been MD. 2005. A pseudoknot in the 3′ non-core region of the *glmS* ribozyme enhances self-cleavage activity. *RNA* **11**:1788–1794.

37. Belasco JG. 2010. All things must pass: contrasts and commonalities in eukaryotic and bacterial mRNA decay. *Nat. Rev. Mol. Cell Biol.* **11**:467–478.

38. Celesnik H, Deana A, Belasco JG. 2007. Initiation of RNA decay in *Escherichia coli* by 5' pyrophosphate removal. *Mol. Cell* **27**:79–90.

39. Bechhofer DH. 2009. Messenger RNA decay and maturation in *Bacillus subtilis*. *Prog. Mol. Biol. Transl.Sci.* **85**:231–273.

40. Condon C, Bechhofer DH. 2011. Regulated RNA stability in the Gram-positives. *Curr. Opin. Microbiol.* **14**:148–154.

41. Collins JA, Irnov I, Baker S, Winkler WC. 2007. Mechanism of mRNA destabilization by the *glmS* ribozyme. *Genes Dev.* **21**:3356–3368.

Chapter 9

Structure-Based Search and In Vitro Analysis of Self-Cleaving Ribozymes

Randi M. Jimenez and Andrej Lupták

Abstract

Detecting functional RNAs is increasingly accomplished through structure-based searches for patterns of conserved secondary structure. With large amounts of new sequencing data becoming available, there is a greater demand for efficient methods of identifying new RNAs. Here we present a method of identifying self-cleaving ribozymes and characterizing the in vitro activity.

Key words: ncRNA, Ribozymes, RNABOB, Oligonucleotides, Self-cleavage

1. Introduction

Structured noncoding RNAs (ncRNAs), ubiquitous to all forms of life, are known to possess a diversity of functions, including regulation of cellular processes and gene expression (1). Noncoding RNA do not benefit from genome annotation because they do not code for proteins. Various methods have lead to the discovery of different ncRNA, but the increasing amount of sequence data demands a computationally efficient ncRNA prediction method. Comparative sequence analysis can be useful in understanding the evolutionary patterns of base changes but are limited in discovering new functional RNA (2). Functional RNAs are conserved in secondary structure as defined by base-paired helical regions joined by sequence-specific single-stranded regions, which can form active sites or binding pockets (3). Strict base-complementarity in paired regions allows base-pair covariation among structured functional RNAs of an identical fold that is otherwise unrecognizable by sequence alignments. Discovering new functional RNAs therefore requires dynamic approaches, which must include structure-based constraints. This chapter will focus on the application of structure-

Jörg S. Hartig (ed.), *Ribozymes: Methods and Protocols*, Methods in Molecular Biology, vol. 848,
DOI 10.1007/978-1-61779-545-9_9, © Springer Science+Business Media, LLC 2012

based searches of genomic ribozymes and complements methods presented in Chapters 2 and 10.

1.1. Ribozymes

RNAs fulfill diverse regulatory and catalytic functions (4, 5). Biological catalytic RNAs include RNase P, group I and II self-splicing introns, self-cleaving ribozymes, and the ribosomal peptidyl transferase (6). Numerous approaches have been used to identify new ribozymes; each method has its own advantages and disadvantages. In vitro selection (7), genomic analysis (8), comparative sequence analysis, and other bioinformatics approaches (9–11) have previously identified new ribozymes. Systematic bioinformatic approaches are increasingly utilized to identify functional RNAs. In particular, structure-based genomic searches have successfully uncovered the widespread phylogenic distribution of HDV-like ribozymes (12), and the hammerhead ribozyme (13, 14).

2. Materials

2.1. Oligo Purification, Transcription, and Self-Cleavage Assay

1. 10× Taq polymerase buffer.
2. 20 mM stocks of all four dNTPs.
3. Ice cold ethanol (100 and 70%).
4. 10× transcription buffer: 400 mM Tris–HCl, pH 7.4; 100 mM dithiothreitol (DTT); 20 mM spermidine; and 1% Triton X-100.
5. 25 mM stocks of each of rCTP, rGTP, rUTP; 2.5 mM stock of rATP.
6. 3 μM 32P-labeled ATP.
7. 2× kinetics buffer: 280 mM KCl; 20 mM NaCl; 20 mM Tris–HCl, pH 7.4.
8. Stop loading buffer: 20 mM EDTA; 8 M urea; 5 mM Tris–HCl, pH 7.4; xylene cyanol and bromophenol blue loading dyes.

2.2. Denaturing PAGE

1. 40% acrylamide stock.
2. 10% ammonium persulfate (APS).
3. TEMED.
4. 10× TBE (Tris-borate and EDTA).
5. Urea.
6. Polyacrylamide gel solution (1 L); final concentrations: 8 M urea, 0.5× TBE, 15% polyacrylamide.
7. 1.5- and 0.8-mm Teflon spacers, wide-toothed and narrow-toothed combs to cast wells, razor blades, and plastic wrap.

8. UV-active fluorescent TLC plates, UV lamp, phosphorimage screens/plates.

9. Loading buffer: 8 M urea; 5 mM Tris, pH 7.4; and xylene cyanol and bromophenol blue loading dyes.

3. Methods

Computationally, single-sequence structure prediction is only partially accurate and multiple sequence alignment concurrent with folding is slow and expensive. Therefore, no structure prediction program alone is reliable in efficient discovery of RNA motifs. Iterative processes featuring multiple search- and structure-based algorithms have previously been described (15–17). Our approach to discovering ribozyme motifs is simple, utilizing the RNABOB program (http://selab.janelia.org/software.html), used for scanning a genome for prescribed RNA secondary structure, and in some cases the RNAfold algorithm of the ViennaRNA package (18, 19), which is used to filter the initial RNABOB output (discussed elsewhere). This approach has been adapted for use on genome searches for a single structured RNA motif of interest at a time. The resulting output sequences are then tested for in vitro activity.

3.1. RNABOB

RNABOB is a fast pattern-searching algorithm written as an implementation of its predecessor RNAMOT (20). The underlying algorithm is a nondeterministic finite state machine similar to that used by UNIX regular expression pattern matching algorithms, but capable of searching for covariance (S. Eddy, unpublished). Implementing RNABOB does not require extensive computer programming knowledge. Descriptors are written as separate text documents (*.des) with the 5′ and 3′ ends of the motif implied by the element ordering (5′→3′). Each element of a motif is given its own designation, i.e., h1, h2, h3, following the syntax of the program as previously described (21). The syntax used for the descriptors is rather powerful, having greater flexibility for complex structures.

RNABOB can search a variety of file formats available through genome databases, with search times varying according to sequence length. The search follows a manner specified by the available options chosen at the command line, and the resulting output can be saved in any text document format. Details regarding implementation can be found in the documentation (rnabob.man, rnabob.ps) accompanying the executables from the downloaded RNABOB source file.

Our structure-based search method requires a POSIX (Portable Operating System Interface for UNIX) compliant operating system.

3.1.1. RNABOB Installation

1. Download RNABOB from the Eddy Lab software website: ftp://selab.janelia.org/pub/software/rnabob/.

2. Expand and archive the tar file in the directory of choice (Optional command line: tar xvzf<filename>).

3. Follow instructions in the install file (INSTALL) to build the program through a computer interface window (shell).

3.1.2. Writing Descriptors

1. Open a new document in a simple text editor (caution: advanced word processing programs, i.e., MS Word, insert hidden characters and should be avoided) and save it with the extension ".des."

2. Divide the motif of interest into a list of individual single stranded elements "s," helical elements "h," or relational elements "r." The "r" designation allows the user to define strict Watson–Crick or non-canonical base pairing for each residue. The "h" designation is general and includes G-U wobble pairs. The top of the descriptor file will be a single line listing each element, separated by a space, starting with the 5′ end. Helical and relational elements are written twice, representing each strand, e.g., h1 and h1′.

3. The subsequent lines should then describe each element individually in terms of the nucleotide content (any IUPAC letter code) as previously described (21), with each line containing three or four fields separated by a tab or blank space. The first field names the element from the topology given in the first line in the file. The second field lists the number of mismatches allowed in that element. The third field is the sequence constraint. The fourth field of a relational element line is the strict base pairing required (T, G, C, A). Helical and relational elements are specified on a single line, with mismatches and primary sequence constraints given as pairs. A colon ":" is used to separate the upstream and downstream regions, left and right, respectively.

4. Sequence-specific constraints can be included in a descriptor to accommodate specific user interests. The addition of these constraints (such as specific Watson–Crick base pairing in stems: i.e., element r5 below) will result in fewer sequence outputs and therefore fewer false-positive returns.

5. Any lines in the descriptor file beginning with "#" are comments and are not interpreted by the program running the descriptor.

   For convenience, descriptor files, sequence files, and output files should be kept near the RNABOB directory to avoid lengthy command lines.

```
# HDV.des

h1 r2 s1 r3 r4 s2 r5 s3 r4' r2' h1' r5' s4 h6 s5 h6'

s6 r3'
```

```
h1  0:0  G:Y

r2  0:0  NNNNN*:*NNNNN TGCA

s1  0    NN[250]

r3  0:1  *NNNNNN:NNNNNN* TGCA

r4  0:0  NNN:NNN TGCA

s2  0    TY

r5  0:0  Y:R TGCA

s3  0    HCG*Y

s4  0    N******

h6  0:1  NNN:NNN

s5  0    NNN[250]

s6  0    NC*RA*
```

The following is an example of a descriptor used to search for a HDV double-pseudoknot motif:

3.2. Genome Databases

The sequence file must contain DNA or RNA sequences and most common file formats are accepted (e.g., FASTA, GenBank, EMBL database). IUPAC degenerate nucleotide codes are accepted.

3.3. Running RNABOB

1. To run a search, open a new shell.
2. Change the directory (unless environment pathways are set to run "rnabob").
3. rnabob usage: rnabob <-options> <descriptor.des> <sequence.fa>

    This displays all output inside the terminal, otherwise it can be sent directly to an output text file ".txt." For example:

```
/Applications/rnabob-2.1/rnabob -Fc Des/HDV.des

Seq/Ribozymes_test.fasta > Outfiles/HDV-

testseq_out.txt
```

This command will run rnabob with "F" and "c" options using the "HDV.des" descriptor and "Ribozymes_test" sequence file. It sends the output into the "Outfiles" folder under the "rnabob-2.1" directory. There are only five options for running RNABOB (described in rnabob.man or rnabob.ps). Fancier output (F) prints the sequence that is matched. Skip (s) gives a single output for a given locus if the descriptor can be satisfied in multiple ways by an ambiguous sequence (degenerate IUPAC, particularly long runs of Ns) that may occur in the sequence file. Complement (c) searches both the sense and antisense strands.

3.4. Structure of Output and Interpreting Output

RNABOB does not provide a graphical representation of the folded RNA but outputs a text representation similar to the top line of the descriptor file. Output will begin with "Waking up rnabob…" which signifies the program is running. Subsequent lines include details about which descriptor file was used, which sequence file was searched, and the options implemented in the search. Each sequence output begins with a header line describing the first and last positions of the motif (seq-f and seq-t respectively), and a description of the sequence, in a format defined by the genome bank from which the file was downloaded. This is particularly helpful should a single file contain sequences of multiple origins. The line following the header contains the sequence which RNABOB returned, matching the descriptor motif with each element separated by a vertical bar. The last line of RNABOB output includes the total number of nucleotides scanned, with the actual number of nucleotides in the sequence file doubled when the "c" option is chosen.

```
Waking up rnabob: version 2.1, March 1996

-----------------------------------------------------

Database file:                  Seqs/Ribozymes_test.fa

Descriptor file:                DesFiles/HDV.des

Complementary strand searched: yes

Filter out overlapping hits:   no

-----------------------------------------------------


  seq-f  seq-t      name      description

 ------ ------ ------------ -----------

    74     140  Mouse_CPEB3

|G|GGGGCC|AC|AGCAGAA|GCG|TT|C|ACGT|CGC|GGCCCC|T|G|TC|AGA|TT

CTGGCGAA|TCT|GCGAA|TTCTGCT|

   146      80 Antisense_Mouse_CPEB3

|G|GGGGCC|AC|AGCAGAA|GCG|TT|C|ACGT|CGC|GGCCCC|T|G|TC|AGA|TT

CTGGCGAA|TCT|GCGAA|TTCTGCT|
```

```
    1      81 gi|269308691|tpg|BK006880.1| TPA_exp:
Anopheles gambiae drz-Agam-1-3

|G|GCTGAC|AAAATCC|TATCTCT|ACC|TC|C|TCGT|GGT|GTCAGC|C|G|G|AA

T|GTGCAGTATCAGTACTGCAT|ATA|CCAAC|AGAGAG|


rnabob run completed.

Total number of bases scanned: 1268
```

The input FASTA file contained five sequences, and RNABOB identified three as matching the HDV motif. In one instance, the mouse CPEB3 gene, the sequence was present in both orientations (sense and antisense), as indicated by the seq-f and seq-t reference locations. The third output originated from an annotated ribozyme in the *Anopheles gambiae* genome, and the header line displays all the pertaining reference information typical of a sequence file downloaded from GenBank.

3.5. Design Constructs

1. Generally, this method follows that previously described (12). Sequences under ~150 nts can generally be ordered through an oligonucleotide synthesis service (such as those offered by Invitrogen), in the form of a single construct under a 25- or 50-nmol size synthesis. Longer sequences may require multiple oligos for a primer extension and PCR amplification, which can be accomplished in one of three ways:

   (a) Mutual primer extension: Forward primers are designed to include the T7 RNA polymerase promoter sequence followed by "gggaga" to facilitate transcription initiation, unless the 5′ end of the construct is G-rich (even a single guanosine may be used as the first transcribed nucleotide; however, we find a purine-rich 5′ terminus transcribes with higher yield). Forward and reverse oligos must overlap by about 20 nts. For example, a 160-nt RNA construct (including promoter and "gggaga") can be divided into two ~100 nt oligos. However, attention should be paid to the outcome of in vitro activity assays, which will be viewed by phosphorimage analysis. In regard to self-cleavage, it is advisable that the difference in length between the RNA precursor and cleaved ribozyme be about 40 nt to promote clear band separation. This can be accomplished by increasing the length of the 5′ end of the construct according to the genomic sequence from the genome file.

   (b) PCR-based gene synthesis: This method follows traditional two-step assembly PCR with all overlapping oligos assembled by self-priming and the product amplified by a pair of primers into the full-length construct.

(c) Alternatively, a one-step PCR gene synthesis approach can also be used (22).

2. Self-cleaving RNA should be transcribed with an inhibitor oligonucleotide complementary to the cleavage site. This method prevents co-transcriptional cleavage, which would eliminate kinetically fast species from the final purified transcript sample, as described previously (12, 23). Optimal lengths for inhibitor oligonucleotide should facilitate a melting temperature of about 42°C, which can be calculated with the help of such web-based tools as Olico Calc: Oligonucleotide Properties Calculator (24). Assuming RNA–DNA hybrids are more stable than DNA–DNA helices, we expect a predicted T_m of 42°C to be sufficient for co-transcriptional inhibition of the ribozyme; however, this T_m can be lowered should the transcription reaction be carried out at lower temperatures.

The following is an example of a HDV ribozyme construct synthesized by mutual priming of two oligos:

drz-Agam-1-1 ribozyme

Ribozyme construct: 5′ gggaacuagc GGC UGA CAA AAU CCU UUC CCA ACC UCC ACG UGG UGU CGG CUG GAU AAU GCA UUA GAA AUG UUG CAU UUA CCA ACU GGG AAGG

AL292: 5′ TTC CCG CGA AAT TAA TAC GAC TCA CTA TA GGG AAC TAG CGG CTG ACA AAA TCC TTT CCC AA

AL293: 5′ CCT TCC CAG TTG GTA AAT GCA ACA TTT CTA ATG CAT TAT CCA GCC GAC ACC ACG TGG AGG TTG GG AAA GGA TTT TGT CAG CC

AL301 (inhibitor): 5′ ATA GGA TTT TGT CAG CC

In this case the leader sequence is only 10 nts long (lower case letters).

3.5.1. Oligo Purification

Oligos can typically be ordered with a variety of purification options (i.e., desalted, PAGE, HPLC, etc.). We recommend purifying all oligos above 25 nts using denaturing PAGE.

1. Resuspend desalted oligo in water to a final concentration of 100 μM. Aliquot half into a 1.5-mL plastic tube and store the rest at –20°C. Add a half-part 8 M urea loading buffer containing xylene cyanol and bromophenol blue loading dyes.

2. Prepare 50 mL of polyacrylamide gel solution of the appropriate percentage for the RNA sample (25).

3. Wash medium gel plates and 1.5-mm spacers and wide-tooth combs with distilled water, then ethanol. With side and bottom spacers in place, clip the plates together with clamps.

4. Add 500 µL of APS to the polyacrylamide gel solution and mix. Pour 4 µL into a 15-mL Eppendorff tube. To this add 10 µL of TEMED to initialize polymerization. Mix and pour into the gel plate assembly to create a plug at the bottom of the gel. Allow ~1 min for polymerization.

5. To the remainder of the gel solution add 50 µL of TEMED, mix, and pour into the gel plate assembly. Insert combs and allow the gel to polymerize completely.

6. Remove clips and combs and move the assembly to an electrophoresis gel box. Add 0.5× TBE buffer to cover the top and bottom of the gel and rinse the wells to remove air bubbles. Load ~10 µL of loading dye into a single well and pre-run the gel for about 20 min at 20 W until dye migration indicates the gel is ready.

7. Turn off the power supply. Rinse the wells and load oligos. Run the gel at 20 W for about 45 min for constructs less than 110 nts.

8. Turn off the power, remove the plate assembly and remove the gel carefully, covering both sides in plastic wrap. Place the gel on top of a fluorescent TLC plate and visualize the bands with a UV lamp. Trace the bands of interest with a marker for excision.

9. Excise each band with a clean razor and cut into pieces. Put the gel pieces into a microcentrifuge filter (spin-X) and add 400 µL of 300 mM KCl. Agitate for 3–4 h.

10. Centrifuge for 1 min at $1,600 \times g$.

11. Discard the microcentrifuge filter with gel pieces. To the elution add 800 µL of cold 100% ethanol and store at –80°C for 20 min minimum.

12. Spin for 10 min at maximum speed and discard the supernatant. Wash the pellet with 1 mL 70% ethanol. Spin down briefly and remove supernatant. Allow to dry.

13. Resuspend the pellet in 30–50 µL of 10 mM Tris–HCl, pH 7.4.

3.5.2. Synthesis of Constructs by Mutual Priming

1. Set up a 100 µL reaction in a PCR tube. Final concentrations should be as follows: 200 µM of each of the four dNTPs; 5 µM of each oligo; 1× Taq buffer and 10 U of Taq DNA polymerase. The primer extension conditions are 94°C for 1 min, followed by 2 cycles of 50°C for 30 s, and 72°C for 2 min as previously described (12).

3.6. Transcription and Purification of Ribozyme Precursor RNA

1. Transcription for catalytic RNA has been described previously (12, 23), allowing the reaction to run for 1 h maximum at the desired temperature. Set up a 20 µL reaction in a 0.5- or 0.2-mL plastic tube. Final concentrations should be as follows: 1× transcription buffer; 2.5 mM each GTP, UTP, and CTP; 250 µM ATP; 4.5 µCi [α-32P]-ATP (Perkin Elmer, Waltham,

MA); 7.75 mM MgCl$_2$; 20 µM inhibitor oligo (specific for construct); 1 U of T7 RNA polymerase; and 0.5 pmol of DNA template. Purify the transcripts by denaturing PAGE gel, stopping electrophoresis when the dye migration indicates the gel is done (for longer constructs, the loading dyes may run completely off the gel). Remove the gel assembly from the gel box and carefully remove one plate. Expose the plastic wrap-covered gel to a phosphorimage screen for about 20 min. Scan the screen using a phosphorimager and print out the image. The bands of interest should be excised and cut into pieces. Put the gel pieces into a spin filter (spin-X) and add up to 150 µL DEPC-treated water. Agitate for 1 h maximum. Centrifuge at maximum speed in a microcentrifuge for 5 min.

2. Prepare a G25 Sephadex bead column by loading ~800 µL suspension of beads onto a spin filter. Centrifuge at 4 kg for 5 min to remove excess liquid.

3. Load the entire RNA elution volume onto the G25 column and centrifuge at 4 kg for 5 min to remove residual urea and EDTA that co-eluted from the gel.

4. Use the flow through for RNA kinetics analysis. Store at –80°C.

3.7. In Vitro Self-Cleavage Assay

1. Ribozyme self-cleavage experiments follow the general procedure previously described (26). The number of time points desired will determine the total reaction volume to make. For example, a 30 µL reaction will supply 3 µL aliquots at about 9 time points accurately. Prepare a 30 µL reaction in a PCR tube. Final concentrations should be as follows: 1× kinetics buffer and the desired reaction concentration of the chosen divalent metal.

2. Set up a cone-bottom well-plate to aliquot time points from the reaction. Each well should contain 3 µL stop buffer containing 20 mM EDTA (or more to chelate all divalent metal ions in the reaction), 5 mM Tris pH 7.4, 8 M urea with xylene cyanol, and bromophenol blue loading dyes at a volume equal to the aliquot taken.

3. It is important not to forget to take a "No Reaction" (NR) RNA aliquot (~0.6 µL) prior to initiating the reaction. This sample should be aliquoted into 3 µL stop buffer on the well plate.

4. The volume of 32P-body labeled RNA used to initiate the reaction should reflect a good radioactive decay signal. Both reaction buffer and RNA should be separately incubated to reaction temperatures prior to use. Initiate the reaction under the desired temperature conditions with a PCR machine set to a constant temperature. (Use a thermocycler with a heated lid to avoid condensation of water on the lid of the tube. Solvent evaporation can drastically change the concentration of solutes when using low volumes.)

5. Remove aliquots in a manner that the reaction is quenched by the stop buffer at the desired time point.

6. Run a denaturing PAGE gel (0.8-mm spacers and small-tooth combs) of the self-cleavage products and expose overnight to a phosphorimage screen. For long exposures it is best to line the back side of the gel with anion exchange paper (DE81, Sigma) and paper towels, or to dry the gel. This step removes excess liquid in the gel which might allow samples to diffuse during storage. Wrap the entire gel in plastic wrap and store the phosphorimage cassette in a refrigerator during exposure. Alternatively, the gel can be first dried and then exposed to an imaging plate.

7. Analyze the gel by Typhoon phosphorimager and ImageQuant software (GE healthcare) to measure band intensities. For self-cleaving ribozymes, the single precursor RNA band will cleave into two visible bands (5′ and 3′ products), which will increase in intensity over time as the self-scission reaction is allowed to proceed. It is important to design constructs in a way which will allow two product bands to be distinguishable by size separation.

3.8. Data Fitting Using Excel

A number of data fitting software are currently available; however, Microsoft Office Excel is commonly used office software. We will therefore describe linear least-squares optimization using Excel. The band intensities of the self-cleavage experiment can be used to solve for the observed rates of cleavage ($k_{obs, cleave}$) using linear regression analysis of a mono- or bi-exponential decay and residuals model fitted by the least-squares approach.

1. Paste the band intensity values into a new Excel spreadsheet. For convenience, this should be done horizontally so that formulas can be pasted directly underneath the corresponding time point. In a convenient location on the spreadsheet, print arbitrary values corresponding to "A," "k_1," and "C" for mono-exponential decay and residuals model or additionally "B" and "k_2" for bi-exponential decay and residuals model.

2. Calculate the fraction of precursor cleaved for each time point by dividing the value of the precursor band intensity by the sum of the precursor and product band(s) intensities at a single time point. The formula can be pasted into subsequent cells without retyping.

3. Using the desired model, calculate for each time point where "t" is the time and "A," "k_1," and "C" are arbitrary values for a mono-exponential model: Fraction intact $= Ae(-k1*t) + C$. This should be done by programming the following into a cell for a mono-exponential decay and residuals model: $=\$P\$40*EXP(-\$Q\$40*B37) + \$R\$40$. Where $\$P\$40$ represents

the value for "A," Q40 represents the value for "k1," B37 represents the time, and R40 represents the value for "C." The dollar sign ensures that the value in that cell is used regardless of where the formula is pasted in the spreadsheet. Therefore, the time value will change as the formula is pasted across the time point columns. This programming is important for utilizing the "Solver" tool, discussed shortly.

4. Calculate the difference at each time point between the model value and the fraction cleaved.

5. Calculate the square of that difference.

6. Calculate the sum of the square differences by programming using the SUM function in a new cell.

7. Solver is an add-in which may need to be loaded into Excel via the Tools menu. Solver finds the optimum value in one cell by adjusting the values in the cells the user specifies. Therefore, this tool can be used to solve the regression of the data points to the model, resulting in the $k_{obs, cleave}$ value for the particular ribozyme.

   (a) In Excel under the Data menu, select Analysis → Solver.

   (b) Set the target cell to the sum of the square differences by selecting the cell containing that value.

   (c) The goal is the minimum value of that selected cell so set "Equal to:" to "min."

   (d) "By changing cells:" should be set to our arbitrary model values (A, k_1, C; P40-R40 in our example).

   (e) Click "Solve." Allow the process to complete. The model value cells should now contain the $k_{obs, cleave}$ value for the particular ribozyme (the value for k_1).

   (f) The "Solve" processing can be visualized if both the calculated model values as well as the fraction cleaved values are plotted vs. time.

8. Alternatively, similar data fitting procedures can be performed in a Google docs spreadsheet (Tools → Solve).

References

1. Sharp, P. A. (2009) The centrality of RNA, *Cell* 136, 577–580.

2. Rivas, E., and Eddy, S. R. (2001) Noncoding RNA gene detection using comparative sequence analysis, *BMC Bioinformatics* 2, 8.

3. Doherty, E. A., and Doudna, J. A. (2000) Ribozyme structures and mechanisms, *Annu Rev Biochem* 69, 597–615.

4. Waters, L. S., and Storz, G. (2009) Regulatory RNAs in bacteria, *Cell* 136, 615–628.

5. Fedor, M. J., and Williamson, J. R. (2005) The catalytic diversity of RNAs, *Nat Rev Mol Cell Biol* 6, 399–412.

6. Cech, T. R. (2009) Evolution of biological catalysis: ribozyme to RNP enzyme, *Cold Spring Harb Symp Quant Biol* 74, 11–16.

7. Salehi-Ashtiani, K., and Szostak, J. W. (2001) In vitro evolution suggests multiple origins for the hammerhead ribozyme, *Nature* 414, 82–84.

8. Cech, T. R. (2002) Ribozymes, the first 20 years, *Biochem Soc Trans* 30, 1162–1166.

9. Bourdeau, V., Ferbeyre, G., Pageau, M., Paquin, B., and Cedergren, R. (1999) The distribution of RNA motifs in natural sequences, *Nucleic Acids Res* 27, 4457–4467.

10. de la Pena, M., and Garcia-Robles, I. (2010) Intronic hammerhead ribozymes are ultraconserved in the human genome, *EMBO Rep* 11, 711–716.

11. de la Pena, M., and Garcia-Robles, I. (2010) Ubiquitous presence of the hammerhead ribozyme motif along the tree of life, *RNA* 16, 1943–1950.

12. Webb, C. H., Riccitelli, N. J., Ruminski, D. J., and Luptak, A. (2009) Widespread occurrence of self-cleaving ribozymes, *Science* 326, 953.

13. Martick, M., Horan, L. H., Noller, H. F., and Scott, W. G. (2008) A discontinuous hammerhead ribozyme embedded in a mammalian messenger RNA, *Nature* 454, 899-U857.

14. Seehafer, C., Kalweit, A., Steger, G., Graf, S., and Hammann, C. (2011) From alpaca to zebrafish: hammerhead ribozymes wherever you look, *RNA* 17, 21–26.

15. Yao, Z. Z., Weinberg, Z., and Ruzzo, W. L. (2006) CMfinder - a covariance model based RNA motif finding algorithm, *Bioinformatics* 22, 445–452.

16. Bengert, P., and Dandekar, T. (2004) Riboswitch finder - a tool for identification of riboswitch RNAs, *Nucleic Acids Research* 32, W154-W159.

17. Turi, A., Loglisci, C., Salvemini, E., Grillo, G., Malerba, D., and D'Elia, D. (2009) Computational annotation of UTR cis-regulatory modules through Frequent Pattern Mining, *BMC Bioinformatics* 10.

18. Hofacker, I. L. (2003) Vienna RNA secondary structure server, *Nucleic Acids Res* 31, 3429–3431.

19. Hofacker, I. L., Fontana, W., Stadler, P. F., Bonhoeffer, L. S., Tacker, M., and Schuster, P. (1994) Fast Folding and Comparison of Rna Secondary Structures, *Monatsh Chem* 125, 167–188.

20. Laferriere, A., Gautheret, D., and Cedergren, R. (1994) An RNA pattern matching program with enhanced performance and portability, *Comput Appl Biosci* 10, 211–212.

21. Riccitelli, N. J., and Luptak, A. (2010) Computational discovery of folded RNA domains in genomes and in vitro selected libraries, *Methods* 52, 133–140.

22. Wu, G., Wolf, J. B., Ibrahim, A. F., Vadasz, S., Gunasinghe, M., and Freeland, S. J. (2006) Simplified gene synthesis: a one-step approach to PCR-based gene construction, *J Biotechnol* 124, 496–503.

23. Chadalavada, D. M., Gratton, E. A., and Bevilacqua, P. C. (2010) The Human HDV-like CPEB3 Ribozyme Is Intrinsically Fast-Reacting, *Biochemistry* 49, 5321–5330.

24. Kibbe, W. A. (2007) OligoCalc: an online oligonucleotide properties calculator, *Nucleic Acids Res* 35, W43–46.

25. Doudna, J. A. (1997) Preparation of homogeneous ribozyme RNA for crystallization, *Methods Mol Biol* 74, 365–370.

26. Salehi-Ashtiani, K., Luptak, A., Litovchick, A., and Szostak, J. W. (2006) A genomewide search for ribozymes reveals an HDV-like sequence in the human CPEB3 gene, *Science* 313, 1788–1792.

Chapter 10

Discovery of RNA Motifs Using a Computational Pipeline that Allows Insertions in Paired Regions and Filtering of Candidate Sequences

Randi M. Jimenez, Ladislav Rampášek, Broňa Brejová, Tomáš Vinař, and Andrej Lupták

Abstract

The enormous impact of noncoding RNAs on biology and biotechnology has motivated the development of systematic approaches to their discovery and characterization. Here we present a methodology for reliable detection of genomic ribozymes that centers on pipelined structure-based searches, utilizing two versatile algorithms for structure prediction. RNArobo is a prototype structure-based search package that enables a single search to return all sequences matching a designated motif descriptor, taking into account the possibility of single nucleotide insertions within base-paired regions. These outputs are then filtered through a structure prediction algorithm based on free energy minimization in order to maximize the proportion of catalytically active RNA motifs. This pipeline provides a fast approach to uncovering new catalytic RNAs with known secondary structures and verifying their activity in vitro.

Key words: Ribozymes, Pseudoknots, RNABOB, RNArobo, RNAfold

1. Introduction

1.1. Structure-Based Searches and RNA Secondary Structure

Structure-based searches allow the user to both define individual elements of the motif of interest and specify the relationship between the individual elements. However, with rising structural complexity, the output of many in silico structure prediction algorithms becomes increasingly convoluted. Benchmarks on prediction accuracy show that RNA structure predicting software

Jörg S. Hartig (ed.), *Ribozymes: Methods and Protocols*, Methods in Molecular Biology, vol. 848,
DOI 10.1007/978-1-61779-545-9_10, © Springer Science+Business Media, LLC 2012

only calculates about 50–70% of the base-pairs correctly (1). Current work hopes to include more biological information to further constrain RNA structure prediction results. The structured functional RNA discovery pipeline presented here, beginning with motif searches, has been utilized with success to identify both HDV-like ribozyme nested-double pseudoknots (2) and hammerhead ribozyme three-way junctions (3, 4).

1.2. RNA Secondary Structure and Helical Inserts

Ribozymes display a variety of tertiary architectures, but can be described as a set of helical elements of variable sequence, often requiring only covariation dictated by Watson–Crick base-pairing, joined by single-stranded regions. Single-stranded segments either form variable connecting regions or sequence-conserved active sites, binding pockets, tertiary contacts, and other specific structural elements. Previously, RNA motif search programs could look for perfect base-pairs or mismatches in secondary structures, but lacked the ability to incorporate potential single nucleotide insertions within helical regions in a convenient manner. We introduce a modification of the RNAMOT program, called RNArobo, a prototype RNA motif search program that allows for insertions within paired regions, a common feature of RNA secondary structures. When integrated into our pipeline for filtering initial structure-based search output, the search yields a higher proportion of functional RNAs with higher structural diversity than RNAMOT or RNABOB searches.

2. Methods

Our structure-based search method requires a POSIX (Portable Operating System Interface for UNIX) compliant operating system as well as a modern version of the Perl programming language freely available at: www.perl.com. Running DotKnot-1.2 locally requires Python programming language available at: http://www.python.org. RNArobo is only the first component of a discovery pipeline wherein further folding constraints (imposed by RNAfold or DotKnot) eliminate unlikely structured functional RNA candidates. The final candidates of this computational sequence filter can be tested in vitro (see Chapter 9).

2.1. In-Depth Motif Searches Using RNArobo

Structured functional RNAs often contain single-base single-stranded insertions within base-paired elements. Descriptors used by RNABOB or RNAMOT programs do not allow such insertions, instead several independent searches must be performed.

In particular, individual descriptors for these searches must cover every possible position of an insertion occurrence within a helical element (and their combinations in case of multiple insertions), making the search impractical, especially for longer RNA motifs. RNArobo introduces an extension to RNABOB descriptors that allows specification of the number of point insertions within each element of the motif. All possible positions for such insertion are considered by RNArobo automatically in a single search.

2.1.1. RNArobo Download

RNArobo is available for any Linux or UNIX workstation at http://compbio.fmph.uniba.sk/rnarobo/.

1. Download the source file. Expand and archive the tar file at a convenient location. (Optional command line: tar–zxvf <file.tat>).

2. To compile RNArobo, change to the RNArobo source directory. Type the command "make" to compile the executable "rnarobo."
   ```
   /Applications/rnarobo-1.9/source/make
   ```

2.1.2. Insertions in Motif Descriptors

RNArobo uses the same descriptor format as for RNABOB. Each line in the descriptor specifies a single structural element using three or four space-separated fields. The first field names the element according to the topology given in the first line of the file (e.g., s1, the first single-stranded region; and r5, the fifth relational element). The second field for single-stranded elements is the number of sequence mismatches allowed in that element, whereas for helical and relational elements, the field specifies mismatch and insertion constraints (e.g., 2:0:1 allows maximum of two sequence mismatches in a defined sequence, zero mispairs, and one insertion; for the purpose of backward compatibility with RNABOB, the third number specifying number or insertions can be omitted). The third field specifies the desired sequence (using IUPAC nomenclature; e.g., HCG*Y, where H stand for any nucleotide except G, Y is a pyrimidine, and * stands for either a single nucleotide or none). The fourth field of a relational element line specifies which nucleotides are allowed to pair nucleotides ACGT, respectively (conventional base-pairing is thus specified as TGCA).

The new feature of RNArobo compared to RNABOB is that in relational elements the single nucleotide insertions are not restricted to a single location, but instead RNArobo systematically searches the sequence allowing for the insertion at every position tolerated within that structural element. The following is an example of a descriptor allowing for an insertion in element r3 of an HDV-like ribozyme.

```
h1 r2 s1 r3 r4 s2 r5 s3 r4' r2' h1' r5' s4 h6 s5 h6'

s6 r3'


h1 0:0    G:Y

r2 0:0    NNNNN:NNNNN TGCA

s1 0      NN[250]

r3 0:0:1  *NNNNNN:NNNNNN* TGCA

r4 0:0    NNN:NNN TGCA

s2 0      TY

r5 0:0    Y:R TGCA

s3 0      HCG*Y

s4 0      NN*****

h6 0:0    NNN:NNN

s5 0      NNN[250]

s6 0      NCNRA*
```

2.1.3. Running RNArobo   Usage for RNArobo is similar to that of RNABOB, following a command line format of:

> rnarobo <-options> <descriptor-file> <sequence-file>

The only available program option is complement (c), which searches both sense and antisense strands.

The following command will run RNArobo with the option "c" using the "HDV-ins.des" descriptor and the "Ribozymes_test.fasta" sequence file. The output will be sent to the "Outfiles" folder under the RNArobo source directory. RNArobo does not filter out overlapping outputs found at a single locus. The output of RNArobo follows the same format as that of RNABOB. Search times for RNArobo are longer than for RNABOB.

```
rnarobo -c Des/HDV-ins.des Seq/Ribozymes_test.fasta >

Outfiles/Results.txt
```

Outfiles/Results.txt:

```
Starting rnarobo: version dev1.9, October 2010

-------------------------------------------------------

Database file:          Ribozymes_test.fa

Descriptor file:            HDV-ins.des

Search order:      R3 R2 R4 S3 R5 H1 S6 S2 H6 S4 S5

S1

Complementary strand searched: yes

----------------------------------------------------

 seq-f  seq-t  name description

------ ------ ------------ ------------

    74    140  Mouse_CPEB3
|G|GGGGCC|AC|AGCAGAA|GCG|TT|C|ACGT|CGC|GGCCCC|T|G|TC|AGA|TT
CTGGCGAA|TCT|GCGAA|TTCTGCT|

    74    139  Mouse_CPEB3
|G|GGGGCC|ACA|GCAGAA|GCG|TT|C|ACGT|CGC|GGCCCC|T|G|TC|AGA|TT
CTGGCGAA|TCT|GCGAA|TTCTGC|

    74    140  Mouse_CPEB3
|G|GGGGCC|AC|AGCAGAA|GCG|TT|C|ACGT|CGC|GGCCCC|T|G|TC|AGA|TT
CTGGCGAA|TCT|GCGAAT|TCTGCT|

   146    80 Antisense_Mouse_CPEB3
|G|GGGGCC|AC|AGCAGAA|GCG|TT|C|ACGT|CGC|GGCCCC|T|G|TC|AGA|TT
CTGGCGAA|TCT|GCGAA|TTCTGCT|

   146    81 Antisense_Mouse_CPEB3
|G|GGGGCC|ACA|GCAGAA|GCG|TT|C|ACGT|CGC|GGCCCC|T|G|TC|AGA|TT
CTGGCGAA|TCT|GCGAA|TTCTGC|

   146    80 Antisense_Mouse_CPEB3
|G|GGGGCC|AC|AGCAGAA|GCG|TT|C|ACGT|CGC|GGCCCC|T|G|TC|AGA|TT
CTGGCGAA|TCT|GCGAAT|TCTGCT|
```

Note that the variable sequences in the last element of the output represent diverse ways to satisfy the structural requirements.

2.2. Parsing Script: Generate a FASTA Format File

FASTA is the standard file format accepted by the majority of web-based structure prediction and sequence alignment software. This text format includes a header line, beginning with a ">" followed by a description of the nucleotide or peptide sequence. Following the header line is the nucleotide or peptide sequence represented in its single-letter code. It is therefore beneficial to have a systematic way of creating FASTA files from the structure-based search outputs of RNABOB or RNArobo. To this end, we use a short Perl script (see Subheading 3; the script is also available through RNArobo download site: http://compbio.fmph.uniba.sk/rnarobo/). It is important that the structure-based search output ".txt" documents are not manually manipulated. Script usage in a shell from the directory containing the perl executable:

perl parse_rnabob.pl <outfile.txt> <element number of interest, or blank for entire sequence>

This script creates a FASTA file format of the RNABOB or RNArobo sequence output (see Chapter 9). The new header line must begin with ">" and contains a simplified sequence description, defined by the script "make header" routine. The subsequent line contains the specified elements (in our example, 14 = h6, 15 = s5, 16 = h6', which together make up the P4-L4 region of the ribozyme) of the RNABOB or RNArobo output sequence with all vertical bars removed.

The following command line parses the "Results.txt" file by creating a FASTA format header and prints only elements 14, 15, and 16 from each sequence. Based on our HDV-like ribozyme descriptor, these three elements correspond to region P4, which should be a stable stem-loop. The result of the parse script is saved as file "HDV-P4.fasta" in the "Outfiles" folder under the rnarobo directory.

```
/Applications/Scripts/perl parse_rnabob.pl /rnarobo-

1.9/Outfiles/Results.txt 14 15 16 > /rnarobo-

1.9/Outfiles/HDV-P4.fasta


> Mouse_CPEB3:74-140

AGATTCTGGCGAATCT

> Mouse_CPEB3:74-139

AGATTCTGGCGAATCT

> Mouse_CPEB3:74-140

AGATTCTGGCGAATCT

> Antisense_Mouse_CPEB3:146-80

AGATTCTGGCGAATCT

> Antisense_Mouse_CPEB3:146-81

AGATTCTGGCGAATCT

> Antisense_Mouse_CPEB3:146-80

AGATTCTGGCGAATCT
```

2.3. Filtering Outputs: Vienna RNAfold Package

To determine the probability of finding a correct fold of simple secondary-structure elements (5) in RNABOB or RNArobo output, we filter the output through the RNAfold algorithm of the ViennaRNA package (6, 7) and calculate the lowest energy conformation of each output sequence. This method of secondary structure prediction is particularly useful to screen for correctly folding stem-loops, and is thus a useful estimate of the stability of peripheral domains. Similarly, FASTA format files can be uploaded to NCBI BLAST (8), UCSC BLAT (9), or multiple sequence alignment tools to acquire information regarding the genomic locus and evolutionary conservation of the RNA motif. This step can also serve as a method of filtering outputs by biological relevance for further biochemical studies.

2.3.1. ViennaRNA Package: Download, Local Run, and Usage

1. Upload FASTA format file to the web-based server: http://rna.tbi.univie.ac.at/cgi-bin/RNAfold.cgi or download the ViennaRNA Secondary Structure package: http://www.tbi.univie.ac.at/RNA/.

2. Expand and archive the tar file in the desired directory. Follow instructions in the install file (INSTALL) to build and configure

the package by typing the appropriate commands in a shell under the directory containing the program files.

3. Running RNAfold can be accomplished in either of two ways.

First, the RNABOB or RNArobo output file can be piped through the parsing script into the RNAfold program. RNAfold options are described in its manual (RNAfold.1) in the "man" folder for the ViennaRNA package. The following command line will take the "Results.txt" RNArobo output, create a temporary FASTA format file, and pipe it into RNAfold, which will calculate the secondary structures for each sequence. The output will appear inside the terminal shell. The "-noPS" option prevents the creation of a PostScript file with secondary structure plots and a "dot plot" of the base-pairing matrix for each result in a separate file.

```
Applications/Scripts/perl parse_rnabob.pl /rnarobo-

1.9/Outfiles/Results.txt | /Applications/ViennaRNA-

1.8.4/Progs/RNAfold -noPS
```

Alternatively, RNAfold can be applied directly to a FASTA file (no spaces are allowed within the sequences). Make sure to initiate the command under the directory containing the Vienna program of interest, in this case RNAfold. The following command line will read the "HDV-P4.fasta" file into the RNAfold program and send the resulting output to the "HDV-P4_fold.fasta" file in Outfiles folder under the rnabob-2.1 directory. The command for this follows the format: RNAfold [-options] <"output.fasta"> fold_output.fasta"

```
Applications/ViennaRNA-1.8.4/Progs/RNAfold -noPS <

/Applications/rnarobo-1.9/Outfiles/ HDV-P4.fasta >

/Applications/rnarobo-1.9/Outfiles/HDV-P4_fold.fasta
```

The RNAfold output can be read in any basic text editor. The output maintains the header and nucleotide sequences of our search, and the third line now describes the sequence secondary structure in a bracket notation followed by the estimated free energy of the structure.

Vienna RNAfold output:

```
> Mouse_CPEB3:74-140

AGAUUCUGGCGAAUCU

(((((((....))))))) ( -4.20)

> Mouse_CPEB3:74-139

AGAUUCUGGCGAAUCU

(((((((....))))))) ( -4.20)

> Mouse_CPEB3:74-140

AGAUUCUGGCGAAUCU

(((((((....))))))) ( -4.20)

> Antisense_Mouse_CPEB3:146-80

AGAUUCUGGCGAAUCU

(((((((....))))))) ( -4.20)

> Antisense_Mouse_CPEB3:146-81

AGAUUCUGGCGAAUCU

(((((((....))))))) ( -4.20)

> Antisense_Mouse_CPEB3:146-80

AGAUUCUGGCGAAUCU

(((((((....))))))) ( -4.20)
```

2.4. DotKnot for Pseudoknot Prediction

Structure prediction approaches based on free energy minimization have a number of drawbacks. Due to the computational complexity of predicting pseudoknots and the limitations of the underlying energy models, most structure prediction algorithms are unable to predict pseudoknots. DotKnot is an efficient pseudoknot prediction program which finds even complex pseudoknots and kissing loops with higher accuracy than other prediction algorithms (10). It works by first evaluating stem regions from the secondary structure probability dot-plot (generated by RNAfold from ViennaRNA package) and then using published pseudoknot loop entropy parameters generates a list of the best locally predicted pseudoknots. This gives DotKnot tremendous utility when filtering for HDV-like ribozymes, as the activity of this class of self-cleaving RNAs is dependent on the formation of the nested double pseudoknot secondary structure. DotKnot-1.2 is freely available as a web server (http://dotknot. csse.uwa.edu.au/), where information regarding acquiring the source files to run DotKnot on local machines can be found.

3. Notes [parse_rnabob.pl]

```perl
#! /usr/bin/perl -w

use strict;

my $USAGE = "$0 <file> <which_element1> <which_element2>

... \n

Parse RNABob output file stored in <file> to fasta format

containing the found regions.


If run without any <which_elements> then it outputs all

elements.

Otherwise, it selects only those elements specified on the

input

(elements are numbered 1,2,... in the order in which they

appear

on the first line of the descriptor).

";

#parse command line arguments

die $USAGE unless @ARGV>=1;

my ($filename, @which) = @ARGV;

#check that all elements of @which are positive integers

foreach my $num (@which) {
```

```perl
    die "Wrong which_element '$num'"

      unless $num=~/^[0-9]+$/ && $num>0;

}

my $file;

open $file, "<$filename" or die "Cannot open $filename";

my $active = 0;

while(my $line = <$file>) {

    chomp $line;

    if ($active) {

     if ($line) {

            #parse line and print fasta header

            print ">",make_header($line),"\n";

                #read another line with sequence

            my $content = <$file>;

            chomp $content;

            my @parts = split '\|',$content;

                #first part should be empty

                die "Wrong format" unless $parts[0] eq "";

                #select parts of the output of interest

            my @subparts;

            if (@which==0) {

             @subparts=@parts;

             shift @subparts;

            } else {
```

```perl
                foreach my $col (@which) {

                        die "Part $col not found" unless exists
$parts[$col];

                    push @subparts,$parts[$col];

                }

            }


            #prints selected parts
            print join("", @subparts), "\n";
        } else {
            # empty line

            $active = 0;
        }
    } else {
        #start of occurrences, end of file header
        if ($line =~ /^ seq-f/) {
            # next line should contain dashed on top of the
table
            $line = <$file>;
            die "Unexpected end of line" unless defined
$line;
            die "Wrong format" unless $line =~ /^[-
\t]*\s*$/;
            $active = 1;
```

```perl
        }

    }

}

sub make_header {

    my ($line) = @_;

    #split line into three parts at whitespace

    my @parts = split " ",$line;

    die "Wrong format" unless @parts>=3;

    #if header contains gi numbers, take onlu the number

    my $header;

    if ($parts[2] =~ /^gi\|[^\|]*\|[^\|]*\|([^\|]*)\|/) {


        $header = $1;

        } else {

        $header = $parts[2];

        }

        return $header.":".$parts[0]."-".$parts[1];

    }
```

Acknowledgments

The authors gratefully acknowledge support from VEGA (1/0210/10) to T.V. at the Faculty of Mathematics, Physics and Informatics, Comenius University; and the Pew Charitable Trusts, the NIH (GM094929-01), and the University of California–Irvine to A.L.

References

1. Eddy, S. R. (2004) How do RNA folding algorithms work?, *Nat Biotechnol* 22, 1457–1458.

2. Webb, C. H., Riccitelli, N. J., Ruminski, D. J., and Luptak, A. (2009) Widespread occurrence of self-cleaving ribozymes, *Science* 326, 953.

3. Seehafer, C., Kalweit, A., Steger, G., Graf, S., and Hammann, C. (2011) From alpaca to zebrafish: hammerhead ribozymes wherever you look, *RNA* 17, 21–26.

4. Jimenez, R. M., Delwart, E., and Luptak, A. (2011) Structure-based search reveals hammerhead ribozymes in the human microbiome, *J Biol Chem* 286, 7737–7743.

5. Riccitelli, N. J., and Luptak, A. (2010) Computational discovery of folded RNA domains in genomes and in vitro selected libraries, *Methods* 52, 133–140.

6. Hofacker, I. L. (2003) Vienna RNA secondary structure server, *Nucleic Acids Res* 31, 3429–3431.

7. Hofacker, I. L., Fontana, W., Stadler, P. F., Bonhoeffer, L. S., Tacker, M., and Schuster, P. (1994) Fast Folding and Comparison of Rna Secondary Structures, *Monatsh Chem* 125, 167–188.

8. Altschul, S. F., Madden, T. L., Schaffer, A. A., Zhang, J., Zhang, Z., Miller, W., and Lipman, D. J. (1997) Gapped BLAST and PSI-BLAST: a new generation of protein database search programs, *Nucleic Acids Res* 25, 3389–3402.

9. Kent, W. J. (2002) BLAT - The BLAST-like alignment tool, *Genome Res* 12, 656–664.

10. Sperschneider, J., and Datta, A. (2010) DotKnot: pseudoknot prediction using the probability dot plot under a refined energy model, *Nucleic Acids Research* 38, e103.

Chapter 11

Crystallographic Analysis of Small Ribozymes and Riboswitches

Geoffrey M. Lippa*, Joseph A. Liberman*, Jermaine L. Jenkins*, Jolanta Krucinska, Mohammad Salim, and Joseph E. Wedekind

Abstract

Ribozymes and riboswitches are RNA motifs that accelerate biological reactions and regulate gene expression in response to metabolite recognition, respectively. These RNA molecules gain functionality via complex folding that cannot be predicted a priori, and thus requires high-resolution three-dimensional structure determination to locate key functional attributes. Herein, we present an overview of the methods used to determine small RNA structures with an emphasis on RNA preparation, crystallization, and structure refinement. We draw upon examples from our own research in the analysis of the leadzyme ribozyme, the hairpin ribozyme, a class I $preQ_1$ riboswitch, and variants of a larger class II $preQ_1$ riboswitch. The methods presented provide a guide for comparable investigations of noncoding RNA molecules including a 48-solution, "first choice" RNA crystal screen compiled from our prior successes with commercially available screens.

Key words: RNA, Riboswitches, Ribozymes, RNA synthesis, Purification, X-ray crystallography, Crystal screen

1. Introduction

Fast sequencing technology has produced myriad new genomes that expand the number of novel noncoding RNA sequences including those of ribozymes and riboswitches (1). To keep pace with ascribing a functional role to these sequences, few approaches are as powerful as X-ray crystallography when it comes to producing an all-atom model of an RNA molecule that serves as framework to support rational analyses. In this chapter, we describe a series of protocols for the preparation, crystallization, and X-ray refinement

---

*Geoffrey M. Lippa; Joseph A. Liberman; Jermaine L. Jenkins contributed equally to this manuscript.

Jörg S. Hartig (ed.), *Ribozymes: Methods and Protocols*, Methods in Molecular Biology, vol. 848,
DOI 10.1007/978-1-61779-545-9_11, © Springer Science+Business Media, LLC 2012

of small ribozymes and riboswitches. Our experience is drawn from: (1) the leadzyme ribozyme—a 24-mer comprising 11-mer and 13-mer synthetic strands (2), (2) the hairpin ribozyme, a 61-mer comprising 12-mer, 13-mer, 17-mer, and 19-mer synthetic strands (3); and (3) a synthetic 33-mer riboswitch that binds the metabolite preQ$_1$ (4). In each of these investigations, the resulting crystal structures were used to gain insight into function. In addition, we present progress on the analysis of an 80-mer preQ$_1$ riboswitch, and its 77-mer variant, produced by in vitro transcription and crystallized from several commercial screens. These results have been compiled with other successful RNA crystallization conditions into a 48-solution "first choice" RNA crystal screen (see Table 1).

Prior to RNA synthesis, we will assume that the target RNA has been validated in terms of enzymatic or ligand-binding activity to define what is necessary and sufficient for function. It has been our experience that the best sequences for crystallization possess conserved core features, e.g., identifiable in RFAM (5), and that these sequences exhibit unique secondary and tertiary folding distinct from competing, nonproductive conformations, e.g., as indicated by HOTKNOTS (6). As a caveat, such programs do not account for the stabilizing effects of transition-state or metabolite binding, but they are a good starting point for the design of crystallization constructs. Such effects can be monitored experimentally by dynamic light scattering (DLS), which has proven useful to assess the compactness and monodispersity of RNA constructs prior to crystallization trials. In addition, we assume herein that RNA phases have been obtained already. Many authoritative reviews have been written about the RNA phase problem. For general approaches on heavy-atom phasing see refs. 7, 8; for molecular replacement see ref. 9. Keel et al. (10) provide an innovative approach for the inclusion of site-specific osmium- or iridium-hexammine binding sites that can be beneficial in experimental phasing. As such, this aspect of the discussion is confined to refinement of the model against X-ray data. Overall, this chapter should be a practical guide to RNA production, crystallization, and refinement with reference to complementary resources to fill in knowledge gaps as necessary.

2. Materials

All water is derived from a NANOpure™ UV/UF system (Thermo Scientific, Asheville, NC) with a resistivity reading of ≥18.1 MΩ. In general all solutions should be 0.22- or 0.45-μm sterile filtered and autoclaved if possible. Specialized reagents and stocks are described below. All steps should use appropriate safety measures to avoid exposure to hazardous chemicals such as ethidium bromide or

Table 1
Summary of "first choice" RNA crystal screening conditions

Number	Precipitant	Salt	Additive 1	Additive 2[a]	Buffer[b]	pH
1	5% PEG 2K	0.10 M KCl	0.005 M MgCl$_2$	Spermine	MES	5.6
2	10% PEG 2K	0.10 M NH$_4$Cl	0.010 M MgCl$_2$	Spermine	Na-cacodylate	6.0
3	15% PEG 2K	0.10 M Li$_2$SO$_4$	0.020 M MgSO$_4$	Spermine	Na-cacodylate	6.5
4[c]	20% PEG 2K	0.20 M NH$_4$Cl	0.010 M Mg(Acetate)$_2$	Spermine	HEPES	7.0
5	25% PEG 2K	0.10 M NaCl	0.001 M Co(NH$_3$)$_6$	Spermine	Tris	7.5
6	5% PEG 8K	0.20 M (NH$_4$)$_2$SO$_4$	0.010 M MgSO$_4$	Spermine	MES	5.6
7[d]	10% PEG 8K	0.10 M KCl	0.020 M MgCl$_2$	Spermine	Na-cacodylate	6.5
8	15% mmePEG 2K	0.20 M NH$_4$Acetate	0.010 M Mg(Acetate)$_2$	Spermine	Na-cacodylate	6.5
9	20% mmePEG 2K	0.20 M NH$_4$Cl	0.010 M MgCl$_2$	Spermine	HEPES	7.0
10[e]	25% mmePEG 2K	0.20 M Li$_2$SO$_4$	0.001 M Co(NH$_3$)$_6$	Spermine	Tris	7.5
11	10% mmePEG 550	0.05 M NH$_4$C$_2$H$_3$O$_2$	0.001 M Co(NH$_3$)$_6$	Spermine	Na-cacodylate	6.0
12	15% mmePEG 550	0.05 M NH$_4$Cl	0.010 M MgCl$_2$	Spermine	Na-cacodylate	6.5
13	20% mmePEG 550	0.05 M KCl	0.020 M MgCl$_2$	Spermine	HEPES	7.0
14	25% mmePEG 550	0.05 M NH$_3$C$_2$H$_3$O$_2$	0.005 M MgCl$_2$	Spermine	Tris	7.5
15[f]	12% mmePEG 8K 12% PEG 200	0.10 M KCl	0.010 M Mg(Acetate)$_2$	Spermine	Na-cacodylate	6.0
16[f]	10% mmePEG 8K 12% PEG 200	0.10 M KCl	0.010 M Mg(Acetate)$_2$	Spermine	Na-cacodylate	6.0
17[f]	11% mmePEG 4K 10% PEG 400	0.050 M KCl	0.015 M Mg(Acetate)$_2$	Spermine	Na-cacodylate	6.0

(continued)

Table 1
(continued)

Number	Precipitant	Salt	Additive 1	Additive 2[a]	Buffer[b]	pH
18[f]	10% mmePEG 8K 15% MPD	0.10 M KCl	0.005 M MgCl$_2$	Spermine	Na-cacodylate	6.0
19	10% PEG 400	0.10 M KCl	0.001 M CoCl$_2$	Spermidine	Na-cacodylate	6.0
20	15% PEG 400	0.10 M NaCl	0.010 M MgCl$_2$	Spermidine	Na-cacodylate	6.5
21	20% PEG 400	0.10 M NH$_4$Cl	0.020 M MgCl$_2$	Spermidine	HEPES	7.0
22	30% PEG 400	0.10 M Mg(Acetate)$_2$			Tris	8.0
23	10% MPD	0.050 M LiSO$_4$	0.010 M MgSO$_4$	Spermidine	Na-cacodylate	6.0
24	15% MPD	0.050 M KCl	0.005 M MgCl$_2$	Spermidine	Na-cacodylate	6.5
25	20% MPD	0.050 M NH$_4$Cl	0.005 M CaCl$_2$	Spermidine	HEPES	7.0
26	25% MPD	0.050 M LiAcetate	0.010 M MgCl$_2$	Spermidine	Tris	7.5
27	30% MPD	0.050 M Mg(Acetate)$_2$		Spermidine	Na-cacodylate	6.0
28	15% MPD	0.10 M NaCl	0.010 M MgSO$_4$	Spermine	Na-cacodylate	6.5
29[c]	20% MPD	0.10 M KCl	0.010 M MgCl$_2$	Spermine	HEPES	7.0
30	25% MPD	0.10 M NaAcetate	0.010 M MgCl$_2$	Spermine	Tris	7.5
31[g]	30% MPD	0.10 M NH$_4$Cl	0.005 M MgSO$_4$	Spermine	Na-cacodylate	6.0
32	1.2 M (NH$_4$)$_2$SO$_4$	0.050 M LiCl	0.001 M Co(NH$_3$)$_6$	Spermine	MES	5.6
33	1.5 M (NH$_4$)$_2$SO$_4$	0.050 M NaCl	0.005 M MgSO$_4$	Spermine	Na-cacodylate	6.0
34[c]	1.8 M (NH$_4$)$_2$SO$_4$	0.010 M MgSO$_4$		Spermidine	Na-cacodylate	6.5
35	2.4 M (NH$_4$)$_2$SO$_4$	0.050 M MgSO$_4$	0.005 M MgSO$_4$	Spermine	HEPES	7.0
36	2.8 M (NH$_4$)$_2$SO$_4$	0.050 M Mg(Acetate)$_2$		Spermine	Tris	7.5

No.	Precipitant	Additive 1	Additive 2	Polyamine	Buffer	pH
37	1.5 M $(NH_4)_2SO_4$		0.005 M $MgSO_4$	Spermidine	Tris	7.5
38	2.0 M $(NH_4)_2SO_4$	2% (v/v) PEG 400	0.001 M $CoCl_2$	Spermidine	HEPES	7.0
39	2.4 M $(NH_4)_2SO_4$		0.010 M $MgSO_4$	Spermidine	Na-cacodylate	6.5
40	2.8 M $(NH_4)_2SO_4$	4% (v/v) PEG 200	0.005 M $CaCl_2$	Spermidine	Na-cacodylate	6.0
41[c]	1.3 M Li_2SO_4	0.001 M $Co(NH_3)_6$		Spermidine	Na-cacodylate	6.0
42[h]	1.5 M Li_2SO_4	2% (v/v) PEG 1K	0.050 $MgSO_4$	Spermine	HEPES	7.5
43[c]	1.8 M Li_2SO_4	0.050 NaCl	0.020 M $MgSO_4$	Spermine	Na-cacodylate	6.0
44	45% Tacsimate™	0.020 M $Mg(Acetate)_2$	0.050 M KCl	Spermine	MES	5.6
45[h]	55% Tacsimate™	0.020 M $MgCl_2$		Spermine	Na-cacodylate	6.5
46[f]	65% Tacsimate™		0.001 M $Co(NH_3)_6$	Spermidine	HEPES	7.0
47	1.5 M 1,6-Hexanediol	0.040 M NH_4Cl	0.010 M $MgCl_2$	Spermine	Na-cacodylate	6.5
48	2.4 M 1,6-Hexanediol	0.040 M $MgSO_4$	0.050 M KCl	Spermidine	Tris	7.5

[a] All concentrations of additive 2 are 0.001 M

[b] All buffer concentrations are 0.050 M

Conditions for diffracting crystals are noted as follows: [a] a 33-mer $preQ_1$ class I riboswitch; [d] an 80-mer class II $preQ_1$ riboswitch without U1A; [c] the minimal hairpin ribozyme; [f] a 77-mer class II $preQ_1$ riboswitch without U1A; [g] the leadzyme ribozyme; and [h] an 80-mer class II $preQ_1$ riboswitch with U1A.

acrylamide, as well as damaging radiation from UV and X-ray sources. Consult material safety data sheets for all chemicals and the manufacturer's documentation for proper instrument operation.

2.1. DNA Template Production with Klenow Fragment

1. 10 mM dNTP solution, cat. #N0447S (New England Biolabs, Ipswich, MA).

2. Klenow fragment, 5,000 U/mL, with 10× reaction buffer, cat. #M0210S (New England Biolabs, Ipswich, MA). 10× Klenow reaction buffer comprises 0.50 M NaCl, 0.10 M Tris–HCl pH 7.9, 0.10 M $MgCl_2$, 0.010 M DTT (New England Biolabs, Ipswich, MA).

3. SeaKem® LE Agarose, cat. #50000 (Lonza, Allendale, NJ).

4. 100% Formamide, cat. #BP228-100 (Fisher Scientific, Pittsburgh, PA).

5. 10 mg/mL Bromophenol blue, cat. #343 (Allied Chemical, Morristown, NJ).

6. 40 mg/mL Ethidium bromide, cat. #BP102-1 (Fisher Scientific, Pittsburgh, PA).

7. 1.0 M NaCl stock solution, cat. #SX0425-3 (EMD Biosciences, San Diego, CA).

8. Agarose gel extraction kit, cat. #28704 (Qiagen Inc., Valencia, CA).

9. Prechilled neat ethanol (100%), cat. #200CSGP (Ultra Pure, LLC, CT).

10. OmniPur® Phenol:Chloroform:Isoamyl Alcohol, 25:24:1, cat. #6810 (EMD Biosciences, San Diego, CA).

11. Required apparatus: a thermostatically controlled water bath, a short/long-wavelength UV transilluminator, a microcentrifuge, and Horizontal Midi-Gel Kit CE 14×10 cm (cat. #MGU-502T, CBS Scientific Comp., Inc., Del Mar, CA).

2.2. RNA Synthesis by T7 Polymerase

1. 1.0 M Tris pH 7.5, Tris base, Molecular Biology Grade, cat. #648310 (CalBioChem, San Diego, CA).

2. 1.0 M DTT, cat. #D9163 (Sigma-Aldrich, St. Louis, MO).

3. 0.5 M Spermidine tetrahydrochloride, cat. #100472 (ICN Biomedical, Irvine, CA).

4. 10% (w/v) Triton X-100, cat. #21568-2500 (Acros, Geel, Belgium).

5. 50% (w/v) PEG 8000, cat. #81268 (Fluka, St. Louis, MO).

6. 1 mg/mL Fraction V, Omnipur BSA solution, cat. #2930 (EMD Biosciences, San Diego, CA).

7. 1.0 M $MgCl_2 \cdot (H_2O)_6$, ACS Grade, cat. #MX0045-1 (EMD Biosciences, San Diego, CA).

8. 0.040 M Ribonucleotide triphosphates (rNTPs) (MP Biomedicals, Solon, OH).

9. Recombinant T7 RNA polymerase, cat. #2085 (Ambion, Inc., Austin, TX) or equivalent.

2.3. Polyacrylamide Gel Electrophoresis Purification of RNA for Crystallography

1. Vertical polyacrylamide gel electrophoresis (PAGE) apparatus, cat. #LSG-400-20-NA (CBS Scientific Company, Inc., Del Mar, CA).

2. TBE running buffer 5× stock is: 54 g of Tris base (final 5× conc. 0.445 M), 27.5 g boric acid (final 5× conc. 0.445 M), 20 mL EDTA from a 0.50 M pH 8.0 stock (final 5× conc. 0.01 M), and sufficient water to bring the volume to 1.0 L (The final pH is 8.3; do not pH the stock).

3. Freshly prepared 8–15% (w/v) acrylamide:bisacrylamide (19:1) with 7.0 M urea to denature RNA. A 0.10 L 15% stock requires the following: 14.25 g acrylamide, 0.75 g bisacrylamide, 42.04 g urea, and 20 mL 5× TBE buffer. Add all solids to a flask and bring the volume to 70 mL. Place the flask on a magnetic stir plate with a heater and mix the solution at 37°C for 1 h. Add 20 mL of 5× TBE and bring the total volume to 0.10 L.

4. Fresh 10% (w/v) ammonium persulfate, cat. #BP179-100 (Fisher Scientific, Pittsburgh, PA).

5. A fresh solution of TEMED (N,N,N',N'-Tetramethylethylenediamine) stock, cat. #T9281-100mL (Sigma-Aldrich).

6. Loading buffer (2×) comprising: 95% formamide, 0.025% (w/v) xylene cyanol, 0.025% (w/v) bromophenol blue, 18 mM EDTA, and 0.025% (w/v) SDS, cat. #AM8547 (Ambion, Austin, TX).

7. Crush-Soak Solution: 20 mM Tris pH 7.5, 2 mM EDTA, and 0.10 M ammonium acetate.

8. Tube Top Filter 50 mL, 0.45 μm cellulose acetate, cat. #430314 (Corning, Inc., Corning, NY).

9. Chromatography Materials include: Toyopearl DEAE-650S, cat. #07472 (Tosoh Bioscience Corp., San Francisco, CA); DEAE wash buffer (20 mM Tris pH 7.5 and 2 mM EDTA); DEAE elution buffer (same as the wash buffer but containing 1.0 M ammonium acetate); desalting chromatography resin is Sephadex G-50 (DNA Grade), cat. #S5897-25G (Sigma-Aldrich, St. Louis, MO); this resin is swollen and run in water.

10. Various ACE Glass, Inc. columns to accommodate ion exchange (20 mm×10 cm) and desalting (15 mm×50 cm) resins. These should be wrapped in aluminum foil and autoclaved for 30 min followed by a dry cycle. Tygon tubing for gravity flow should be treated similarly, but first wet the inside of the tube to prevent damage from heating.

11. Fluor-coated TLC plate, cat. #AM10110 (Applied Biosystems, Carlsbad, CA).

12. Sterile scalpel blade and metal spatula.

13. A lyophilizer is used for the final desalting step although this can be substituted with a speedvac operated without heating.

2.4. Sparse Matrix Screening by the Hanging Drop Method

With RNA in hand the search for crystallization conditions begins by screening against commercial kits designed for nucleic acids such as: the Natrix and Nucleic Acid Mini Screen (Hampton Research, Aliso Viejo, CA), as well as the JBScreen Nuc-Pro (MiTeGen, Ithaca, NY). For liquid handling, we prefer a robotic system that operates in 96-well format. If lead crystallization conditions are identified, we then proceed to optimization using larger, manual setups (see Subheading 2.5). Materials for robotic crystal screening include:

1. A Mosquito™ robot (TTP Labtech, Melbourn, UK) or equivalent (see Note 1).

2. 96-Well plates such as SpectraPlate-96HB, cat. #P12-106 (PerkinElmer Life and Analytical Sciences, Waltham, MA).

3. Clear seal film, cat. #HR4-523 (Hampton Research Inc., Aliso Viejo, CA).

4. Seal and sample aluminum foil lids, cat. #538619 (Beckman Coulter, Brea, CA).

5. ViewDrop 96-well plate seals for hanging drop, 25 per pack, cat. #4150-05100 (TTP Labtech).

6. Crystallization screen such as Natrix HT, cat. #HR2-131 (Hampton Research Inc.).

7. 8-Position 2- and 5-μL micro-reservoir strips for 9-mm pitch, 50 per pack, cat. #4150-03110 (TTP Labtech).

8. 8-Channel pipettor, P-200, cat. #89079-944 (VWR, Radnor, PA).

2.5. Materials for Manual Optimization of RNA Crystals

1. VDX 24-well hanging drop plate, cat. #HR3-140 (Hampton Research Inc.).

2. Siliconized glass circular cover slides, 22 mm diameter, cat. #HR3-231 (Hampton Research Inc.).

3. White petrolatum jelly cat. #S80117 (Sigma-Aldrich), which is heated to liquification on a hot plate, poured inside a 10-cc syringe with the narrow end parafilmed, and chilled on ice. The plunger is restored to extrude a column of jelly for sealing the VDX plate.

2.6. Materials for the Refinement of RNA Crystal Structures

1. An Intel-based computer running Mac OSX 10.5.8 or higher. Representative systems include: a MacBook Pro laptop with a 2.6 GHz Intel Core 2 Duo processor, 4 GB 667 MHz DDR2

SDRAM, and a 0.5Tb HD connected to a Zalman Trimon ZM-M220W stereo monitor; and a Quad Mac tower running OSX 10.6.5 with a 2.66 GHz Quad-Core Intel Xeon processor, 13 GB 1,066 MHz DDR3 RAM, an NVidea Quadro FX 4800 graphics card (for quad buffered stereo), and 3× 1.5Tb HDs.

3. Methods

3.1. Introduction to the Synthesis of RNA by Chemical and Enzymatic Approaches

Chemically synthesized oligonucleotides are appropriate for the production of short RNA sequences. We have had success crystallizing synthetic sequences ranging from 11 to 33 nt. Such strands are produced by solid-phase chemical synthesis (11). Major advantages include rapidity of production, accommodation of nonstandard modifications (3, 12, 13), and attainment of relatively high yields. The feasibility of synthesis can be evaluated by the heuristic: yields(%)=coupling efficiency$^{(\text{chain size }-1)} \times 10$, where coupling efficiency depends on the nature of the nucleotide monomer (e.g., DNA, RNA, or chemical variants thereof) as well as the chemistry employed in coupling. Traditional RNA phosphoramidites were developed for DNA synthesis, and employ 5′-DMT and 3′ β-cyanoethyl (CE) protecting groups (14). This approach can couple with 98% efficiency at each step, but leaves a relatively labor-intensive workup (7, 15). More recently, the combination of a 5′-silyl ether and an acid labile 2′-OH orthoester provides superior protection chemistry identifiable as 2′-ACE phosphoramidites that exhibit >99% coupling efficiency and a relatively facile workup (16). It has been our experience that the former approach offers more choices of nonstandard nucleotide modifications (e.g., Glen Research, Sterling VA), and there are more opportunities to partner with a company that will incorporate "homemade" phosphoramidites into an RNA strand as part of a custom synthesis (12). By contrast, the latter approach is commercially more costly but the yields of longer oligonucleotides are superior. In the end the user must choose what is best for his or her application.

When longer RNA strands are needed in milligram quantities, chemical synthesis becomes impractical due to poor yields and high costs. In vitro transcription using short, double-stranded DNA templates derived from Klenow fragment fill-in offers an alternative to plasmid-based approaches that require cloning and DNA isolation from bacteria. Klenow allows rapid and routine production of RNA transcripts (limited to ~140 bases in part by poor yields of DNA primer and template strands), which is ideal for screening numerous different small RNA constructs for crystallization. In this method, two DNA strands are ordered in which one encodes the desired "template" for the RNA sequence to be transcribed, whereas the complementary strand harbors a 5′ T7 RNA polymerase promoter site in a "primer" strand. Due to yield limitations associated with the synthesis of very large DNA oligonucleotides, the approach utilizes strands that are relatively short (<70 nt). By design the primer and template strands exhibit complementary base pairing with a T_M of 50–60°C in a central region of the duplex (see Fig. 1a). Klenow fragment is used to fill-in the primer-template pair yielding double-stranded DNA that can be used by T7 polymerase to synthesize RNA. Herein we describe this process, which adheres to the concept that the RNA construct itself is a

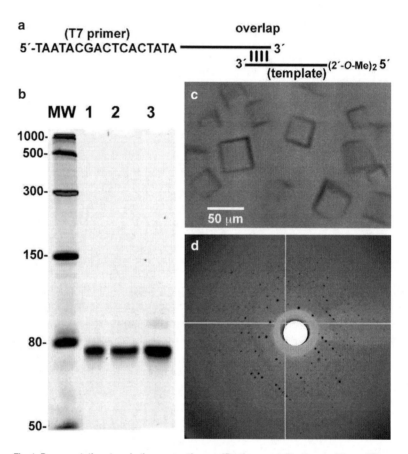

Fig. 1. Representative steps in the preparation, purification, crystallization, and X-ray diffraction of a riboswitch aptamer domain. (a) Schematic diagram of the starting T7 polymerase primer and template that are used for Klenow fragment extension into double-stranded DNA. (b) Representative purity of an 80-mer RNA riboswitch that binds preQ$_1$. The transcribed product was purified by denaturing PAGE and analyzed here on an 8% gel with 7.0 M urea. The sample was suspended in 2× loading buffer. RNA in *lanes 1–3* were loaded in 0.055, 0.110, and 0.500 mg quantities, respectively. (c) Cubic crystals of the RNA from (b) generated by robotic screening using condition 42 of Table 1. (d) X-ray diffraction from a representative 80-mer crystal from (c) using the 22-micrometer microfocus beam at station A1 of the Cornell High Energy Synchrotron Source. The image edge is 5 Å resolution. The crystal was cryoprotected from a 1:1 mixture of paratone-N and silicone oil.

variable in crystal screening (17). As a further consideration, the template strand is synthesized to include tandem 2'-*O*-methyl ribose nucleotides at the 5'-end of the DNA (18). This modification reduces untemplated transcription whereby extra nucleotides are added at the 3'-end of the RNA product, thus lowering purity and potentially confounding crystallization (18). The reader is directed to papers by Uhlenbeck for template and primer design, as well as the factors that influence a strong start by the RNA polymerase, such as use of the initial sequence 5'-GGC (see ref. 19 and references therein).

3.2. DNA Template Production with Klenow Fragment

1. Primer and template DNAs are dissolved in water to 100 pM/μL and used as needed or stored at –20°C.

2. To 460 μL of water, add 60 μL of 10× Klenow reaction buffer, 40 μL of primer, and 40 μL of template with mixing. The solution is heated to 95°C and cooled quickly to 4°C less than the T_M of the overlapping sequence. The solution is held here for 20 min, then cooled quickly to 37°C prior to the addition of the Klenow reaction mixture.

3. The Klenow reaction mixture is prepared by adding 20 μL of 10× Klenow buffer and 36 μL of 10 mM dNTPs to 134 μL of water. After mixing, 10 μL of Klenow enzyme is added and the mixture is warmed to 37°C immediately before addition to the DNA solution.

4. The Klenow reaction mixture from step 3 is added to the DNA primer-template mixture with gentle mixing and then incubated at 37°C for 1 h.

5. The dsDNA product is purified via agarose gels. A volume of 0.70 mL of formamide is added to the synthesized DNA, heated to 70°C for 10 min, and then cooled quickly on ice. A trace amount of bromophenol blue is added, and the sample is loaded onto a 10 cm long, 2.5–3% (w/v) agarose gel containing sufficient ethidium bromide to attain 2 μg/mL. DNA mass markers can be loaded if desired. The running buffer is also made 2 μg/mL in ethidium bromide. The power supply is set to 6 V/cm and electrophoresis continues until the bromophenol blue travels 80% of the gel length.

6. The DNA is visualized under long-wavelength UV light and the correct DNA band is excised. The DNA is extracted and purified using an agarose extraction kit.

7. The DNA is phenol/chloroform extracted as follows. A 0.10 mL mixture comprising phenol/chloroform (1:1) is added per 0.10 mL of DNA eluted from the gel extraction kit. The extraction solution is shaken vigorously, and then microcentrifuged at $9,000 \times g$ for 2 min. The upper layer contains DNA and is removed, followed by addition of 0.30 mL of chloroform alone.

This solution is shaken vigorously and microcentrifuged. The upper layer is removed and the chloroform extraction step is repeated again. After centrifugation, the upper layer of DNA is removed.

8. Ethanol precipitation of the DNA proceeds by adding 15 μL of 1.0 M NaCl and 0.30 mL of prechilled, 100% ethanol and holding the solution at −70°C for 1 h or −20°C overnight. The solution is then microcentrifuged at $15,000 \times g$ for 10 min. The supernatant is removed carefully and the pellet is washed with 70% ethanol and dried on the bench top or by a speedvac without heating.

9. The DNA concentration is estimated by dissolving the pellet in 0.10 mL water. The optical density is measured at 260 nm where 1.0 OD unit represents 50 μg/mL DNA.

3.3. RNA Synthesis by Bacteriophage T7 RNA Polymerase

1. Initial in vitro transcription reactions on a trial scale (50–100 μL) are conducted in 1.5-mL microcentrifuge tubes to determine optimal conditions.

2. Each reaction is prepared at 22°C to avoid precipitation of solution components such as DTT or spermidine hydrochloride.

3. Each trial reaction contains the following components: 8.3 μg DNA template per 1 mL reaction, 0.075 M Tris pH 7.5, 0.010 M DTT, 0.002 M spermidine, 0.01% Triton X-100, 4% PEG 8K, 50 μg/mL BSA, 0.030 M $MgCl_2$, 4 mM rNTP, and 30 μg/mL of T7 RNA polymerase (see Note 2).

4. The mixture should be incubated at 37°C for no more than 24 h.

5. After incubation, the reaction is centrifuged at $14,000 \times g$ for 10 min before removing the soluble phase for gel purification.

6. The products can be analyzed by denaturing PAGE electrophoresis on a small scale. If necessary, the reaction can be optimized by changing the amount of input $MgCl_2$, T7 RNA polymerase, or rNTPs (see Note 3).

3.4. Purification of RNA for Crystallization

Once a satisfactory trial-scale RNA reaction has been established, the conditions should be scaled up to a 1–10 mL size. The goal is to produce ≥1 mg of pure RNA for crystallization trials. Robotic screening methods (see Subheading 3.9) make it feasible to conduct 96 crystallization trials with as little as 150 μg of pure RNA, but we aspire to set up multiple 96-well plates with concentrated material. The purity and homogeneity of the RNA is a key factor in crystallization, and the primary limitation of chemical or enzymatic synthesis is incomplete polymerization (i.e., failure sequences). A secondary problem unique to in vitro transcription is the presence of $n+1$ or $n+2$ untemplated nucleotides that are added to the 3′-end. For purification, we prefer reverse phase HPLC for RNA strands ≤20 nt; procedures for this are described elsewhere (7, 15). For longer strands—or sequences with significant secondary structure—it is necessary to use denaturing PAGE. The electric field separates the various polymer lengths by charge, and the appropriate chain is excised from the gel. We describe this method as applied to RNA strands to be used in crystallization (see Fig. 1b, c).

3.5. First Anion Exchange Purification (DEAE)

1. The day before PAGE purification, prepare an Ace Glass column by pouring a 3 mL slurry of resin.

2. Flow the wash buffer over the resin (ten column volumes). The wash packs the resin while bringing it to the correct ionic strength.

3. Add the reaction(s) from Subheading 3.3 directly to the resin. Save the flow through.

4. Add wash buffer (five bed volumes). Save the wash volume (which may contain unbound RNA, but more likely contains mononucleotides).

5. Add elution buffer (ten column volumes). Collect eluate in small aliquots (0.5–1.0 mL), testing the absorption of each fraction at 260 nm for the presence of RNA.

6. Plot the OD vs. fraction number to determine the appropriate pooling of the RNA.

7. Add three volumes of 100% ethanol to the pooled RNA and place at −20°C for ~15 h.

8. Spin sample at $14,000 \times g$ for 25 min to harvest the RNA as a precipitated pellet.

9. Gently decant the supernatant into a separate tube.

10. Wash the remaining RNA pellet with 70% ethanol to remove trace salts.

11. Allow the pellet to dry on the bench for 30 min. The pellet can be stored at −20°C.

12. Suspend the RNA by gentle pipetting or vortexing in water.

3.6. Denaturing Polyacrylamide Gel Electrophoresis

1. The polyacrylamide gel is semi-preparative with 0.75–3.0-mm Teflon spacers depending on the amount of RNA to load and the separation requirements. This will use about 0.150 L of acrylamide for the thickest spacer. Pour 0.150 L of acrylamide stock into a 0.50-L Erlenmeyer flask. Add 0.90 mL of 10% ammonium persulfate. Add 0.150 mL of TEMED to the gel solution. A comb with a single lane is preferred to evenly distribute the sample. The gel can be allowed to polymerize for 16 h, but precautions must be taken to avoid excessive drying (such as wrapping the exposed acrylamide in plastic with a damp paper towel).

2. Pre-run the gel at 200 V (with a starting power of 15 W) and proceed until the wattage level becomes nearly constant (2–3 h depending on the size of the gel).

3. Mix the RNA with an equal amount of 2× loading buffer and heat 70°C for 5 min before plunging on ice. Load the dissolved RNA onto the gel distributing evenly along its length. Usually there will be 100–200 OD 260 nm units loaded per gel.

4. Run the gel at 200 V until the sample completely enters the gel, then increase to 400 V. The longer the gel runs the better the resolution and it will be easier to isolate the desired RNA sequence, which should be the most prominent band.

5. Remove the correctly sized band by UV shadowing. First place the gel between two sheets of plastic wrap and then position the gel atop a fluor-coated TLC plate that fluoresces under UV light. In a darkened room, expose the gel to UV light. The major product will appear as a dark band on a green background. Use a Sharpie® pen to mark the boundaries of the band to excise—work quickly to avoid UV damage to the RNA.

6. With the lights on, carefully excise the band with a sterile scalpel (removing the plastic wrap) and cut the band into short strips (1–2 cm in length). Place the acrylamide fragments into a pre-weighted 50-mL tube. The gel should be pulverized further with a sterile metal spatula.

7. Add crush-soak solution to the 50-mL tube; soak volume = 3× the gel mass.

8. Place the sealed tube in a shaker at 50–100 rpm at 22°C for no more than 16 h.

9. Filter the solution using a 50-mL Tube Top Filter to remove the acrylamide. The recovered acrylamide can be soaked again ≤8 h.

3.7. Second Anion Exchange Purification (DEAE)

1. A fresh column is prepared as in Subheading 3.5.

2. Load the combined, filtered solutions from Subheading 3.6, step 9 directly to the resin making sure to collect the flow through.

3. Repeat steps 4–11 of Subheading 3.5.

3.8. Gel Filtration Chromatography (Desalting)

1. Water is added to the G50 resin, which can be swollen and degassed by autoclaving. The room temperature slurry is poured into a 15 mm×50 cm sterile column to fill 95% of the column length. The resin is allowed to pack by flowing water through by gravity for 6 h.

2. The RNA from Subheading 3.7 in a volume ≤1 mL is layered gently onto the G50 resin.

3. The column should be flowing while loading and 1.0 mL fractions can be collected.

4. When the RNA is done loading, gently layer the G50 column with 0.5 mL volumes of water until the RNA has entered more than 3 cm into the column. 5 mL of water is added and the column is sealed for a gravity feed.

5. Measure the OD 260 nm of each fraction and plot the outcome. The graph should be a single, sharp peak. Shoulders or doublets may be signs of overloading or improper loading by disruption of the sample plate during the addition of water.

6. Pool the appropriate fractions in a 20-mL tube and measure the OD 260 nm.

7. Freeze the solution in N_2 (*l*) and make two holes in the cap with a hot needle.

8. Place sample on a lyophilizer until the RNA is completely dry. The sample should be a fluffy, white powder with no signs of oil, crystallinity, or discoloration, which are indications of incomplete desalting that necessitate repeating the G50 step. Lyophilized RNA can be stored at –20°C.

3.9. Robotic Crystal Screening Using the Mosquito™

1. Use a multichannel pipette to dispense 75 μL of each screening condition into the well of a 96-well plate.

2. Position the plate below the pipette assembly of the robot.

3. Choose a 2- or 5-μL sample strip based on the size of the desired hanging drop. We prefer 250–350 nL drops, which require a 5-μL strip whereas smaller drops use a 2-μL strip.

4. Pipette 5 μL for 350 nL drops into each of the receptacles of the sample strip. In general, dispense 30% more RNA solution per well than actually needed to allow for evaporation.

5. After pipetting the RNA, start the robot immediately. Runs can be programmed according to the manufacturer's documentation.

6. When complete, invert the 96-condition hanging drop appliqué, align with the guide arrow, and press firmly with the aluminum block to seal the plate.

7. Peel off the protective anti-scratch layer of the appliqué for viewing.

8. Store the plate at 22 or 4°C for 3–5 days before checking the drops.

9. Record your observations for each condition using a consistent scoring system.

10. Promising crystals may take a few days to months to appear. Although it is possible to recover crystals from the 96-well plate by cutting the plastic appliqué encompassing the desired drop, it is often necessary to produce larger crystals under optimized conditions.

3.10. Manual Optimization of RNA Crystals Through Grid Screening

Initial screens may yield light to heavy precipitate, phase separation, microcrystals, showers of crystals, needles, or a few large crystals suited for diffraction tests. Under most circumstances, there is room for improvement. After all, high-throughput screens are designed to cover a broad number of conditions, but these often require a follow-up "grid" screen in which a single variable is changed per experiment to identify factors that best optimize crystal growth and diffraction. In general, a suitable crystal should be single with sharp edges and free of defects (see Fig. 1c); dimensions should be >100 μm on each edge. In this section, we describe the general approaches for establishing manual screens, which are often restricted to incremental changes in pH (±0.25 U), salt (±20%), or precipitant (±20% of hit) based on initial hits from robotic screening (see Note 4). Temperature, RNA concentration, and the RNA sequence itself should not be overlooked as optimization variables (see Note 5). Methods are as follows:

1. Grease the circular opening of each well on a 24-well plate with a petrolatum-filled 10-cc syringe leaving a small gap to release air when the cover slide is pressed in place.

2. Pipette each precipitating agent into its respective reservoir. The total volume per reservoir is 1.0 mL and the final buffer concentration is 50 mM.

3. Pipette 1.7 μL of RNA solution to the center of a glass cover slide. Gently add 1.7 μL of reservoir solution to the RNA drop; do not mix the drop contents.

4. Invert the cover slide and place it onto the corresponding well solution using pressure to seal the well. Use a pipet tip to press down firmly on the cover slip edges to remove air gaps in the grease.

5. Store the crystallization plate in a temperature-controlled environment (see Note 6). Check the plates 3–5 days later and in 1-week intervals thereafter. Record the observations carefully every time the drops are inspected.

6. Single crystals can be harvested and cryoprotected as described (20, 21).

7. For suggestions on how to improve crystallization constructs to produce better crystals, see Note 7.

3.11. Refinement of RNA Crystal Structures

The ideal diffraction pattern should be from a single lattice, and each reflection should be nearly circular, and free of splitting and diffuse scattering; combinations of reflections should comprise distinct elliptical lunes (see Fig. 1d). For background on X-ray data collection, processing, and reduction see refs. 22, 23. To facilitate model building and refinement, the resolution of the diffraction pattern should exceed 3.2 Å resolution. Previously we noted there are many authoritative descriptions of RNA phasing and X-ray structure determination, and we will not describe this process here. Instead we will assume that the experimental phase problem has been addressed, and that the problem is one of refinement whereby the model $|F_{calc}|$ is manipulated to optimize agreement with the observed data $|F_{obs}|$ while maintaining agreement with stereochemical restraints. A suitable program for RNA refinement is the PHENIX (Python-based Hierarchal ENvironment for Xtallography) software suite (24). Here we describe aspects of refinement that are pertinent to RNA crystallography. We will assume that the user is running PHENIX from a UNIX-like operating system X-window, but we point out where the GUI offers a distinct advantage.

3.12. Importing Data and Adding a Test Set

1. Importing structure factor amplitudes (hkl, I_{obs}, $I\sigma$). There are a variety of X-ray data reduction programs. PHENIX can import or write data in various formats including: SCALEPACK (.sca), CCP4 (.mtz), CNS (.cns), and SHELX (.hkl).

2. Convert intensities to amplitudes using the following command line arguments:

   % phenix.reflection_file_converter mystructure.sca --write-mtz-amplitudes --mtz-root-label=FOBS --mtz=mystructure

3. To check your structure factors before starting refinement, use Phenix Xtriage. This will perform a quick analysis of your data and determine a probable solvent content, twinning analysis, Wilson scaling, and anomalous signal:

   % phenix.xtriage mystructure.mtz n_bases=61 log=mystructure_xtriage.log

4. To generate a new test set add "--generate-r-free-flags=True" to the argument above.

5. To preserve R_{free} flags from a previous structure, use the "Reflection file editor" under the Reflection tools tab in the PHENIX GUI as follows: (1) if unit cell dimensions for both structures are not identical, one file must be converted into a format that does not include the unit cell dimensions, e.g., XPLOR or CNS; (2) use the GUI to import the reflection file to be refined against (i.e., "mystructure.mtz") as well as the file that contains the desired R_{free} flags (i.e., "previous_structure.mtz"); (3) highlight mystructure.mtz in the top window then click "use symmetry from selected file"; (4) in the "Input arrays" window highlight the "Data type" to be moved to the "Output arrays window" by using the "+" icon below the window; this may include: (a) FOBS, SIGFOBS, etc. from mystructure.mtz and (b) the R_{free} flag from previous_structure.mtz; (5) click the "R-free flag generation …" button and in the window that opens deselect "Generate R-free flags if not already present"; then click "OK"; (6) verify that the "Extend

existing R-free arrays(s) to full resolution range" option is selected and type the high and low-resolution values for your my_structure.mtz in the "high and low resolution values" boxes; (7) check that "Unit cell" and "Space group" values are correct; (8) in the "Output file" box, enter the correct path and file name for the new data file with R_{free} flags, e.g., "mystructure-Rfree.mtz"; and (9) click the "Run" button.

3.13. Preparation of a Protein Data Bank File with Ligands for PHENIX Refinement

1. PHENIX will be unable to perform refinement if any atom, metal, or ligand names in your protein data bank (PDB) file are unrecognized. Such nonstandard atoms or residues must be identified before refinement.

2. To view the coordinates and a cif (crystallographic information file) file for a particular ligand that is part of a larger structure (e.g., found in the PDB's Ligand Expo or a downloaded cif file) use the following commands:

   % phenix.reel --chemical-components=ATP

   % phenix.reel ATP.cif

3. To optimize and examine every atom in the file:

   % phenix.reel starting_model.pdb --do-all

4. The PHENIX routine "ready_set" adds hydrogens to nucleic acids and ligands in a PDB file in preparation for refinement. It also checks all the atom names and three-letter identifier codes against the PHENIX monomer library and the PDB's ligand database, which can result in new cif files or ".edits" file (also called "edits" file) for unknown identifiers. To run this command:

   % phenix.ready_set mypdb.pdb cif_file_name=myligand.cif

   Default behavior is to add hydrogens to ligands that have less than one quarter of their possible hydrogens but not to optimize_ligand_geometry. Novel linkages should also be found by ready_set and written to a custom edits file for bond restraint. This file can be written by the user to specify any custom link, bond angle, or bond length.

   The format for the edits file for a 2'–5' phosphodiester linkage (12, 25) is:

```
refinement.geometry_restraints.edits {

  bond {

    action = *add

    atom_selection_1 = name O2´ and chain A and resname 3DA and resseq 5

    atom_selection_2 = name P and chain A and resname Gr and resseq 6

    symmetry_operation = x,y,z

    distance_ideal = 1.6

    sigma = 0.020

    slack = None

  }

}

refinement.geometry_restraints.edits {

  angle {

    action = *add

    atom_selection_1 = name O2´ and chain A and resname 3DA and resseq 5

    atom_selection_2 = name P and chain A and resname Gr and resseq 6

    atom_selection_3 = name OP2 and chain A and resname Gr and resseq 6

    angle_ideal = 109

    sigma = 4

  }

}
```

5. Metal coordination spheres can be written to an edits file using:

   % phenix.metal_coordination mystructure.pdb

6. Phenix.eLBOW (electronic Ligand Builder and Optimization Workbench) can be used to produce geometry restraints files for novel ligands. The program accepts various file input file formats (PDB, SMILES, Mol) and writes out a PDB and a cif file. Default behavior is to do a simple force field optimization

of the input coordinates but alternative force fields such as AM1 are available:

% phenix.elbow ligand.pdb

7. To use eLBOW to process all unknown ligands in a pdb file:

% phenix.elbow starting_model.pdb --do-all

8. The user has the option to not optimize and keep input geometry:

% phenix.elbow --final_geometry ligand.pdb

9. We recommend that all ligands output by phenix.elbow be checked visually as quality control.

10. Initial coordinates should be assigned a global B-factor before refinement (see Note 8).

3.14. Rigid-Body Refinement

When solving a Fourier problem in which the starting model (i.e., the phase source) and F_{obs} are significantly different in terms of subtle rotations and translations of the contents in the respective asymmetric units, it is possible to reposition the starting model by treating the entire model as a rigid body or to divide various known domains into groups that can be refined as independent (rigid) bodies. Rigid-body refinement can commence as follows:

1. Execute the first round of rigid-body refinement treating each whole molecule in the asymmetric unit as a separate rigid body:

% phenix.refine mystructure-Rfree.mtz starting_model.pdb
    strategy=rigid_body output.prefix=model-rigid1

This creates a rigid-body refined PDB file called model-rigid1_001.pdb. If there are any ligands or unusual nucleotides, then include the correct cif file after the input PDB file on the command line. Progress can be checked by examining the R_{work}, R_{free}, rmsd bond lengths, and rmsd bond angles located in the header of the output PDB file.

2. Multiple rounds of rigid-body refinement can be performed for RNA structures with multiple domains or when there is more than one molecule in the asymmetric unit. The method is fast and should be restricted to low-resolution data, which is sufficient to rotate and translate each rigid-body starting model to match the unknown structure to be refined. Various statements for parsing the structure into independently refined rigid-body domains may appear as follows:

% phenix.refine mystructure-Rfree.mtz starting_model.pdb
    strategy=rigid_body sites.rigid_body="chain A or chain B"
    sites.rigid_body="chain C or chain D."

Here chains A and B are treated as one group, and C and D are treated as a separate group.

3. The different chains can be treated as independent rigid bodies by the statement:

% phenix.refine mystructure-Rfree.mtz starting_model.pdb strategy=rigid_body sites.rigid_body="chain A" sites. rigid_body="chain B" sites.rigid_body="chain C" sites. rigid_body="chain D."

4. The net effect should be a lowering of R_{free} and R_{work} by global repositioning of the starting model, which supplies the initial phases.

3.15. Refinement of Individual Atomic Coordinates and B-Factors

1. To perform the default three cycles of bulk solvent correction and scaling, as well as refinement of coordinates and ADP (atomic displacement parameters), use the following command:

% phenix.refine mystructure-Rfree.mtz starting_model.pdb

When complete PHENIX writes several files including: (1) an updated coordinate file that includes the refined structure with a header containing a summary of the refinement statistics; (2) a new MTZ file that includes electron density map coefficients compatible with interactive graphics programs such as COOT (26); (3) the log file of the refinement run; (4) a refinement parameters file (.eff); (5) a geometry restraints file (.geo); (6) a defaults file (.def) that includes all the previous commands, geometry edits, cif file locations, and modifications. This file can start the next round of refinement using the command:

% phenix.refine mystructure_002.def

2. To perform different combinations of refinement steps, one can string commands together using "strategy":

% phenix.refine mystructure-Rfree.mtz starting_model.pdb strategy=rigid_body+individual_sites+group_adp

3. To set the number of macrocycles performed:

% phenix.refine mystructure-Rfree.mtz starting_model.pdb main. number_of_macro_cycles=6

4. Improvements in the radius of convergence are also achievable by simulated annealing (see Note 9).

3.16. TLS Refinement

1. Translation, libration, screw-axis (TLS) refinement allows correction for anisotropic motion of atomic groups or domains, which can improve the model agreement with experimental data. To invoke TLS refinement type:

% phenix.refine mystructure-Rfree.mtz starting_model.pdb refine.adp.tls="chain A" refine.adp.tls="chain B" refine.adp. tls="chain C" refine.adp.tls="chain D"

2. Alternatively a "tls_group_selections.params" file can be used that specifies the TLS groups. This option will add ANISOU

records to the refined pdb for atoms in the TLS groups. As such, the total B-factor includes the sum contribution of B-factor_tls and B-factor_individual values.

3.17. Optimizing Refinement Target Weights

1. As a starting point, it is best to determine a weighting scheme that gives the lowest R_{free}:

   % phenix.refine Data-Rfree.mtz model.pdb optimize_wxc=true optimize_wxu=true

   The command wxc=true will activate the optimization of X-ray and stereochemistry weighting, and wxu true will activate optimization of X-ray data and ADP weighting. Issuing this command will generate the lowest possible R-values, but may result in unacceptably high bond length and angle rmsd values, so examine the log file and the refined model output.

2. Once an initial set of weights is calculated from step 1, it is possible to manually set the weight of the stereochemical and ADP restraint terms during refinement if R_{work} and R_{free} values are not satisfactory. To target a specific weight value enter:

   % phenix.refine Data-Rfree.mtz model.pdb target_weights. wxc_scale=X target_weights.wxu_scale=Y, where X determines the contribution of X-ray/stereochemical restraints and Y determines the X-ray/ADP contribution. The default values for wxc_scale and wxu_scale are $X = 0.5$ and $Y = 1.0$, respectively. Increasing either value will loosen restraints.

3. If the stereochemical restraints appear too tight, then the weighting of stereochemistry in refinement can be adjusted as follows:

   % phenix.refine Data-Rfree.mtz model.pdb wxc_scale=X

   The same course of action can be taken for the ADP weight (wxu) if desired.

3.18. Use of Coot for Interactive Model Building

1. COOT is a program for interactive refinement of macromolecular coordinates based primarily on manual fitting of electron density maps. Here we assume the user is reading in a PDB file from PHENIX-based refinement and its corresponding MTZ map coefficients described in Subheading 3.15. Although RNA autobuilding is possible in PHENIX, cycles of PHENIX-refinement are often punctuated by manual intervention, which can improve the radius of convergence in which a model is trapped in a local energy minimum.

2. Start by opening an .mtz file in COOT. This can be done by choosing the "Auto Open MTZ ..." under the "File" tab of the Black window GUI. This option creates both $2mF_o - DF_c$ and $mF_o - DF_c$ electron density maps.

3. A full review of Coot capabilities is provided elsewhere (26). However, the following commands are worth reviewing since they are invoked frequently during RNA model building: (1) "Real Space Refine Zone"—fits model to electron density map selected; (2) Regularize Zone—normalize geometry to ideal values; (3) "LSQ Superpose ..." located under the "Calculate" tab—allows pair-wise atomic superpositions between structures; (4) "Check/Delete Waters ..."—under the "Validate" tab; (5) also under the "Validate" tab, the commands "Density fit analysis," "ADP variance," and "MOLPROBITY Analysis" (27); (6) "Get PDB and Map using EDS ..." under the "File" tab—retrieves the requested PDB coordinates and electron density maps from the Electron Density Sever (28) if they are available.

4. Notes

1. A robotic screening system is not essential but has the advantage of improving throughput with a relatively small amount of sample. Several fee-for-service screening facilities are available such as the University of Rochester Structural Biology and Biophysics Facility (http://www.urmc.rochester.edu/Structural-Biology-Biophysics/), or the batch-under-oil 1536-well screen at the Hauptman-Woodward Medical Research Institute, Inc. (Buffalo, NY) (http://www.hwi.buffalo.edu/faculty_research/crystallization.html). As an alternative, manual screening by use of the reagents in Table 1 represents a starting point.

2. Two factors are worth immediate consideration if difficulties arise in transcription. These include the concentration of $MgCl_2$ and activity of T7 polymerase. The amount of $MgCl_2$ will determine the extent of the reaction since it is depleted during the polymerization as an insoluble complex with inorganic pyrophosphate. T7 polymerase should be fresh or stored at $-20°C$ in 20% glycerol under reducing conditions with DTT.

3. If it becomes apparent that the transcription products are a mixture of truncated and full-length species, it is possible that the polymerase is failing at segments of identical bases. One possible solution is to add 10% more of the nucleotide present in the repeat region.

4. The following variables should be taken into account when choosing or designing crystallization screens for RNA: (1) pH range: 5.6–8.0 (suggested buffers: Na-cacodylate, MES, HEPES, Tris); (2) monovalent salts of chloride: Li^+, Na^+, K^+, NH_4^+; acetate: Li^+, Na^+, K^+, NH_4^+; sulfate: Li^+, NH_4^+; (3) divalent cations (5–20 mM): Mg^{2+}, Ca^{2+}, Mn^{2+}, and Co^{2+}; (4) trivalents: 0.5–5.0 mM $Co(NH_3)_6Cl_3$; (5) organic solvents: 5–40%

(v/v) 2-methyl-2,4-pentanediol (MPD); and 1.5–2.4 M 1,6-hexanediol; (6) poly(ethylene) glycol (PEG): 5–30% (w/v) PEG 400, monomethyl ether (MME) PEG 550, MME PEG 2000, MME PEG 5000, PEG 3350, PEG 6000, and PEG 8000; (7) polyamine additives: 1–2 mM spermine and spermidine; (8) chelating agents: 0.1–0.5 mM EDTA; and (9) organic acid salts: 20–60% Tacsimate™.

5. The higher concentration of RNA, the more likely it is to crystallize. Ideally, the RNA will be concentrated to its solubility limit prior to initiating crystallization trials. Thus far, most of our ribozymes and riboswitches have crystallized in a range between 0.25 and 0.70 mM. Establishing the desired concentration of RNA may be as simple as resuspending the lyophilized RNA in a buffer of choice (e.g., 10 mM Na-cacodylate pH 6.0), or concentrating by centrifugation. For example, the $preQ_0$ metabolite was added to our riboswitch aptamers under dilute conditions, then concentrated (4). For the crystals in Fig. 1c, dilute U1A was added dropwise to semi-concentrated RNA to attain a molar ratio of 1:1.05 before concentrating.

6. Like protein crystallization, RNA crystal growth can be sensitive to vibration and temperature. Screening is conducted in a low-vibration environment with a constant temperature ($\pm1°C$) in the range of 4–35°C. Plates are checked for crystal growth using a microscope at the temperature of screening. We typically choose 20 and 4°C to store plates during crystallization trials.

7. There are several documented approaches to improving RNA crystals if initial crystallization trials are unsuccessful or crystals exhibit poor diffraction. These include: adding U1A (29, 30), changing helix length (31), trying a different species (4), adding blunt or sticky ends (7), replacing nonconserved stem-loops with GNRA tetraloops, or using a transition-state analog for ribozymes (12, 25).

8. To set an overall ADP value (e.g., based on Wilson scaling):

```
% phenix.refine mystructure-Rfree.mtz starting_model.pdb
    set_b_iso=25
```

9. To perform simulated annealing:

```
% phenix.refine mystructure-Rfree.mtz starting_model.pdb
    simulated_annealing=true
```

Acknowledgments

We thank Profs. Harold C. Smith and Clara L. Kielkopf for sharing their expertise on RNA. We thank Jason Salter for assistance with diffraction analysis, as well as the staff of MacCHESS and SSRL for help with X-ray data collection. This work was supported in part by NIH grants GM063162 and RR026501 to J.E.W. MacCHESS is supported by NSF award DMR-0225180 and NIH/NCRR award RR01646SSRL. SSRL is operated by Stanford on behalf of the U.S. DOE. The SSRL Structural Molecular Biology Program is supported by the DOE, and by NIH/NCRR and NIGMS.

References

1. Wedekind, J. E. (2011) in *Met. Ions Life Sci.: Structural and Catalytic Roles of Metal Ions in RNA*, eds. A. Sigel, H. Sigel, R. Sigel, Royal Society of Chemistry, London, pp. 299–345.

2. Wedekind, J. E. and McKay, D. B. (2003) Crystal structure of the leadzyme at 1.8Å resolution: metal ion binding and the implications for catalytic mechanism and allo site ion regulation. *Biochemistry* **42**, 9554–9563.

3. Alam, S., Grum-Tokars, V., Krucinska, J., Kundracik, M. L. and Wedekind, J. E. (2005) Conformational heterogeneity at position U37 of an all-RNA hairpin ribozyme with implications for metal binding and the catalytic structure of the S-turn. *Biochemistry* **44**, 14396–14408.

4. Spitale, R. C., Torelli, A. T., Krucinska, J., Bandarian, V. and Wedekind, J. E. (2009) The structural basis for recognition of the PreQ0 metabolite by an unusually small riboswitch aptamer domain. *J Biol Chem* **284**, 11012–11016.

5. Griffiths-Jones, S., Moxon, S., Marshall, M., Khanna, A., Eddy, S. R. and Bateman, A. (2005) Rfam: annotating non-coding RNAs in complete genomes. *Nucleic Acids Res* **33**, D121–124.

6. Ren, J., Rastegari, B., Condon, A. and Hoos, H. H. (2005) HotKnots: heuristic prediction of RNA secondary structures including pseudoknots. *RNA* **11**, 1494–1504.

7. Wedekind, J. E. and McKay, D. B. (2000) Purification, crystallization, and X-ray diffraction analysis of small ribozymes. *Methods Enzymol.* **317**, 149–168.

8. Golden, B. L., Gooding, A. R., Podell, E. R. and Cech, T. R. (1996) X-ray crystallography of large RNAs: heavy-atom derivatives by RNA engineering. *RNA* **2**, 1295–1305.

9. Robertson, M. P. and Scott, W. G. (2008) A general method for phasing novel complex RNA crystal structures without heavy-atom derivatives. *Acta Crystallogr D Biol Crystallogr* **D64**, 738–744.

10. Keel, A. Y., Rambo, R. P., Batey, R. T. and Kieft, J. S. (2007) A general strategy to solve the phase problem in RNA crystallography. *Structure* **15**, 761–772.

11. Beaucage, S. L. and Reese, C. B. (2009) Recent advances in the chemical synthesis of RNA. *Curr Protoc Nucleic Acid Chem* **Chapter 2**, Unit 2 16 11–31.

12. Torelli, A. T., Spitale, R. C., Krucinska, J. and Wedekind, J. E. (2008) Shared traits on the reaction coordinates of ribonuclease and an RNA enzyme. *Biochem Biophys Res Commun* **371**, 154–158.

13. Spitale, R. C., Volpini, R., Mungillo, M. V., Krucinska, J., Cristalli, G. and Wedekind, J. E. (2009) Single-atom imino substitutions at A9 and A10 reveal distinct effects on the fold and function of the hairpin ribozyme catalytic core. *Biochemistry* **48**, 7777–7779.

14. Sinha, N. D., Biernat, J., McManus, J. and Koster, H. (1984) Polymer support oligonucleotide synthesis XVIII: use of beta-cyanoethyl-N,N-dialkylamino-/N-morpholino phosphoramidite of deoxynucleosides for the synthesis of DNA fragments simplifying deprotection and isolation of the final product. *Nucleic Acids Res* **12**, 4539–4557.

15. Spitale, R. C. and Wedekind, J. E. (2009) Exploring ribozyme conformational changes with X-ray crystallography. *Methods* **49**, 87–100.

16. Hartsel, S. A., Kitchen, D. E., Scaringe, S. A. and Marshall, W. S. (2005) RNA oligonucleotide synthesis via 5'-silyl-2'-orthoester chemistry. *Methods Mol Biol* **288**, 33–50.

17. Ferre-D'Amare, A. R., Zhou, K. and Doudna, J. A. (1998) A general module for RNA crystallization. *J Mol Biol* **279**, 621–631.

18. Sherlin, L. D., Bullock, T. L., Nissan, T. A., Perona, J. J., Lariviere, F. J., Uhlenbeck, O. C. and Scaringe, S. A. (2001) Chemical and enzymatic synthesis of tRNAs for high-throughput crystallization. *RNA* **7**, 1671–1678.

19. Milligan, J. F. and Uhlenbeck, O. C. (1989) Synthesis of small RNAs using T7 RNA polymerase. *Methods Enzymol* **180**, 51–62.

20. Garman, E. F. and Doublie, S. (2003) Cryocooling of macromolecular crystals: optimization methods. *Methods Enzymol* **368**, 188–216.

21. Garman, E. F. and Owen, R. L. (2006) Cryocooling and radiation damage in macromolecular

crystallography. *Acta Crystallogr D Biol Crystallogr* **62**, 32–47.

22. Arndt, U. W. and Wonacott, A. J., *The Rotation method in crystallography: Data collection from macromolecular crystals*, Elsevier/North-Holland, New York, 1977.

23. Otwinowski, Z. and Minor, W. (1997) Processing of X-ray Diffraction Data Collected in Oscillation Mode. *Methods in Enzymology* **276**, 307–326.

24. Adams, P. D., Afonine, P. V., Bunkoczi, G., Chen, V. B., Davis, I. W., Echols, N., Headd, J. J., Hung, L. W., Kapral, G. J., Grosse-Kunstleve, R. W., McCoy, A. J., Moriarty, N. W., Oeffner, R., Read, R. J., Richardson, D. C., Richardson, J. S., Terwilliger, T. C. and Zwart, P. H. (2010) PHENIX: a comprehensive Python-based system for macromolecular structure solution. *Acta Crystallogr D Biol Crystallogr* **66**, 213–221.

25. Torelli, A. T., Krucinska, J. and Wedekind, J. E. (2007) A comparison of vanadate to a 2'-5' linkage at the active site of a small ribozyme suggests a role for water in transition-state stabilization. *RNA* **13**, 1052–1070.

26. Emsley, P., Lohkamp, B., Scott, W. G. and Cowtan, K. (2010) Features and development of Coot. *Acta Crystallogr D Biol Crystallogr* **66**, 486–501.

27. Chen, V. B., Arendall, W. B., III, Headd, J. J., Keedy, D. A., Immormino, R. M., Kapral, G. J., Murray, L. W., Richardson, J. S. and Richardson, D. C. (2010) MolProbity: all-atom structure validation for macromolecular crystallography. *Acta Crystallogr D Biol Crystallogr* **66**, 12–21.

28. Kleywegt, G. J., Harris, M. R., Zou, J. Y., Taylor, T. C., Wahlby, A. and Jones, T. A. (2004) The Uppsala Electron-Density Server. *Acta Crystallogr D Biol Crystallogr* **60**, 2240–2249.

29. Ferre-D'Amare, A. R. and Doudna, J. A. (2000) Crystallization and structure determination of a hepatitis delta virus ribozyme: use of the RNA-binding protein U1A as a crystallization module. *J Mol Biol* **295**, 541–556.

30. Ferre-D'Amare, A. R. (2010) Use of the spliceosomal protein U1A to facilitate crystallization and structure determination of complex RNAs. *Methods* **52**, 159–167.

31. MacElrevey, C., Spitale, R. C., Krucinska, J. and Wedekind, J. E. (2007) A posteriori design of crystal contacts to improve the X-ray diffraction properties of a small RNA enzyme. *Acta Crystallogr. D Biol. Crystallogr.* **63**, 812–825.

Chapter 12

Functional Dynamics of RNA Ribozymes Studied by NMR Spectroscopy

Boris Fürtig, Janina Buck, Christian Richter, and Harald Schwalbe

Abstract

Catalytic RNA motifs (ribozymes) are involved in various cellular processes. Although functional cleavage of the RNA phosphodiester backbone for self-cleaving ribozymes strongly differs with respect to sequence specificity, the structural context, and the underlying mechanism, these ribozyme motifs constitute evolved RNA molecules that carry out identical chemical functionality. Therefore, they represent ideal systems for detailed studies of the underlying structure–function relationship, illustrating the diversity of RNA's functional role in biology. Nuclear magnetic resonance (NMR) spectroscopic methods in solution allow investigation of structure and dynamics of functional RNA motifs at atomic resolution. In addition, characterization of RNA conformational transitions initiated either through addition of specific cofactors, as e.g. ions or small molecules, or by photo-chemical triggering of essential RNA functional groups provides insights into the reaction mechanism. Here, we discuss applications of static and time-resolved NMR spectroscopy connected with the design of suitable NMR probes that have been applied to characterize global and local RNA functional dynamics together with cleavage-induced conformational transitions of two RNA ribozyme motifs: a minimal hammerhead ribozyme and an adenine-dependent hairpin ribozyme.

Key words: NMR spectroscopy, Caged compound, RNA folding, Self-cleaving ribozymes, Time-resolved NMR

1. Introduction

1.1. NMR Studies on Self-Cleaving Ribozyme RNA Motifs

Self-cleaving ribozyme RNA motifs include the hammerhead (HHR) (1, 2), the hairpin (3–6), the hepatitis delta virus (3–8), the Varkud satellite (9, 10), and the *glmS* ribozymes (11). These motifs are naturally occurring RNA sequences that are involved in different cellular processes including processing of satellite or virus RNAs or regulation of gene expression. For example, the tobacco ringspot virus satellite possesses two catalytic RNA motifs, the

Jörg S. Hartig (ed.), *Ribozymes: Methods and Protocols*, Methods in Molecular Biology, vol. 848,
DOI 10.1007/978-1-61779-545-9_12, © Springer Science+Business Media, LLC 2012

HHR and the hairpin ribozyme, that participate in a "rolling cycle" mechanism of its own replication. The small self-cleaving ribozymes display sequence lengths between 40 and 200 nucleotides (nt) and can adopt a variety of different secondary and tertiary structures. Hydrolysis of the RNA phosphodiester backbone follows an *in line* transesterification mechanism (see Subheading 3.2). The global architectures of different ribozymes, their respective active site conformation pre- and post-cleavage and conformational dynamics during catalysis have been extensively characterized using diverse biochemical and biophysical methods.

Given the dynamic nature of ribozymes and their compatible molecular size, nuclear magnetic resonance (NMR) spectroscopic methods were particularly valuable for describing structure and dynamics of these catalytic RNA motifs. Various NMR parameters that report on structure and dynamics yielded insights into the functional aspects of small self-cleaving ribozymes, e.g. the HHR and hairpin ribozyme. Early reports on base-pairing interactions examined by NMR spectroscopy revealed the role of G:A-mismatch pairs within the structure of HHR (12). Conformational changes in the active site, while the surrounding stems remain stable, were shown to be fundamental to catalytic activity (13). Analysis of NMR residual dipolar coupling data revealed relative motions of helices for conformational states pre- and post-cleavage (14). The dynamic nature of HHR and its essential functional interactions with divalent ions could also be monitored by using direct NMR reporter signals introduced within the sequence, such as ^{15}N of ^{19}F (15, 16), or by secondary effects of phosphothioates on the chemical shifts of naturally occurring ^{31}P-NMR signals of the RNA backbone (17, 18). For the minimal hairpin ribozyme, 3D structures of individual RNA domains (A and B) were solved by NMR spectroscopy (19, 20) proposing a full-length model of the ribozyme motif in solution (20).

1.2. Minimal Hammerhead Ribozyme

For many years, the minimal hammerhead ribozyme (mHHR), representing the smallest motif (~40-nt) derived from naturally occuring ribozymes, served as a model system to study conformational rearrangements and dynamics associated with RNA catalysis. Its structure–function relationship has been examined by a multitude of biochemical and biophysical techniques. mHHR hydrolyses the 5′-3′-phosphodiester linkage in a sequence-specific manner (21). The catalytic activity was found to be dependent on the presence of divalent ions that can, however, be replaced by high concentrations of monovalent ions (22). Furthermore, the optimal pH value for the chemical step of the reaction was determined to be between pH ~6–8. Within the central 3-way junction, comprising the ribozyme active site, the global structure of 14 nucleotides is

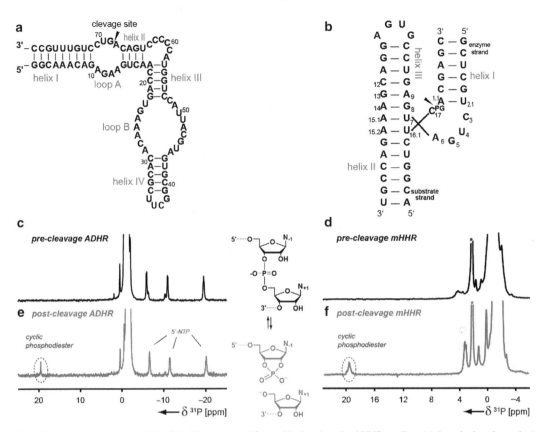

Fig. 1. Secondary structure models of the ribozyme motifs used in the described NMR studies; (**a**) the adenine-dependent hairpin ribozyme and (**b**) the minimal hammerhead ribozyme; ^{31}P-NMR spectra of ribozyme motifs in the pre-cleaved conformation (**c**) for ADHR (in the absence of adenine) and (**d**) for mHHR (with photo-protected 2′-OH nucleotide); ^{31}P-NMR spectra in the post-cleaved conformation (**e**) for ADHR (after incubation with adenine for 20 h) and (**f**) for mHHR (after deprotection through ~1 s laser pulse and 4 h incubation time). The ^{31}P-NMR resonance at $\delta(^{31}\text{P})\sim19.5$ ppm characteristic for the 2′,3′-cyclic phosphodiester reaction product is highlighted. The *middle* panel shows a schematic representation of the transesterification mechanism. The *asterisk* in (**d, f**) indicates the buffer resonance.

conserved which includes an essential triplet N15.1-U16.1-H17 (N for all nucleotides, H for all nucleotides except G) on the 5′-side of the scissile phosphodiester bond (see Fig. 1b). Early mHHR crystal structures delineated the overall 3D-architecture of the RNA; yet, the detailed cleavage mechanism could not be explained. Both nucleotides adjacent to the active site adopt conformations incompatible with the proposed nucleophilic *in line* attack. FRET studies, electric bifrigence, and NMR spectroscopy revealed that the helices in the ribozyme motif experience substantial angular movements and adopt different global shapes depending on the ionic conditions. These findings already led to the notion that dynamic rearrangements within the RNA were at the basis of the catalytic reaction.

However, in the native sequence context, in full-length hammerhead ribozyme motifs, additional long-range interactions could be identified remote from the cleavage site. The elongated ribozyme motif (eHHR) shows an enhanced cleavage rate that is ~10^3 times higher than the rate of its minimal counterpart (mHHR) (23). The eHHR crystal structure illustrates that the remote structural elements compact the overall fold and render the catalytic core to be *in line* for the cleavage reaction. Comparison of ribozyme structures of different sequence context and correlation with their reactivity as well as in conjunction with the data provided by our time- and site-resolved NMR experiments (24) reveals the following model for mHHR activity: the crystal structure of mHHR represents the ground state of the ribozyme motif with a low-energy dynamic orientation of helices. Due to diffuse electrostatic shielding by Mg^{2+}-ions, the helices adjacent to the three-way junction exert relative motions towards each other. These helical motions are coupled to the conformation of the catalytic core that undergoes conformational exchange. A sub-population of these conformations can undergo an *in line* attack required for the hydrolysis step to occur. In the extended hammerhead ribozyme, the interhelical motions are fixed by long-range helix–helix interactions that increase fortuitously the reactive conformation within the catalytic core.

1.3. Adenine-Dependent Hairpin Ribozyme

RNA catalysis has been studied in the adenine-dependent hairpin ribozyme (ADHR), representing a synthetic ribozyme motif that undergoes self-cleavage after addition of the exogenous cofactor adenine. The "inactive" ribozyme ADHR was designed to analyze the role of cofactors in RNA catalytic mechanisms (25). Based on the wild-type RNA sequence of the minimal catalytic hairpin ribozyme motif of the tobacco ringspot virus satellite, this ribozyme mutant was selected in vitro in a SELEX procedure (5). The 80-nt ribozyme motif represents the largest RNA for which time-resolved NMR studies could describe cleavage kinetics thus far (26). Analysis of the NMR data reveals that ADHR represents a ribozyme motif with high activation barriers towards cleavage-competent states that are only infrequently reached and presumably exhibit short lifetimes. However, if an active conformation is populated, the chemical reaction step can proceed. Thus, for ADHR conformational dynamics depict the rate-limiting step in catalysis. The cofactor adenine can restore essential catalytic function. However, the ribozyme motif remains also additionally dependent on cations. The multiple factor dependence of catalysis renders ADHR an interesting model system to dissect catalytic contributions to ribozyme activity.

2. Materials

1. For experimental details concerning chemical synthesis of 2′-OH photo-protected and specifically ^{13}C-labelled nucleoside building blocks and solid-phase RNA synthesis, we refer to following reviews (27–33).

2. ^{1}H,^{13}C-HSQC (34), HCC-TOCSY (35), HCCH-E.COSY (36), and ^{13}C-edited-NOESY (37) spectra were recorded on a Bruker AV800MHz spectrometer equipped with a 5-mm z-axis gradient TXI-HCN cryogenic probe at 293 K.

1D ^{31}P-NMR spectra and 3D HCP (38) experiments were recorded by using an AV600MHz Bruker NMR spectrometer equipped with a 5-mm z-axis gradient TCI-HCP cryogenic probe and an AV300MHz Bruker NMR spectrometer equipped with a 5-mm z-axis gradient BBO-RT probe.

1. NMR kinetic experiments of the mHHR cleavage reaction were recorded at different temperatures on a Bruker AV800MHz spectrometer equipped with a 5-mm z-axis gradient TXI-HCN cryogenic probe (T~288–298 K). RNA samples had a concentration of ~100 μM and were prepared in phosphate buffer pH ~7.5 containing additional NaCl (25 mM) and MgCl$_2$ (10 mM).

2. Reaction initiation was maintained photo-chemically with an in situ laser setup. In the laser installation, light of 334–380 nm wavelength (total power ~4.5 W) of a CW argon ion laser (Spectra-Physics, Darmstadt, Germany) is directly coupled into the NMR tube via a quartz fibre connected to a quartz tip insert (39–41).

3. Laser illumination of the NMR sample is achieved via spectrometer-controlled coupling optics by TTL connection to beam-shutter drivers.

4. Processing and analysis of the experimental data was performed using the software FELIX (Accelrys).

5. Non-linear fitting of the kinetic traces was achieved using the program Sigma Plot.

2.4. Divide-and-Conquer Strategy for Imino Proton-Based Characterization of Adenine-Dependent Hairpin Ribozyme Conformations

1. For ADHR-specific SELEX procedures and RNA synthesis by using T7-RNA polymerase in vitro transcription, we refer to Meli et al. (25) (unlabelled rNTPs: Fermentas, St. Leon-Rot, Germany; ^{15}N-labelled rNTPs: Silantes, Munich, Germany).

2. Unlabeled model hairpin RNAs were purchased from Dharmacon (Boulder, CO, USA).

3. RNA buffer conditions: HEPES (40 mM), pH ~7.5, MgCl$_2$ (6 mM, if not otherwise stated). NMR spectra were recorded in H$_2$O/D$_2$O (9:1).

4. NMR experiments were performed using Bruker NMR spectrometers (Bruker, Rheinstetten, Germany) AV900MHz, AV800MHz, AV700MHz, and AV600MHz with 5-mm z-axis gradient TXI-HCN cryogenic probes.

5. NMR resonance assignment of RNAs was performed by applying the following standard NMR experiments: ^{1}H,^{15}N-HSQC, 2D-^{1}H,^{1}H-NOESY (42), HNN-COSY (43), and 3D-^{1}H,^{1}H,^{15}N-NOESY-HSQC (44) with WATERGATE water suppression (45) or *jump–return–echo* pulse sequences (46).

6. NMR data acquisition and processing was done with the software TOPSPIN (Bruker, Karlsruhe, Germany).

2.5. Imino Proton Resonances as Reporter Signals to Follow the Catalytic Reaction of the Adenine-Induced Cleavage of ADHR

1. Kinetic NMR experiments were performed at following experimental conditions: ADHR (220 μM), adenine (5.6 mM), 298 K (buffer conditions see Subheading 2.4). A minimum RNA concentration of ~200 μM was experimentally determined for this 80-nt RNA, in order to obtain a reliable sensitivity of a single 1D spectrum of the kinetic NMR experiment at the particular time resolution.

2. The NMR data were recorded as pseudo-2D experiments with a time resolution of ~4.25 min/1D spectrum (128 scans per 1D spectrum).

3. The cofactor adenine was added manually to the NMR sample, which resulted in a reaction dead time of ~2 min. However, the NMR spectrum directly after cofactor addition was congruent with the cofactor-free RNA spectrum, proving evidence that the resulting dead time through manual addition is not relevant for our kinetic investigation of the cleavage reaction under the applied experimental conditions (see Note 1).

4. Kinetic NMR data were analyzed with the programs TOPSPIN (Bruker, Karlsruhe, Germany) and Sigma Plot.

5. The reaction time constants k (min^{-1}) for individual RNA imino proton signals ($\delta(^{1}$H)~9–15 ppm) were obtained, extracting their intensities over a period of ~20 h. The extracted data were corrected by the zero point of the kinetics, normalized by the averaged values of the last 20 residual data

points, and fitted with a monoexponential function. Half-life values ($t_{1/2}$ (min)) could be obtained by using the formula for a first-order process ($t_{1/2} = \ln2/k$). The error stated for k is the fitting error.

3. Methods

3.1. General Strategy for Sequence Specifically Labelled and Photo-Protected RNA for (Real-Time) NMR Studies

NMR spectroscopic studies of nucleic acids generally suffer from the low chemical shift dispersion of NMR resonances due to the small number of individual building blocks with similar chemical constitution. Therefore, the choice of RNA constructs, functional relevant reporter signals, the RNA isotope labelling scheme, and/or the scientific question to be studied by NMR spectroscopic methods are of importance, especially for larger RNAs (>50nt). Incorporation of selective reporter signals, namely NMR active spin-1/2 nuclei such as ^{13}C or ^{15}N, greatly enhances the ability to perform studies with site or even atomic resolution. Available methods to synthesize selectively labelled RNAs include either biochemical approaches by in vitro transcription—using nucleotide-specific isotopic labelled rNTPs—or chemical synthesis on solid phase—using sequence-specific labelled building blocks and/or ligation strategies. Biochemical labelling strategies enable introduction of nucleotide-specific reporter signals within the RNA sequence. Next to uniformly labelling of the RNA, also introduction of individual ^{13}C- or ^{15}N-labelled rNTPs can be obtained through in vitro transcription. Another possibility is to completely label certain segments of larger RNAs and incorporate those into the full-length molecule via ligation techniques (47–49). In conjunction with NMR-filter experiments that select for either ^{1}H-^{12}C or ^{1}H-^{13}C (accordingly, ^{1}H-^{14}N or ^{1}H-^{15}N) sites within an RNA, such labelling schemes are beneficial in larger RNAs in order to resolve signal overlap (50).

We investigated local dynamics and cleavage-associated conformational changes of the active site residues (C17 and A1.1) of mHHR. These nucleotides were known to be involved in the underlying chemical reaction step of ribozyme catalysis. Therefore, the two residues adjacent to the scissile phosphodiester bond were ^{13}C-labelled within the ribose moiety. Such labelling scheme could be achieved by solid-phase RNA synthesis (24). In order to enable in situ catalytic cleavage, a photo-labile protecting group was covalently attached to the reactive 2'-OH group of nucleotide C17 (see Fig. 2). With this photo-protected nucleotide, the ribozyme is cleavage-incompetent and thus, remains uncleaved until laser-induced removal of the protecting group. In situ photolytic deprotection within the NMR spectrometer allows timed initiation of the chemical reaction that can be followed by time-resolved NMR. The ribozyme motif is synthesized in two complementary RNA strands, denoted as enzyme strand and substrate strand (the

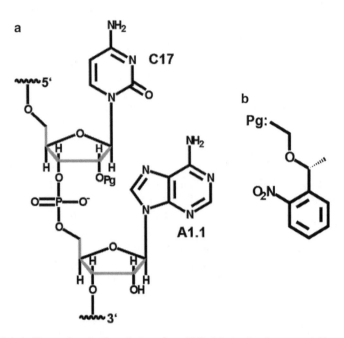

Fig. 2. Labelling and protection strategy for mHHR; (**a**) structural representation of the two core nucleotides, ^{13}C-labelled atoms are highlighted in grey; as indicated, the nucleophilic C17 2'-OH is photo-protected (Pg) by the npeom group in order to inhibit premature nucleophilic attack of the phosphate backbone; (**b**) constitution of the npeom protecting group.

latter containing the labelled nucleotides). With this labelling scheme only resonances stemming from A1.1 and C17 can be observed in ^{13}C-edited NMR experiments both, under static conditions (see Subheadings 3.1 and 3.2) and in laser-triggered time-resolved NMR experiments (see Subheadings 3.3).

3.2. Selective ^{13}C-Labelling of Ribose Atoms as Reporter Signals for Local Conformational Dynamics Within mHHR Cleavage Site

Local conformational changes of the nucleotides adjoining the cleavage site in mHHR (C17 and A1.1) are one important aspect that determines the rate of the chemical reaction. In order to link structural changes to reaction dynamics, the conformation of these nucleotides has to be described in detail. The sequence-specific ^{13}C-labelling allows measurement of chemical shifts as an NMR parameter sensitive to structural changes within the RNA. Additionally, homo- and heteronuclear scalar J-couplings can be determined as NMR parameters illustrating local conformational dynamics of the respective sugar moieties. Furthermore, the changes in structure and dynamics of the catalytic site can be delineated in dependence of external factors such as Mg^{2+} or pH. For mHHR in the pre-cleaved state, the sugar moieties around the cleavage site undergo a fast equilibrium between C2'- and C3'-endo conformations (~50:50% based on measurements of $J_{H1'H2'}$ and $J_{H3'H4'}$ couplings). Addition of Mg^{2+} leads to an ensemble of conformations where dynamic transitions between energetically similar conformations occur on the millisecond timescale (based

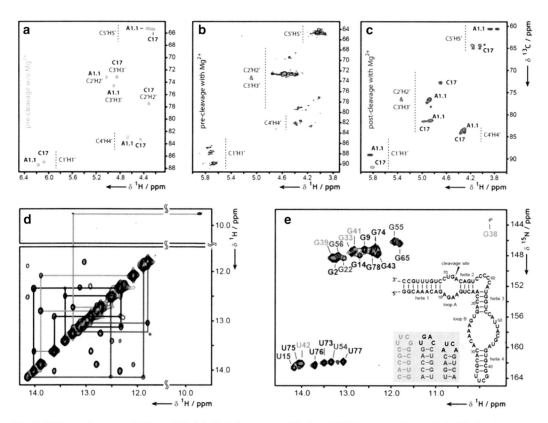

Fig. 3. NMR spectroscopy of differentially labelled ribozyme motifs; (a–c) HSQC spectra of selectively ^{13}C-ribose labelled mHHR in three different states (293 K, 800 MHz); (a) pre-cleavage photo-protected Mg^{2+}-free state; (b) pre-cleavage photo-protected state with 10 mM Mg^{2+}; (c) post-cleavage state with 10 mM Mg^{2+} after photo-initiated cleavage reaction; (d–e) imino proton NMR resonance assignment of the 80-nt ^{15}N-labelled adenine-dependent hairpin ribozyme; (d) ^{1}H,^{1}H-NOESY spectrum of ADHR (without Mg^{2+}, 293 K, 600 MHz, 150 ms mixing time); lines represent resolved imino-imino proton connectivities with colour coding according to full-length secondary structural elements that are mimicked by model hairpin RNAs (see (e), asterisk: NOE detectable in ^{1}H,^{1}H-NOESY with different mixing time); (e) overlay of ^{1}H,^{15}N-HSQC spectra of full-length ADHR (black) and model hairpin RNAs with annotated resonance assignment of ADHR. Model hairpin 1/mimic of helix 4 is colour coded in orange, model hairpin 2/mimic of helix 3 in blue, and model hairpin 3/mimic of helix 2 in red; inset: secondary structures of ADHR and model hairpin RNAs.

on the measurement of signal line-shapes in heteronuclear spectra). In the post-cleavage state, a homogenous conformation is observed for both nucleotides (see Fig. 3).

3.3. ^{31}P-NMR Applied to Characterize the Specific 2′,3′-Cyclic Phosphodiester Ribozyme Cleavage Product and Local Functional Conformational Changes Within the Active Site

The catalytic mechanism of self-cleavage in naturally occurring RNA ribozymes includes hydrolysis of the RNA phosphodiester backbone (51). This hydrolysis is supposed to proceed via a trans-esterification mechanism, for which the *in line* orientation of the reactive centres is the structural pre-requisite for the cleavage mechanism to proceed. The chemical reaction step of the cleavage process involves nucleophilic attack of the vicinal 2′-hydroxyl on the scissile phosphate, resulting in RNA-ends with a 2′,3′-cyclic phosphodiester and a 5′-hydroxyl group on the 5′- and 3′-site, respectively. Here, we illustrate that ^{31}P-NMR spectroscopy offers the possibility (1) to specifically characterize cleavage of the RNA phosphodiester backbone and (2) to directly observe conformational changes within the active site.

1. Comparison of 1D ^{31}P-NMR spectra of ADHR and the mHHR in the respective pre-cleaved conformation and post-cleavage shows an additional upcoming signal at $\delta(^{31}P) \sim 19.5$ ppm following catalytic cleavage, respectively (see Fig. 1). This ^{31}P chemical shift can be assigned to a 2′,3′-cyclic phosphodiester (52) and is well separated from the ^{31}P-NMR resonances corresponding to the RNA phosphodiester backbone at $\delta(^{31}P) \sim -1.5$ ppm. Thus, this signal represents a direct reporter of one characteristic cleavage product within the active site in the proposed catalytic cleavage mechanisms (see Note 2).

2. NMR correlation experiments that determine the chemical shifts of the bound phosphorous atoms (3D HCP) allow a description of the phosphodiester backbone conformation. The ^{31}P-chemical shifts can be used as a direct measure of the conformations within the RNA backbone angles α and ζ (53). Based on the dependence of chemical shifts on external factors such as temperature or ionic strength preferred metal-binding sites can be inferred (see Note 3).

3.4. Selective ^{13}C-Labelling Enables Monitoring of mHHR Active Site Cleavage Reaction at Atomic Resolution

The catalytic cleavage reaction of mHHR was characterized by real-time NMR experiments. The kinetic traces were recorded in a pseudo-3D dataset. ^{13}C-filtered 1D-^{1}H experiments (128 experiments, with 16 s time resolution) were recorded and provide the blank experimental data prior to cleavage followed by irradiation with a laser pulse of ~1 s to initiate the reaction. After reaction initiation through deprotection of the photo-protected 2′-OH of nucleotide C17, 7 planes with 128 experiments each were recorded. The resulting pseudo-3D dataset (consisting of 1 (pre-cleavage) + 7 (post-cleavage) planes of 128 1D spectra each; (per kinetic point 16 transients were averaged)) was Fourier-transformed with 4,096 data points in the direct dimension after multiplication with a squared cosine window function, resulting in 1,024 ^{13}C-filtered 1D-^{1}H spectra. These were submitted to phasing a polynomial baseline correction and peak integration of the ribose proton signals. The kinetic traces were extracted as an integral over the respective peak half-height over time. For mHHR, starting from the dynamic ensemble of slowly inter-converting conformers in the presence of Mg^{2+}, for all analyzed peaks identical build-up curves showing biphasic behaviour were detected (see Note 4; Fig. 4).

3.5. Divide-and-Conquer Strategy for Imino Proton-Based Characterization of Adenine-Dependent Hairpin Ribozyme Conformations

RNA imino proton signals are direct reporters of base-pairing interactions, for which chemical shifts are sensitive to structural transitions and changes in intra- and intermolecular interactions and were thus here chosen as reporter signals. Structural characterization based on RNA imino protons involves the NMR resonance assignment, including isotope labelling of this sizable RNA. For the 80-nt ADHR, imino proton resonance assignment was

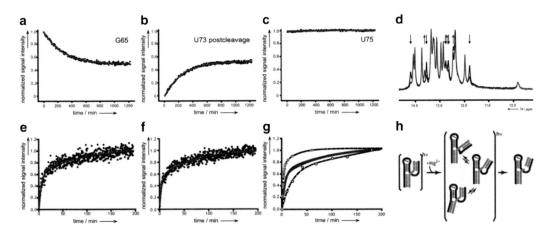

Fig. 4. NMR spectroscopic detected kinetics of ribozyme cleavage reactions; (a–d) time-resolved NMR studies of cofactor-induced cleavage of ADHR; (a–c) normalized signal intensity vs. time (min) of imino proton signals G65 and U73 of ADHR post-cleavage with monoexponential fit (*red solid line*) and U75 (linear fit) extracted from time-resolved NMR kinetic experiments; (d) overlay of 1D NMR spectra of imino proton resonances of ADHR pre-cleavage (without adenine, *black*) and post-cleavage (after incubation with adenine, *red*). *Arrows* indicate signals that are upcoming (*red*) or decreasing (*black*) upon cofactor addition; (e–h) time-resolved NMR studies of mHHR cleavage reaction; (e–f) kinetic traces of two signals extracted from the real-time ^{13}C-filtered NMR experiment conducted on selectively labelled mHHR, the traces of the reaction are best fitted by biexponential behaviour (*red lines*), (g) comparison with kinetic data taken from literature (see text) and based on biochemical assays; here also biexponential behaviour can be fitted reliably; (h) schematic representation of the reaction sequence as a result of static and real-time NMR experiments for mHHR.

performed by standard homo- and heteronuclear NMR assignment strategies (53, 54), including ^{1}H,^{15}N-HSQC (55), 2D-^{1}H,^{1}H-NOESY, HNN-COSY (43), and 3D-^{1}H,^{1}H,^{15}N-NOESY-HSQC (44) experiments. Many of the sequential NOE-based imino proton connectivities for ADHR could be resolved (see Fig. 3). However, due to the size of the RNA construct and its intrinsic characteristics including less structured loop and bulge regions, exclusively canonical base-pairing motifs in the helical regions and exchange processes, analysis of NMR spectra was affected by reduced spectral dispersion, discontinuity of NOE connectivities, and lack of unique assignment starting points for the four individual helices and/or line broadening. These factors led to incomplete sequential connectivities and hindered straightforward resonance assignment.

The *divide-and-conquer* strategy, an approach successfully applied here, is based on the fact that the chemical shift of individual resonances is substantially dependent on sequential and structural context (see Note 5). Thus, the full-length ADHR construct was divided in shorter helical segments, so-called model hairpin RNAs mimicking individual structural elements. These model hairpin RNAs were analyzed separately and the characteristics of their NMR spectra (chemical shifts and NOE connectivities of imino protons) could then reliably be correlated with the spectra of the full-length RNA and thus confirm and complete the resonance assignment of ADHR (see Fig. 3). Based on the imino proton resonance assignment, pre- and post-cleaved ribozyme conformations could be characterized and the cofactor-induced catalytic mechanism could be analyzed (see Subheading 3.5).

3.6. Imino Proton Resonances as Reporter Signals to Follow the Catalytic Reaction of the Adenine-Induced Cleavage of ADHR

RNA conformational changes upon cleavage were analyzed investigating RNA imino proton resonances of ADHR after addition of the catalytically essential cofactor adenine. Comparison of NMR spectra before and after catalytic cleavage reveals significant differences indicating that several imino proton resonances can serve as reporter signals to follow the reaction time course (see Fig. 4). The kinetics associated with the cleavage process occur on a timescale of minutes, and analysis of the NMR signals post-cleavage shows ~52% signal decay, in line with biochemical studies that also observe an equilibrium of ligation and cleavage (25). The kinetics observed for all RNA residues reveals comparable time constants (k (min^{-1})) with individual variations for different residues from the averaged reaction time constant $\Delta k_{mean} < 9\%$. Under the experimental conditions described here, the averaged half-life of the cleavage reaction is $t_{1/2} \sim 187.2 \pm 11.1$ min. Based on the imino proton resonance assignment (see Subheading 3.4), the signals affected during the catalytic cleavage could be assigned to nucleotides positioned around the cleavage site. However, two different cleavage-dependent effects could be detected on the imino proton signals. (1) The imino proton signals in helix 2 (G14, U15, G65) strongly decay in intensity, suggesting these base–pairing interactions to be absent in the post-cleaved conformation. (2) The signals of nucleotides in helix 1 (G9, U73, G74) also decay in intensity but in contrast, resonate at a different chemical shift in the RNA conformation post-cleavage. Interestingly, the neighbouring nucleotide in helix 1 U75 is not affected upon cleavage (see Fig. 4); here, neither chemical shift nor intensity changes are detectable, indicating this structural region to be uniformly stable. In conclusion, the cleavage-dependent structural effects, observed on the imino proton resonances, are likely due to the conformational changes in the structural region nearby the active site, for which a structural rearrangement can be linked with the conformational changes upon catalytic cleavage.

4. Notes

1. The in situ reaction initiation by addition of the cofactor adenine within the NMR spectrometer can be achieved using a rapid mixing device (56, 57) that would enable NMR kinetic data acquisition with much shorter reaction dead times (~ms-s).

2. As ^{31}P-NMR spectra provide unique measures of the active site cleavage reaction, the kinetics of the ribozyme reaction could also be characterized by time-resolved ^{31}P-NMR spectroscopy.

3. Selective ^{13}C-labelling within the sugar moiety in conjunction with HCP correlation experiments can also be used to measure additional parameters such as $^{n}J_{CP}/^{n}J_{HP}$ scalar couplings or $\Gamma_{CH,31P-CSA}$ cross-correlated relaxation rates in order to determine the RNA backbone angles at higher precision.

4. To further characterize the structural changes during catalysis, additional NMR parameters such as ^{1}H-^{13}C-RDCs could be measured in a time-resolved fashion.

5. Model hairpin RNA constructs basically depict spectral characteristics of a full-length RNA construct, if the respective isolated structural region reflects a similar conformation than in the full-length RNA construct. Additional long-range tertiary contacts in the full-length construct might render comparison of spectral characteristics impossible.

Acknowledgments

We thank Marie-Christine Maurel, Stefan Pitsch, Li Yan-Li, Jacques Vergne, and Philipp Wenter for fruitful collaborations. The work was funded by the DFG. H.S. is member of the DFG-funded cluster of excellence: macromolecular complexes.

References

1. Forster, A.C. and R.H. Symons, *Self-cleavage of virusoid RNA is performed by the proposed 55-nucleotide active site*. Cell, 1987. **50**(1): p. 9–16.

2. Prody, G.A., et al., *Autolytic processing of dimeric plant virus satellite RNA*. Science, 1986. **231**(4745): p. 1577–80.

3. Fedor, M.J., *Structure and function of the hairpin ribozyme*. J. Mol. Biol., 2000. **297**(2): p. 269–91.

4. Feldstein, P.A., J.M. Buzayan, and G. Bruening, *Two sequences participating in the autolytic processing of satellite tobacco ringspot virus complementary RNA*. Gene, 1989. **82**(1): p. 53–61.

5. Hampel, A. and R. Tritz, *RNA catalytic properties of the minimum (–)sTRSV sequence*. Biochemistry, 1989. **28**(12): p. 4929–33.

6. Haseloff, J. and W.L. Gerlach, *Sequences required for self-catalysed cleavage of the satellite RNA of tobacco ringspot virus*. Gene, 1989. **82**(1): p. 43–52.

7. Been, M.D. and G.S. Wickham, *Self-cleaving ribozymes of hepatitis delta virus RNA*. Eur. J. Biochem., 1997. **247**(3): p. 741–53.

8. Wu, H.N., et al., *Human hepatitis delta virus RNA subfragments contain an autocleavage activity*. Proc. Natl. Acad. Sci. USA, 1989. **86**(6): p. 1831–5.

9. Lilley, D.M., *The Varkud satellite ribozyme*. RNA, 2004. **10**(2): p. 151-8.

10. Saville, B.J. and R.A. Collins, *A site-specific self-cleavage reaction performed by a novel RNA in Neurospora mitochondria*. Cell, 1990. **61**(4): p. 685–96.

11. Winkler, W.C., et al., *Control of gene expression by a natural metabolite-responsive ribozyme*. Nature, 2004. **428**(6980): p. 281–6.

12. Katahira, M., et al., *Formation of sheared G:A base pairs in an RNA duplex modelled after ribozymes, as revealed by NMR*. Nucleic Acids Res, 1994. **22**(14): p. 2752–9.

13. Simorre, J.P., et al., *A conformational change in the catalytic core of the hammerhead ribozyme upon cleavage of an RNA substrate.* Biochemistry, 1997. **36**(3): p. 518–25.

14. Bondensgaard, K., E.T. Mollova, and A. Pardi, *The global conformation of the hammerhead ribozyme determined using residual dipolar couplings.* Biochemistry, 2002. **41**(39): p. 11532–42.

15. Hammann, C., D.G. Norman, and D.M. Lilley, *Dissection of the ion-induced folding of the hammerhead ribozyme using 19 F NMR.* Proc Natl Acad Sci U S A, 2001. **98**(10): p. 5503–8.

16. Tanaka, Y., et al., *NMR spectroscopic analyses of functional nucleic acids-metal interaction and their solution structure analyses.* Nucleic Acids Symp Ser (Oxf), 2005(49): p. 51–2.

17. Suzumura, K., et al., *Significant change in the structure of a ribozyme upon introduction of a phosphorothioate linkage at P9: NMR reveals a conformational fluctuation in the core region of a hammerhead ribozyme.* FEBS Lett, 2000. **473**(1): p. 106–12.

18. Osborne, E.M., et al., *The identity of the nucleophile substitution may influence metal interactions with the cleavage site of the minimal hammerhead ribozyme.* Biochemistry, 2009. **48**(44): p. 10654–64.

19. Colmenarejo, G. and I. Tinoco, Jr., *Structure and thermodynamics of metal binding in the P5 helix of a group I intron ribozyme.* J Mol Biol, 1999. **290**(1): p. 119–35.

20. Butcher, S.E., F.H. Allain, and J. Feigon, *Solution structure of the loop B domain from the hairpin ribozyme.* Nat. Struct. Biol., 1999. **6**(3): p. 212–6.

21. Hertel, K.J., D. Herschlag, and O.C. Uhlenbeck, *Specificity of hammerhead ribozyme cleavage.* EMBO J, 1996. **15**(14): p. 3751–7.

22. Curtis, E.A. and D.P. Bartel, *The hammerhead cleavage reaction in monovalent cations.* RNA, 2001. **7**(4): p. 546–52.

23. De la Pena, M., S. Gago, and R. Flores, *Peripheral regions of natural hammerhead ribozymes greatly increase their self-cleavage activity.* EMBO J., 2003. **22**(20): p. 5561–70.

24. Fürtig, B., et al., *NMR-spectroscopic characterization of phosphodiester bond cleavage catalyzed by the minimal hammerhead ribozyme.* RNA Biol., 2008. **5**(1): p. 41–8.

25. Meli, M., J. Vergne, and M.C. Maurel, *In vitro selection of adenine-dependent hairpin ribozymes.* J. Biol. Chem., 2003. **278**(11): p. 9835–42.

26. Buck, J., et al., *NMR-spectroscopic characterization of the adenine-dependent hairpin ribozyme.* Chembiochem, 2009. **10**: p. 2100–10.

27. Wenter, P., et al., *Short, synthetic and selectively 13C-labeled RNA sequences for the NMR structure determination of protein-RNA complexes.* Nucleic Acids Res, 2006. **34**(11): p. e79.

28. Kawashima, E. and K. Kamaike, *Synthesis of stable-isotope (C-13 and N-15) labeled nucleosides and their applications.* Mini-Reviews in Organic Chemistry, 2004. **1**(3): p. 309–332.

29. Milecki, J., *Specific labelling of nucleosides and nucleotides with C-13 and N-15.* Journal of Labelled Compounds & Radiopharmaceuticals, 2002. **45**(4): p. 307–337.

30. van Buuren, B.N., et al., *NMR spectroscopic determination of the solution structure of a branched nucleic acid from residual dipolar couplings by using isotopically labeled nucleotides.* Angew Chem Int Ed Engl, 2004. **43**(2): p. 187–92.

31. Lagoja, I.M. and P. Herdewijn, *Chemical synthesis of C-13 and N-15 labeled nucleosides.* Synthesis-Stuttgart, 2002(3): p. 301–314.

32. Milecki, J., et al., *The first example of sequence-specific non-uniformly C-13(5) labelled RNA: Synthesis of the 29mer HIV-1 TAR RNA with C-13 relaxation window.* Tetrahedron, 1999. **55**(21): p. 6603–6622.

33. Quant, S., et al., *Chemical Synthesis of C-13-Labeled Monomers for the Solid-Phase and Template Controlled Enzymatic-Synthesis of DNA and Rna Oligomers.* Tetrahedron Letters, 1994. **35**(36): p. 6649–6652.

34. Palmer, A.G., et al., *Sensitivity Improvement in Proton-Detected 2-Dimensional Heteronuclear Correlation Nmr-Spectroscopy.* Journal of Magnetic Resonance, 1991. **93**(1): p. 151–170.

35. Kay, L.E., et al., *A Gradient-Enhanced Hcch Tocsy Experiment for Recording Side-Chain H-1 and C-13 Correlations in H2o Samples of Proteins.* Journal of Magnetic Resonance Series B, 1993. **101**(3): p. 333–337.

36. Schwalbe, H., et al., *Determination of a Complete Set of Coupling-Constants in C-13-Labeled Oligonucleotides.* Journal of Biomolecular NMR, 1994. **4**(5): p. 631–644.

37. Schleucher, J., et al., *A General Enhancement Scheme in Heteronuclear Multidimensional Nmr Employing Pulsed-Field Gradients.* Journal of Biomolecular NMR, 1994. **4**(2): p. 301–306.

38. Marino, J.P., et al., *A 3-Dimensional Triple-Resonance H-1,C-13,P-31 Experiment - Sequential through-Bond Correlation of Ribose Protons and Intervening Phosphorus Along the Rna Oligonucleotide Backbone.* Journal of the American Chemical Society, 1994. **116**(14): p. 6472–6473.

39. Buck, J., et al., *Time-resolved NMR-spectroscopy: ligand-induced refolding of riboswitches.* Methods of Molecular Biology (Riboswitches, A. Serganov (ed.)), 2009. **540**: p. 161–71.

40. Wenter, P., et al., *Kinetics of photoinduced RNA refolding by real-time NMR spectroscopy.* Angew. Chem. Int. Ed. Engl., 2005. **44**(17): p. 2600–3.

41. Wenter, P., et al., *A caged uridine for the selective preparation of an RNA fold and determination of its refolding kinetics by real-time NMR.* Chembiochem, 2006. 7(3): p. 417–20.

42. Jeener, J., et al., *Investigation of Exchange Processes by 2-Dimensional Nmr-Spectroscopy.* Journal of Chemical Physics, 1979. **71**(11): p. 4546–4553.

43. Dingley, A.J. and S. Grzesiek, *Direct observation of hydrogen bonds in nucleic acid base pairs by internucleotide $^2J_{NN}$ couplings.* J. Am. Chem. Soc., 1998. **120**(33): p. 8293–7.

44. Zhang, O.W., et al., *Backbone H-1 and N-15 Resonance Assignments of the N-Terminal Sh3 Domain of Drk in Folded and Unfolded States Using Enhanced-Sensitivity Pulsed-Field Gradient Nmr Techniques.* Journal of Biomolecular Nmr, 1994. **4**(6): p. 845–858.

45. Liu, M., et al., *Improved WATERGATE pulse sequences for solvent suppression in NMR spectroscopy.* J. Magn. Reson., 1998. **132**: p. 125–9.

46. Sklenar, V. and A. Bax, *A new water suppression technique for generating pure-phase spectra with equal excitation over a wide bandwidth.* J. Magn. Reson., 1987. **75**: p. 378–83.

47. Duss, O., et al., *A fast, efficient and sequence-independent method for flexible multiple segmental isotope labeling of RNA using ribozyme and RNase H cleavage.* Nucleic Acids Res, 2010. **38**(20): p. e188.

48. Lapham, J. and D.M. Crothers, *RNase H cleavage for processing of in vitro transcribed RNA for NMR studies and RNA ligation.* Rna-a Publication of the Rna Society, 1996. 2(3): p. 289–296.

49. Xu, J., J. Lapham, and D.M. Crothers, *Determining RNA solution structure by segmental isotopic labeling and NMR: Application to Caenorhabditis elegans spliced leader RNA 1.* Proceedings of the National Academy of Sciences of the United States of America, 1996. **93**(1): p. 44–48.

50. Buck, J., et al., *Time-resolved NMR methods resolving ligand-induced RNA folding at atomic resolution.* Proc. Natl. Acad. Sci. USA, 2007. **104**(40): p. 15699–704.

51. Fedor, M.J. and J.R. Williamson, *The catalytic diversity of RNAs.* Nat. Rev. Mol. Cell Biol., 2005. **6**(5): p. 399–412.

52. Gorenstein, D.G., *Nucleotide conformational analysis by 31P nuclear magnetic resonance spectroscopy.* Annu Rev Biophys Bioeng, 1981. **10**: p. 355–86.

53. Fürtig, B., et al., *NMR spectroscopy of RNA.* Chembiochem, 2003. **4**(10): p. 936–62.

54. Wijmenga, S.S. and B.N.M. van Buuren, *The use of NMR methods for conformational studies of nucleic acids.* Prog. Nucl. Magn. Reson. Spec., 1998. **32**: p. 287–387.

55. Bodenhausen, G. and D.J. Ruben, *Natural abundance Nitrogen-15 NMR by enhanced heteronuclear spectroscopy.* Chem. Phys. Lett., 1980. **69**: p. 185–9.

56. Mok, K.H., et al., *Rapid sample-mixing technique for transient NMR and photo-CIDNP spectroscopy: applications to real-time protein folding.* J. Am. Chem. Soc., 2003. **125**(41): p. 12484–92.

57. Manoharan, V., et al., *Metal-Induced Folding of Diels-Alderase Ribozymes Studied by Static and Time-Resolved NMR Spectroscopy.* Journal of the American Chemical Society, 2009. **131**(17): p. 6261–6270.

Chapter 13

Deoxyribozyme-Based, Semisynthetic Access to Stable Peptidyl-tRNAs Exemplified by tRNA^Val Carrying a Macrolide Antibiotic Resistance Peptide

Dagmar Graber, Krista Trappl, Jessica Steger, Anna-Skrollan Geiermann, Lukas Rigger, Holger Moroder, Norbert Polacek, and Ronald Micura

Abstract

We present a protocol for the reliable synthesis of non-hydrolyzable 3′-peptidyl-tRNAs that contain all the respective genuine nucleoside modifications. The approach is exemplified by tRNA^Val-3′-NH-VFLVM-NH$_2$ and relies on commercially available *Escherichia coli* tRNA^Val. This tRNA was cleaved site-specifically within the TΨC loop using a 10–23 type DNA enzyme to obtain a 58 nt tRNA 5′-fragment which contained the modifications. After cleavage of the 2′,3′-cyclophosphate moiety from the 5′-fragment, it was ligated to the 18 nt RNA-pentapeptide conjugate which had been chemically synthesized. By this methodology, tRNA^Val-3′-NH-VFLVM-NH$_2$ is accessible in efficient manner. Furthermore, we point out that the approach is applicable to other types of tRNA.

Key words: Dephosphorylation, DNA enzyme, Enzymatic ligation, T4 RNA ligase, tRNA

1. Introduction

Macrolide antibiotics, such as erythromycin, bind close to the entrance of the ribosomal exit tunnel, adjacent to the peptidyl transferase center. Thereby, they hinder elongation of the nascent polypeptide chain (1). It has been shown that translation of short peptides can render the ribosome resistant to macrolide antibiotics and that the amino acid sequence and size of the so-called resistance peptides are critical for their activity. Furthermore, a significant correlation has been observed between different peptide consensus sequences and structurally distinct macrolide antibiotics to which resistance is conferred (2–9). This suggests a direct interaction between the peptide, the drug, and the ribosome. Since the peptide

Jörg S. Hartig (ed.), *Ribozymes: Methods and Protocols*, Methods in Molecular Biology, vol. 848,
DOI 10.1007/978-1-61779-545-9_13, © Springer Science+Business Media, LLC 2012

alone does not confer resistance, a likely scenario is that translation of these peptides is necessary to expel the macrolide antibiotic from the ribosome.

To date, there is little experimental data available addressing this hypothesis. In particular, structural characterization of a ribosome containing a macrolide antibiotic in complex with a tRNA bearing an antibiotic resistance peptide is lacking. To reveal structural details of these interactions would represent a major contribution in terms of understanding the resistance mechanism. For such an undertaking, 3'-peptidyl-tRNA derivatives are required that are stable under the conditions encountered during ribosome crystallographic setups or ribosome probing experiments (see Fig. 1). To achieve such tRNA derivatives, we have recently introduced a flexible approach that tailors natural tRNAs by DNA enzymes in combination with enzymatic ligation using T4 RNA ligase (10, 11).

Most tRNAs contain nucleoside modifications up to position 58 while the following 18 nucleotides at the 3'-terminus remain unmodified. Therefore, we set out for a strategy that delivers the tRNA 5'-fragment with all modifications from wild-type tRNA whereas the 3'-fragment representing the "artificial" component stems from chemical synthesis. In our case, this component is an RNA-peptide conjugate that comprises a stable amide instead of the natural ester linkage. Such conjugates are directly accessible by solid-phase synthesis according to a published procedure (12).

Enzymatic ligation of two tRNA fragments with T4 RNA ligase gives high yields if the site of ligation is positioned within the anticodon loop (see Fig. 2, site I) (13). For our approach, however, the identification of an appropriate ligation site within the TΨC loop (see Fig. 2, sites III or IV) was mandatory as mentioned above. Figure 2 provides an overview of initial experiments using unmodified *E. coli* or *Saccharomyces cerevisiae* tRNA^{Phe} fragments and summarizes the ligation yields that have been achieved. Clearly, ligation between nucleosides 58 and 59 (site IV) turned out to fulfill the requirement of high yields. Other ligation sites residing more upstream in the TΨC loop (site III) or in the variable loop (II) resulted in significantly lower yields (see Fig. 2). Encouraged by these initial insights, we optimized 10–23 type deoxyribozyme-catalyzed cleavage of natural tRNAs between positions 58 and 59 to obtain the tRNA 5'-fragment with all nucleoside modifications and a 2',3'-cyclophosphate moiety. Subsequent dephosphorylation with T4 polynucleotide kinase provided the free 2' and 3' hydroxyl groups of the tRNA 5'-fragment which was then applied in the enzymatic ligation reaction to the synthetic RNA-peptide conjugate.

Exemplarily, this step-by-step protocol describes (1) tRNA^{Val} cleavage, (2) dephosphorylation, and (3) enzymatic ligation to yield tRNA^{Val}-3'-NH-VFLVM-NH$_2$.

**E.coli ribosomes,
23S rRNA**

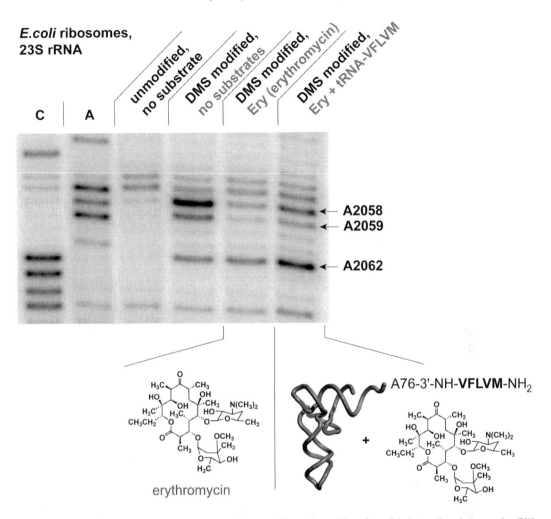

Fig. 1. Non-hydrolyzable 3'-peptidyl-tRNAs are promising candidates for probing of possible interactions between the tRNA peptide chain, the macrolide antibiotic, and the peptidyl transferase center (PTC): preliminary results on chemical probing by dimethylsulfate (DMS) and primer extension analysis of *Escherichia coli* 23S rRNA indicate that the pentapeptide moiety of tRNA^{Val}-3'-NH-VFLVM-NH$_2$ is properly organized within the ribosomal PTC and enters the ribosomal tunnel as indicated by an increased DMS modification at A2062 (9, 25, 26); at the same time, the characteristic erythromycin footprint at A2058 is decreased, but still significantly present when the tRNA conjugate has been added. This may either indicate a conformational rearrangement of the ternary complex or partial release of the drug together with residual protection of A2058 through the peptide chain. In this preliminary experiment, the tRNA^{Val} moiety did not contain the natural nucleoside modifications.

2. Materials

*2.1. Cleavage
of tRNA^{Val}*

1. *E. coli* tRNA^{Val} (*Sigma Aldrich*, R2645): 10 U.

2. 10–23 DNA enzyme: 5'-d(<u>GGT-GGG-TGA-TGA-CGG-GAG</u>-GCT-AGC-TAC-AAC-G<u>AC-GAA-CCG-CCG-ACC-CCC-T</u>)-3' was purchased from *IDT* (grade: desalted) and used without further purification (see Note 1).

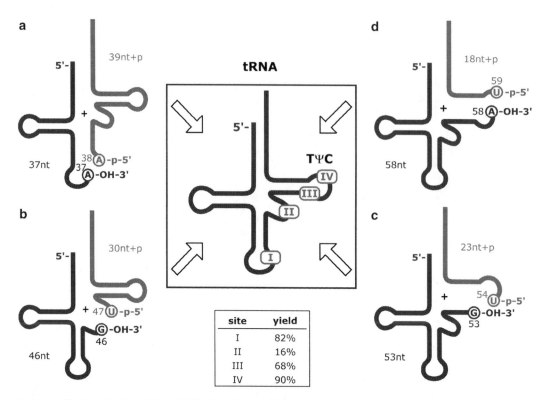

Fig. 2. Identification of potential high-yield ligation sites for tRNA assembly by T4 RNA ligase using chemically synthesized RNA strands. (**a**) Ligation at site I (A37/A38; *E. coli* tRNA[Phe]) yielded 82%, (**b**) site II (G46/U47; *Saccharomyces cerevisiae* tRNA[Phe]) 16%, (**c**) site III (G53/U54; *S. cerevisiae* tRNA[Phe]) 68%, and (**d**) site IV (A58/U59, *E. coli* tRNA[Phe]) 90%. The RNA strands were prepared by automated solid-phase synthesis; the acceptor strands with the 3′-hydroxyl group are denoted in *black* and the donor strands with the 5′-monophosphate in *grey*; the nature of terminal ribonucleotides is denoted.

3. Cleavage buffer: 50 mM Tris–HCl (pH 7.5), 10 mM $MgCl_2$, 10 mM DTT. Dissolve Tris (*Sigma-Aldrich*, 605 mg, 5 mmol) in 90 mL nanofiltered water and adjust to pH 7.5 with HCl_{conc}. Dissolve magnesium chloride hexahydrate (*Sigma-Aldrich*, 203 mg, 1 mmol) and DTT (*Sigma-Aldrich*, 154 mg, 1 mmol) in that solution and finally add nanofiltered water to obtain a final volume of 100 mL; store at 4°C.

4. Water bath and thermostat.

2.2. Dephosphorylation of the tRNA[Val] 5′-Fragment

1. T4 PNK (*Promega*), 10 U/μL in storage buffer (*Promega*): 20 mM Tris–HCl (pH 7.5), 25 mM KCl, 2 mM DTT, 0.1 mM EDTA, 0.1 μM ATP, 50% (v/v) glycerol.

2. 10× T4 PNK reaction buffer: 700 mM Tris–HCl (pH 7.6 at 25°C), 100 mM $MgCl_2$, 50 mM DTT.

3. Vivaspin centrifugal concentrator (*Sartorius Stedim Biotech*).

4. Thermostat.

2.3. Enzymatic Ligation to Yield tRNA*Val*-3′-NH-VFLVM-NH$_2$

1. Acceptor RNA strand: the tRNA 5′-fragment was obtained from the previous steps.

2. Donor RNA strand: the RNA-peptide conjugate 5′-p-UCC-CGU-CAU-CAC-CCA-CCA-3′-NH-VFLVM-NH$_2$ was prepared by solid-phase synthesis according to ref. (12).

3. DNA splint oligonucleotide: 5′-d(TGG-TGG-GTG-ATG-ACG-GCC-GCC-GAC-CCC-CTC-CTT-GTA-AGG)-3′ was purchased from *IDT* (grade: desalted) and was used without further purification (see Note 1).

4. T4 RNA ligase (*Fermentas*), 10 U/μL in storage buffer: 10 mM Tris–HCl (pH 7.5), 1 mM DTT, 50 mM KCl, 0.1 mM EDTA, 50% (v/v) glycerol.

5. 10× T4 RNA ligase reaction buffer (*Fermentas*): 500 mM HEPES-NaOH (pH 8.0 at 25°C), 100 mM MgCl$_2$, 100 mM DTT.

6. 10× ATP (*Fermentas*): 10 mM.

7. 10× BSA (*Fermentas*): 1 mg/mL.

8. Thermostat.

2.4. Purification

1. Phenol solution: A phenol/chloroform/isoamyl alcohol (25/24/1, v/v/v; *Sigma-Aldrich*) solution is extracted 3 times with water and stored in a dark bottle under a water layer at 4°C.

2. Chloroform/isoamyl alcohol (24/1, v/v) solution.

3. HPLC system.

4. Anion exchange column DNAPac PA-100 (*Dionex*): 4×250 mm for HPLC analysis or 9×250 mm for HPLC purification.

5. 250 mM Tris–HCl, pH 8.0, buffer stock solution: Dissolve Tris (*Sigma-Aldrich*, 30.29 g, 250 mmol) in 900 mL nanofiltered water. Adjust to pH 8.0 with HCl$_{conc}$ and add nanofiltered water to obtain a final volume of 1,000 mL.

6. Eluant A: 25 mM Tris–HCl (pH 8.0), 6 M urea. Dissolve urea (*Roth*, 360 g, 6 mol) and 100 mL 250 mM Tris–HCl in 1,000 mL nanofiltered water. Filter the solution through a cellulose acetate filter (*Sartorius Stedim Biotech*, 0.2 μm pore size).

7. Eluant B: 25 mM Tris–HCl (pH 8.0), 6 M urea, 500 mM NaClO$_4$. Dissolve urea (*Roth*, 360 g, 6 mol), sodium perchlorate monohydrate (*Sigma-Aldrich*, 70.23 g, 500 mmol), and 100 mL 250 mM Tris–HCl in 1,000 mL nanofiltered water. Filter the solution through a cellulose acetate filter (*Sartorius Stedim Biotech*, 0.2 μm pore size).

8. Sep-Pak® Plus C18 environmental cartridges (*Waters*).

9. 1 M Triethylammonium bicarbonate (TEAB) buffer: Pour triethylamine (*Fluka*, 101.2 g, 1 mol) in 800 mL nanofiltered water and bubble through CO$_2$ until pH 8 is obtained. Adjust

the volume to 1,000 mL with nanofiltered water. Store the buffer solution at 4°C and dilute to 0.15 M just before use.

10. CH_3CN, CH_3CN/H_2O (1/1, v/v), H_2O; HPLC-grade acetonitrile (*Acros Organics*), nanofiltered water purified on a Millipore-Q System.

2.5. Mass Spectrometry

1. LC-ESI MS system.

2. XTerra®MS C18 (*Waters*): particle size 2.5 μm, 1.0×50 mm.

3. Eluant A: 8.6 mM Et_3N, 100 mM 1,1,1,3,3,3-hexafluoroiso-propanol in H_2O (pH 8.0). Dissolve triethylamine (*Fluka*, 435 mg, 4.3 mmol) in 450 mL nanofiltered water, add 1,1,1,3,3,3-hexafluoroisopropanol (*Fluka*, 8.4 g, 50 mmol) and adjust to a final volume of 500 mL with nanofiltered water. Degas for 5 min by ultrasonication and store at 4°C.

4. Eluant B: methanol (HPLC-grade, *Acros Organics*).

5. 20 mM EDTA solution: Dissolve ethylenediaminetetraacetic acid disodium salt dihydrate (*Fluka*, 74 mg, 0.2 mmol) in 10 mL nanofiltered water.

3. Methods

3.1. Cleavage of tRNA^Val^

3.1.1. General Remarks

For the deoxyribozyme-catalyzed cleavage of tRNA^Val^, a synthetic 10–23 type DNA enzyme (14) with 17 nt long recognition arms (see Fig. 3a, b) is applied. The site of tRNA cleavage has been placed between A58 and U59 for two reasons. First, these nucleosides are highly conserved also in other tRNA species and they represent an established dinucleotide sequence for cleavage by 10–23 DNA enzymes (5′-...AU...-3′) (15, 16). Second, they provide ideal nucleobase termini (purine at the 3′-end of the acceptor strand and pyrimidine at the 5′-end of the donor strand) for high-yield enzymatic ligation using T4 RNA ligase later on (17, 18).

3.1.2. Cleavage Protocol

1. 5 nmol of *E. coli* tRNA^Val^ and 10 nmol of 10–23 DNA enzyme were lyophilized to dryness, in two separate microtubes (see Note 2).

2. Dissolve tRNA and DNA in 62.5 μL cleavage buffer each, and incubate at room temperature for 30 min. Vortex and centrifuge the solutions twice, to ensure complete redissolving (see Note 3).

3. Combine the solutions by adding the DNA enzyme to the tRNA, vortex the mixture for 15 s, and centrifuge for 20 s at room temperature. Take a start sample of 2.5 μL for HPLC analysis.

4. Perform the cleavage reaction using thermal cycling rounds: 1 min at 90°C, 5 min at 60°C, 10 min at 35°C, 10 min at 25°C (three rounds), then 5 min at 60°C, and 80 min at 25°C (see Note 4).

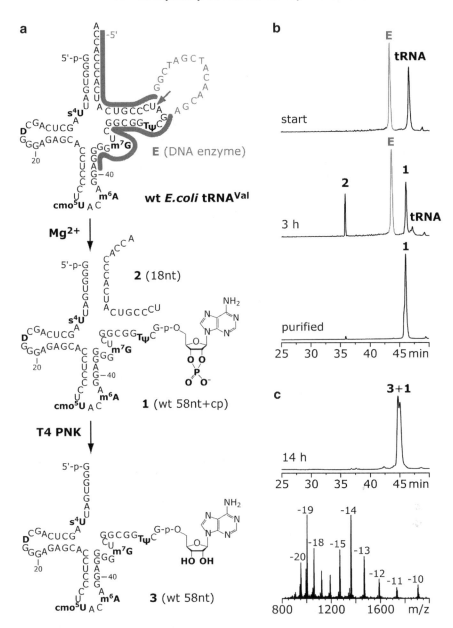

Fig. 3. Cleavage of *E. coli* tRNA^{Val} and dephosphorylation of the tRNA 5′-fragment obtained. (**a**) Reaction scheme: tRNA and tRNA fragments (*black*), 10–23 DNA enzyme (*grey*) for site-specific cleavage between A58/U59 (*arrow*). The cleavage efficiency was 88% and required temperature cycles. After purification of the tRNA 5′-fragment 1, the cyclophosphate moiety was removed by incubation with T4 polynucleotide kinase (PNK). (**b**) Anion-exchange HPLC analysis of the cleavage reaction. The individual traces reflect the sample composition at the beginning (start), at the end (3 h), and after HPLC purification (purified). (**c**) Anion-exchange HPLC analysis of the dephosphorylation reaction using T4 polynucleotide kinase (PNK) shows 50% conversion after 14 h (*upper trace*). Complete product formation of 3 after a reaction time of 28 h was confirmed by LC-ESI MS analysis (*bottom trace*; m/z (calc) = 19,049, m/z (found) = 19,047 ± 5). s^4U 4-thiouridine; *D* dihydrouridine; cmo^5U uridine-5-oxyacetic acid; m^6A N^6-methyladenosine; m^7G 7-methylguanosine; *T* 5-methyluridine; *Ψ* pseudouridine.

5. Take a reaction sample of 2.5 μL for HPLC analysis after each thermal cycling round (see Note 5).

6. Stop the cleavage reaction by freezing the solution in liquid nitrogen (see Note 6).

3.1.3. Purification by Anion-Exchange HPLC

1. Prior to purification, analyze the start and reaction samples taken before (see Subheading 3.1.2, step 5) by anion-exchange chromatography. We recommend the use of *Dionex* DNAPac® PA-100 column (4×250 mm) at 80°C, flow rate: 1 mL/min, gradient: 0–60% eluant B in eluant A within 45 min, UV detection at 260 nm (see Note 7).

2. Purify the tRNA fragment by anion-exchange chromatography. Take aliquots of the reaction solution to avoid column over-loading and start with 20 μL to optimize the gradient.

   We recommend the use of *Dionex* DNAPac® PA-100 column (9×250 mm) at 80°C, flow rate: 2 mL/min, gradient: 28–40% eluant B in eluant A within 20 min, UV detection at 260 nm (see Note 8).

3. Collect and combine product containing fractions (see Note 9).

4. Attach the C18 Sep-Pak® cartridge to a 10 mL polypropylene syringe barrel and treat consecutively with 2× 10 mL CH_3CN, 3× 10 mL CH_3CN/H_2O (1/1, v/v), 3× 10 mL H_2O, and equilibrate with 3× 10 mL 0.15 M TEAB buffer (see Note 10).

5. Dilute product containing fractions with an equal volume of 0.15 M TEAB buffer and load the solution on the conditioned cartridge.

6. Wash with 10 mL TEAB buffer, and then with 3× 5 mL H_2O.

7. Elute product with 3× 10 mL CH_3CN/H_2O (1/1, v/v) into a round bottom flask and remove the solvent in vacuo.

8. Dissolve the product in 1 mL of nanofiltered water (see Note 11).

9. Determine the yield of isolated product as units of optical density at 260 nm (OD_{260}) by UV spectroscopy at room temperature. To calculate the oligonucleotide concentration according to *Lambert Beer's* law, approximate the extinction coefficient ε of the product as the sum of the individual extinction coefficients of the nucleotides contained.

3.2. Dephosphorylation of the tRNA^{Val} 5′-Fragment

3.2.1. General Remarks

The tRNA^{Val} 5′-fragment which is obtained by 10–23 DNA enzyme cleavage carries a 2′,3′-cyclophosphate group. This terminus cannot be used as substrate for the intended enzymatic ligation to the synthetic 3′-peptidylamino-RNA conjugate using T4 RNA ligase. Therefore, T4 polynucleotide kinase (T4 PNK), which is known to have 3′-phosphatase activity to release the cyclophosphate, is applied and provides the terminal ribose moiety with free 2′-OH and 3′-OH groups (19) (see Fig. 3a, c).

3.2.2. Dephosphorylation Protocol

1. Take 4.2 nmol of tRNA 5'-fragment carrying the 2',3'-cyclo-phosphate moiety out of the stock solution and load it on a Vivaspin centrifugal concentrator (see Note 12).

2. Treat the RNA sample consecutively with an equal volume of nanofiltrated water, 5× T4 PNK reaction buffer, 2.5× T4 PNK reaction buffer, 1.25× T4 PNK reaction buffer, and 1× T4 PNK reaction buffer. Centrifuge after each treatment at room temperature (see Note 13).

3. Adjust the volume to 266 μL with 1× T4 PNK reaction buffer, vortex the solution for 5 s, and centrifuge for 10 s.

4. Start the dephosphorylation reaction by adding 14 μL of T4 PNK (10 U/μL in storage buffer) to the solution. Vortex carefully (tip gently) and centrifuge for up to 6 s if necessary (see Note 14).

   In the dephosphorylation reaction the final RNA concentration is 15 μM and the enzyme concentration is 0.5 U/μL in a final volume of 280 μL.

5. Incubate the reaction solution for 14 h at 37°C.

6. Add again 14 μL of T4 PNK (10 U/μL in storage buffer) to the solution (final concentration of 1.0 U/μL) and incubate for further 14 h at 37°C (see Note 15).

3.2.3. Purification by Phenol Extraction

1. Add an equal volume of phenol solution to the reaction sample, vortex the mixture for 30 s, and centrifuge for 1 min at room temperature.

2. Transfer the upper aqueous layer in a second microtube and repeat step 1.

3. Transfer the upper aqueous layer in a third microtube and add an equal volume of chloroform/isoamyl alcohol (24/1, v/v) solution, vortex the mixture for 30 s, and centrifuge for 1 min at room temperature.

4. Transfer the upper aqueous layer in a fourth microtube and repeat the previous step to avoid accumulation of phenol in the product containing aqueous layer.

5. Transfer the upper aqueous layer in a fifth microtube.

6. To maximize the overall yield two, successive extraction steps might be performed as outlined: Add half the volume of nano-filtered water to the organic layer in microtube one, vortex the mixture for 30 s, and centrifuge for 1 min at room temperature. Transfer the aqueous layer to the phenol solution of the second microtube and repeat vortexing and centrifugation. Transfer the aqueous layer to the chloroform/isoamyl alcohol solution of the third microtube and after vortexing and centrifugation transfer it to the organic layer of the fourth microtube.

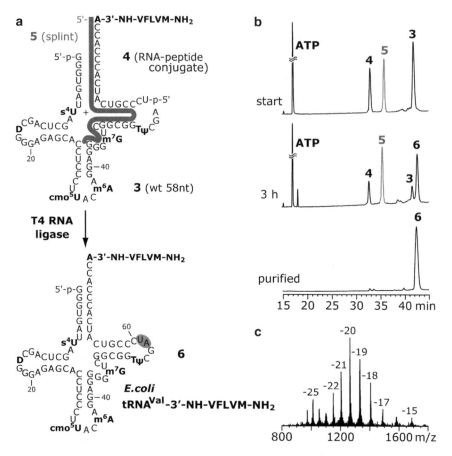

Fig. 4. Enzymatic ligation of tRNA^{Val}-3′-VFLVM-NH$_2$ 6 using T4 RNA ligase and a DNA splint. (a) Reaction scheme: tRNA fragment 3 and RNA-peptide conjugate 4 (*black*), DNA splint (*grey*) for the formation of a stable pre-ligation complex. (b) Anion-exchange HPLC analysis of the ligation reaction with the sample composition at the beginning (start), at the end (3 h), and after HPLC purification (purified). (c) Characterization of the 3′-peptidyl-tRNA product 6 by LC-ESI mass spectrometry (*m/z* (calc) = 25,287, *m/z* (found) = 25,286 ± 5).

Repeat vortexing and centrifugation and add the upper layer to the aqueous solution of the fifth microtube.

7. Lyophilize the aqueous solution to a volume of 100 μL.

3.3. Enzymatic Ligation to Yield tRNA^{Val}-3′-NH-VFLVM-NH$_2$

3.3.1. General Remarks

The ligation of the tRNA^{Val} 5′-fragment to the synthetic conjugate, 5′-p-UCC-CGU-CAU-CAC-CCA-CCA-3′-NH-VFLVM-NH$_2$, relies on T4 RNA ligase. A 39 nt DNA splint is used to form a sufficiently stable pre-ligation complex that brings the 5′-phosphate of the donor strand (synthetic conjugate) in close proximity to the 3′-OH group of the acceptor strand (58 nt tRNA fragment) (see Fig. 4). This is a prerequisite for efficient ligation with T4 RNA ligase (17, 18, 20–24).

3.3.2. Ligation Protocol

1. Take equimolar amounts (2 nmol) of donor and acceptor strands and DNA splint from their respective stock solutions, mix in a microtube, and lyophilize to dryness in vacuo.

2. Dissolve the sample in 36.2 μL nanofiltered water and incubate at room temperature for 30 min. Vortex and centrifuge the solution twice, to ensure complete resuspension.

3. Heat the solution to 90°C for 2 min and then allow to cool to room temperature (~21°C) within 15 min.

4. Add 2.5 μL 10× ATP to the solution, vortex the mixture for 5 s, and centrifuge for 10 s at room temperature. Take a start sample of 2 μL for HPLC analysis.

5. Add 5 μL of 10× T4 RNA ligase reaction buffer and 5 μL of 10× BSA to the solution, vortex the mixture for 5 s, and centrifuge for 10 s.

6. Start the reaction by adding 1.3 μL of T4 RNA ligase (10 U/μL in storage buffer) to the solution. Vortex carefully (tip gently) and centrifuge for 6 s if it's necessary (see Note 14).

   In the ligation reaction the final oligonucleotide concentration is 40 μM for each fragment and the final ligase concentration is 0.25 U/μL in a final volume of 50 μL.

7. Incubate the reaction solution for 3 h at 37°C (see Note 16).

8. Purify the reaction solution by phenol extraction, analogue as described in Subheading 3.2.3; except step 7: Lyophilize the aqueous solution to a volume of approximately 1 mL.

9. Isolate the ligation product by anion-exchange HPLC, analogue as described in Subheading 3.1.3. Different HPLC purification parameters have to be used: *Dionex* DNAPac® PA-100 column (9× 250 mm) at 60°C, flow rate: 2 mL/min, gradient: 31–44% eluant B in eluant A within 20 min, UV detection at 260 nm.

3.4. Characterization by Mass Spectrometry

1. Take 100–160 pmol of the RNA stock solution and lyophilize to dryness.

2. Dissolve the RNA in 30 μL of 20 mM EDTA solution.

3. Analyze the sample by LC-ESI mass spectrometry (negative ion mode) at 21°C using following conditions: flow rate: 30 μL/min; gradient: 0–100% eluant B in eluant A within 20 min; UV detection at 254 nm. Prior to each injection, column equilibration was performed by purging with buffer A for 30 min.

4. Notes

1. Check the thermodynamic stability of DNA/tRNA secondary structures using the freely available software packages "RNAfold (Webserver)" or "Mfold". Melting temperatures of the tRNA/DNA duplexes should be in a range of 60–70°C.

2. To avoid difficulties in redissolving the tRNA, this procedure can be replaced by a concentration step using a Vivaspin

centrifugal concentrator, followed by a washing step with cleavage buffer.

3. Due to the separate handling, a defined starting point of the cleavage reaction is obtained.

4. Denaturation at 90°C is performed in a water bath. 60°, 35°, and 25°C are obtained using a thermostat. Indicated minutes referred to the reaction period at the corresponding temperature; the sample is kept in the thermostat throughout all thermostatization steps (60°–35°C in 5 min and 35°–25°C in 2 min).

5. Analyze the sample during the cleavage reaction to monitor the progress.

6. At this point the sample might be stored at –20°C for several days.

7. If insufficient cleavage is observed, vary the temperature cycles or design longer DNA enzyme recognition arms.

8. Be aware that the gradient is dependent on the column age and the oven temperature and might significantly differ for individual HPLC setups. Therefore, no absolute values of gradients are provided here.

9. Store the product containing fractions at 4°C until they are desalted all together.

10. Applying pressure, which results in a flow rate not exceeding one drop per second, is recommended. Avoid running the cartridge dry.

11. Stock solutions stored at –20°C are stable for several months.

12. Be sure to select a model with a molecular weight cutoff of at least 50% smaller than the molecular size of the RNA sample.

13. High concentrations of ammonium ions, phosphate ions, KCl, or NaCl inhibit the PNK enzyme and decrease the dephosphorylation rate and/or lead to incomplete conversion.

14. Always keep the enzymes (T4 PNK and T4 RNA ligase) at –20°C and tip gently before use.

15. This step is performed to achieve quantitative conversion.

16. To monitor the progress, take a reaction sample of 2.5 μL every hour and analyze it after phenol extraction on HPLC.

Acknowledgments

We thank the Austrian Science Fund FWF (P21641, I317 to R.M.; Y315 to N.P.) and the Ministry of Science and Research (GEN-AU project "Non-coding RNAs" P0726-012-012 to R.M.; P110420-012-012 to N.P.) for funding.

References

1. Mankin, A. S. (2008) Macrolide myths. *Curr. Opin. Microbiol.* **11**, 414–421.

2. Tenson, T., Mankin, A. S. (2006) Antibiotics and the ribosome. *Mol. Microbiol.* **59**, 1664–1677.

3. Vimberg, V., Xiong, L., Bailey, M., Tenson, T., Mankin, A. S. (2004) Peptide-mediated macrolide resistance reveals possible specific interactions in the nascent peptide exit tunnel. *Mol. Microbiol.* **54**, 376–385.

4. Tenson, T., Mankin, A. S. (2001) Short peptides conferring resistance to macrolide antibiotics. *Peptides* **22**, 1661–1668.

5. Tripathi, S., Kloss, P. S., Mankin, A. S. (1998) Ketolide resistance conferred by short peptides. *J. Biol. Chem.* **273**, 20073–20077.

6. Tenson, T., Xiong, L., Kloss, P., Mankin, A. S. (1997) Erythromycin resistance peptides selected from random peptide libraries. *J. Biol. Chem.* **272**, 17425–17430.

7. Dam, M., Douthwaite, S., Tenson, T., Mankin, A. S. (1996) Mutations in domain II of 23S rRNA facilitate translation of a 23S rRNA-encoded pentapeptide conferring erythromycin resistance. *J. Mol. Biol.* **259**, 1–6.

8. Tenson, T., DeBlasio, A., Mankin, A. S. (1996) A functional peptide encoded in the Escherichia coli 23S rRNA. *Proc. Natl. Acad. Sci. USA* **93**, 5641–5646.

9. Ramu, H., Vázquez-Laslop, N., Klepacki, D., Dai, Q., Piccirilli, J., Micura, R., Mankin, A. S. (2011) Nascent peptide in the ribosome exit tunnel affects functional properties of the A-site of the peptidyl transferase center. *Mol. Cell* **41**, 321–330.

10. Graber, D., Moroder, H., Steger, J., Trappl, K., Polacek, N., Micura, R. (2010) Reliable semi-synthesis of hydrolysis-resistant 3'-peptidyl-tRNA conjugates containing genuine tRNA modifications. *Nucl. Acids Res.* **38**, 6796–6802.

11. Steger, J., Graber, D., Moroder, H., Geiermann, A.-S., Aigner, M., Micura, R. (2010) Efficient access to nonhydrolyzable initiator tRNA based on the synthesis of 3'-azido-3'-deoxyadenosine RNA. *Angew. Chem. Int. Ed.* **49**, 7470-7472.

12. Moroder, H., Steger, J., Graber, D., Fauster, K., Trappl, K., Marquez, V., Polacek, N., Wilson, D. N., Micura, R. (2009) Non-hydrolyzable RNA-peptide conjugates: a powerful advance in the synthesis of mimics for 3'-peptidyl tRNA termini. *Angew. Chem. Int. Ed.* **48**, 4056–4060.

13. Sherlin, L. D., Bullock, T. L., Nissan, T. A., Perona, J. J., Lariviere, F. J., Uhlenbeck, O. C., Scaringe, S. A. (2001) Chemical and enzymatic synthesis of tRNAs for high-throughput crystallization. *RNA* **7**, 1671–1678.

14. Santoro, S. W., Joyce, G. F. (1997) A general purpose RNA-cleaving DNA enzyme. *Proc. Natl. Acad. Sci. USA* **94**, 4262–4266.

15. Höbartner, C., Silverman, S. K. (2007) Recent advances in DNA catalysis. *Biopolymers* **87**, 279–292.

16. Silverman, S. K., Baum, D. A. (2009) Use of Deoxyribozymes in RNA Research. *Meth. Enzymol.* **469**, 95–117.

17. Persson, T., Willkomm, D. K., Hartmann, R. K. (2005) T4 RNA ligase. In *Handbook of RNA Biochemistry*, Eds. Hartmann, R. K., Bindereif, A., Schön, A., Westhof, E., Wiley-VCH, Weinheim, Germany, pp 53–74.

18. Arn, E. A., Abelson, J. (1998) RNA ligases: Function, mechanism, and sequence conservation. In *RNA Structure and Function*, Eds. Simons, R. W., Grunberg-Manago, M., CSHL Press, New York, USA, pp. 695–726.

19. Schürer, H., Lang, K., Schuster, J., Mörl, M. (2002) A universal method to produce in vitro transcripts with homogeneous 3' ends. *Nucl. Acids Res.* **30**, e56.

20. Lang, K., Micura, R. (2008) The preparation of site-specifically modified riboswitch domains as an example for enzymatic ligation of chemically synthesized RNA fragments. *Nat. Protoc.* **3**, 1457–1466.

21. Höbartner, C., Micura, R. (2004) The chemical synthesis of selenium-modified oligoribonucleotides and their enzymatic ligation leading to an U6 snRNA stem-loop segment. *J. Am. Chem. Soc.* **126**, 1141–1149.

22. Höbartner, C., Rieder, R., Kreutz, C., Puffer, B., Lang, K., Polonskaia, A., Serganov, A. Micura, R. (2005) Syntheses of RNAs with up to 100 nucleotides containing site-specific 2'-methylseleno labels for use in X-ray crystallography. *J. Am. Chem. Soc.* **127**, 12035–12045.

23. Rieder, R., Höbartner, C., Micura, R. (2009) Enzymatic ligation strategies for the preparation of purine riboswitches with site-specific chemical modifications. In *Riboswitches*, Ed. Serganov, A. *Meth. Mol. Biol.* **540**, 15-24.

24. Rieder, R., Lang, K., Graber, D., Micura, R. (2007) Ligand-induced folding of the *adenosine deaminase* A-riboswitch and implications on riboswitch translational control. *ChemBioChem* **8**, 896–902.

25. Vázquez-Laslop, N., Ramu, H., Klepacki, D., Kannan, K., Mankin, A. S. (2010) The key function of a conserved and modified rRNA residue in the ribosomal response to the nascent peptide. *EMBO J.* **29**, 3108–3117.

26. Hansen, L. H., Mauvais, P., Douthwaite, S. (1999) The macrolide-ketolide antibiotic binding site is formed by structures in domains II and V of 23 S ribosomal RNA. *Mol. Microbiol.* **31**, 623–631.

Chapter 14

Probing Functions of the Ribosomal Peptidyl Transferase Center by Nucleotide Analog Interference

Matthias D. Erlacher and Norbert Polacek

Abstract

The ribosome is a huge ribonucleoprotein complex in charge of protein synthesis in every living cell. The catalytic center of this dynamic molecular machine is entirely built up of 23S ribosomal RNA and therefore the ribosome can be referred to as the largest natural ribozyme known so far. The in vitro reconstitution approach of large ribosomal subunits described herein allows nucleotide analog interference studies to be performed. The approach is based on the site-specific introduction of nonnatural nucleotide analogs into the peptidyl transferase center, the active site located on the interface side of the large ribosomal subunit. This method combined with standard tests of ribosomal functions broadens the biochemical repertoire to investigate the mechanism of diverse aspects of translation considerably and adds another layer of molecular information on top of structural and mutational studies of the ribosome.

Key words: Ribosome, In vitro reconstitution, Protein synthesis, 23S rRNA, Translation fidelity, Nucleotide analog interference

1. Introduction

Protein synthesis is a complex and energy-intensive process. About 50% of the cell's energy resources are dedicated to translation (1). The ribosome is the central protagonist of protein biosynthesis and translates the genetic information, stored in the form of messenger RNA (mRNA) codons, into the corresponding amino acid sequences of proteins. The amino acids are delivered to the ribosome as aminoacyl-tRNAs complexed with elongation factor Tu (EF-Tu) and GTP. This ternary complex is recognized by the ribosome's A-site in a codon-dependent manner, a reaction called decoding, and the amino acid is subsequently incorporated into the growing nascent peptide chain (2). This process is performed by

Jörg S. Hartig (ed.), *Ribozymes: Methods and Protocols*, Methods in Molecular Biology, vol. 848,
DOI 10.1007/978-1-61779-545-9_14, © Springer Science+Business Media, LLC 2012

the ribosome with an impressing speed and accuracy. By catalyzing 15–20 peptide bond/s (3), the error rate is in the range of 10^{-3} to 10^{-4} (4–6). Another remarkable feature of the ribosome is its catalytic center. The active site is composed of 23S ribosomal RNA (rRNA) only (7) and therefore makes the ribosome the largest so far known ribozyme (8). The peptidyl transferase center (PTC), where the two principal chemical reactions of protein synthesis are catalyzed, namely peptide bond formation and peptidyl-tRNA hydrolysis, is packed with universally conserved nucleotides of domain V of the 23S rRNA (7, 9).

Although high-resolution crystal structures revealed the three-dimensional architecture of the active site and many biochemical and mutational approaches were applied to elucidate the mechanism of translational processes, molecular details of the reactions are still not fully understood at the molecular level (reviewed in refs. (1, 10)). Standard mutational studies seemed to be insufficient to unravel the contribution of specific 23S rRNA functional groups of PTC residues to ribosomal catalysis. Therefore, a novel in vitro genetics approach was established that allows the manipulation of 23S rRNA residues at the functional group or even single atom level (11, 12). Key to this approach is the use of circularly permuted 23S rRNA (cp-23S rRNA) transcripts in an in vitro assembly reaction of the *Thermus aquaticus* large ribosomal subunit (50S). The cp-23S rRNA is constructed in such a way that its new ends are positioned near the site of interest whereas a short sequence gap is introduced around the 23S rRNA residue under investigation. This missing rRNA fragment is chemically synthesized to contain any desired nucleotide analog and is provided in trans during the 50S in vitro reconstitution approach (see Fig. 1).

This procedure, also referred to as "gapped-cp-reconstitution," has so far been successfully applied to study transpeptidation (10, 13, 14), peptidyl-tRNA hydrolysis (15), EF-G GTPase activation (16, 17), and tRNA translocation (18). While the gapped-cp-reconstituted has revealed the importance of 23S rRNA backbone groups for ribosomal catalysis, the ribose 2'-OH at A2451 for peptide bond formation, and the ribose at A2602 for peptidyl-tRNA hydrolysis, the contribution of the universally conserved PTC nucleobases to protein synthesis remained obscure (10). It is possible that the nucleobases in the PTC are critical for processes distinct from catalysis, such as the fast and productive accommodation of the aminoacyl-tRNA 3'-ends into the 50S A-site after the acceptor end has been released from EF-Tu. Indeed contribution of 23S rRNA nucleotides for tRNA accommodation and translation fidelity has been proposed (6, 19). Here, we report on the procedure to generate chemically engineered ribosomes carrying nucleotide modifications in the PTC to study translation fidelity.

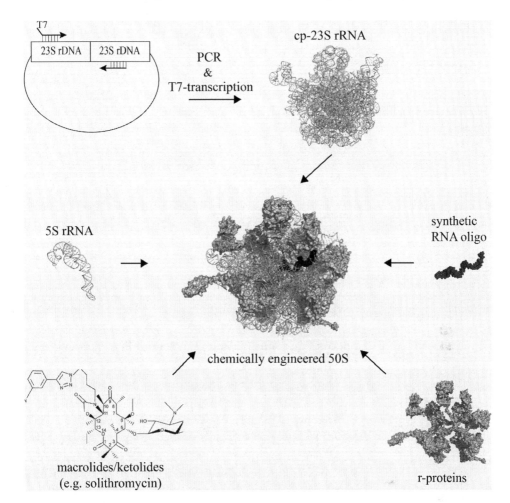

Fig. 1. Flowchart for the preparation of gapped-cp-reconstituted ribosomes. The cp-23S rRNA is produced by in vitro transcription from the cp-23S rDNA template using T7 RNA polymerase. The cp-23S rDNA template is generated by PCR (PCR primers are depicted as *arrows*) using a plasmid carrying tandemly repeated 23S rRNA genes from *T. aquaticus*. The cp-23S rRNA is designed so that a short sequence gap is generated encompassing the residue under investigation. The 5S rRNA is transcribed from a linearized vector and combined with the cp-23S rRNA, the compensating synthetic RNA oligo, and the total protein fraction of *T. aquaticus* 50S ribosomal subunits (TP50) in the presence of an antibiotic from the macrolide/ketolide family to generate chemically engineered 50S subunits by in vitro reconstitution.

2. Materials

2.1. PCR Reaction to Generate the Circularly Permuted 23S rDNA

1. PCR buffer (10×): 100 mM Tris/HCl (pH 9.5), 200 mM $(NH_4)_2SO_4$, 15 mM $MgCl_2$, 10 mM DTT, 0.05% Nonidet P-40 (see Notes 1 and 2).

2.2. Transcription

1. RiboMAX Large Scale RNA Production System-T7 (Promega).

2. Sephadex G-50 Fine (GE Healthcare) is suspended in 0.3 M NaOAc pH 5.5 and subsequently autoclaved and stored at room temperature (RT).

3. NaOAc (Roth) is dissolved in water to a concentration of 3 M and the pH is set to 5.5 using glacial acetic acid (Roth) and kept at RT.

4. Phenol/Choloform/Isoamylalcohol (25/24/1) (Roth) stored at 4°C.

2.3. Agarose Gel

1. TAE buffer (10×): 24.2 g Tris, 6.6 g NaOAc, and 1.85 g of EDTA are dissolved in water and the pH is adjusted to ~8 using glacial acetic acid and stored at RT.

2. Running buffer (1× TAE): The 10× buffer is diluted to 1× and 20 μL of a 10 mg/mL ethidium bromide solution is added to 500 mL buffer and stored at RT.

3. 1% (w/v) agarose gel is made by using the running buffer containing ethidium bromide.

2.4. 50S In Vitro Reconstitution

1. Reconstitution buffer (10×): 100 mM Tris/HCl (pH 7.4), 2 M NH$_4$Cl, 20 mM MgCl$_2$, 1 mM EDTA, 25 mM β-mercaptoethanol in water.

2. Spermidine solution (70 mM): spermidine (Fluka) is dissolved in water.

3. Antibiotics (6 mM): solithromycin (CEMPRA), telithromycin (LGM Pharma), azithromycin (Fluka), or roxythromycin (Fluka) are dissolved in 100% ethanol to a final concentration of 6 mM (see Note 3).

4. RNA oligomers: Commercially available modified and unmodified RNA oligonucleotides were obtained from Dharmacon or Microsynth. Particular nonnatural RNA building blocks were synthesized and incorporated using the 2'-O-TOM chemistry (20, 21).

2.5. Puromycin Reaction

1. HSM buffer (10×): 200 mM Hepes pH 7.6, 0.5 mM spermine, 40 mM β-mercaptoethanol.

2. Puromycin solution: 2.91 mg puromycin is dissolved in 500 μL 1× binding buffer and the pH is adjusted to 7.5 using KOH.

3. Binding buffer (10×): 200 mM Hepes pH 7.6, 1.5 M NH$_4$Cl$_2$, 40 mM β-mercaptoethanol, 20 mM spermidine, 0.5 mM spermine, 60 mM MgCl$_2$.

4. Ac-Phe-tRNA[Phe]: the peptidyl-tRNA analog can be prepared according to published protocols (11).

5. Poly(U) messenger RNA analog: poly(U) was purchased from Sigma and dissolved in water to a concentration of 20 mg/mL.

6. Scintillation cocktail: Rothiszint eco plus (Roth).

2.6. Poly(U)-Directed Poly(Phe) Synthesis and Misincorporation Assay

1. Translation buffer (10×): 200 mM Hepes pH 7.6, 45 mM MgAc$_2$, 1,500 mM NH$_4$Ac, 40 mM β-mercaptoethanol, 20 mM spermidine, 0.5 mM spermine.

2. ATP solution (100 mM): ATP disodium salt (Roche) is dissolved in 100 mM Tris pH 9.4.

3. GTP solution (100 mM): GTP disodium salt (Roche) is dissolved in 100 mM Tris pH 9.4.

4. Acetylphosphate (300 mM): acetylphosphate is dissolved in water.

5. L-phenylalanine (5 mM): phenylalanine (ICN Biomedicals) is dissolved in water.

6. L-phenylalanine [ring-^{14}C] purchased from Hartmann Analytic.

7. tRNA^{Phe} and tRNA^{bulk}: purified tRNA^{Phe} (Sigma) and tRNA^{bulk} (Sigma) are dissolved in water.

8. L-leucine (500 μM): leucine (Sigma) is dissolved in water.

9. L-[^{3}H] leucine purchased from Hartmann Analytic.

10. S100: The *E. coli* post-ribosomal supernatant (S100) preparation is performed according to ref. (22) and stored in aliquots at –80°C.

11. Native *E. coli* or *T. aquaticus* ribosomal subunits are prepared via 10–40% sucrose gradients and stored in small aliquots at –80°C (23).

12. Glass microfibre filters GF/C: obtained from Whatman.

3. Methods

The herein described method allows the design and generation of a gapped circularly permuted 23S rRNA and the subsequent reconstitution to functional 50S ribosomal subunits from *T. aquaticus*. Using a PCR reaction new 5′ and 3′ ends, and consequently a sequence gap, can be introduced into the 23S rDNA at the position desired (see Fig. 1). This PCR product is then used as template for a T7 RNA polymerase-driven transcription to generate cp-23S rRNA. The cp-23S rRNA is subsequently combined with in vitro transcribed 5S rRNA, the total proteins of the large ribosomal subunit (TP50), and a synthetic RNA oligomer (20, 21). The synthetic RNA is designed to fill the gap of the cp-23S rRNA and carries the desired nucleotide modification. The 50S subunits can be reassociated with highly purified small ribosomal subunits to form 70S particles. These particles can then be tested, e.g., in the puromycin reaction for their capability to catalyze transpeptidation or in translation assays such as the poly(U)-directed poly(Phe) synthesis to monitor their overall performance in

Table 1
Activities of chemically engineered ribosomes carrying nonnatural nucleotide analogs at A2451

Modification	Peptide bond formation $(k_{rel})^a$	Poly(Phe) synthesis $(k_{rel})^a$
Wild type	1.00	1.00
2'-deoxyribose abasic	<0.01[b]	0.02[d]
Ribose abasic	0.53[b]	0.71[d]
3-deazaadenosine	0.77[b]	0.65
2'-deoxyadenosine	0.11[b]	0.11
2'-fluoroadenosine	0.02[c]	0.06[d]
2'-aminoadenosine	1.02[c]	0.73[d]

[a] The initial rates in peptide bond formation and poly(U)-dependent poly(Phe) in vitro translation were determined from experimental points in the linear range of the reactions. All rates (k_{rel}) were normalized to the rate observed with gapped-cp-reconstituted ribosomes carrying the synthetic wild-type 26-mer in the PTC, which was taken as 1.00
[b] Data taken from (11)
[c] Data taken from (14)
[d] Data taken from (12)

protein synthesis (see Table 1). In addition the translation fidelity of reconstituted particles carrying modifications in the PTC can be investigated using a misincorporation assay based on the poly(Phe) synthesis (see Fig. 2). Depending on the design of the cp-23S rRNA, the reconstituted particles can show varying activities in different assays for testing ribosomal functions (12).

3.1. PCR to Generate the Gapped-cp-rDNA

1. To generate the template for the transcription of the gapped-cp-23S rRNA, a PCR is performed using the plasmid pCPTaq23S carrying a tandem repeat of the 23S rDNA gene as template (11). The PCR primers define the desired new 5' and 3' ends of the cp-23S rDNA (see Fig. 1). The forward primer carries the T7 promoter sequence (TAATACGACTCACTATA) to allow subsequent in vitro transcription.

2. The PCR reaction contains 1 µM forward and reverse primers, respectively, 100–150 ng of the pCPTaq23S plasmid, 200 µM of each dNTP, and 1–2 µL of self-prepared Taq DNA polymerase. The volume of the reaction is usually 100 µL.

3. Cycle the PCR reaction using the following conditions: (step 1) 5' at 92°C, (step 2) 25" at 92°C, (step 3) 25" between 50 and 68°C (depends on the Tm of the primers used), (step 4) 3' 40" at 72°C. steps 2–5 are repeated 29 times.

4. The PCR product is verified on a 1% agarose gel and subsequently purified using a PCR purification kit (Quiagen).

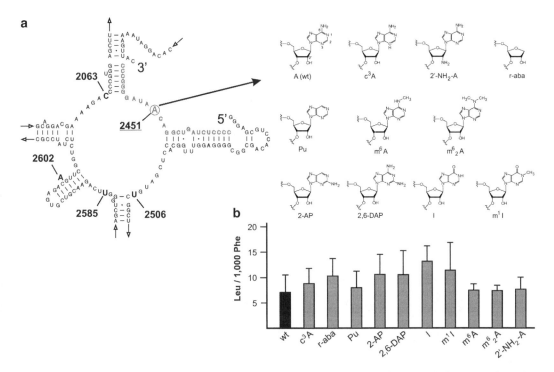

Fig. 2. Translation fidelity of ribosomes carrying nonnatural modifications at the PTC nucleotide A2451. (a) Secondary structure of the PTC of gapped-cp-reconstituted 50S subunits showing the newly introduced endpoints of the cp-23S rRNA (5′ and 3′, respectively). The chemically synthesized 26-nucleotide RNA, which compensates for the missing rRNA fragment, is shown in *gray*. Inner core PTC residues are indicated and nucleotide A2451 is highlighted. The chemical structures of nucleotide analogs that were introduced at 23S rRNA position 2451 are depicted on the *right*. (b) The translation fidelity was elucidated by measuring the amount of the misincorporated [³H]-leucine into a poly([¹⁴C]Phe) chain during a poly(U)-directed in vitro translation assay. The error rates were calculated from at least three independent experiments and are given as [³H]Leu per 1,000 incorporated [¹⁴C]Phe residues. None of the tested nucleoside analogs at A2451 (as well as those introduced at A2602 or U2506; data not shown) significantly altered the error rate of the chemically engineered ribosomes. Tested nucleoside analogs: c^3A 3-deaza-adenosine; $2′-NH_2-A$ 2′-amino-adenosine; *r-aba* ribose-abasic analog; *Pu* purine; m^6A N6-methyladenosine; m^6_2A N6,N6-dimethyladenosine; *2-AP* 2-aminopurine; *2,6-DAP* 2,6-diaminopurine; *I* inosine; m^1I 1-methylinosine.

3.2. Transcription of cp-23S rRNA and 5S rRNA

1. The *T. aquaticus* 5S RNA is transcribed in vitro from a plasmid carrying the 5S rRNA gene (23) and the gapped cp-23S rRNA from the PCR product described in Subheading 3.1.

2. The transcription is performed in 100 μL using a Promega kit according to the provided manual. After 3–4 h of incubation at 37°C, the template DNA is digested using 8 U of RQ1 DNAse (provided with the kit) for 15 min at 37°C.

3. The transcript is purified by gravity flow using a self-prepared Sephadex G50 column (2 mL) and 0.3 M NaOAc as running buffer. The fractions containing the RNAs are pooled and after phenol/chloroform/isoamylalcohol extraction, precipitated with 3 volumes of EtOH.

4. The concentration of the RNA is determined by measuring the OD_{260} using a NanoDrop spectrophotometer, and the quality is checked by running an aliquot on a 1% agarose gel (see Note 4).

3.3. Reconstitution of the 50S Ribosomal Subunits (the Gapped-cp-Reconstitution)

1. A reconstitution reaction containing 10 pmol cp-23S rRNA is performed in 1× reconstitution buffer, in the presence of 10 pmol 5S rRNA, 20–50 pmol synthetic RNA oligomer (amount needed varies depending on the cp-23S rRNA construct), 6 mM spermidine, 500 µM solithromycin (see Note 3), 1.5–5 µL TP50 (prepared as described in ref. (11)). The TP50 preparation needs to be tested and titrated to obtain optimal ribosomal activities (see Note 5). 10 pmol of cp-23S rRNA is usually reconstituted to 50S subunits in a total volume of 11.9 µL.

2. It is recommended to use a thermocycler for the following incubation process: 2′ at 70°C, 40″ at 65°C, 40″ at 60°C, 40″ at 55°C, 40″ at 50°C, 60′ at 44°C. After the incubation at 44°C the $MgCl_2$ concentration has to be raised to 20–30 mM (depending on the used cp-23S rRNA construct) by adding 1 µL of the respective $MgCl_2$ stock solution, before the reconstitution is incubated 30′ at 60°C. After the 60°C step, the procedure can be interrupted (see Note 6).

3. In the final step, small ribosomal subunits from *T. aquaticus* or *E. coli* (depends on the functional assay performed afterward; see below) are added to achieve 70S ribosomes. The ratio of reconstituted 50S particles to native 30S particles is typically between 10:1 and 10:4 (see below). The reassociation step is carried out at 40°C for 10 min whereas the small ribosomal subunits are simply added directly to the reconstitution reaction.

3.4. Puromycin Reaction

1. The reassociated 70S ribosomes (13.9 µL; containing 10 pmol reconstituted 50S subunits and 4 pmol native *T. aquaticus* 30S) are combined with 0.8 pmol Ac[^{3}H]Phe-tRNA^{Phe} (15,000 cpm/pmol), 40 µg of poly(U) mRNA analog, 1.4 µL 10× HSM buffer in a total volume of 27.8 µL for 15 min at 37°C.

2. 6.2 µL of 10.7 mM puromycin solution is added to the reaction and incubated at 37°C for 1–120 min.

3. The reaction is terminated by adding 4.5 µL 10 M KOH and incubated for 5 min at 37°C.

4. After neutralization of the reaction with 90 µL of 1 M KH_2PO_4, 1 mL of cold ethyl acetate is added to extract the reaction product Ac[^{3}H]Phe-puromycin by vortexing for 1 min.

5. The samples were centrifuged at $18,900 \times g$ for 1 min and 0.8 mL of the organic phase is transferred into a liquid scintillation vial.

6. 5 mL of scintillation cocktail is added and the reaction product counted via liquid scintillation counting (see Note 5).

3.5. Poly(U)-Directed Poly(Phe) Synthesis

1. For one in vitro translation reaction, gapped-cp-reconstituted 50S subunits that were assembled from 20 pmol cp-23S rRNA are reassociated with 2 pmol of native *E. coli* 30S subunits (see Note 7).

2. The reassociated 70S ribosomes are subsequently precipitated with three volumes of EtOH at –80°C for 1 h and pelleted by centrifugation for 25 min at 4°C at $18,900 \times g$.

3. The ribosomal precipitate is resuspended in 8 µL translation buffer (1×). 25 µg of poly(U) mRNA analog is added and the volume is increased to 15 µL with 1× translation buffer.

4. This so-called binding reaction is incubated at 37°C for 15 min.

5. The charging reaction, which contains 8 mM ATP, 4 mM GTP, 12 mM acetylphosphate, 50 µg of deacylated tRNA^{Phe}, and 2–5 µL of S100, is set up separately. The amount of S100 varies depending on the quality of the extraction preparation and needs to be tested. 1 nmol of labeled phenylalanine is added whereas the specific activity of Phe should be between 200 and 400 cpm/pmol. The charging reaction is incubated for 2 min at 37°C before it is combined with the binding reaction (see above) and incubated at 42°C for 5–180 min.

6. To stop the poly(Phe) synthesis reaction, 10 µL of BSA (10 µg/µL) is added (see Note 6) and 2 mL of 5% TCA is used for hot TCA precipitation of the reaction products. Therefore, the samples are incubated at 95°C for 15 min. After cooling the reaction for 15 min on ice, the sample is filtered through a glass fiber filter, washed with 3 mL 5% TCA and 2 mL of ethyl ether/ethanol, and dried on room temperature.

7. The filter is transferred into a scintillation tube, 5 mL of scintillation cocktail is added and the product is measured by liquid scintillation counting (see Note 5).

3.6. Translation Fidelity Assay

1. This reaction is based on the poly(U)-directed poly(Phe) synthesis assay (4). Leucine is the classic amino acid to be misincorporated under these conditions since leucine is encoded by the UUA/G codons and therefore differs from the phenylalanine codon only in the wobble position. Instead of tRNA^{Phe}, an *E. coli* tRNA^{bulk} preparation is used, that contains a mixture of all cellular tRNAs. Moreover, radioactively labeled L-leucine is added to the reaction. Different radioactive labels for leucine and phenylalanine have to be used to elucidate the amounts of incorporated amino acids independently. The misincorporation of leucine into the poly(Phe) peptide chain is an indicator

for the translation fidelity of the ribosomes (4). The binding reaction is in principal carried out as described in Subheading 3.5, with the exception of the volume of the reaction. In this described case, the volume needs to be reduced to 13 μL.

2. The charging reaction differs from the above-described protocol in the use of tRNA^{bulk} (50 μg) instead of tRNA^{Phe} and the addition of leucine. To be able to reliably detect the misincorporated L-leucine, the specific activity has to be rather high. In this procedure the use of [^{3}H]-labeled leucine with a specific activity of 10,000 cpm/pmol is recommended. The volume of the charging reaction is 12 μL (see Note 8).

3. The incubation of the reaction was carried out at 42°C for 120 min. The reaction was stopped and filtered as described in step 6 of Subheading 3.5.

4. For measuring the overall translation performance and the amount of misincorporated leucine, the scintillation counter has to be programmed to measure both radioactive labels separately.

4. Notes

1. All solutions should be prepared in deionized water with a resistivity of 18.2 MΩ and is referred as water throughout this text.

2. The buffers are usually stored at –20°C if not otherwise mentioned.

3. In principal, all four mentioned macrolide antibiotics stimulate the 50S reconstitution efficiency. However, solithromycin showed the strongest effects.

4. If the yield of transcription reactions is low, the addition of the sequence GGATCC to the 5′ end of the T7-promotor is recommended. This extended sequence might enhance the transcription efficiency.

5. In case of high background activities in the puromycin reaction or the poly(Phe) synthesis assay, the added native 30S particles should be additionally purified since they might contain minute amounts of native 50S subunits. Therefore, the 30S subunits should be applied on a 10–40% sucrose gradient to get rid of contaminating native 50S particles. If this does not improve the results, the TP50 preparation could be a reason for it. It might be beneficial to test the ribosomal proteins in the absence of transcribed rRNAs to check for possible contaminations with native 23S rRNA.

6. After shock freezing using liquid nitrogen, the samples can be stored at this point for several months at −80°C.

7. For the poly(Phe) assay the use of 30S ribosomal subunits from *E. coli* is recommended. 30S particles from *T. aquaticus* show significantly lower activities during in vitro translation.

8. The specific activities of labeled L-leucine and L-phenylalanine for the misincorporation assay possibly need adjustment depending on the processivity of the used construct and the reaction conditions (e.g., Mg^{2+}-concentrations). For a detailed description of poly(Phe) and misincoporation assays for native particles see ref. 4.

Acknowledgments

We thank Ronald Micura and his team for a fruitful collaboration and for continuously providing top-quality synthetic RNA oligos. Prabhavathi Fernandes from Cempra Pharmaceuticals (Chapel Hill, NC, USA) is acknowledged for providing solithromycin. Nina Clementi and Anna Chirkova are thanked for experimental advice and comments on the manuscript. In addition we are grateful for constant support from Alexander Mankin, Knud Nierhaus, Wolfgang Piendl, and Alexander Hüttenhofer. Work in our laboratory is funded by grants from the Austrian Science Foundation FWF (Y315 to N.P. and P22658-B12 to M. E.) and the Austrian Ministry of Science and Research (GenAU project consortium "non-coding RNAs" D-110420-012-012 to N.P).

References

1. Polacek, N., and Mankin, A. S. (2005) The ribosomal peptidyl transferase center: structure, function, evolution, inhibition. *Crit. Rev. Biochem. Mol.* **40**, 285–311.

2. Wilson, D. N., Blaha, G., Connell, S. R., Ivanov, P. V., Jenke, H., Stelzl, U., Teraoka, Y., and Nierhaus, K. H. (2002) Protein synthesis at atomic resolution: mechanistics of translation in the light of highly resolved structures for the ribosome. *Curr. Protein Pept. Sci.* **3**, 1–53.

3. Katunin, V. I., Muth, G. W., Strobel, S. A., Wintermeyer, W., and Rodnina, M. V. (2002) Important contribution to catalysis of peptide bond formation by a single ionizing group within the ribosome. *Mol. Cell* **10**, 339–346.

4. Szaflarski, W., Vesper, O., Teraoka, Y., Plitta, B., Wilson, D. N., and Nierhaus, K. H. (2008) New features of the ribosome and ribosomal inhibitors: non-enzymatic recycling, misreading and back-translocation. *J. Mol. Biol.* **380**, 193–205.

5. Zaher, H. S., and Green, R. (2009) Fidelity at the molecular level: lessons from protein synthesis. *Cell* **136**, 746–762.

6. Wohlgemuth, I., Pohl, C., and Rodnina, M. V. (2010) Optimization of speed and accuracy of decoding in translation. *EMBO J.* **29**, 3701–3709.

7. Nissen, P., Hansen, J., Ban, N., Moore, P. B., and Steitz, T. A. (2000) The structural basis of ribosome activity in peptide bond synthesis. *Science* **289**, 920–930.

8. Chirkova, A., Erlacher, M. D., Micura, R., and Polacek, N. (2010) Chemically engineered ribosomes: A new frontier in synthetic biology. *Curr. Org. Chem.* **14**, 148–161.

9. Ban, N., Nissen, P., Hansen, J., Moore, P. B., and Steitz, T. A. (2000) The complete atomic structure of the large ribosomal subunit at 2.4 A resolution. *Science* **289**, 905–920.

10. Erlacher, M. D., and Polacek, N. (2008) Ribosomal catalysis: The evolution of mechanistic concepts for peptide bond formation and peptidyl-tRNA hydrolysis. *RNA Biol.* **5**, 5–12.

11. Erlacher, M. D., Lang, K., Shankaran, N., Wotzel, B., Huttenhofer, A., Micura, R., Mankin, A. S., and Polacek, N. (2005) Chemical engineering of the peptidyl transferase center reveals an important role of the 2′-hydroxyl group of A2451. *Nucleic Acids Res.* **33**, 1618–1627.

12. Erlacher, M. D., Chirkova, A., Voegele, P., and Polacek, N. (2011) Generation of chemically engineered ribosomes for atomic mutagenesis studies on protein biosynthesis. *Nature Prot.* **6**, 580–592.

13. Erlacher, M. D., Lang, K., Wotzel, B., Rieder, R., Micura, R., and Polacek, N. (2006) Efficient ribosomal peptidyl transfer critically relies on the presence of the ribose 2′-OH at A2451 of 23S rRNA. *J. Am. Chem. Soc.* **128**, 4453–4459.

14. Lang, K., Erlacher, M., Wilson, D. N., Micura, R., and Polacek, N. (2008) The role of 23S ribosomal RNA residue A2451 in peptide bond synthesis revealed by atomic mutagenesis. *Chem. Biol.* **15**, 485–492.

15. Amort, M., Wotzel, B., Bakowska-Zywicka, K., Erlacher, M. D., Micura, R., and Polacek, N. (2007) An intact ribose moiety at A2602 of 23S rRNA is key to trigger peptidyl-tRNA hydrolysis during translation termination. *Nucleic Acids Res.* **35**, 5130–5140.

16. Clementi, N., Chirkova, A., Puffer, B., Micura, R., and Polacek, N. (2010) Atomic mutagenesis reveals A2660 of 23S ribosomal RNA as key to EF-G GTPase activation. *Nat. Chem. Biol.* **6**, 344–351.

17. Clementi, N., and Polacek, N. (2010) Ribosome-associated GTPases: The role of RNA for GTPase activation. *RNA Biol.* **7**, 521–527.

18. Chirkova, A., Erlacher, M. D., Clementi, N., Zywicki, M., Aigner, M., and Polacek, N. (2010) The role of the universally conserved A2450-C2063 base pair in the ribosomal peptidyl transferase center. *Nucleic Acids Res.* **38**, 4844–4855.

19. Sanbonmatsu, K. Y. (2006) Alignment/misalignment hypothesis for tRNA selection by the ribosome. *Biochimie* **88**, 1075–1089.

20. Wachowius, F., and Hobartner, C. (2010) Chemical RNA modifications for studies of RNA structure and dynamics. *Chembiochem* **11**, 469–480.

21. Micura, R. (2002) Small interfering RNAs and their chemical synthesis. *Angew. Chem. Int. Ed. Engl.* **41**, 2265–2269.

22. Bommer, U. A., Burkhardt, N., Jünemann, R., Spahn, C. M. T., Triana-Alonso, F. J., and Nierhaus, K. H. (1997) Ribosomes and polysomes. In: Graham, J., and Rickwood, D., (eds) Subcellular fractionation: A practical approach, IRL Press, Washington DC.

23. Khaitovich, P., Tenson, T., Kloss, P., and Mankin, A. S. (1999) Reconstitution of functionally active Thermus aquaticus large ribosomal subunits with in vitro-transcribed rRNA. *Biochemistry* **38**, 1780–1788.

Chapter 15

Single Molecule FRET Characterization of Large Ribozyme Folding

Lucia Cardo, Krishanthi S. Karunatilaka, David Rueda, and Roland K.O. Sigel

Abstract

A procedure to investigate the folding of group II intron by *single molecule Fluorescence Resonance Energy Transfer* (smFRET) using total internal reflection fluorescence microscopy (TIRFM) is described in this chapter. Using our previous studies on the folding and dynamics of a large ribozyme in the presence of metal ions (i.e., Mg^{2+} and Ca^{2+}) and/or the DEAD-box protein Mss116 as an example, we here describe step-by-step procedures to perform experiments. smFRET allows the investigation of individual molecules, thus, providing kinetic and mechanistic information hidden in ensemble averaged experiments.

Key words: Single molecule, FRET, Group II introns, Folding, TIRF microscopy, Dwell times, DEAD-box helicases

1. Introduction

In the last 2 decades, bio- and nanotechnologies have focused on the development of methods capable of analyzing the structure/function correlation of biomolecules at the single molecule level (1–7). Such methods aim to overcome some of the limitations of bulk experiments. Ensemble averaged experiments yield an average dynamic behavior from a large number of molecules, thus any unsynchronized, dynamic information about short-lived or low-populated intermediates can often be lost or misinterpreted. Instead, observing the behavior of single molecules in real time allows the direct analysis of individual folding pathways of different molecules within a large ensemble, including short-lived intermediates, without need for synchronization. Thus, the distribution of different behaviors, rather than the average of these distributions, originating from the contribution of each single observed molecule, can be determined by single molecule experiments.

Jörg S. Hartig (ed.), *Ribozymes: Methods and Protocols*, Methods in Molecular Biology, vol. 848,
DOI 10.1007/978-1-61779-545-9_15, © Springer Science+Business Media, LLC 2012

Ribonucleic acids (RNAs) are highly versatile systems with a large conformational diversity and intricate folding pathways, which are regulated by metal ions and often by other molecules (mainly proteins) in vivo (8–11). The analysis with "traditional" methods, such as NMR, CD, UV melting, or electrophoresis techniques, provides essential information regarding RNA folding (12–16), but studies of single molecules of RNA performed in the last years have helped to uncover new insights that bulk experiments could not reveal (7).

Single molecule fluorescence resonance energy transfer (smFRET) (17–26) is one of the most widely used methods to observe the folding of single RNAs (7, 17, 27–32). In a standard FRET experiment, the biomolecule of interest is labeled with a donor–acceptor fluorophore pair. Upon excitation of the donor, the efficiency of the energy transfer from the donor to the acceptor is dependent on the distance between the two fluorophores (see Note 1) (1). Thus, FRET can be used as a molecular ruler in the 2–10 nm distance range, making it an ideal tool to study the structural dynamics and function of biomolecules.

The interference of background signals from various fluorescence sources is one of the main issues in detecting single fluorophores. This can be minimized by measuring highly diluted samples (pM) and using methods that permit the excitation of very small sample volumes (μm^3). Two strategies are commonly employed: confocal microscopy and total internal reflection fluorescence microscopy (TIRFM) (see Note 2) (33, 34). Here, we focus on the application of TIRFM to detect FRET of single molecules of RNA, using the example of the *Sc*.ai5γ group II intron ribozyme from *Saccharomyces cerevisiae*. The folding pathway of this large RNA was characterized by smFRET revealing a new folding paradigm for this large RNA and showing that increasing amounts of Mg^{2+} not only fold the RNA but also increase the dynamic behavior of the single domains (35). The addition of small amounts of Ca^{2+} instead leads to the formation of two distinct subpopulations (36) whereas the DEAD-box protein Mss116 can substitute for a large part of Mg^{2+} and stabilizes the active state (37).

Group II introns belong to the class of large phosphoryltransfer ribozymes together with group I introns and RNase P RNA. Although the sequence conservation in group II introns is very low, their secondary structure is greatly conserved, usually divided into six subunits, identified as domains D1-D6 (38). D1 is the largest domain and provides the scaffold for the docking of other domains, while D5 is the most conserved domain and comprises a large part of the catalytic core. The natural *S. cerevisiae* (*Sc*.) ai5γ group IIB intron (yeast mitochondrial intron residing in the *cox1* gene, ~900 nb) is one of the best characterized introns of this category (38). Its folding (and consequently its activity) requires high ionic strength in vitro, but splicing in vivo is also assisted by protein cofactors (39–41). The folding dynamics of this intron has been investigated by smFRET by annealing two short DNA strands (15–20 nts) each functionalized with either Cy3 or Cy5 (17) (see Note 3). The synthetic Cy3-DNA and Cy5-DNA oligonucleotides were annealed to a modified version of Sc.ai5γ group II intron named Sc.D135-L14 (see Fig. 1) that includes D1, D3, and D5 as well as two 15 nts loops within D1 and D4 whose sequences are complementary to the DNA oligos (see Note 4). Additionally, the 3′-end of the intron is elongated with a sequence complementary to a third DNA strand functionalized with biotin at its 5′-end.

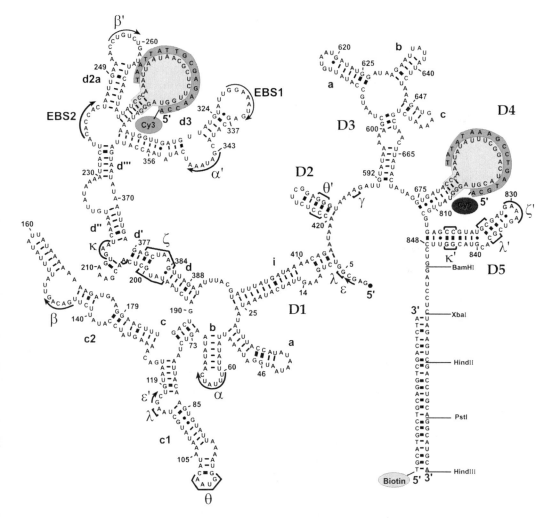

Fig. 1. Secondary structure of the Sc.D135-L14 ribozyme originating from the wild-type Sc.ai5γ group II intron from *Saccharomyces cerevisiae*. In Sc.D135-L14 D2 is reduced to a hairpin, D6 is deleted, and two modular loops (L1 and L4) are inserted to permit annealing of DNA-Cy3 and DNA-Cy5 (highlighted in dark). The 3′-end of the intron is elongated with a sequence suitable for the annealing of a biotinylated DNA (also in dark) in order to immobilize the construct on a strepta-vidin coated slide. The numbering corresponds to the wild-type Sc.ai5γ and tertiary contacts (Greek letters) as well as exon/intron binding sequences (EBS and IBS) are indicated.

Such, immobilization of the RNA can be achieved on the surface of a streptavidin-coated quartz slide via the strong streptavidin–biotin interaction.

Before advancing to the smFRET experiments, the correct (and optimal) annealing of Cy3-, Cy5-, and biotin-DNAs must be confirmed by native gel electrophoresis. Furthermore, all parameters influencing R_0 (and consequently also FRET efficiency) are constants (see Note 1) known for every pair of dyes, except κ^2 that is the factor related to the reciprocal orientation of fluorophores dipoles in the space. The fluorophores are usually assumed to freely rotate without conformational constrictions, meaning that an average value of $\kappa^2 = 2/3$ can be applied in calculating FRET efficiency. Fluorescence anisotropy measurements (18, 42) can be performed to confirm that donor and acceptor are indeed rotating freely, which is the case for Sc.D135-L14 (35).

The immobilized and fluorophore-carrying RNA/DNA complex is analyzed on a slide containing a home-built microfluidic chamber placed on an inverted microscope (see Fig. 2a, b) (17, 33). To achieve total internal reflection and visualize the single molecules, the laser excitation beam reaches the slide through

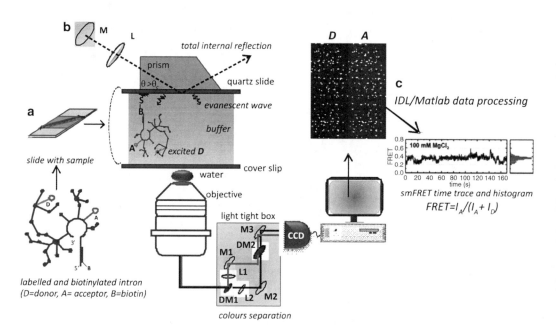

Fig. 2. smFRET TIRF setup: (**a**) The Cy3/Cy5-labeled and biotinylated Sc.D135-L14 ribozyme is loaded into the microfluidic chamber of a self made slide coated with streptavidine. (**b**) The mirror M and the lens L focus the excitation laser beam to the slide with an angle larger than θ_c to achieve total internal reflection. The resulting evanescent wave excites the molecules within 100–150 nm from the slide surface. The emitted light of the donor D and acceptor A is collected through the objective into the light-tight box where the signals from D and A are separated and detected by the CCD camera. (**c**) Single molecules detected through donor and acceptor emission are visible as bright dots in two parallel images. (**d**) Data analysis yields a FRET time trajectory of each single molecule (the shown time trace of Sc.D135-L14 at 100 mM MgCl$_2$ adapted from ref. (35)).

a quartz prism placed over the slide itself (see Fig. 2b). The beam is totally reflected without penetrating below into the sample if the incidence angle is larger than the critical angle (θ_c) (see Note 5). Such, an evanescent wave is created which diffuses only shortly (100–200 nm) below the quartz/solution interface and only the molecules present in that small volume are excited (see Note 6). The fluorescence emission from the sample is collected through an objective and directed into a light-tight box containing a set of dichroic mirrors and lenses that separate the donor and acceptor wavelengths (33). The two signals are then simultaneously detected with a CCD camera as two individual images. The acquisition software allows observation of both the donor and acceptor channel in real time displaying the single molecules as tiny bright dots (see Fig. 2c). Each pair of dots (from the donor and acceptor channels) is analyzed with the home-built software to determine the corresponding time trajectories of their smFRET efficiencies. The apparent FRET efficiency (43) is calculated as:

$$FRET = \frac{I_A}{I_A + I_D},$$

where I_A and I_D are the emission intensities of acceptor and donor, respectively, as integrated from each pair of dots. The image of immobilized fluorescent microspheres (beads) is used as calibration tool to map the two channels. This control is recommended in order to achieve a perfect correlation between two signals in the two channels relative to the same single molecule (33).

In the following we first describe the preparation of *Sc.* D135-L14 ribozyme by in vitro transcription, including the native gel and fluorescent anisotropy control experiments. The smFRET section then includes the description of slide preparation, the execution of smFRET experiment, and the data analysis. As an example, the smFRET experiments and analysis of the *Sc.*D135-L14 group II intron construct in dependence of $MgCl_2$ concentration or in the presence of the DEAD-box protein Mss116 are described.

2. Materials

All chemicals used for preparing buffers and stock solutions are at least puriss p.a. and purchased from usual suppliers. Buffers and solutions are prepared using double distilled autoclaved H_2O (ddH_2O) and subsequently filtered using 0.2-μm sterile filters (Filtropur S syringe filters for volumes up to 100 mL or Steritops Express™PLUS bottle top filters for larger volumes). Polyacrylamide gels are prepared using AccuGel™ 29:1 (acrylamide:bisacrylamide; 40% w/v) stabilized solution from National Diagnostic (UK).

Glassware and consumables (Eppendorfs, Falcon tubes, pipette tips, etc.) must be either autoclaved or bought as sterile and DNase/RNase-free items.

2.1. General Stock solutions

The following stock solutions should be at hand in order to prepare the buffers and solutions listed below:

1 M Tris–HCl, pH 7.5 (2-Amino-2-hydroxymethyl-propane-1,3-diol·HCl); 1 M EDTA, pH 8.0 (ethylenediamine-N,N,N',N'-tetraacetic acid); 5 M NaOH; 500 mM HEPES, pH 7.5 (4-(2-hydroxyethyl)-1-piperazineethanesulfonic acid); 100 mM MOPS, pH 6.0 (3-(N-morpholino)propanesulfonic acid); 5 M NaCl.

2.2. Buffers and Solutions for RNA Sc.D135-L14 Transcription

1. *Stock solutions of nucleoside 5′-triphosphates* (*NTPs*). Adenosine 5′-triphosphate (ATP; GE Healthcare), guanosine 5′-triphosphate (GTP; GE Healthcare), cytidine 5′-triphosphate (CTP; GE Healthcare), and uridine 5′-triphosphate (UTP; Acros-Brunschwig). About 130 mg of NTP are dissolved in 800 μL of H_2O and 20 μL of 1 M Tris–HCl each. The pH is adjusted with freshly prepared 5 M NaOH to pH 7.0 and the solution filled up to 1 mL with ddH_2O. The exact concentration of the freshly prepared NTP solutions is determined by UV spectroscopy (see Note 7). Aliquots of ~200 μL each are then stored at –20°C and used for transcription within a few months.

2. *pT7D135-L14 plasmid stock solution.* This plasmid encodes the D135-L14 sequence and was stored at a concentration of ~0.3 mg/mL in ddH_2O at –20°C. Standard digestion with *Hind*III (or any other specific restriction enzyme) followed by phenol–chloroform extraction is done before in vitro transcription with T7 RNA polymerase. The concentration of the cut plasmid is determined based on its absorbance at 260 nm (double-stranded DNA average extinction coefficient $\varepsilon = 0.020$ μg/mL/cm).

3. *10× Transcription buffer* (*pH 7.5*), *5 mL.* 400 mM Tris–HCl (pH 7.5), 400 mM DTT (1,4-Dithio-DL-threitol), 100 mM spermidine, 200 mM $MgCl_2$ and 0.1% Triton X-100.

4. *T7 RNA polymerase.* Homemade T7 RNA polymerase (44, 45) or commercially available T7 polymerase can be used.

5. 100 mL cold ethanol (–20°C) and 10 mL 5 M NaCl.

2.3. Buffers and Solutions for Polyacrylamide Gel Electrophoresis (PAGE)

1. *Denaturing electrophoresis buffer.* 1× TBE buffer (~700 mL for preparative gel) prepared from 10× TBE buffer (Tris-borate-EDTA; 0.89 M Tris, 0.89 M boric acid pH 8.3, and 20 mM Na_2EDTA; from National Diagnostic, UK).

2. *Denaturing gel loading buffer, 10 mL.* 8 M Urea, 2 mM Tris (pH 7.5), 20 mM EDTA, 0.02% xylene cyanol, and 0.02% blue bromophenol (store at 4°C).

3. *Denaturing 5% polyacrylamide gel electrophoresis (PAGE) gel solution, 500 mL.* 62.5 mL of AccuGel™ 29:1 (acrylamide:bisacrylamide; 40% w/v), 210 g Urea (ultrapure grad, from EUROBIO, France), and 50 mL 10× TBE (Tris-borate-EDTA) buffer (from National Diagnostic, UK) are diluted with ddH₂O to 500 mL. Gel solutions are best prepared one day prior to use, filtered after preparation, and stored in the dark at 4°C. For a 18% gel use 225 mL AccuGel™.

4. *Native gel electrophoresis buffer, 5× stock, 300 mL, pH 7.4.* 330 mM HEPES, 170 mM Tris–HCl (pH 7.5), 15 mM magnesium acetate. The pH of the solution is adjusted to 7.4 by adding 5 M NaOH. Electrophoresis buffer is used at 1×.

5. *Native gel loading buffer, 1 mL* 60% glycerol.

6. *Native 6% PAGE gel solution, 100 mL.* 16 mL of AccuGel™ 29:1 (40% w/v), and 20 mL 5× native gel buffer diluted with ddH₂O to 100 mL total.

7. *Gel casting solutions.* 500 μL of 10% Ammonium persulfate and 50 μL TEMED (tetramethylethylenediamine) for each 100 mL of gel casting solution (either denaturing or native gel).

8. *Elution buffer (pH 6.0), 50 mL:* 10 mM MOPS, 1 mM EDTA, 250 mM NaCl. The buffer should be kept at 4°C and protected in alumina foil after preparation.

2.4. Buffers and Solutions for smFRET Experiments

1. *Cy3 and Cy5 labeled DNA and T-Biotin-DNA oligonucleotides* are purchased from Microsynth, Balgach (Switzerland) or from HHMI Biopolymer/Keck Foundation Biotechnology Resource Laboratory, Yale University, New Haven CT (USA), purified by gel 18% denaturing PAGE, redissolved in 200 μL of ddH₂O and stored in the dark at –20°C. Concentrations are measured by UV–vis, using the "nearest neighbor" based method for T-Biotin-DNA (see Note 8), and the following molar extinction coefficients for the fluorophore-DNA oligos: $\varepsilon_{Cy3\text{-}DNA}$ (550 nm) 150,000/M/cm, $\varepsilon_{Cy5\text{-}DNA}$ (647 nm) 250,000/M/cm.

2. *Catalase* from bovine liver (crystalline suspension in water containing 0.1% thymol, by Sigma) and *Glucose Oxidase* Type VII (Sigma-Aldrich).

3. 0.5 mL of *1 mg/mL Biotinylated BSA solution* (ImmunoPure Biotinylated Bovine Serum Albumine by Fisher Thermo Scientific) and 0.5 mL of *0.2 mg/mL Streptavidin solution* (from *Streptomyces avidinii*, by Invitrogen). Both solutions are stored at 4°C.

4. *Fluorescent beads.* FluoroSpheres® carboxylate-modified microspheres (0.2 μm, red fluorescent, 2% in distilled water, 2 mM azide) from Invitrogen are used. Keep at 4°C and protected from light (see Note 9).

5. *5× Reaction buffer (pH 6.9), 10 mL*: 400 mM MOPS, 2.5 M KCl. Keep at 4°C and cover with aluminum foil.

6. *T50 buffer (pH 7.5), 1 mL*: 50 mM Tris–HCl, 50 mM NaCl. If proteins are also present in the reaction mixture, i.e., in the described case Mss116, the following additional buffers and solutions have to be prepared

7. *Solutions for preparation of PEG-coated slides*: 3-aminopropyl-triethoxysilane (Vectabond reagent, Vector Laboratories, Inc., Burlingame, CA), biotin polyethylene glycol succinimidyl carboxymethyl (BIO-PEG-SCM, 3,400/5,000 MW, Laysan Bio. Inc., Arab, AL), and methoxy polyethylene glycol succinimidyl carboxymethyl (m-PEG-SCM, 5,000 MW, Laysan Bio. Inc., Arab, AL).

8. *5× Reaction buffer for protein experiments (pH 7.5), 5 mL*: 200 mM MOPS, 500 mM KCl. Keep at 4°C and cover with aluminum foil.

9. *1× Reaction buffer for protein experiments (pH 7.5), 5 mL*: 40 mM MOPS, 100 mM KCl and 8 mM $MgCl_2$. Keep at 4°C covered with aluminum foil.

10. *10% Sugar-buffer for protein experiments (pH 7.5), 5 mL*: 10% D-glucose (w/v), 40 mM MOPS, 100 mM KCl and 8 mM $MgCl_2$. Keep at 4°C covered with alumina foil.

11. *100 mM ATP (pH 7), 1 mL.*

3. Methods

3.1. RNA Transcription and Purification

For a more detailed description see also Subheading 3.1 of Chapter 16.

1. 5 mL in vitro transcription of *Sc*.D135-L14: Mix HindIII digested pT7D135-L14 plasmid (7.5 μg/mL final concentration in 5 mL) with 1× transcription buffer, ATP, CTP, GTP, and UTP (final concentration of 5 mM in 5 mL for each NTP) and add ddH_2O up to 4.9 mL. Aliquot the reaction solution into five different Eppendorf tubes (980 μL in each tube) and add 20 μL of T7 RNA polymerase to each tube (see Note 10). Shake the reaction mixtures for 4–5 h at 37°C and 300 rpm (overnight is also possible for shorter RNAs).

2. Spin down the insoluble magnesium pyrophosphate resulting from the transcription ($1,500 \times g$ for 5 min), combine the supernatants into a falcon tube, add 5 M NaCl (final concentration of 250 mM NaCl), and mix with cold ethanol (3× volume). Store at –20°C (at least 6 h, better overnight) or, alternatively, at –80°C for 1 h.

3. Centrifuge at 4°C and $13,000 \times g$ for 40 min. Separate the white pellets from the supernatant by decantation and dissolve the pellets in as little of ddH_2O as possible (typically ~2 mL for a 5 mL transcription). Keep the solution on ice.

4. Add an equal volume of denaturing loading buffer and purify by denaturing 5% PAGE. Usually the product solution of a 5 mL in vitro transcription is split onto two preparative gels (28×42 cm with 1.5 mm spacers).

5. The RNA bands are located by UV-shadowing, excised and extracted from the gel by the crush-and-soak method: Crush the gel into a fine slurry by smashing it through a 10-mL syringe that has been melted at the tip and puckered again with a needle. Transfer the gel into falcon tubes and soak and shake it in Elution Buffer (3× the volume of the gel) for 4 h at 4°C.

6. Separate the eluted RNA from the gel by centrifugation and precipitate with 3 volumes of cold ethanol. Store at -80°C for 1 h and centrifuge at 4°C and $13,000 \times g$ for 40 min.

7. After separation from the supernatant, the white pellets are vacuum-dried in a concentrator ("speed-vac"), redissolved in as little ddH_2O as possible and stored at –20°C.

8. Measure the absorbance at 260 nm and use the following equation valid for large RNAs: $conc = A/(\#NTP/100)$, with $\#NTP = 637$ for the D135-L14 RNA construct.

3.2. Control Experiments

3.2.1. Native Gel Electrophoresis

Native gel electrophoresis experiments are performed to verify the correct annealing of the Cy3- and Cy5-DNAs to the RNA under the conditions used in the smFRET experiments.

1. Using the oligonucleotide stock solutions (see Subheading 2.4), prepare seven samples of 9 μL each with 1× Reaction buffer containing (a) Sc.D135-L14 (2 μM), (b) Cy3-DNA (15 μM), and (c) Cy5-DNA (15 μM), with the following combinations: (1) only (a), (2) only (b), (3) only (c), (4) (b)+(c), (5) (a)+(b), (6) (a)+(c), (7) (a)+(b)+(c). Concentrations given above are the final concentrations. Use dark (brown) Eppendorf tubes for samples preparation.

2. Heat to 90°C for 45 s.

3. Add 1 μL of 1 M $MgCl_2$ (100 mM final concentration) right after heating to 90°C and incubate at 42°C for 10–15 min.

4. Add 8 μL of loading native buffer to each sample and perform electrophoresis using the 6% native gel (17×24 cm) at 4°C for 2 h (15 W). Use a molecular imaging scanner to visualize the bands containing Cy3 and Cy5 (we use a Typhoon molecular scanner) (see Note 11).

5. Additionally, the gel is stained in a GelRed™ bath (5 μL of GelRed™, Biotium Inc. Hayward CA, in 100 mL ddH₂O) for 15 min followed by 5 min in ddH₂O and detected by UV-shadowing at 260 nm to visualize possible bands not containing Cy3 and Cy5.

3.2.2. Fluorescence Anisotropy Measurements

Fluorescence anisotropy measures the rotational diffusion of fluorescent molecules. Linking a fluorophore to a large biomolecule may affect its ability to rotate freely, thus affecting the values of κ^2 and R_0. We performed fluorescence anisotropy measurements (18, 42) to confirm that Cy3 and Cy5 are freely rotating when bound to the DNA annealed to Sc.D135-L14 (35) and that $\kappa^2 = 2/3$ can be assumed (see Note 1).

Prepare four samples containing Cy3-DNA (0.15 μM) and/or Cy5-DNA (0.5 μM, total volume of 200 μL each) in the presence and absence of Sc.D135-L14 (0.3 μM) and fold in 100 mM MgCl₂ as described above (see Subheading 3.2.1).

1. Anisotropy experiments are conducted in a cuvette (100 μL) to measure fluorescence polarization using polarized filters on the spectrophotometer. Fluorescence intensities (I) of polarized excitation and emission in vertical (v, 0°) and horizontal (h, 90°) positions are recorded in all four possible combinations I_{vv}, I_{vh}, I_{hv}, and I_{hh}. The anisotropy value r is calculated as described (42):

$$r = \frac{I_{vv} - gI_{vh}}{I_{vv} + 2gI_{vh}} \quad \text{with} \quad g = \frac{I_{hv}}{I_{hh}}.$$

If the observed anisotropy values for the free fluorophores are similar to those of the fluorophores attached to DNA in the Sc. D135-L14 RNA–DNA complex is an indication of the free rotation ability of fluorophores in a complex. It is noteworthy that, due their short fluorescent lifetimes, the fluorescent anisotropies of Cy3 and Cy5 alone in solution are high compared to other commonly used fluorophores such as fluoresceine or rhodamine.

3.3. smFRET Experiments

3.3.1. Instruments

The folding of Sc. D135-L14 ribozyme under optimal in vitro splicing conditions (500 mM KCl, 80 mM MOPS pH 6.9 and 100 mM MgCl₂) or under near-physiological conditions (100 mM KCl, 40 mM MOPS pH 7.5 and 8 mM MgCl₂) was monitored using a prism-based total internal reflection fluorescence (TIRF) inverted microscope (IX-71, Olympus, Center Valley, PA). Such a setup has recently been excellently described in detail (33) and is thus only shortly summarized here (see Fig. 2).

In order to obtain total reflection of the laser beam (532 nm, 3 mW, CrystaLaser GCL-532-L, Reno, NV), the angle of incidence (θ) on the slide should be larger than the critical angle (θ_c). When $\theta > \theta_c$, the laser beam totally reflects and creates an evanescent wave at the slide-solution interface that can penetrate a few 100 nm to excite the fluorophore-labeled samples (see Fig. 2b). The laser beam is introduced through a quartz Pellin-Broca prism (CVI Melles-Griot, Albuquerque, NM) by a mirror (Newport, Irvine, CA). The angle of incidence can be controlled by adjusting the height of the mirror. The reflected laser beam is focused to the prism using a BK7 lens with a 100 mm focal length (Newport, Irvine, CA).

The donor and acceptor emissions are collected through an inverted microscopic objective and transferred into a light-sealed box through a slit (see Fig. 2b). The first dichroic mirror (DM 1, 635DCXR, Chroma, Rockingham, VT) physically separates the donor and acceptor intensities and allows them to pass through lenses (L1 and L2, 200 mm focal length and 2.0 in. diameter) that amplify the image. The second dichroic mirror (DM 2) recombines the separated donor and acceptor emission signals as side-by-side images onto a high quantum yield CCD camera (Ixon+, DV-897E, Andor, South Windsor, CT). The CCD camera amplifies the signals by the highest electron multiplication (EM) gain to maximize the signal-to-noise ratio and transfers the digitalized frames to a computer for data analysis (see Fig. 2c).

3.3.2. Preparation of Microscope Slides with a Microfluidic Chamber

Standard microscope quartz slide ($76 \times 25 \times 1$ mm, Finkenbeiner Inc., Waltham, MA, USA) must be modified to create a microfluidic chamber, in which the sample is loaded and immobilized. The slides must be thoroughly cleaned and protected from dust and any type of contamination. A very detailed description has recently been published (33) and is repeated here in slightly different words.

1. Drill two holes into the quartz slides as shown in Fig. 3. Use a hand drill (Dremel 300-N, Racine, WI) held by a work-station (Dremel 220-01, Racine, WI) and diamond drill bits (1.0 mm diameter, Kingsley North, Norway, MI). With a marker, spot the positions of the holes. Drill through the slide placed over a sustaining cylindrical ring immersed in a basin filled with enough water to submerge the slide. The diamond drill bit must penetrate the slide very slowly to avoid slide rupture. Use each diamond bit for 5–6 times only.

2. Clean the slides with a thick paste of powder detergent (Alconox, VWR) and water. Rub slides thoroughly with fingers for at least 20 s. Rinse with water and keep wiping with fingers to

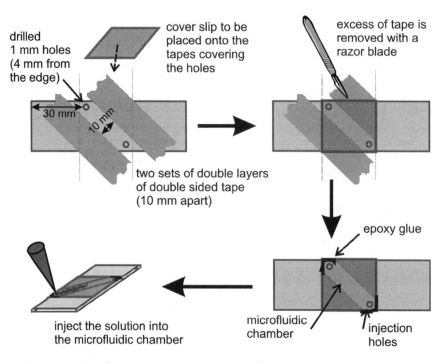

drilled 1 mm holes (4 mm from the edge)

cover slip to be placed onto the tapes covering the holes

30 mm

10 mm

two sets of double layers of double sided tape (10 mm apart)

excess of tape is removed with a razor blade

epoxy glue

microfluidic chamber

injection holes

inject the solution into the microfluidic chamber

Fig. 3. Schematic representation of a slide preparation with the microfluidic channel, into which the sample is loaded (see also text in Subheadings 3.3.2 and 3.3.3).

make sure that all the detergent is removed. Rub and rinse with ethanol and again with ddH$_2$O. The slides should look perfectly clean at the end of this procedure.

3. Place the slides in a beaker containing 100 mL of autoclaved water, 20 mL of 30% ammonium hydroxide, and 20 mL of 30% hydrogen peroxide (work under fume hood). Boil the solution for 20 min, gently stir with a magnetic bar that does not hit the slides and make sure that the whole surface of the slides is submerged in the solution.

4. Use tweezers to take the slides out of the solution, rinse each of them with water, dry them with a Bunsen burner flame (see Note 12), and place them on a metal railing in order to keep the slide uncontaminated (especially the central part of the slides that will hold the fluidic chambers).

5. When the slide cools down, place two stripes of twin-sided adhesive tape parallel to the line defined by the two holes (see Fig. 3). Keep about 6–10 mm distance between the two stripes and precisely place two additional stripes of tape on top of the first ones.

6. Carefully place a coverslip (Microscope cover slides 24 × 24 mm, Huber & Co. AG, Reinach, Switzerland) on the tapes, well centered on the slide covering the two holes. Using the top

part of the tweezers apply some pressure at the corners to ensure that the coverslip sticks well to the slide and is water tight.

7. Cut off the overhanging tape with a razor blade and quickly apply epoxy glue to seal the corners (see Fig. 3, see Note 13). When the glue is dry (about 15–20 min) store each slide in a sterilized container (i.e., 50-mL falcon tubes, see Note 14).

3.3.3. Preparation of PEG-Coated Slides for Protein Experiments

In order to minimize nonspecific binding of proteins to the slide surface, polyethyleneglycol (PEG) is used as a passivating agent in single molecule experiments (37, 46). The procedure below explains the preparation of PEG-coated quartz slides for single molecule experiments with Mss116 and D135-L14 RNA.

1. Clean the slides as described in the first four steps of Subheading 3.3.2.

2. After drying the slides with a flame, place the slides and coverslips into separate glass coplin jars, filled with 1 M potassium hydroxide (KOH), and sonicate for ~1 h.

3. Rinse the slides and coverslips first with distilled water then with methanol. Fill the jars with methanol and sonicate again for ~1 h.

4. Clean a beaker with methanol for aminopropylsilation. Mix 1 mL of 3-aminopropyltriethoxysilane reagent (Vectabond) kept in room temperature for ~1 h with 100 mL methanol and 5 mL glacial acetic acid.

5. Remove methanol and fill the coplin jars containing the slides and coverslips with the previously prepared 3-aminopropyltriethoxysilane solution.

6. Incubate the slides and coverslips for 10 min and sonicate for 1 min. After sonication, the slides and coverslips should be incubated again for another 10 min.

7. Decant the aminopropylsilane mixture into the appropriate waste container and rinse the slides first with methanol and then with double distilled water.

8. Rinse the slides and coverslips again with methanol and dry them using nitrogen or argon.

9. Prepare the PEGylation buffer by dissolving 84 mg of sodium bicarbonate in 10 mL double distilled water and filter to sterilize.

10. Prepare PEGylation reaction solution (five slides): Mix 4–8 mg of BIO-PEG-SCM, ~80 mg of m-PEG-SCM and 320 μL of the bicarbonate PEGylation buffer in 1-mL centrifuge vial. Vortex the solution to dissolve the PEG and then, centrifuge at 13,000 g for 1 min to remove bubbles.

11. In order to perform PEGylation, the slides should be placed in clean PEGylation reaction containers (use a clean pipette tip

box and add water in the bottom of the container to maintain a humid environment).

12. Place 70 μL of PEGylation reaction solution onto the surface of each slide and slowly place the coverslips onto the slides covering the solution without creating any bubbles between the slide and the coverslip.

13. Close the containers and incubate overnight at room temperature in a dark place to allow the PEGylation reaction to occur.

14. Rinse slides and coverslips with ddH$_2$O and dry with nitrogen or argon as previously explained.

15. Assemble the slides and coverslips following the steps 5–7 as explained in the Subheading 3.3.2.

3.3.4. Beads Slide Preparation

The slides containing the bead solution do not need to have drilled holes. These slides can be completely sealed and reused for several months when necessary.

1. Rinse a cleaned slide (see procedure described in Subheading 3.3.2) with methanol and dry it under a flux of N$_2$. Place double layers of twin-sided adhesive tape parallel to each other and parallel to the long side of the slide about 6–10 mm apart (see Fig. 4a). Place the coverslip (Microscope coverslips 24 × 24 mm) over the center of the slide creating a microfluidic channel that will hold the beads solution. Cut off the excess of tape with a razor blade.

2. Load 50–70 μL of 0.5 M MgCl$_2$ into the chamber with a pipetman. Prepare 1/2,500 diluted solution of FluoroSpheres® carboxylate-modified fluorescent beads in water and inject 50–70 μL of this solution into the chamber from the opposite side than used to load the MgCl$_2$ solution.

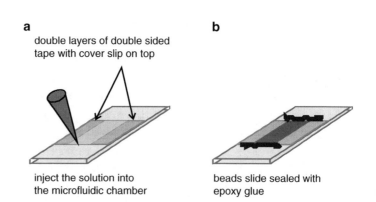

a
double layers of double sided tape with cover slip on top

inject the solution into the microfluidic chamber

b
beads slide sealed with epoxy glue

Fig. 4. Preparation of beads slides. (**a**) The two sets of double sticky tape are placed parallel to the slide and covered with a coverslip. The solutions are filled from the opposite ends and the microfluidic chamber is sealed with epoxy glue (**b**) (see also text in Subheading 3.3.4).

3. Check if it is possible to see the formed beads using the single molecule setup and whether the image quality and number of beads are satisfactory; (see also Subheading 3.3.5). Seal the microfluidic chamber with epoxy glue (see Fig. 4b, Note 15).

3.3.5. smFRET Analysis of Sc.D135-L14 in 100 mM MgCl₂, Sample Preparation and Slide Loading

The metal ion-dependent folding of *Sc*.D135-L14 was studied using single molecule fluorescence experiments (35). Here, we explain the sample preparation and slide loading to monitor the folding of D135-L14 as an example case for such studies.

1. Prepare the pipette tips for loading: Cut off 2–3 mm of the cone end of the 200-µL pipette tip with a blade razor. Ensure that the tip tightly fits into the injection hole so that the solutions do not leak during loading but flush out only from the opposite hole (see Fig. 3, Note 16).

2. Prepare a fresh stock of oxygen scavenging solution by mixing ~50 µL (powder volume) of glucose oxidase, 12.5 µL catalase (see Note 17), and 100 µL of T50 buffer. This solution can be stored at 4°C, but used for no more than 3 days.

3. Prepare 10 mL of 1× Reaction buffer by diluting 2 mL of 5× Reaction buffer in 8 mL of ddH₂O (autoclaved).

4. Prepare a 10% sugar/buffer solution by dissolving 100 mg D-(+)-glucose in 1 mL (total volume) of 1× Reaction buffer. Pass the solution through a 0.2-µm filter.

5. Prepare 10 µL stock solutions each of *Sc*.D135-L14 (2.5 µM), Cy3-DNA, Cy5-DNA, and T-Biotin-DNA oligos (100 µM each). Keep the solutions on ice and the dye-DNA oligo solutions protected from light.

6. Prepare *solution A*: in a dark 200-µL Eppendorf, mix 2.5 µL of ddH₂O, 2 µL of 5× Reaction buffer, 1 µL of β-mercaptoethanol (1, 47) (see Note 18), 2 µL of *Sc*.D135-L14 stock solution, and 0.5 µL of each Cy3-DNA, Cy5-DNA, and T-Biotin-DNA stock solutions. Vortex, spin down, and heat at 90°C for 45 s. Add 1 µL of 1 M MgCl₂ and incubate at 42°C for 15–20 min. Solution A now contains 10 µL of folded and labeled 0.5 µM *Sc*. D135-L14, 5 µM of each DNAs, 1× Reaction buffer, and 100 mM MgCl₂ concentration.

7. Start loading the slide: inject ~80 µL of biotinylated BSA solution and incubate for ~10 min to allow for uniform absorption to the slide surface (see Notes 19 and 20).

8. Meanwhile, prepare the oxygen scavenging system (OSS) by mixing 2 µL of the stock oxygen scavenging solution, 2 µL β-mercaptoethanol, and 196 µL of the 10% sugar/buffer solution. Incubate for 15–20 min before use to activate the scavenger system.

9. Wash the microfluidic chamber with ~200 μL of T50 buffer to remove the excess of BSA and inject ~200 μL of streptavidine solution. Incubate for ~10 min to optimize biotin binding.

10. In the meanwhile prepare *solution B*: Mix 1 μL *solution A* and 1 μL β-mercaptoethanol in 98 μL 1× Reaction buffer.

11. Prepare *solution C* (the final sample solution to be injected): Mix 1 μL *solution B* and 2 μL β-mercaptoethanol in 197 μL 1× Reaction buffer. The RNA is now ~25 pM.

12. Turn off the lights (see Note 21): Wash the microfluidic chamber with ~200 μL of 1× Reaction buffer and inject ~200 μL *solution C*. Allow binding of the biotinylated RNA to the streptavidin for 7–10 min and inject ~200 μL OSS. Equilibrate for ~5 min before taking measurements.

3.3.6. Sample Preparation and Slide Loading for the Analysis of Sc.D135-L14 in the Presence of Mss116

Protein-mediated folding of *Sc*.D135-L14 can be studied using single molecule fluorescence experiments with the DEAD-box protein Mss116 and ATP (37). This section explains the sample preparation and slide loading to monitor the folding of D135-L14 in the presence of Mss116 and ATP under near-physiological conditions.

1. Prepare *solution A* by mixing 2.7 μL of ddH$_2$O, 2 μL of 5× Reaction buffer, 1 μL of β-mercaptoethanol, 2 μL of *Sc*. D135-L14 stock solution, and 0.5 μL of each Cy3-DNA, Cy5-DNA, and T-Biotin-DNA stock solutions as explained in the Subheading 3.3.5. Heat-anneal the sample for 45 s at 90°C and incubate at 30°C for 15–20 min after addition of 0.8 μL of 100 mM MgCl$_2$ to assure proper annealing and folding. At the end of this step we obtain a 10 μL solution containing 0.5 μM D135-L14, 5 μM of all DNA oligos, 100 mM KCl, 40 mM MOPS pH 7.5, and 8 mM MgCl$_2$.

2. Load the PEG-coated slide with ~80 μL of streptavidin solution and incubate for ~10 min to allow binding of streptavidin to the slide surface.

3. Meanwhile, prepare *solution B* and *solution C* as described above (step s 10–11 in Subheading 3.3.5).

4. Wash the microfluidic chamber with ~200 μL of 1× Reaction buffer and inject ~200 μL of *solution C* (step 3). Incubate for 7–10 min for complete RNA binding.

5. Prepare the OSS with 25 nM Mss116 and 1–2 mM ATP by mixing 2 μL of stock oxygen scavenging solution and 2 μL of β-mercaptoethanol using 10% sugar-buffer solution for protein experiments (final volume 200 μL). Store this solution at 4°C until ready to inject.

6. Inject 200 μL of the OSS containing 25 nM Mss116 and 1–2 mM ATP (step 5) and incubate for 5 min.

3.3.7. Calibration of
the Experimental Setup

In order to obtain the best possible single molecule data, first the instrument needs to be calibrated using the fluorescent beads slide (see Subheading 3.3.4). This allows to build a map that matches a molecule's signal in the donor channel with its corresponding signal in the acceptor channel.

1. Switch on the CCD camera and the acquisition software and cool the camera to –80°C.

2. Place a drop of ddH$_2$O on the high numerical aperture water immersion objective (60×) and place the fluorescent beads slide on the slide holder over the objective with the coverslip facing the objective (see Note 22). Secure the slide with stage clamps.

3. Position the prism onto the slide adding a drop of refraction index-matched immersion oil between slide and prism.

4. Switch the laser on and look for the fluorescent beads through the microscope's eye piece using the appropriate filters. Focus the image (see Note 23) and center the laser beam using the focusing lens placed in front of the prism. The set up is aligned when individual bright dots from isolated beads are visible (see Note 24).

5. To visualize the image onto the 512×512 pixel EM-CCD camera, switch to the side port. Focus and center the signal in the observation window using the focusing lens until single beads are clearly visible as bright dots in the donor and acceptor channels (see Note 25). Search for at least 50 sharp and well-focused beads, while avoiding large bright dots that may correspond to bead aggregates.

6. Save ~30 frames to be further analyzed to generate the image map (see Note 26 and Subheading 3.3.8).

3.3.8. Performing the Single Molecule FRET Experiment

After the instrument has been calibrated (see Subheading 3.3), proceed with the single molecule analysis of the RNA. The slides with the reaction solution are prepared as described (see Subheading 3.3.5). Always work in the dark when handling the fluorophore-labeled RNA.

1. Five minutes after adding the OSS (step 12, Subheading 3.3.5) place the slide on the microscope's slide holder and visualize the single molecules as described above (steps 1–5 in Subheading 3.3.7).

2. An EM gain level of 400 is typically used to visualize the Cy3-Cy5 fluorophore pair (see Note 27).

3. Move the stage in order to observe a good set of single molecules and record a set of frames at the desired frame rate. Let the measurement run until photobleaching of the fluoro-

phores is observed (photobleaching time depends on the laser intensity used).

4. Move the stage to another area of single molecules on the same slide to record a new set of frames. From a good slide, eight to ten movies can be collected yielding a large distribution of single molecules for data analysis.

3.3.9. Data Analysis

To obtain single molecule time trajectories, the location of single molecule peaks in the donor and acceptor channels is mapped using the image obtained from the immobilized fluorescent beads as a calibration map. The emission intensities of the donor and acceptor fluorophores for every recoded frame are obtained by integrating the corresponding peaks after background subtraction.

Since single molecule experiments can generate significantly large amounts of data, the analysis of single molecule FRET data requires the following specific criteria:

(a) Anti-correlated donor and acceptor emission intensities.

(b) Single-step photobleaching of fluorophores.

(c) Stable emission intensities corresponding to single fluorophore.

Single molecule trajectories of the D135-L14 ribozymes provide valuable information about the folding of introns under different reaction conditions (35, 37). Each single molecule FRET trajectory represents the folding behavior of an individual RNA molecule exhibiting different FRET states corresponding to different structural conformations (Fig. 5a). In order to determine the distribution of different conformational states under given reaction conditions, FRET histograms can be constructed using FRET trajectories from more than 100 molecules, or until convergence (Fig. 5b). Therefore, FRET histograms represent the general effect of different reaction conditions on D135-L14 RNA folding. Under equilibrium conditions, the relative heights of peaks in the FRET histograms can be used to determine the relative stabilities of different conformational states.

In addition, FRET trajectories can be used to obtain valuable kinetic and mechanistic information about the D135-L14 RNA folding pathway using dwell times in each conformational state. The dwell time is the amount of time a molecule spends in a given state before switching to the next state. Folding rate constants of D135-L14 can be determined by fitting the dwell-time distribution of each state with an exponential decay function (Fig. 5c, d). Alternatively, the complex single molecule trajectories can be analyzed using a Hidden Markov Model (HMM) for unbiased estimation of the number of distinct FRET states and the folding

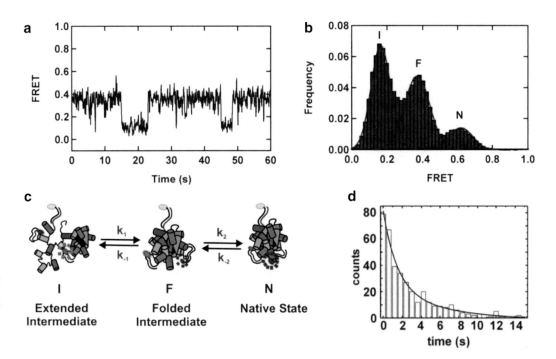

Fig. 5. Single molecule data analysis. (**a**) A typical FRET trajectory showing the behavior of a single D135-L14 ribozyme in the presence of Mss116 and ATP under near-physiological conditions (8 mM $MgCl_2$, 100 mM KCl and 40 mM MOPS pH 7.5). (**b**) FRET histogram showing the distribution of three structural conformations: the extended intermediate state (I), the folded intermediate state (F) and the native state (N). (**c**) D135-L14 minimal folding pathway showing folding rate constants k_1, k_{-1}, k_2, and k_{-2}. (**d**) A typical dwell-time distribution fit to a single exponential decay to obtain a rate constant for a given transition. Figures are adapted from refs. (35, 37).

rate constants among those states (37, 48). In addition, the order in which different conformational states appear in the trajectories also enables to distinguish between obligatory folding intermediates from off pathway intermediates.

The resulting folding rate constants for the D135-L14 ribozyme can then be used to calculate RNA folding free energy diagrams, which reveal the ribozyme's folding pathways (37).

3.3.10. Final Remarks

Single molecule FRET allows to study the folding and dynamics of even large RNAs and RNA protein complexes, at a so far unprecedented resolution and provides information that is generally not accessible by bulk experiments. Monitoring the behavior of hundreds of single molecules in real time over several minutes reveals an astonishing diversity in behavior of these individual systems, e.g., short-lived intermediate folding states that are unknown from ensemble averaged experiments. From the individual time trajectories, rate constants of the single folding steps, binding events, possibly catalysis, as well as thermodynamic data can be calculated making this method a very powerful tool widely applicable in modern biophysical research.

Using the example of the Cy3/Cy5 labeled D135-L14 ribozyme derived from the yeast mitochondrial group II intron *Sc.* ai5γ group II intron, we here describe all steps starting from RNA transcription and isolation to the recording of single molecule time trajectories and data evaluation as well as the slight modifications applied when investigating RNA protein interactions, e.g., in the Mss116-D135-L14 system. Every RNA behaves slightly differently, folds under different conditions, and uses different cofactors for catalysis. In addition, the here described setup and methodology are used specifically in our groups and thus depend on each other. Generally, smFRET is not "black-box" technique, but requires hands-on optimization of the mostly self assembled setups as well as adjusting the methodology. Hence, every system has to be thoroughly optimized in order to get the most valuable data, but smFRET allows to readily implementing one's own ideas and directions thus contributing to this fast moving field of biophysical analysis.

4. Notes

1. The efficiency of energy transfer is given by $E_{FRET} = 1/(1+(R/R_0)^6)$, where R is the distance between the two fluorophores and R_0 the Förster radius at which 50% of energy transfer efficiency is observed:

$$(R_0)^6 = 8.8 \times 10^{-28} k^2 n^{-4} Q_0 J, \quad J = \int f_D(\lambda)\varepsilon_a(\lambda)\lambda^4 d\lambda$$

where κ^2 is the factor describing the orientation of dipoles, n the refractive index of the medium, Q_0 the donor quantum yield in the absence of the acceptor, $f_D(\lambda)$ the fluorescence intensity of the donor, and $\varepsilon_a(\lambda)$ the molar extinction coefficient of the acceptor at the same wave length λ.

2. Confocal microscopy is mostly used for the detection of freely diffusing molecules in solution with low time resolutions because of the avalanche photodiode (ADP) and photomultipliers that are used in this technique (34). In contrast, TIRFM allows the excitation of smaller volumes, which limit the effect of noise background although the signal is detected with charge-couple device (CCD) cameras that work with higher time resolutions.

3. To investigate short RNA sequences (<60–80 nts) by FRET, the two fluorophores are covalently attached by direct insertion during chemical synthesis. In contrast, larger constructs are obtained by in vitro transcription by T7 RNA polymerase,

which prevents the insertion of modified nucleotides in the middle of a sequence.

4. The underlying construct Sc.D135 containing only D1, D3, and D5 is the best characterized group II intron construct that contains all necessary units for folding and catalytic activity (38). Control experiments have shown that also Sc.D135-L14 retains equivalent catalytic activity (35).

5. θ_c depends on the indexes of refraction of the quartz slide and buffer solution. The right incidence angle is controlled by regulating the height and distance of the mirror M, as described in detailed in ref (33).

6. This setup is known as prism-based TIRF. Alternatively in the objective-based TIRF method, the evanescent wave can be generated through the objective (33). Here, the alignment of the laser beam is more complicate and this strategy is used only when the presence of the prism over the slide is an impediment for the type of experiment to perform.

7. Extinction coefficients of the nucleotides (M/cm) are: ε_{ATP} (260 nm) = $15.4 \cdot 10^3$, ε_{UTP} (260 nm) = $10.0 \cdot 10^3$, ε_{CTP} (270 nm) = $9.0 \cdot 10^3$, ε_{GTP} (249 nm) = $13.7 \cdot 10^3$.

8. The extinction coefficients ε can be estimated using web-based tools, e.g., http://biophysics.idtdna.com/UVSpectrum.html or http://www.owczarzy.net/abstr11.htm. These calculation tools are based on the "nearest neighbor" method meaning that they take into account not only the extinction of each nucleobase, but also the identity of the neighboring bases (49, 50).

9. Approximate fluorescence excitation/emission of these beads is 580/606 nm. Alternatively, crimson fluorescent Fluoro-Spheres® carboxylate-modified microspheres (625/645 nm), 0.2 μm, 2% in distilled water, 2 mM azide, from Invitrogen, can be used.

10. The concentration of homemade T7 is usually not determined. We concentrate T7 as much as possible (45). Hence, the optimal amount needs to be optimized in transcription trials on an analytical scale (50 μL transcription). Transcription efficiency strongly varies depending on the RNA sequence, plasmid concentration, T7 batch, $MgCl_2$, and NTP concentration (see also Chapter 16, Subheading 3.1).

11. An image of such a native gel experiment is reported in Fig. S2 of the Supporting Information of ref. (35).

12. The slides will break easily if held for too long over the flame. Move the slide with the tweezers toward the flame and slowly move it back. Repeat this for about 10 s until the slide is dry.

13. Use just the minimal amount of glue to seal the open corners, since excess epoxy glue penetrates into the microfluidic chamber by capillary forces and plugs the holes.

14. Slides can be stored for about 6–8 weeks. Used slides can be recuperated: Soften the glue by boiling the slides in ddH$_2$O for 20 min or until the glue turns yellowish. Use a razor blade to remove the coverslip and the glue and proceed with the cleaning as described in the beginning of Subheading 3.2.

15. Calibration of the instrument can be performed with the same beads slide for many weeks as the beads solution in MgCl$_2$ is photostable.

16. The same tip can be used for all the injections described below.

17. Collect the supernatant of the suspension. Alternatively aliquot 16 µL of suspension in an Eppendorf tube, centrifuge for 30 s, and take 12.5 µL of the supernatant.

18. β-mercaptoethanol quenches the triplet state of the dyes, which helps to achieve a steady light emission minimizing any blinking effects (1, 47).

19. Ideally, both sample preparation and injection into the slide should be carried out in the room set up for single molecule experiments. Keep shut the black curtains of the room during the sample injection, and work close to the microscope thus the slide can be easily transferred for analysis after the injection.

20. Each solution is slowly and completely injected, carefully avoiding formation of bubbles and allowing the excess of solution to flow out of the opposite hole. After injection remove the tip from the injection hole by keeping down the plunger button of the pipetman to avoid sucking out the solution again.

21. The injection of *solution C* must occur in a dark room assisted only by low power LED lights. We use USB notebook led lights (5V, 48mA) plugged using a 2-port USB charger.

22. Make sure that the slide is resting on the holder and not on the lens of the objective.

23. Take care not to jam the objective into the slide, cracking the slide or, even worse, scratching the lens of the objective.

24. If the image cannot be found, it is recommended to remove the filter and use a low-magnification objective (e.g., 10×) to pre-align the laser beam on the image center. Then switch back to the high numerical aperture objective. Make sure that still a drop of water is on the lens.

25. Usually no or only a very low EM gain level (~20) is required to visualize the fluorescent beads.

26. We use the "Run Till Abort" acquisition mode of the Labview program that allows the saving of each picture frame of the movie recorded by the CCD camera at the respective frame rate. This movie containing the sequence of frames is saved to the hard drive as one large *.pma file, which is just a binary file containing all saved frames. This file is then processed with IDL (ITT VIS) scripts to extract the single molecules time trajectories.

27. It is crucial to turn off *all* lights when the EM gain function is enabled, to avoid fast aging of the CCD camera.

Acknowledgements

Financial support by the University of Zürich, an ERC Starting Grant 2010 (259092-MIRNA to R.K.O.S.), as well as the National Science Foundation (MCB0747285 to D.R.) and the National Institutes of Health (R01 GM085116 to D.R.) is gratefully acknowledged.

References

1. Hinterdorfer P., Oijen A. v. (2009) Handbook of Single-Molecule Biophysics. Springer, New York.

2. Cornish P. V., Ha T. (2007) A survey of single-molecule techniques in chemical biology. *ACS Chem Biol* **2**, 53–61.

3. Greenleaf W. J., Woodside M. T., Block S. M. (2007) High-resolution, single-molecule measurements of biomolecular motion. *Annu Rev Biophys Biomol Struct* **36**, 171–190.

4. Kulzer F., Orrit M. (2004) Single-molecule optics. *Annu Rev Phys Chem* 55, 585-611.

5. Kapanidis A. N., Strick T. (2009) Biology, one molecule at a time. *Trends Biochem Sci* **34**, 234–243.

6. Moerner W. E. (2007) New directions in single-molecule imaging and analysis *Proc Natl Acad Sci U S A* **104**, 12596–12602.

7. Karunatilaka K. S., Rueda D. (2009) Single-molecule fluorescence studies of RNA: A decade's progress. *Chem Phys Lett* **476**, 1–10.

8. Fedorova O., Solem A., Pyle A. M. (2010) Protein-facilitated folding of group II intron ribozymes. *J Mol Biol* **397**, 799–813.

9. Lilley D. M. J. (2005) Structure, folding and mechanisms of ribozymes. *Curr Opin Struct Biol* **15**, 313–323.

10. Thirumalai D., Hyeon C. (2009) Theory of RNA folding: from hairpins to ribozymes. In: Walter N. G., Woodson S. A., Batey R. T. (ed) Non-Protein Coding RNAs. Springer, Heidelberg Germany

11. Lilley D. M. J., Eckstein F. (2008) Ribozymes and RNA catalysis. RCS Publishing, Cambridge UK.

12. Lee M. K., Gal M., Frydman L., Varani G. Real-time multidimensional NMR follows RNA folding with second resolution. *Proc Natl Acad Sci U S A* **107**, 9192–9197.

13. Solomatin S., Herschlag D. (2009) Methods of site-specific labeling of RNA with fluorescent dyes. *Methods Enzymol* **469**, 47–68.

14. Johannsen S., Korth M. M. T., Schnabl J., Sigel R. K. O. (2009) Exploring metal ion coordination to nucleic acids by NMR. *Chimia* **63**, 146–152.

15. Schlatterer J. C., Brenowitz M. (2009) Complementing global measures of RNA folding with local reports of backbone solvent accessibility by time resolved hydroxyl radical footprinting. *Methods* **49**, 142–147.

16. Woodson S. A., Koculi E. (2009) Analysis of RNA folding by native polyacrylamide gel electrophoresis. *Methods Enzymol* **469**, 189–208.

17. Roy R., Hohng S., Ha T. (2008) A practical guide to single-molecule FRET. *Nat Methods* **5**, 507-516.

18. Ha T. (2001) Single-molecule fluorescence resonance energy transfer. *Methods* **25**, 78–86.

19. Uphoff S., Holden S. J., Le Reste L., Periz J., van de Linde S., Heilemann M., Kapanidis A. N. Monitoring multiple distances within a single molecule using switchable FRET. *Nat Methods* 7, 831–U890.

20. Blanco M., Walter N. G. (2010) Analisis of complex single molecule FRET time trajectories. *Methods Enzymol* 472, 153–178.

21. Ditzler M. A., Aleman E. A., Rueda D., Walter N. G. (2007) Focus on function: single molecule RNA enzymology. *Biopolymers* 87, 302–316.

22. Greenfeld M., Herschlag D. (2010) Measuring the energetic coupling of tertiary contacts in RNA folding using Single Molecule Fluorescence Resonance Energy Transfer. *Methods Enzymol* 472, 205–220.

23. Kapanidis A. N., Weiss S. (2009) Single-molecule FRET analysis of the path from transcription initiation to elongation. In: Buc V. H., Strick T. (ed) RNA polymerases as molecular motors. RCS Publishing, Cambridge UK

24. Abelson J., Blanco M., Ditzler M. A., Fuller F., Aravamudhan P., Wood M., Villa T., Ryan D. E., Pleiss J. A., Maeder C., Guthrie C., Walter N. G. (2010) Conformational dynamics of single pre-mRNA molecules during in vitro splicing. *Nat Struct Mol Biol* 17, 504–U156.

25. Bokinsky G., Zhuang X. W. (2005) Single-molecule RNA folding. *Acc Chem Res* 38, 566–573.

26. Rueda D., Bokinsky G., Rhodes M. M., Rust M. J., Zhuang X. W., Walter N. G. (2004) Single-molecule enzymology of RNA: Essential functional groups impact catalysis from a distance. *Proc Natl Acad Sci U S A* 101, 10066–10071.

27. Rueda D., Guo Z. J., Karunatilaka K. (2009) Splicing Mechanisms: Lessons from Single-Molecule Spectroscopy. *J Biomol Struct Dyn* 26, 48.

28. Weiss S. (2000) Measuring conformational dynamics of biomolecules by single molecule fluorescence spectroscopy. *Nature Struct Biol* 7, 724–729.

29. Zhuang X. W. (2005) Single-molecule RNA science. *Annu Rev Biophys Biomol Struct* 34, 399–414.

30. Ha T., Zhuang X. W., Kim H. D., Orr J. W., Williamson J. R., Chu S. (1999) Ligand-induced conformational changes observed in single RNA molecules. *Proc Natl Acad Sci U S A* 96, 9077–9082.

31. Zhuang X. W., Kim H., Pereira M. J. B., Babcock H. P., Walter N. G., Chu S. (2002) Correlating structural dynamics and function in single ribozyme molecules. *Science* 296, 1473–1476.

32. Aleman E. A., Lamichhane R., Rueda D. (2008) Exploring RNA folding one molecule at a time. *Curr Opin Chem Biol* 12, 647–654.

33. Zhao R., Rueda D. (2009) RNA folding dynamics by single-molecule fluorescence resonance energy transfer. *Methods* 49, 112-117.

34. Vukojevic V., Heidkamp M., Ming Y., Johansson B., Terenius L., Rigler R. (2008) Quantitative single-molecule imaging by confocal laser scanning microscopy. *Proc Natl Acad Sci U S A* 105, 18176–18181.

35. Steiner M., Karunatilaka K. S., Sigel R. K. O., Rueda D. (2008) Single-molecule studies of group II intron ribozymes. *Proc Natl Acad Sci U S A* 105, 13853–13858.

36. Steiner M., Rueda D., Sigel R. K. O. (2009) Ca2+ Induces the Formation of Two Distinct Subpopulations of Group II Intron Molecules. *Angew Chem Int Ed Engl* 48, 9739–9742.

37. Karunatilaka K. S., Solem A., Pyle A. M., Rueda D. (2010) Single-molecule analysis of Mss116-mediated group II intron folding. *Nature* 467, 935–939.

38. Fedorova O., Zingler N. (2007) Group II introns: structure, folding and splicing mechanism. *Biol Chem* 388, 665–678.

39. Sigel R. K. O. (2005) Group II intron ribozymes and metal ions - A delicate relationship. *Eur J Inorg Chem* 2281–2292.

40. Solem A., Zingler N., Pyle A. M. (2006) A DEAD protein that activates intron self-splicing without unwinding RNA. *Mol Cell* 24, 611–617.

41. Sigel R. K. O., Pyle A. M. (2007) Alternative roles for metal ions in enzyme catalysis and the implications for ribozyme chemistry. *Chem Rev* 107, 97–113.

42. Lakowicz J. R. (2006) Principles of fluorescence spectroscopy. Springer, New York.

43. Pljevaljcic G., Millar D. P., Deniz A. A. (2004) Freely diffusing single hairpin ribozymes provide insights into the role of secondary structure and partially folded states in RNA folding. *Biophys J* 87, 457–467.

44. Davanloo P., Rosenberg A. H., Dunn J. J., Studier F. W. (1984) Cloning and expression of the gene for bacteriophage-T7 RNA polymerase. *Proc Natl Acad Sci U S A.* 81, 2035–2039.

45. Gallo S., Furler M., Sigel R. K. O. (2005) In vitro transcription and purification of RNAs of different size. *Chimia* 59, 812–816.

46. Lamichhane R., Solem A., Black W., Rueda D. (2010) Single-molecule FRET of protein-nucleic acid and protein-protein complexes: Surface passivation and immobilization. *Methods* 52, 192–200.

47. Rasnik I., McKinney S. A., Ha T. (2006) Nonblinking and longlasting single-molecule fluorescence imaging. *Nat Methods* **3**, 891–893.

48. McKinney S. A., Joo C., Ha T. (2006) Analysis of single-molecule FRET trajectories using hidden Markov modeling. *Biophys J* **91**, 1941–1951.

49. SantaLucia J. J. (1998) A unified view of polymer, dumbbell, and oligonucleotide DNA nearest-neighbor thermodynamics. *Proc Natl Acad Sci USA.* **95**, 1460–1465.

50. Cavaluzzi M. J., Borer P. N. (2004) Revised UV extinction coefficients for nucleoside-5′-monophosphates and unpaired DNA and RNA. *Nucleic Acids Res* **32**, e13.

Chapter 16

Metal Ion–RNA Interactions Studied via Multinuclear NMR

Daniela Donghi and Roland K.O. Sigel

Abstract

Metal ions are indispensable for ribonucleic acids (RNAs) folding and activity. First they act as charge neutralization agents, allowing the RNA molecule to attain the complex active three dimensional structure. Second, metal ions are eventually directly involved in function. Nuclear magnetic resonance (NMR) spectroscopy offers several ways to study the RNA–metal ion interactions at an atomic level. Here, we first focus on special requirements for NMR sample preparation for this kind of experiments: the practical aspects of in vitro transcription and purification of small (<50 nt) RNA fragments are described, as well as the precautions that must be taken into account when a sample for metal ion titration experiments is prepared. Subsequently, we discuss the NMR techniques to accurately locate and characterize metal ion binding sites in a large RNA. For example, ^2J-[^{1}H,^{15}N]-HSQC (heteronuclear single quantum coherence) experiments are described to qualitatively distinguish between different modes of interaction. Finally, part of the last section is devoted to data analysis; this is how to calculate intrinsic affinity constants.

Key words: Ribonucleic acids, Metal ions, In vitro transcription, Nuclear magnetic resonance spectroscopy, Chemical shift perturbation, Affinity constants

1. Introduction

Ribonucleic acids are not simple working copies of the genome, but rather play crucial roles in a large number of biological functions (1). Since the discovery of catalytically active RNAs in the early 80s, i.e., ribozymes (2, 3), the research aimed at understanding structure and function of RNAs has remarkably increased.

Due to their polyanionic nature, RNAs are intrinsically associated with metal ions (4). In general, metal ions are diffusely bound to RNAs, acting primarily as charge screening agents. However, it has been shown that around 10% of the total negative charge is compensated by metal ions that are site-specifically bound (5). In order to be active, RNA molecules need to adopt complex three dimensional structures, stabilized by a large number and variety of tertiary contacts. In this folding process, localized pockets

Jörg S. Hartig (ed.), *Ribozymes: Methods and Protocols*, Methods in Molecular Biology, vol. 848,
DOI 10.1007/978-1-61779-545-9_16, © Springer Science+Business Media, LLC 2012

of accumulated negative charge are formed, that strongly bind positively charged ions, that is Na^+, K^+, Ca^{2+}, and Mg^{2+}, the most abundant solvated metal ions in cells (5). In particular, Mg^{2+} has been shown to perform both a structural and a catalytical role. For example, the presence of two Mg^{2+} ions in the catalytic core of group I and group II introns, as revealed by recent X-ray structures, strongly supports the hypothesis of a two metal ion mechanism for the catalysis performed by these two classes of large ribozymes (6, 7). Hence, the study of the interaction between metal ions and RNA has a double significance: it allows to characterize and understand the structural arrangement as well as the functioning of these large biomolecules, both being tightly interlinked.

NMR is a powerful technique able to provide information on biologically relevant molecules at the atomic level (8). Mostly, RNAs comprising 20–40 nucleotides are studied by NMR, but recently also NMR structures of RNAs in the size range of 25–40 kDa have been described (8, 9). For a detailed description on the use of NMR for RNA structural studies please refer to Chapter 12 of this volume. The study of RNA–metal ion interactions by NMR is challenging, as the exchange rate of metal ions in RNA–metal systems is equal or faster than the NMR time-scale. The resulting line broadening consequently hampers evaluation and characterization of the exact interaction.

The biologically most relevant metal ion in RNA biochemistry is Mg^{2+}, which is commonly considered as spectroscopically silent (5). It is however worth mentioning that Mg^{2+} is NMR active, and, despite its low sensitivity, low natural abundance, and its large quadrupolar moment, a few reports have been published dealing with the use of this "exotic" nucleus in the study of NTPs–Mg^{2+} as well as nucleic acids–Mg^{2+} interactions (for a recent review, see ref. 10). For example, by ^{25}Mg NMR line shape analysis of RNAs titrated with isotopically enriched Mg^{2+} it was possible to estimate the numbers and the association constants of weakly bound Mg^{2+} to tRNAs (11). However, in order to localize and quantify RNA–metal ion interactions by NMR, ^1H, ^{13}C, ^{15}N, and ^{31}P are the nuclei normally applied. They all have spin-½ and are straightforward to record and study. Moreover, the now easy and cost-affordable access to ^{13}C and ^{15}N isotope labeled RNAs (12) overcomes the difficulties intrinsically related to the low receptivity and abundance of these nuclei. Nevertheless, it is worth noting that most of the biologically relevant metal ions are NMR active, but show low sensitivity and spins larger than ½, resulting in broad signals not always easy to record.

RNA building blocks contain numerous metal ion binding atoms (see Fig. 1): phosphoryl oxygens, guanosine O6, uracil O4, and purine N7 are the most important ones. The metal ions can coordinate either directly through inner-sphere interactions

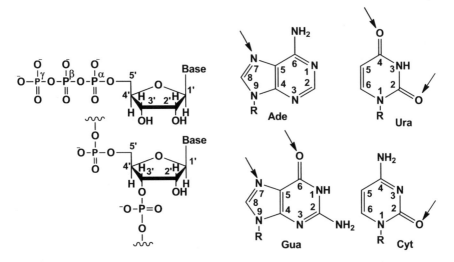

Fig. 1. General NTPs structure with atomic numbering schemes (*left top*) and nucleotide structure within RNA with the bridging 5′- and 3′-phosphates groups (*left bottom*). The chemical structures of the four most abundant natural nucleobases in RNA are given on the *right* together with their numbering scheme: adenine (Ade), guanine (Gua), uracil (Ura), and cytosine (Cyt). The *arrows* indicate the most common coordination sites for metal ions.

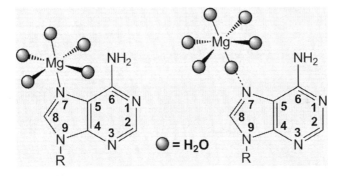

Fig. 2. Different coordination modes of metal ions to nucleic acids. Inner-sphere (*left*) and outer-sphere (*right*) coordination of [Mg(H$_2$O)$_6$]$^+$ to adenine-N7 is depicted.

or through outer-sphere interactions, i.e., mediated through the hydrogen bonds of the coordinated water molecule (see Fig. 2) (5). It is worth noting that the coordination properties of metal ions to single nucleotides can be directly translated to large nucleic acids (5).

Many studies have been devoted to the interaction between RNA and metal ions others than Mg^{2+} to better characterize this interaction. For example, Mn^{2+} is used in paramagnetic line broadening studies to probe for inner-sphere interaction while [Co(NH$_3$)$_6$]$^{3+}$, with its kinetically inert amines, can mimic the outer-sphere interaction of $[Mg(H_2O)_6]^{2+}$, yielding NOE cross peaks between the amine ligands and RNA protons (13, 14). Also, d^{10} have been applied: for example, strong perturbations in ^{15}N, ^{13}C and ^{1}H chemical shifts were observed upon titration of a hammerhead ribozyme metal binding site model with

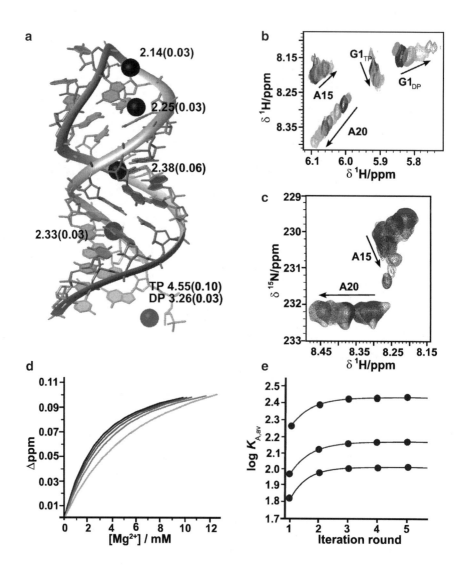

Fig. 3. NMR solution structure (a) of the truncated domain 6 of the yeast mitochondrial group II intron *Sc*.ai5γ (D6-27) used for the metal ions binding experiments described in Notes 38–40 (20, 21). The five Mg^{2+} ions and their intrinsic log K_a values are indicated (20). Overlaid selected regions of [^{1}H,^{1}H]-NOESY (b) and ^{2}J-[^{1}H,^{15}N]-HSQC (c) experiments show chemical shift perturbation within D6-27 upon Mg^{2+} titration (20, 21). Signals that broaden with increasing Mg^{2+} indicate direct coordination nearby whereas a "simple" change in chemical shift "just" indicates a structural change induced by neighboring Mg^{2+}-binding. The change of a 1:1 fit (d) and of the log $K_{a,av}$ values of a RNA with three binding sites (e) upon the iterative calculation procedure of intrinsic affinity constants is schematically indicated. Panel (a) prepared with MOLMOL (35) using PDB ID 2AHT. Figure 3c is adapted from Erat et al. (21).

Cd^{2+}: a high-field shift of about 20 ppm for ^{15}N, and a low-field shift of 0.38 and 2.3 ppm for ^1H and ^{13}C, respectively, confirm a direct coordination of Cd^{2+} to the N7 of a specific guanosine (15). When NMR active metal ions are used, $^{15}N–M^{n+}$ direct scalar coupling constants are evidence of the interaction, as shown recently in a DNA duplex containing Ag^+ (16). $^2J_{NN}$ metal mediated couplings have been reported for Hg^{2+} mediated thymine–thymine basepairs (17). However, the chemical exchange within RNA–metal ions system mostly prevents direct observation of J-coupling (18). Lastly, also ^{31}P chemical shift perturbation can be used to explore RNA–metal ion interactions (19).

The Mg^{2+}–RNA interaction is characterized by a general intrinsic lability and relatively low stability. Nevertheless, NMR titration experiments of RNA molecules with Mg^{2+} allow a detailed examination of simultaneous metal ion binding to various sites, as well as offer the possibility to calculate intrinsic binding affinities (20). In this protocol we first concentrate on the practical aspects of RNA sample preparation for such experiments. Then $[^1H,^1H]$-NOESY (Nuclear Overhauser Effect spectroscopy) and ^2J-$[^1H,^{15}N]$-HSQC experiments are described to follow Mg^{2+} binding by both ^1H and ^{15}N chemical shift perturbation, resulting in complementary information. Perturbation of proton chemical shifts is caused by either electronic effects of metal ion binding, structure perturbation caused by metal binding in the vicinity or a combination of both. Often it is necessary to first identify the binding pocket in order to obtain quantitative information on the binding constants. On the other hand, observation of the directly involved nucleus is indispensable to study the way of interaction: for example ^{15}N resonances allow to discriminate between inner- and outer-sphere interaction modes (21) (see Fig. 2). Finally in Notes 38–40, the interaction between Mg^{2+} and the truncated but catalytically active D6-27 from the yeast mitochondrial Sc.ai5γ group II intron (see Fig. 3) is described in more details as a practical example (20, 21).

2. Materials

All solutions must be prepared by using autoclaved bi-distilled water (18.2 MΩ cm resistivity at 25°C), that is additionally filtered through sterile 0.2-μm Filtropur syringe filters (Sarstedt) or Steritop™ bottle top filter units (Millipore). To this we refer as "ultrapure water" throughout the text. All NMR tubes, reaction vessels, pipette tips, etc. used must be RNase free. All chemicals are purchased from either Fluka-Sigma-Aldrich or Brunschwig Chemie, at puriss p.a. or biograde, if not otherwise specified.

2.1. In Vitro Transcription, Polyacrylamide Gel Electrophoresis Purification and Electroelution

1. Stock solutions of nucleosides 5′-triphosphate (NTPs): adenosine 5′-triphosphate (ATP; GE Healthcare), guanosine 5′-triphosphate (GTP; GE Healthcare), cytidine 5′-triphosphate (CTP; GE Healthcare), uridine 5′-triphosphate (UTP; Acros-Brunschwig). To prepare 1 mL of an about 200 mM solution, weigh roughly 120 mg of dried NTPs, add 20 μL of 1 M Tris(2-Amino-2-hydroxymethyl-propane-1,3-diol)–HCl (pH 7.5) (dissolve 60.57 g in 500 mL of ultrapure water, adjust the pH with HCl and store at room temperature) and 800 μL ultrapure water and mix well. Adjust the pH to 7.0 with freshly prepared 5 M NaOH solution, and fill up to 1 mL (see Note 1). The exact concentration of the freshly prepared NTP solutions is determined by UV spectroscopy (see Note 2). Prepare 1–2 mL of NTP solutions, divide in 300–400 μL aliquots and store them at –20°C. Use the NTP solutions within a few months.

2. NaOH and HCl solutions at different concentration: prepare 10 mL each of 5 M NaOH and 1 M HCl solutions. Lower concentrated stock solutions are obtained by dilution (1 M, 100 mM, 10 mM, 1 mM). Store at room temperature.

3. Transcription buffer (5×): 200 mM Tris–HCl (pH 7.5), 200 mM DTT (Dithiothreitol) and 10 mM spermidine (N-(3-aminopropyl)butane-1,4-diamine) in ultrapure water. For 1 mL of solution, mix 200 μL of 1 M Tris–HCl (pH 7.5), 200 μL of 1 M DTT (154.25 mg in 1 mL of ultrapure water, store at –20°C), 5 μL of 2 M spermidine (290.5 mg in 1 mL of ultrapure water, store at –20°C) and 595 μL of ultrapure water. Prepare 10–20 mL of transcription buffer solution and store it in 1 mL aliquots at –20°C. If properly stored, they can be used indefinitely.

4. Stock solutions of the two strands each of the double-stranded DNA template comprising the T7 promoter region plus the coding region (22), prepared at a concentration between 50 and 200 μM (checked with an UV spectrophotometer). The DNA strands are normally purchased from commercial suppliers (e.g., Microsynth) and purified by polyacrylamide gel electrophoresis (PAGE) before usage (see Notes 3–5). Store at –20°C.

5. Ultrapure 1 M MgCl$_2$ solution in water, as purchased from the commercial supplier (see Note 6). Store at room temperature.

6. Homemade bacteriophage T7 RNA polymerase in 30% glycerol (see Note 7). Store at –20°C.

7. 100 and 70% ethanol. Store at room temperature.

8. 5 M NaCl in ultrapure water. Weigh 2.9 g of NaCl in a 12-mL plastic tube and add up to 10 mL total volume of ultrapure water. Mix vigorously. Store at room temperature.

9. Gel solution in ultrapure water, 7 M urea (Eurobio), 89 mM TBE (Tris/Borate/EDTA-ethylenediaminetetraaceticacid), and 5–20% of polyacrylamide starting from a stock solution of AccuGel™ (40% w/v; 29:1 acrylamide/bisacrylamide), depending on the size of the RNA under investigation (see Note 8). For example, to prepare 1 L of 18% gel solution, weigh 420 g of urea in a 1-L beaker, add 100 mL of 10× TBE (see below) and 450 mL of 40% AccuGel™ 29:1 solution. Fill up to 1 L with ultrapure water. Mix with a magnetic stirrer for half an hour at 50–60°C to dissolve urea. Pass through a sterile filter and keep in a 1-L screwed cap autoclaved bottle. Store in the dark at 4°C.

10. TBE running buffer (1×). Dilute 100 mL of commercially available 10× TBE (0.89 M Tris–Borate pH 8.9, and 20 mM Na_2EDTA) to 1 L with ultrapure water. Transfer to a 1-L screwed cap bottle, mix and autoclave it. Prepare 3–5 L of running buffer. Store at room temperature.

11. TEMED (N,N,N',N'-tetramethyl-ethane-1,2-diamine), as purchased from the commercial supplier. Store at 4°C.

12. Ammonium persulfate (APS) solution in ultrapure water, 10% (w/v). Prepare 10 mL and store at 4°C. Use within a few weeks.

13. Denaturing urea loading buffer. 11.8 M Urea, 0.083% dyes, 8.3% sucrose, 4.2 mM Tris–HCl (pH 7.5), 0.83 mM EDTA (pH 8.0). Prepare 50 mL of loading buffer: weigh 35.4 g of urea in a 50-mL falcon tube, add 4.15 g of sucrose, 0.21 mL of 1 M Tris–HCl (pH 7.5), 0.415 mL of 0.1 M EDTA (pH 8), (dissolve 37.2 mg of $Na_2EDTA·2H_2O$ in 1 mL of ultrapure water and adjust the pH), 2.075 mL of each dyes (starting from a 2% solution of dye). Fill up with ultrapure water. Store at 4°C (see Note 9).

2.2. Preparation, Purification, and Concentration of the NMR Samples

1. 1 M KCl solution in ultrapure water (dissolve 37.25 g of KCl in 500 mL of ultrapure water, store at room temperature), and ultrapure water for the washing and concentration of the RNA samples.

2. For NMR sample preparation: 100% D_2O (Armar Chemicals, see Note 10), 1 M KCl in D_2O (74.55 mg in 1 mL), 1 mM Na_2EDTA in D_2O (3.36 mg in 1 mL, and further dilute 100 µL to 1 mL), DCl and NaOD in D_2O at different concentration (normally from 0.1 to 100 mM, starting from concentrated solution purchased from commercial suppliers) to adjust the pH of the sample. Prepare 1 mL aliquots of each needed reagents and store at 4°C.

3. For metal ion titration: 100 mM $MgCl_2$ in D_2O (95.21 mg of ultra pure water free $MgCl_2$ in 10 mL of D_2O). The exact

concentration of the stock solution is determined by potentiometric pH titration employing EDTA (23). Store at room temperature.

3. Methods

It is important to always use latex gloves during sample preparation and manipulation. Human skin is a rich source of RNases and other contaminations that cause RNA degradation. If the sample tends to easily degrade, it is advisable to prepare fresh buffers and solution each time.

3.1. In Vitro Transcription, Polyacrylamide Gel Electrophoresis Purification and Electroelution

1. Slowly defreeze the NTPs, the transcription buffer and the two DNA strands solutions on ice. Mix the NTPs (see Note 11) (final concentration of 5 mM) with the transcription buffer (final concentration of 1×), 0.01% (w/v) Triton X-100 (polyethylene glycol p-(1,1,3,3-tetramethylbutyl)-phenyl ether) and ultrapure water in a 12-mL plastic tube. According to the efficiency of the transcription (that depends on the RNA sequence), 5 or 10 mL transcription is recommended in order to obtain enough RNA for the preparation of 1–3 NMR samples (ranging from 0.5 to 1 mM concentration in a typical 250–300 μL volume). Mix gently after each addition. Add the optimal amount of $MgCl_2$ (around 20–40 mM) (see Note 6) and the two complementary DNA strands (around 1 μM) (see Note 12). Aliquot the solution in 1.5-mL plastic tubes, each containing up to 1 mL transcription solution. Only at this point add the polymerase (stored at –20°C until its usage) to each transcription tube (see Note 13). For a practical example of the needed amount of transcription reagents, see Table 1. Incubate the transcription tubes for 4–12 h at 37°C, shaking at 300 rpm (see Notes 13 and 14).

2. If the transcription works well, a white precipitate due to $Mg_2P_2O_7$ (magnesium pyrophosphate) formation is usually visible. Spin down the magnesium pyrophosphate and transfer 2.5 mL each of the supernatant in 12-mL centrifuge compatible plastic tubes. Precipitate the solutions with 3 volumes of 100% EtOH and 1/20 volume of 5 M NaCl, either overnight at –20°C or at –80°C for 1 h. Thereafter centrifuge the frozen tubes at 4°C for 30 min at $13,000 \times g$, collect the RNA precipitate and dry it in vacuum for few minutes.

3. Prepare for the gel electrophoresis. Wipe the gel plates, spacers and combs with ultrapure water and 70% EtOH, assemble them on a horizontal surface and fix them with clamps. For a 5 mL transcription, it is advisable to prepare two big gels

Table 1
Practical example of the composition of a 5 mL transcription solution

Reagents	Starting concentration (mM) (stock solutions)	Final concentration (mM) (transcription solution)	Needed amount (µL) (transcription solution)
Transcription buffer	5×	1×	1000.0
GTP	150	5	166.7
ATP	160	5	156.3
CTP	140	5	178.6
UTP	150	5	166.7
Triton X-100	10%	0.01%	5.0
Template DNA strand	70	1×10^{-3}	35.7
Top strand DNA	50	1×10^{-3}	50.0
MgCl$_2$	1,000	35	175.0
RNA T7 polymerase	100×	1×	50.0
Ultrapure water			3016.0

The third column lists the final concentration of each reagent in the transcription reaction. In the fourth column the added volume of each reagent is given. Note that the efficiency of T7-RNA polymerase and thus the added amount must be checked for every new batch

(45×30 cm with 1.5-mm spacers). For each gel, transfer about 200 mL of gel solution (prepared the day before and stored at 4°C) to an autoclaved beaker, add 1 mL of APS and 120 µL of TEMED. Gently mix the solution with a plastic syringe and pour it into the preassembled glass plates. Try to avoid bubbles by gently tapping the glass with your knuckles during pouring. When the plates are filled place the combs. The gel will take around 1 h to polymerize.

4. After polymerization remove the combs and place the gels in the vertical electrophoresis gel apparatus. Fix them to the upper chambers with clamps. Fill the upper and the lower chambers with 1× TBE running buffer. Pre-run the gels for 10 min. Dissolve the dried RNA in a total volume of 0.5–1 mL of ultrapure water, add the same amount of denaturing urea loading buffer and load the sample. Use a maximum of 1.5 mL of sample for each gel. Run the gels. For example, 18% gels can be run overnight at low power (10–15 W per gel) or during the day at higher power (30–40 W per gel) (see Note 15).

5. Stop the electrophoresis when the RNA reaches the lower third of the gel. Disassemble the gel with the use of a thin spatula and place it on a piece of cellophane. Put it under a 260 nm

UV lamp in order to visualize the RNA sample. Excise it from the gel by cutting the band in 3–8 mm pieces with a scalpel.

6. Recover the RNA sample by electroelution at 200 V at room temperature into the smallest volume possible using an electroelution system with BT1 and BT2 membranes (Whatman® Elutrap System). Normally the electroelution procedure is repeated 2–3 times, each lasting for 1–2 h. The collected solution is precipitated with EtOH and NaCl, as described above. Check by UV-shadowing if all the RNA has been recovered from the gel.

7. Centrifuge the solutions containing RNA (30 min, $13,000 \times g$, at 4°C), collect the RNA pellets and dry them in vacuum.

3.2. Preparation, Purification and Concentration of the NMR Samples

The sample for metal titration NMR measurements must contain only RNA (and the needed amount of salt for folding, see below). All buffers used during transcription (especially Tris, which is a good chelating ligand for metal ions (24)) as well as water soluble acrylamide impurities must be carefully and completely removed.

1. The isolated RNA is purified by centrifugation (Vivaspin 2-mL ultrafiltration devices, Sartorius Stedim biotech) with several washing steps employing first 1 M KCl and then ultrapure water (see Notes 16 and 17). For example, for a 300 µL of a 0.5 mM sample, the washing procedure must be repeated at least 5 times with 1 M KCl at pH 7.5–8 and 5 times with ultrapure water at room temperature at the speed recommended by ultrafiltration devices suppliers. Each time the volume of the sample should be reduce from 2 mL to about 300 µL. Centrifugation time can vary from 10 to 30 min according to the sample concentration.

2. After several rinsing cycles, collect the sample and check its concentration by UV measuring the absorption at 260 nm. According to its concentration and intended usage divide the sample into aliquots and lyophilize them (see Notes 18 and 19).

3. Dissolve the previously lyophilized RNA in 100% D_2O, in the presence of the needed amount of KCl (see Notes 20 and 21). The ideal sample concentration is around 0.5–1 mM (see Note 22). Add 4–8 µL of 1 mM EDTA in D_2O (up to 10 µM concentration) in order to chelate transition metal ion impurities that may be present in the sample. Adjust the pH using DCl and/or NaOD solutions (see Notes 23 and 24) and check again the concentration by UV.

4. Transfer the sample into a 5-mm Shigemi tube, ideally in 300 µL of total volume (see Notes 25 and 26).

3.3. NMR Measurements: [¹H,¹H]-NOESY and ²J-[¹H,¹⁵N]-HSQC Experiments

The sample is now ready for the NMR measurements. To obtain a good resolution at reasonable experimental time, the spectrometers used for *bio*-NMR experiments normally have magnetic fields of 500 MHz or higher and are equipped with z axis pulsed field gradient cryoprobes™ (Bruker) or cold probes (Varian/Agilent). Even if the experiments here described are performed in D_2O it is important to find the right conditions for water suppression.

The scope of the experiments described below is to observe the chemical shift changes of ¹H and ¹⁵N upon metal ion titration and use these changes to describe qualitatively and quantitatively the studied RNA–metal ion system.

1. Record a series of ¹H-NMR spectra at different temperatures and KCl concentrations (see Note 20), evaluate the line-width of the resonances and set the conditions which are best. Normally, the temperatures used are between 293 and 310 K.

2. If sugar and aromatic proton chemical shift changes are followed, [¹H,¹H]-NOESY experiments are best suited (see Fig. 3b, Note 27). Usually, only the H1' are followed as all other sugar protons suffer from a strong overlap. To observe N7–H8 chemical shift changes, ²J-[¹H,¹⁵N]-HSQC experiments (see Figs. 3c and 4, Notes 28–30) with a ¹⁵N labeled RNA sample should be recorded.

3. After recording the first experiments at 0 mM M^{n+} concentration start the metal ion titration: a defined amount of MgCl$_2$ solution in D_2O (normally 1 μL additions are performed) is laid on the inner NMR tube wall. The tube is covered with parafilm and the solution gently mixed (see Note 31). The tube can also be centrifuged with a NMR tube table centrifuge. Strong line broadening of ¹⁵N resonances is often observed already in the presence of 2.5 mM Mg^{2+} while, in the case of ¹H resonances, a larger concentration range of up to 10–15 mM Mg^{2+} is possible (21). Depending on which nucleus is observed, perform 0.5 or 1 mM addition steps (see Notes 32 and 33).

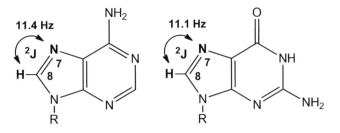

Fig. 4. ¹H,¹⁵N long-range coupling between H8 and N7 of adenine and guanosine nucleobases as observed by the ²J-[¹H,¹⁵N]-HSQC experiment (see Fig. 3c).

4. Repeat the NMR experiments without changing parameters at each titration steps. Only the 90° pulse and the field homogeneity must be checked after every addition, as well as the number of scans (see Notes 27, 28 and 34).

5. When the titration experiment is finished remove the Mg^{2+} as soon as possible. In order to do that, purify the sample with PAGE electrophoresis, followed by electroelution (see above) (see Note 35). Clean the Shigemi tube with ultrapure water, 0.1 M EDTA and EtOH. Alternatively, add an excess of EDTA to the NMR sample to chelate magnesium and purify it by centrifugation (see above).

3.4. Data Treatment and Evaluation of Affinity Constants

In order to obtain meaningful data as many ^1H and ^{15}N resonances of the studied RNA sample must be attributed, e.g., by the sequential walk strategy, as described elsewhere (8, 25). Perform at least 10–12 titration steps in order to obtain a reliable fit of the experimental data and accurate affinity constants.

3.4.1. Qualitative Evaluation of ^1H Data

1. Map the ^1H chemical shift variations ($\Delta\delta$) of as many peaks as possible based on the [^1H,^1H]-NOESY experiments and tabulate them.

2. Calculate the $\Delta\delta$ values at each step of Mg^{2+} addition and plot them in a bar diagram to qualitatively obtain a first picture of Mg^{2+} binding and Mg^{2+}-induced structure perturbation. The observed $\Delta\delta$ values, both upfield and downfield, are normally in the range 0.05–0.25 ppm. Usually, an upfield shift is due to an increase in stacking interaction attributable to metal ion binding nearby, while the downfield shift can be attributed to a decrease in stacking interactions and/or the electron-pulling effect of metal ion coordination. The plot of $\Delta\delta$ vs. Mg^{2+} concentration offers a qualitative, easy and clear picture of metal ion binding and allows to divide the RNA molecule in few binding regions. This is a good starting point for the following steps of data analysis consisting in the quantitative evaluation of the binding.

3.4.2. Estimating the Affinity Constants

1. Plot the chemical shifts of the observed protons vs. the increasing total Mg^{2+} concentration.

2. Fit the data to Eq. 1, corresponding to a 1:1 binding model, using for example a Levenberg–Marquardt nonlinear least square fit algorithm, to calculate the affinity constant K_{Ai} of Mg^{2+} toward a binding site i in the RNA molecule (see Note 36). This can be done with software like Origin®, Matlab® or similar mathematical programs yielding first preliminary affinity constants. In Eq. 1, $[RNA_i]_{tot}$ corresponds to the total RNA concentration and $[M^{n+}]_{tot}$ to the total increasing magnesium concentration. δ_{obs} is the chemical shift of the observed proton

while $\delta_{RNA}i$ and $\delta_{RNA \cdot M}$ are the chemical shifts of the unbound and completely bound species, respectively.

$$\delta_{obs} = \delta_{RNAi} + (\delta_{RNA \cdot M} - \delta_{RNAi}) \frac{[M^{n+}]_{tot} + [RNA_i]_{tot} + \frac{1}{K_{Ai}} - \sqrt{\left([M^{n+}]_{tot} + [RNA_i]_{tot} + \frac{1}{K_{Ai}}\right)^2 - 4[M^{n+}]_{tot}[RNA_i]_{tot}}}{2[RNA_i]_{tot}} \tag{1}$$

3. Group the binding constants estimated with Eq. 1 into binding sites according to their similar values and neighborhood. The qualitative plot of $\Delta\delta$ thereby helps to define binding regions.

4. Use Mn^{2+} line broadening experiments to confirm the different binding sites within the molecule (see Note 37).

5. Calculate averaged affinity constants $K_{A,av1i}$ by taking either the arithmetic or weighted mean of the respective K_{Ai} values for each binding site.

3.4.3. Determination of the Intrinsic Affinity Constants

Usually more than one Mg^{2+} coordination site with similar affinity is present in RNA. Consequently, the various binding sites are filled in parallel and the amount of Mg^{2+} available for each site is smaller than the total Mg^{2+}. The average affinity constants obtained as described in Subheading 3.4.2 do not take this effect into account. To determine the intrinsic affinity constant for a given site, $[M^{n+}]_{tot}$ in Eq. 1 must be substituted by the actual available Mg^{2+} concentration (20).

1. Use the obtained $K_{A,av1i}$ values to calculate the concentration of bound Mg^{2+}, $[M^{n+}]_{bound,i}$, at each binding site with Eq. 2.

$$[M^{n+}]_{bound,i} = \frac{(K_{A,av1i}[M^{n+}]_{tot} + K_{A,av1i}[RNA_i]_{tot} + 1) - \sqrt{(-(K_{A,av1i}[M^{n+}]_{tot} + K_{A,av1i}[RNA_i]_{tot} + 1))^2 - 4K_{A,av1i}^2[M^{n+}]_{tot}[RNA_i]_{tot}}}{2K_{A,av1i}} \tag{2}$$

2. Calculate the actual available Mg^{2+} concentration, $[M^{n+}]_{avail,i}$, using Eq. 3, where $[M^{n+}]_{bound}$ corresponds to the sum of the concentrations of Mg^{2+} bound to all binding sites.

$$[M^{n+}]_{avail,i} = [M^{n+}]_{tot} - \sum [M^{n+}]_{bound} + [M^{n+}]_{bound,i} \tag{3}$$

3. Fit again the chemical shift data vs. the corrected concentration $[M^{n+}]_{avail,i}$ at each binding site using Eq. 1. New affinity constants are obtained and again averaged for each individual binding site, to give a second set of average affinity constants $K_{A,av2i}$ (see Fig. 3d).

4. Insert these new affinity constants in Eqs. 2 and 3 to calculate an improved value for $[M^{n+}]_{bound,i}$ and $[M^{n+}]_{avail,i}$ for each individual binding site (see Fig. 3d, e).

5. Repeat the procedure iteratively until no changes in $K_{A,av}$ are observed anymore within the error limit (see Fig. 3d, e).

6. Plot the $K_{A,av}$ values for each iteration step vs. the corresponding iteration round and fit them to an asymptotic function in order to obtain the final intrinsic affinity constant for each binding site (see Fig. 3e).

   Naturally, the above described calculation procedure is not restricted to Mg^{2+} binding but can be applied to any system that contains multiple binding sites for the same ligand at similar affinities. In Notes 38 and 39 details of this procedure for an example case are given.

3.4.4. Qualitative
Evaluation of ¹⁵N Chemical
Shift Changes

Proton chemical shift changes in such experiments can have different causes, i.e., direct metal ion coordination or conformational changes in the RNA structure due to metal ion binding close by. Extending the investigations into the ^{15}N dimension, often allows to distinguish between these two cases and possibly even to identify the exact coordinating ligand.

1. Record ^2J-$[^1H,^{15}N]$-HSQC experiments during a metal ion titration (see Note 40). To obtain a qualitative picture overlay to compare the behavior of the single resonances (see Fig. 3c). You can also map both ^1H and ^{15}N chemical shifts as described in Subheading 3.4.1.

2. Analyze the shifts according to the following qualitative rules: In case shift changes in either ^{15}N or both dimensions and a severe line broadening are observed, this can be taken as evidence for a direct (inner-sphere) Mg^{2+}–N coordination (A15 in Fig. 3c) based on the fast ligand exchange rate of Mg^{2+}. If only the ^1H but not in the ^{15}N chemical shift is affected (A20 in Fig. 3c) structural changes are the cause. If both ^1H and ^{15}N signals are shifted but without line broadening outer-sphere interaction is probably the cause (see Note 41).

4. Notes

1. In order to minimize dilution, first add 5 M NaOH and then add NaOH of lower concentration (1 M, 100 mM, and 10 mM) to reach the desired pH. HCl solution at different concentration can be used to decrease the pH again if too much base has been added.

2. Use the following extinction coefficients of the nucleotides $(M^{-1}cm^{-1})$: $\varepsilon_{ATP}(260\,nm) = 15.4 \times 10^3, \varepsilon_{UTP}(260\,nm) = 10.0 \times 10^3, \varepsilon_{CTP}(270\,nm) = 9.0 \times 10^3, \varepsilon_{GTP}(249\,nm) = 13.7 \times 10^3$.

3. In principle, only the T7 promoter region must be double-stranded. However, our experience shows that higher transcription yields are reached by using fully double stranded-DNA oligonucleotides.

4. For RNA sequences comprising more than 50 nucleotides, synthetic double stranded-DNA oligonucleotides are usually substituted by cut plasmids. Nevertheless, RNA samples used for NMR experiments are mostly shorter than 40–50 nucleotides.

5. The efficiency of the T7 RNA polymerase depends critically on the first six nucleotides of the transcription. Most importantly, a 5′-guanosine is a prerequisite. 5′-GGGAGA, 5′-GGGAUC, and 5′-GAGCGG are known to be good starting sequences. For a collection of good starting sequences, see also ref. 26.

6. Note, $MgCl_2$ is not a titrimetric standard. The $MgCl_2$ concentration is crucial for the transcription yield and must be optimized for each sequence. In order to do so, transcription trials are performed on 50-μL scale. The efficiency of the transcription is checked with an analytical gel. In our experience, a 35 mM concentration is optimal for transcription of relatively short RNAs (<50 nt).

7. Bacteriophage T7 RNA polymerase does not do a primer extension but rather starts from a single GTP, which makes it useful for in vitro transcriptions. However, T7 tends to add one or more random nucleotides at the 3′-end of the run-off transcription, leading to problems with 3′ homogeneity. A way to circumvent 3′- and 5′-inhomogeneous ends can be found in refs. (22, 27). T7 can be obtained commercially, but for the large scale of NMR transcriptions it is advisable to express it oneself. An easy-reading protocol on T7 RNA polymerase preparation and its use for in vitro transcription has been published recently (22).

8. For fragments comprising 30–50 nucleotides use 18% polyacrylamide gel solution. For longer fragments, use lower percentage of polyacrylamide (28). Safety note: unpolymerized acrylamide is a neurotoxin. To avoid skin contact, wear protection gloves and glasses as well as a lab coat while handling.

9. The dyes normally used are xylene cyanol (XC) and bromophenol blue (BB), whose migration in polyacrylamide denaturing gels is known (28). For very short RNAs also Orange G can be used, which runs with a dinucleotide.

10. 100% refers normally to a minimum of 99.98% D_2O isotope enrichment. Buy small bottles (10 mL maximum) and store them at room temperature. Repeated opening and closing of the bottle causes an increasing of H_2O content.

11. Labeled nucleotides (^{13}C and/or ^{15}N) are nowadays normally purchased from a commercial supplier (e.g., Silantes GmbH). In the past, such nucleotides had prohibitive prices and different routes for their synthesis were proposed (8, 12). In the last few years more competitive prices made it possible to overcome the need of synthesizing them for most applications.

12. As for the $MgCl_2$ concentration (see Note 6), the amount of DNA used in transcription must be optimized. This is best done by performing 50 μL transcription trials in the presence of different amount of DNA and evaluating the transcription efficiency with an analytical gel. In our experience the best concentration of DNA is between 0.6 and 1.2 μM.

13. The amount of T7 RNA polymerase to be used depends on its efficiency and stock concentration, which must be checked any time a new polymerase batch is prepared. Again this is best done by 50 μL transcription trials with varying T7 concentrations and evaluated on analytical gels. The optimal duration of the transcription depends on the RNA sequence and must be checked independently. Longer sequences tend to degrade if too long transcription times are used.

14. During transcription, metal ion assisted partial hydrolysis of the terminal NTP seems to occur, leading to a mixture of a 5′ terminal tri- and diphosphate groups. Once the sample is prepared no further change in the ratio of di- and triphosphate terminal groups seems to occur (20).

15. A $n/n+1$ nucleotide resolution can be reached by using the right percentage of acrylamide, not overloading the gel, and performing long runs. Other separation techniques based on chromatographic methods are also known but will not be discussed here (29).

16. Before usage, the filter devices must be rinsed following the instructions of the supplier. Choose the filter devices according to the molecular weight of the sample. For a good recovery of RNA the membrane cutoff should be at least 50% smaller than the molecular weight of the RNA. For RNAs used for NMR, which normally range from 7 to 20 kDa, membranes with 3 kDa cutoff are advisable.

17. This point is very critical in order to prepare a sample useful for metal titration studies: Tris used in the transcription buffer is a good chelating ligand for metal ions (24) and must be completely removed to obtain reliable values of affinity constants.

18. The lyophilized RNA can be stored in a closed vial at −20°C up to several months without risk of degradation.

19. If hairpins are investigated, it is important to verify that a hairpin and not a duplex is formed in solution. To avoid duplex

formation and any possible association events one possibility is to dilute the RNA sample with up to 100 mL of ultrapure water, denature the RNA at 80–90°C in a water bath, and cool it again on ice. It is better using glassware instead of plastic vessels which can release plasticizers into RNA sample upon heating. Concentrate the sample by centrifugation and lyophilization.

20. KCl binds unspecifically to RNA and is needed to ensure the proper RNA secondary structure. Its concentration normally ranges from 50 to 100 mM.

21. Often phosphate buffers are used in NMR samples. This must be strictly avoided when performing metal ion titrations, as phosphate is a good ligand for metal ions (30).

22. Aggregates can form at higher concentration, which hamper the quality of the spectra. Aggregates can be detected by NMR diffusion experiments (DOSY) or spectroscopically by dynamic light scattering (DLS) (16).

23. The ideal pH is close to neutrality or slightly acidic. In our experience, the best pH range is between 6.7 and 7.2. RNA is sensitive to alkaline conditions, at which it undergoes base catalyzed hydrolysis.

24. As the NMR solution is prepared in D_2O, 0.4 log units have to be added to the pH-meter reading in order to get the correct pD value (31).

25. Shigemi tubes are special NMR microtubes, whose magnetic susceptibility is matched to a specific solvent, and which are ideal for measuring small sample amounts. They consist of an outer tube and a plunger. The solution is poured into the tube and the plunger is then carefully placed on top of the solution. Care should be taken to avoid bubbles at the interface between the plunger and the sample.

26. A 200 µL volume is the minimum but to have a good field homogeneity it is advisable to use at least 300 µL. If you have only little RNA, 300 µL 0.5 mM sample is better than 200 µL 0.75 mM sample.

27. Depending on the sample concentration, 5–8 h for a [^{1}H,^{1}H]-NOESY spectrum are required. With increasing amounts of Mg^{2+}, the peaks get broader and more scans are needed in order to obtain a good signal to noise ratio. Consequently the last points of the titration can each take up to 12 h. As the homogeneity of the system is fundamental in order to analyze the data, it is advisable to finish the titration within a reasonable time. i.e., one needs at least 4–5 consecutive days of measuring time. The same mixing time should be used in all the experiments (200–250 ms) (8).

28. This experiment is useful to follow the chemical shift variation of N7 of the purine nucleobases. It uses the ^2J coupling constant between N7 and H8 (32) (see Fig. 4) and allows to contemporary follow N7 and H8 chemical shift perturbation upon Mg^{2+} titration (21). As in the case of the [^1H,^1H]-NOESY (see Note 27), these experiments can take up to 8–10 h at high Mg^{2+} concentration.

29. The samples used for these experiments can be fully or partly ^{13}C, ^{15}N, or only ^{15}N labeled. In case a ^{13}C, ^{15}N labeled sample is used, ^{13}C and ^{15}N need to be simultaneously decoupled. Samples can be prepared in which only adenosines or only guanosines are labeled. Although time consuming, the use of different samples containing only one kind of labeled nucleotide at a time allows a straightforward and unambiguous attribution of the peaks.

30. In our experience ^2J-[^1H,^{15}N]-HSQC experiments comprising a selective 180° ^{15}N IBurp2 shaped pulse give good results (21).

31. If folding needs to start from single stranded-RNA, the solution is transferred to a 1.5-mL plastic tube. The solution is warmed up using a heating plate, $MgCl_2$ is added, the solution mixed by a vortex mixer, and cooled down on ice before transferred back into the NMR tube. Such, mixing is most effective but usually leads to a considerable loss of sample after a few titration steps.

32. Here, we describe the titration using $MgCl_2$ but with other metal ions it works accordingly. For example $MnCl_2$ and $CaCl_2$ can be used. In the case of Cd^{2+} it is important to use $Cd(ClO_4)_2$ and replace KCl with $NaClO_4$ because $CdCl^+$ is a rather stable complex in solution.

33. Mg^{2+} binding is responsible for both chemical shift changes and signal line broadening due to the ligand exchange rate on the NMR time-scale. Nevertheless, evaluating both the proton chemical shift changes and their line broadening allows to depict an accurate picture of the metal ion binding sites (33).

34. High concentration of salt increases the pulse length and may hamper proper tuning and matching.

35. M^{2+} ions tend to facilitate hydrolysis, resulting in RNA degradation. After performing the metal ion titration make sure that you have all the data needed and desalt the sample before starting the data analysis.

36. Chemical shifts of aromatic and sugar H1′ protons are usually best dispersed and provide the best data. Also imino protons can be investigated (in 90% H_2O/10% D_2O) but can often not be properly fit to a 1:1 binding model because of solvent exchange phenomena (34).

37. These experiments are performed following the above described procedure (20): [^{1}H,^{1}H]-NOESY experiments are recorded at each titration steps, from 0 to 200 µM Mn^{2+} concentration, in 10–20 µM addition steps. MnCl$_2$ should be added in smaller steps than with Mg^{2+} because strong line broadening is observed already at very low Mn^{2+} concentration (13). Again, aromatic and H1′ sugar protons are evaluated.

38. The method was used to evaluate the intrinsic affinity constants of Mg^{2+} to D6-27 (see Fig. 3) (20). A 0.85 mM sample (100 mM KCl, 10 µM EDTA, pD = 6.7 in 100% D$_2$O) was titrated with increasing amounts of Mg^{2+} (0–12 mM in 11 steps) and [^{1}H,^{1}H]-NOESY experiments were collected at any step at 303 K. The ^{1}H changes were plotted against Mg^{2+} concentration and (if possible) fit to Eq. 1 with a Levenberg–Marquardt nonlinear least square regression. Preliminary affinity constants $K_{A,est}$ were calculated using the total concentration of Mg^{2+} in solution and grouped into five individual binding sites on the basis of their value, their proximity, and Mn^{2+} line broadening to determine averaged $K_{A,av}$ values for each binding site. These values were inserted in Eqs. 2 and 3 to calculate the actual Mg^{2+} concentration. The corrected Mg^{2+} concentrations were then used in Eq. 1 in the next iteration step to obtain improved $K_{A,est}$ values. This iterative procedure was repeated until the calculated intrinsic affinity constants did not change anymore within the experimental errors (see Fig. 3d, e). The final intrinsic affinity constants as well as the five binding sites to which they refer are shown in Fig. 3a (20).

39. Particular attention needs to be paid when evaluating the affinity constants at the 5′ terminus. As stated in Note 14, 5′-tri- and 5′-diphosphate terminal groups are usually present after transcription, both of which are good Mg^{2+} binding sites (20). First, the exact ratio (based on NOE cross peaks ratio or ^{31}P signals) must be determined. Second, in contrast to all the other binding sites, whose pK_a values are far away from neutrality, at pD 6.7 (at which titration of D6-27 were performed) competition between D$^+$ and Mg^{2+} occurs (NTP, pK_{a/H_2O} = 6.50 ± 0.05, and NDP, pK_{a/H_2O} = 6.40 ± 0.05) (20). The apparent affinity constants at 5′-termini can then be corrected using Eq. 4. K_{a/D_2O} represents the acidity constant of the tri-/diphosphate in D$_2$O, calculated according to Eq. 5.

$$\log K_{A,DP/TP} = \log K^{app}_{A,DP/TP} + \log\left(1 + \frac{[D^+]}{K_{a/D_2O}}\right) \qquad (4)$$

$$pK_{a/D_2O} = 1.015 pK_{a/H_2O} + 0.45 \qquad (5)$$

40. For example, this method was used to study the interaction between D6-27 (see Fig. 3c) and Mg^{2+}. Different samples were prepared, with either only guanosines or only adenosines ^{13}C and ^{15}N labeled (concentration ranging from 0.5 to 1.57 mM, 50 mM KCl, 10 μM EDTA, pD = 6.7). The samples were titrated with Mg^{2+} from 0 to 2.5 mM, in 0.5 mM steps. At any step, ^2J-$[^1H,^{15}N]$-HSQC experiments were recorded, at 298 K. The ^{15}N chemical shift variation observed upon Mg^{2+} titration were small (lower than 2 ppm) but allowed to give insights into the nature of the interaction (21).

41. If enough titration steps have been performed and the data is of high enough quality also a quantitative evaluation is possible as described in Subheadings 3.4.2 and 3.4.3.

Acknowledgments

Financial support by the Seventh European Community Frame-work Program (Marie-Curie Post-doctoral Fellowship to D. D. and ERC Starting Grant 2010 259092-MIRNA to R. K. O. S.), the Swiss National Science Foundation (Project 200021-124834/1 to R. K. O. S.), and the University of Zürich is gratefully acknowledged.

References

1. Gesteland, R. F., Cech, T. R., Atkins, J. F. (2006) The RNA World, Cold Spring Harbor Press.

2. Kruger, K., Grabowski, P. J., Zaug, A. J., Sands, J., Gottschling, D. E., Cech, T. R. (1982) Self-splicing RNA: autoexcision and autocyclization of the ribosomal RNA intervening sequence of Tetrahymena. Cell 31, 147–157.

3. Guerrier-Takada, C., Gardiner, K., Marsh, T., Pace, N., Altman, S. (1983) The RNA moiety of ribonuclease P is the catalytic subunit of the enzyme. Cell 35, 849–857.

4. Sigel, R. K. O. (2005) Group II intron ribozymes and metal ions - a delicate relationship. Eur J Inorg Chem 12, 2281–2292.

5. Freisinger, E., Sigel, R. K. O. (2007) From nucleotides to ribozymes – a comparison of their metal ion binding properties. Coord Chem Rev 251, 1834–1851.

6. Stahley, M. R., Strobel, S. A. (2005) Structural evidence for a two-metal-ion mechanism of group I intron splicing. Science 309, 1587–1590.

7. Toor, N., Rajashankar, K., Keating, K. S., Pyle, A. M. (2008) Structural basis for exon recognition by a group II intron. Nature Struct Mol Biol 15, 1221–1222.

8. Flinders, J., Dieckmann, T. (2006) NMR spectroscopy of ribonucleic acids. Prog Nucl Mag Res Spec 48, 137–159.

9. Davis, J. H., Tonelli, M., Scott, L. G., Jaeger, L., Williamson, J. R., Butcher, S. E. (2005) RNA helical packing in solution: NMR structure of a 30 kDa GAAA tetraloop-receptor complex. J Mol Biol 351, 371–382.

10. Ronconi, L., Sadler, P. J. (2008) Applications of heteronuclear NMR spectroscopy in biological and medicinal inorganic chemistry. Coord Chem Rev 252, 2239–2277.

11. Reid, S. S., Cowan, J. A. (1990) Biostructural chemistry of Magnesium ion characterization of the weak binding sites on transfer RNAphe(Yeast) - Implications for conformational change and activity. Biochemistry 29, 6025–6032.

12. Lu, K., Miyazaki, Y., Summers, M. F. (2010) Isotope labeling strategies for NMR studies of RNA. J Biomol NMR 46, 113–125.

13. Gonzalez, R. L., Jr., Tinoco, I., Jr. (2001) Identification and characterization of metal ion

binding sites in RNA. *Methods Enzymol* 338, 421–443.

14. Johannsen, S., Korth, M. M. T., Schnabl, J., Sigel, R. K. O. (2009) Exploring Metal Ion Coordination to Nucleic Acids by NMR. *Chimia* 63, 146–152.

15. Tanaka, Y., Taira, K. (2005) Detection of RNA nucleobase metalation by NMR spectroscopy. *Chem Commun*, 2069–2079.

16. Johannsen, S., Megger, N., Bohme, D., Sigel, R. K. O., Muller, J. (2010) Solution structure of a DNA double helix with consecutive metal-mediated base pairs. *Nature Chem* 2, 229–234.

17. Tanaka, Y., Oda, S., Yamaguchi, H., Kondo, Y., Kojima, C., Ono, A. (2007) ^{15}N,^{15}N J-coupling across Hg(II): Direct observation of Hg(II)-mediated T-T base pairs in a DNA duplex. *J Am Chem Soc* 129, 244–245.

18. Tanaka, Y., Ono, A. (2008) Nitrogen-15 NMR spectroscopy of N-metallated nucleic acids: insights into N-15 NMR parameters and N-metal bonds. *Dalton Trans*, 4965–4974.

19. Maderia, M., Horton, T. E., DeRose, V. J. (2000) Metal interactions with a GAAA RNA tetraloop characterized by ^{31}P NMR and phosphorothioate substitutions. *Biochemistry* 39, 8193–8200.

20. Erat, M. C., Sigel, R. K. O. (2007) Determination of the intrinsic affinities of multiple site-specific Mg^{2+} ions coordinated to domain 6 of a group II intron ribozyme. *Inorg Chem* 46, 11224–11234.

21. Erat, M. C., Kovacs, H., Sigel, R. K. O. (2010) Metal ion-N7 coordination in a ribozyme branch domain by NMR. *J Inorg Biochem* 104, 611–613.

22. Gallo, S., Furler, M., Sigel, R. K. O. (2005) In vitro transcription and purification of RNAs of different size. *Chimia* 59, 812–816.

23. Sigel, H., Zuberbühler, A. D., Yamauchi, O. (1991) Comments on potentiometric pH titrations and the relationship between pH-meter reading and hydrogen ion concentration. *Anal Chim Acta* 255, 63–72.

24. Sigel, H., Griesser, R. (2005) Nucleoside 5′-triphosphates: Self-association, acid-base, and metal ion-binding properties in solution. *Chem Soc Rev* 34, 875–900.

25. Erat, M. C., Sigel, R. K. O. (2005) Structure determination of catalytic RNAs and investigations of their metal ion-binding properties. *Chimia* 59, 817–821.

26. Milligan, J. F., Groebe, D. R., Witherell, G. W., Uhlenbeck, O. C. (1987) Oligoribonucleotide synthesis using T7 RNA polymerase and synthetic DNA templates. *Nucleic Acids Res* 15, 8783–8798.

27. Erat, M. C. (2011) *in* "Advances in Biochemical Spectroscopy" (Pascal, S. M., and Dingley, A., Eds.), pp. 488, IOS Press, Amsterdam, The Netherlands.

28. Sambrook, J., Russel, D. W. (2001) Molecular Cloning: A Laboratory Manual, Cold Spring Harbor Laboratory Press, Cold Spring Harbor.

29. Anderson, A. C., Scaringe, S. A., Earp, B. E., Frederick, C. A. (1996) HPLC purification of RNA for crystallography and NMR. *RNA* 2, 110–117.

30. Schnabl, J., Sigel, R. K. O. (2010) Controlling ribozyme activity by metal ions. *Curr Op Chem Biol* 14, 269–275.

31. Glasoe, P. K., Long, F. A. (1960) Use of glass electrodes to measure acidities in deuterium oxide. *J Phys Chem* 64, 188–190.

32. Wijmenga, S. S., van Buuren, B. N. M. (1998) The use of NMR methods for conformational studies of nucleic acids. *Prog Nucl Mag Res Sp* 32, 287–387.

33. Erat, M. C., Zerbe, O., Fox, T., Sigel, R. K. O. (2007) Solution structure of the domain 6 from a self-splicing group II intron ribozyme: A Mg^{2+} binding site is located close to the stacked branch adenosine. *ChemBioChem* 8, 306–314.

34. Sigel, R. K. O., Sashital, D. G., Abramovitz, D. L., Palmer III, A. G., Butcher, S. E., Pyle, A. M. (2004) Solution structure of domain 5 of a group II intron ribozyme reveals a new RNA motif. *Nat Struct Mol Biol* 11, 187–192.

35. Koradi, R., Billeter, M., Wüthrich, K. (1996) MOLMOL: A program for display and analysis of macromolecular structures. *J Mol Graphics* 14, 29–32 & 51–55.

Chapter 17

Analysis of Catalytic RNA Structure and Function by Nucleotide Analog Interference Mapping

Soumitra Basu, Mark J. Morris, and Catherine Pazsint

Abstract

Nucleotide analog interference mapping (NAIM) is a quick and efficient method to define concurrently, yet singly, the importance of specific functional groups at particular nucleotide residues to the structure and function of an RNA. NAIM can be utilized on virtually any RNA with an assayable function. The method hinges on the ability to successfully incorporate, within an RNA transcript, various $5'-O-(1\text{-thio})$nucleoside analogs randomly via in vitro transcription. This could be achieved by using wild-type or Y639F mutant T7 RNA polymerase, thereby creating a pool of analog doped RNAs. The pool when subjected to a selection step to separate the active transcripts from the inactive ones leads to the identification of functional groups that are crucial for RNA activity. The technique can be used to study ribozyme structure and function via monitoring of cleavage or ligation reactions, define functional groups critical for RNA folding, RNA–RNA interactions, and RNA interactions with proteins, metals, or other small molecules. All major classes of catalytic RNAs have been probed by NAIM. This is a generalized approach that should provide the scientific community with the tools to better understand RNA structure–activity relationships.

Key words: RNA structure and function, Ribozyme, Nucleotide analog interference mapping

1. Introduction

Biochemical studies that rely on chemical modifications using footprinting or interference mapping have been extensively used to derive information on RNA structure and function (1, 2). NAIM is a technique that blends the phosphorothioate modification interference with the inherent simplicity of phosphorothioate-based RNA sequencing to generate information on critical functional groups within an RNA (see Fig. 1; (3–6)). A major difference between NAIM and site-directed mutagenesis is that in NAIM the smallest mutated unit is a single functional group, whereas in site-directed mutagenesis the whole nucleotide must be changed. This allows NAIM to generate data at the functional group level, often furnishing details at the atomic

Jörg S. Hartig (ed.), *Ribozymes: Methods and Protocols*, Methods in Molecular Biology, vol. 848,
DOI 10.1007/978-1-61779-545-9_17, © Springer Science+Business Media, LLC 2012

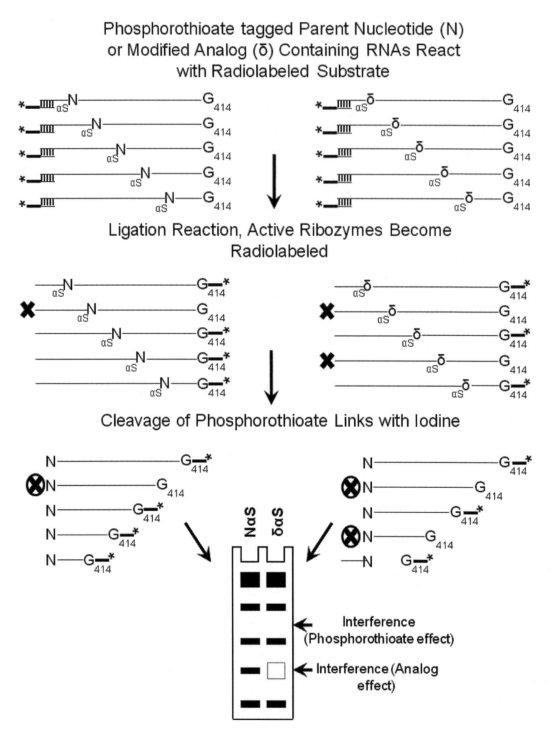

Fig. 1. Schematic of nucleotide analog interference mapping. In this example, the active ribozymes performs a ligation reaction and covalently transfers the radioactive tag to its 3′-end from a radiolabeled substrate. NαS indicates the parent nucleotide, and δαS the modified nucleotide analog.

level, such as contribution of a single hydrogen bond. NAIM has been used extensively to study almost all major classes of catalytic RNAs, including, group I introns (7, 8), group II intron (9), RNase P (10), hairpin ribozyme (11), HDV ribozyme (12), and the varkud satellite (VS) ribozyme (13). Additionally, it has been used to study metal ion and small-molecule interactions with RNA. For example, both monovalent and divalent interactions with RNA have been probed with NAIM (14–17). The technique has also been used to study small-molecule interactions with RNA in case of the *glm*S ribozyme (18).

The first step to perform NAIM is to randomly incorporate the α-phosphorothioate tagged nucleotides with a functional group substitution or deletion into an RNA molecule by in vitro transcription. This produces a pool of transcripts with substitutions of analogs at different nucleotide positions along the RNA molecule (see Fig. 1). The analog substitutions will determine whether the particular ribozymes (or any other RNA) among the pool are still functional or are inactive depending on the importance of the substituted or deleted functional group at that exact position. The ribozyme function is generally a cleavage or a ligation reaction, in either case the active population must be radioactively labeled. Because the analogs are incorporated randomly, sites of incorporation are precisely mapped, by cleaving the phosphorothioate tag with iodine which is followed by polyacrylamide gel electrophoresis (PAGE) to separate the fragments (5). The RNA is then visualized by autoradiography. Each band on the sequencing gel represents the position of a particular nucleotide. A band missing from the sequence ladder will signify that the activity of the RNA was lost due to the specific analog substitution at that particular site (see Fig. 1; (19)). The ability to determine the importance of each nucleotide and its functional groups in such an efficient and high-throughput manner is why NAIM can be effectively applied to study the structure and function of ribozymes.

In this chapter, we will outline the general methods for synthesis of the phosphorothioate-tagged modified nucleotide analogs, incorporation of the analogs into the *Tetrahymena* group I intron ribozyme, NAIM experiment performed on one of the engineered version of the catalytic RNA from *Tetrahymena*, and interpretation and quantitation of the data.

2. Materials

2.1. Reagent Preparation

2.1.1. Synthesis of α-Phosphorothioate-Tagged Nucleotide Triphosphates

The four parent 5′-*O*-(1-thio) nucleoside triphosphates are available from Life Technologies, CA. These compounds and several of the nucleotide analogs described in this chapter are also available from Glen Research (Sterling, VA). Many of the modified nucleosides can be obtained from Sigma (St. Louis, MO).

1. 100 mg unprotected nucleoside.

2. 50 mL anhydrous pyridine.

3. 2 mL triethylphosphate (TEP).

4. $PSCl_3$.

5. Trioctylamine.

6. 0.5 M LiCl in H_2O.

7. Tributylammonium pyrophosphate (TBAP).

8. Triethyl amine (TEA).

9. 50 mM Triethyl ammonium bicarbonate (TEAB).

10. n-Propanol.

11. Ammonium hydroxide.

12. Argon.

13. Cellulose chromatographic plate (for TLC).

14. DEAE Sephadex A-25.

2.1.2. In Vitro Transcription with α-Phosphorothioate Analogs

1. Linearized plasmid DNA template (1 μg/μL).

2. 10× Transcription buffer: 400 mM Tris–HCl, pH 7.5, 40 mM spermidine, 100 mM dithiothreitol (DTT), 150 mM $MgCl_2$, and 0.5% Triton X-100.

3. 10× Nucleotide triphosphates (NTP's): 10 mM each of NTP, pH 7.5. Individual NTP is weighed to make approximately 1 mL of 100 mM of stock solution in DEPC treated ddH_2O. The solution is adjusted to pH 7.5 and filtered through a 22-μm syringe filter. Absorbance is measured on a UV spectrophotometer and the exact concentration is calculated using the appropriate extinction coefficient. 10× NTP mixture is prepared by diluting the stock to a solution containing 10 mM of each of the NTPs.

4. Nucleotide analogs: Often a diastereomeric mixture is used, unless the S_p isomer is purified from the R_p by HPLC. The T7 RNA polymerase only recognizes the S_p from a diastereomeric mixture and during incorporation there is a reversal of configuration converting it to an R_p.

5. T7 RNA polymerase and Y639F mutant version of the enzyme.

6. Urea, premade polyacrylamide solution, ammonium persulfate, and TEMED (N,N,N',N'-Tetramethylethylenediamine).

7. 5 M NaCl.

8. 100 mM Tris–EDTA, pH 7.0 (TE).

9. 100 and 70% ethanol.

10. Formamide loading buffer: 90% formamide, 100 mM Tris–HCl, pH 7.0, 22 mM EDTA, 0.001% bromophenol blue, and 0.001% xylene cyanol (0.22-μm filter sterilized).

1. Purified RNA.

2. Alkaline phosphatase, calf intestinal (CIP).

3. T4 polynucleotide kinase (PNK).

4. Yeast PolyA polymerase.

5. Phenol/chloroform (Tris buffered).

6. 100 and 70% Ethanol.

7. 100 mM TE.

8. [γ-^{32}P]-ATP (6,000 Ci/mmol) and [α-^{32}P]cordycepin triphosphate (5,000 Ci/mmol), PerkinElmer.

9. Kodak X-OMAT Film.

10. 5 M NaCl.

3. Methods

3.1. Synthesis of NTPαS

The 5'-*O*-(1-thio) nucleoside triphosphates are prepared from unprotected nucleosides on a scale of approximately 50–200 mg of starting material depending on the availability. The synthesis involves a one-pot, two-step reaction, adopted from the method developed by Arabashi and Frey (20) (see Note 1).

100 mg of nucleoside precursor is transferred to a 25-mL round-bottomed flask, and co-evaporated with 3×10 mL of anhydrous pyridine. The nucleoside is further co-evaporated with (1–2 mL) of toluene to remove any residual pyridine. After drying, the flask is capped with a septum, and the inside purged with argon.

The dry nucleoside is dissolved in a minimal volume of triethyl phosphate (TEP). The volume is kept less than 1 mL. Nucleosides that have difficulty becoming soluble are heated with an air-gun while constantly stirring the mixture (see Note 2).

The syntheses of purine analogs are slightly different than pyrimidines. To the purine analog 1.1 equivalent of trioctylamine (Aldrich) (MW 353.7, density 0.816 g/mL, approximately 176 μL depending upon the molecular weight of the analog) and 1.1 equivalent of PSCl$_3$ (Aldrich) (approximately 45 μL, MW 169.4, density 1.67 g/mL) are added. The reaction is stirred for 30–60 min under argon. The progress of the reaction is monitored by analyzing a small portion of the reaction (100–200 μL) via thin-layer chromatography (TLC). To drive the reaction forward additional amounts of trioctylamine and PSCl$_3$ may be added. The reaction is quenched with water, and the products are resolved by cellulose TLC (Aldrich) in a solvent system of 0.5 M LiCl in water. The products on the TLC plate are visualized by handheld UV light. The monophosphate migrates slower than the unreacted starting material.

We took a slightly modified approach to synthesize the 5'-O-(1-thio-1, 1-dicloro)phosphoryl nucleoside intermediate on our way to prepare the triphosphates of the pyrimidine analogs (21). The nucleoside is dried as described previously. The dry nucleoside (about 100 mg) is dissolved in minimum amount of TEP. The solution is cooled to 0°C and trioctylamine (1.2 equivalents) and collidine (Sigma) (1.2 equivalents, MW 121.8, density 0.917 g/mL) are added under anhydrous conditions. PSCl$_3$ (about 1.5 equivalents) is added dropwise to the cooled solution, and allowed to react for 30 min before warming to room temperature and continuing the reaction at that condition for another 45 min. Monitoring the progress of the reaction and other steps are similar to the processes used for purine analogs.

To the 5'-O-(1-thio-1, 1-dicloro)phosphoryl nucleoside formed at the first step a TBAP (Sigma) (TBAP, MW 451.5, 4 equivalents of TBAP per mole of PSCl$_3$) solution is added, and stirred for 30 min. The TBAP solution is prepared by dissolving in TEP (0.1 g/mL) by mild heating with an air-gun. A small portion (100–200 μL) of the reaction is removed and quenched by adding a few drops of triethylamine (TEA) (Aldrich). The TEA precipitates the phosphates, which is collected by centrifugation, dissolved in 50 mM TEAB, and the products are analyzed by silica TLC using a solvent system of n-propanol:ammonium hydroxide:water (6:3:1) v/v. TLC analysis shows at least two products, both of which migrate slower than the unreacted nucleoside precursor and the nucleoside monophosphate. One of the products is the desired triphosphate and the other is the cyclic triphosphate, which hydrolyzes to the linear triphosphate after several hours at room temperature. After the reaction is over, the triphosphates are precipitated by adding about 50 equivalents of TEA (MW 101.19, density 0.726 g/mL, calculated per mole of the nucleoside precursor). The resulting precipitate is collected by centrifugation and dissolved in 5 mL of 50 mM TEAB and kept overnight at room temperature. The cyclic intermediate of the nucleoside triphosphate gets hydrolyzed under this condition. The nucleoside triphosphate is purified by DEAE Sephadex chromatography using a linear gradient from 50 to 800 mM TEAB in a total volume of 1 L (22). For guanosine analogs using a gradient from 50 to 1 M TEAB in a total volume of 1 L yields a better result. The nucleotide typically elutes at approximately 0.6 M TEAB. The fractions (about 15 mL each) are checked for the presence of the nucleotide by UV measurements (A$_{260}$). Positive fractions are pooled as a batch of 2–3 and lyophilized. The reaction products are characterized by ^{31}P NMR and mass spectrometry. The ^{31}P NMR spectra of the correct nucleotide shows three resonance peaks at 42–43 (α), –5–10 (γ), and –20–25 (β) ppm relative to an 85% phosphoric acid standard (23). The nucleotide amount is quantitated by measuring the UV absorbance and the identity is established by mass spectrometry.

3.2. Incorporation of Phosphorothioate Nucleotide Analogs via In Vitro Transcription

One of the basic requirements of NAIM is to incorporate the nucleotide analogs randomly within the RNA, which is usually accomplished enzymatically via T7 RNA polymerase (24, 25). Certain analogs especially the minor groove modified analogs are incorporated better by the Y639F mutant version of T7 RNA polymerase (26, 27). The incorporation level is chosen to generate sufficient signals within the detectable range and also to limit cooperative interference due to too many substitutions. Usually the incorporation level is kept at about 5% to fulfill the above criteria (8, 28). To achieve the roughly 5% incorporation level the ratio of analog to parent NTP that are used is shown in Table 1. These conditions are selected based upon a transcription condition where each of the NTP except the parent is present at 1 mM. The concentration of the parent nucleotide may need to be varied to create a suitable ratio with the modified analog to achieve 5% incorporation level (see Table 1) (see Note 3).

The reactions are performed according to standard procedures for in vitro transcription. The template may be derived from plasmid linearized with the suitable restriction enzyme or may be a synthetic DNA template. The linearized plasmid is phenol/chloroform extracted and then ethanol precipitated. The precipitated DNA is resuspended in a volume of TE to make the stock 1 mg/mL. If using a synthetic template (typically used for transcription of small RNAs) DNAs longer than 35–40 nucleotides must be purified by denaturing PAGE before using them for transcription (29). A typical transcription reaction will contain buffer, plasmid DNA template (0.05 μg/μL), T7 RNA polymerase (5 units/μL), inorganic pyrophosphatase (0.001 units/μL), nucleotide analog, and NTPs (see Table 1). The reactions are incubated for 3 h at 37°C. The reaction is purified by 7 M urea denaturing PAGE. The RNA transcripts are visualized by UV shadowing, excised from the gel with a clean razor blade, and eluted overnight in TE, 250 mM NaCl at 4°C. The eluted RNA is ethanol precipitated, and resuspended in 100 μL of TE. The RNA is quantitated by measuring the absorbance at 260 nm.

3.2.1. Optimization of Analog Incorporation

The incorporation of nucleotide analogs at particular sites within an RNA transcript or the overall incorporation level may vary, and therefore it is necessary to normalize the incorporation efficiency (see Note 4). The incorporation efficiency is determined by resolving the cleavage products of ^{32}P-5′-end labeled RNA transcripts containing the parental nucleotide or the nucleotide analog by denaturing PAGE and comparing individual band intensities of the parental nucleotide containing lane with the corresponding bands of the nucleotide analog lane. Twenty picomoles of RNA are treated with 200 units of CIP for 1 h at 37°C to remove the 5′-terminal phosphate group. The reaction is phenol/chloroform extracted and the RNA ethanol precipitated. The RNA is 5′-radiolabeled by incubating it with

Table 1
Conditions for incorporation of nucleotide analogs into RNA for interference mapping experiments

Analog δTPαS (S$_p$ isomer only) (except ones with*)	δTPαS (mM)	Parent NTP (mM)	T7 RNA polymerase (WT or Y639F)	References
AαS	0.05	1.0	WT	(28)
7dAαS	0.05	1.0	WT	(35)
m^{6}AαS	0.1	1.0	WT	(35)
c^{3}AαS	2.0	0.5	WT	(40)
FormαS	0.1	1.0	WT	(40)
n^{8}AαS	0.5	1.0	WT	(40)
DAPαS	0.025	1.0	WT	(8)
PurαS	2.0	1.0	WT	(35)
2APαS	0.5	1.0	WT	(35)
dAαS	0.75	1.0	Y639F	(35)
^{OMc}AαS	2.0	0.2	Y639F	(35)
^{F}AαS	0.25	1.0	Y639F	(42)
7F7dAαS	0.05	1.0	WT	(39)
2F7dAαS	0.2	1.0	WT	(39)
2FAαS	0.2	1.0	WT	(39)
^{SH}AαS	0.2	1.0	Y639F	(37)
GαS	0.05	1.0	WT	(28)
IαS	0.1	1.0	WT	(8)
m^{2}GαS	0.75	0.5	Y639F	(35)
dGαS	0.25	1.0	Y639F	(44)
7dGαS	0.05	1.0	WT	(41)
S^{6}GαS (4 mM Mn^{2+})	0.25	1.0	WT	(14)
^{OMc}GαS	2.0	0.1	Y639F	(3)
^{SH}GαS	0.25	1.0	Y639F	(37)
CαS	0.05	1.0	WT	(28)
n^{6}CαS	0.5	0.5	WT	(21)
F^{5}CαS	0.5	1.0	WT	(21)
ΨiCαS	0.05	1.0	WT	(21)

(continued)

Table 1
(continued)

Analog δTPαS (S$_p$ isomer only) (except ones with*)	δTPαS (mM)	Parent NTP (mM)	T7 RNA polymerase (WT or Y639F)	References
ZαS	0.5	0.5	Y639F	(43)
m^{4}CαS* (1 mM Mn^{2+})	2.0	0.1	Y639F	Unpublished
2PyαS*	1.0	1.0	Y639F	Unpublished
dCαS	0.75	1.0	Y639F	(11)
^{SH}CαS	0.2	1.0	Y639F	(37)
^{OMe}CαS	2.0	0.05	Y639F	(3)
UαS	0.05	1.0	WT	(28)
dUαS	0.25	1.0	Y639F	(44)
^{F}UαS	0.25	1.0	Y639F	(44)
m^{5}UαS	0.5	1.0	WT	Unpublished
^{OMe}UαS	2.0	0.1	Y639F	(3)
^{SH}UαS	0.2	1.0	Y639F	(37)
n^{6}UαS* (0.5 mM Mn^{2+})	1.0	0.1	Y639F	Unpublished
c^{3}2PyαS* (1 mM Mn^{2+})	1.5	0.5	Y639F	Unpublished
$\psi\alpha$S	0.05	1.0	WT	Unpublished

Note: All the analogs are incorporated at a level of approximately 5%, except c^{3}AαS (about 1%) and ^{OMe}AαS (about 1–2%)

* The analogs are diastereomeric mixture (both R$_p$ and S$_p$ isomers present)

10 units of T4 PNK and 50 mCi of [γ-^{32}P]ATP for 1 h at 37°C. The reaction is passed through a ProbeQuant G-50 spin-column (GE Healthcare, Pittsburgh, PA) to remove unincorporated radionucleotides. The labeled RNA is purified by denaturing PAGE, visualized by autoradiography, the associated band is cut off the gel, and eluted into 1% SDS in TE overnight at 4°C (see Note 5). To remove the SDS and precipitate the RNA the eluent is extracted with equal volume of phenol/chloroform and ethanol precipitated. The labeled RNAs containing equal amount of counts are mixed with two volumes of loading buffer, reacted with 1/10th volume of I$_2$ in ethanol (50 mM), and heated at 90°C for 2 min (5). The iodine selectively cleaves the phosphorothioate linkages, which are the sites of parent or analog incorporation at about 10–15% efficiency. The cleavage products are resolved on a preheated denaturing polyacrylamide gel. To control for the presence of any nonspecific cleavage, an equal amount of labeled RNAs, which are not treated with iodine

are also loaded on the same gel. Large RNAs are resolved on gels of different acrylamide concentration and are run for different lengths of time to resolve the maximum portion of the molecule. For the *Tetrahymena* intron, which is about 400 nucleotides long, 5 and 6% denaturing polyacrylamide gels are used. The 5% gels are typically run for 4 h 15 min and 3.5 h at 75 W, whereas the 6% gels are run for 2.5 and 1 h at identical power. This procedure can resolve more than 98% of about 400 nucleotides of the *Tetrahymena* intron. The gels are dried and exposed to phosphorimager screens (Kodak, Rochester, NY). The intensities of the bands are quantitated by Bio-Rad Quantity One software. If a Molecular Dynamics machine is used for scanning the gel, the ImageQuant software is used for quantitation.

3.3. Selection of Active RNA

A primary requirement for successful NAIM assay is to be able to distinguish between an active from inactive population within a nucleotide analog substituted RNA pool (19). Broadly, there are two selection principles that have been used. The first approach requires physically separating the active variants from the inactive. This can be achieved by using a native gel-mobility, filter binding, or denaturing PAGE, and have been successfully employed to study folding of ribozyme domain, RNA–protein interactions, and ribozyme cleavage activity (12, 14, 18, 30). The other approach is where a ribozyme-mediated ligation activity is employed, that selectively labels active variants from the pool of randomly doped RNAs (8).

One of the most used assays for NAIM is the 3′-exon ligation reaction executed by the L-21 G414 version of the *Tetrahymena* group I intron, which is the reverse of the second step of intron splicing (see Fig. 2a; (31–33)). In this reaction, the terminal guanosine (G414) attacks the oligonucleotide substrate mimicking the 5′–3′ ligated exon and transfers the 3′-exon to the 3′-end of the intron to attach covalently. By using a 3′-end radiolabeled substrate, the active variants among the intron pool selectively label themselves. A typical ligation reaction with the L-21 G414 intron is performed as follows. The RNAs are prefolded by incubating in reaction buffer at 50°C for 10 min. After a quick spin the RNAs are further heated for 2 min at 50°C at which time the substrate dT(−1)S [CCCUC(dT) AAAAA], radiolabeled at 3′-end by yeast poly(A) polymerase (Affymetrix, Cleveland, OH) and [α-^{32}P] cordycepin (34) is dissolved in the same buffer and heated at 50°C for 2 min. After 2 min, dT(−1)S substrate, equal in volume to the ribozyme reaction, is added with proper mixing. The reaction is stopped after 10 min by adding two volumes of loading buffer. The reaction is then split into two fractions and to one of the fractions I$_2$-ethanol solution is added (1/10th volume) to cleave the sulfur containing linkages. Both reactions are heated to 90°C for 2 min and loaded onto a 5 or 6% denaturing polyacrylamide gel (see Fig. 2b), and after the completion of the run the gel is dried and visualized by exposing the dried gel to a phosphorimager plate followed by scanning.

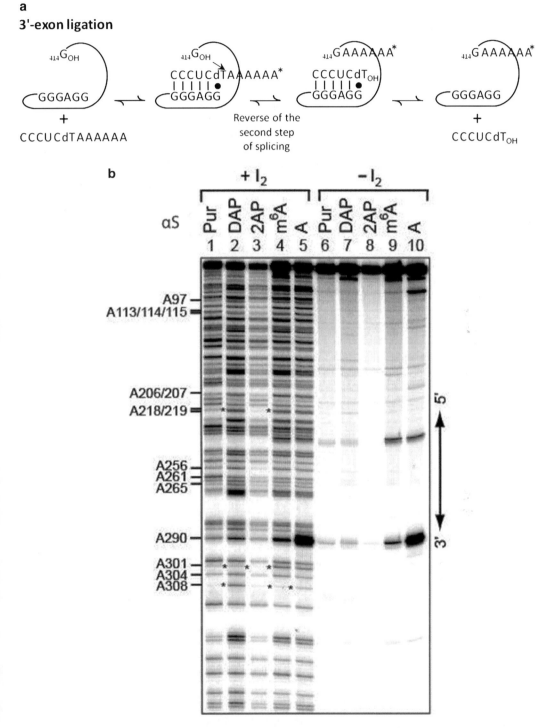

Fig. 2. Interference mapping on the *Tetrahymena* group I intron catalytic RNA. (**a**) Schematic of the 3′-exon ligation reaction for the L-21 G414 version of the group I intron with the 3′-radiolabeled substrate dT(−1)S. In the reaction the terminal guanosine (G414) makes a nucleophilic attack on the substrate that mimics the 5′-3′ ligated exon, and covalently transfers the 3′-exon portion on the 3′-end of the ribozyme. (**b**) Autoradiogram of the cleaved sequences of the ligated products which are doped with a subset of adenosine analogs. Some of the sites of interference are identified, and a few of those

c

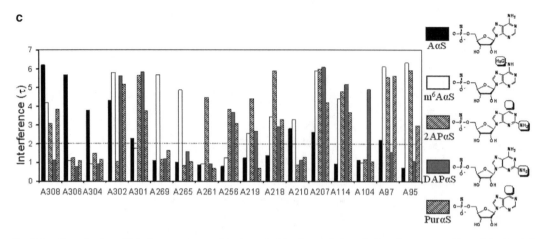

Fig. 2. (continued) sites of interferences are marked by *asterisks. Lanes 1–5* are iodine cleavage reactions of the ligated products. *Lanes 6–10* are the corresponding lanes that were not treated with iodine, to control for any nonspecific cleavage. The particular analog corresponding to each lane is marked on *top* of the autoradiogram. A290 is a very consistent nonspecific cleavage site and thus is always uninformative. Under our assay condition A302 and A306 show very strong phosphorothioate effect and therefore are uninformative. (**c**) A quantitative depiction of the interference data derived from (**b**) using histograms. The 5′-radiobaled transcripts were also considered (as per Eq. 1) during calculation of the τ values.

3.4. Data Quantitation

Quantitation of each band intensity from the iodine cleavage sequence ladder is required to identify sites of interference as a result of analog substitution. The bands corresponding to cleavage products from the 5′-radiolabeled unselected RNA is compared with those from the selected RNA to control for incorporation differences at different sites, sites of phosphorothioate interference, and sites of nucleotide analog interference. Individual bands on the gels are selected, and the area under the peak corresponding to band intensity, is calculated using the QuantityOne (Bio-Rad) software. Interference at each site is calculated by substituting the value of individual band intensities into the following equation:

$$\text{Interference } (\tau) = (N\alpha S_{\text{selected}}) / (\delta\alpha S_{\text{selected}}) \div (N\alpha S_{\text{unselected}}) / (\delta\alpha S_{\text{unselected}}) \ (1)$$

where, $N\alpha S$ is the parent phosphorothioate nucleotide and the $\delta\alpha S$ is the modified analog of that nucleotide.

The equation normalizes the interference values for any site-specific incorporation differences or due to the phosphorothiaote tag. For the sites of noninterference the τ value is expected to be approximately 1. To normalize the data, the average interference value for all positions that are within at least 2–3 standard deviations of the mean is calculated and the interference value at each position is divided by this value (see Fig. 2c). The resulting set of values mostly hover around 1.0 ± 0.2. The τ value of above 2 is considered as interference, τ values below 0.5 are enhancement and values around 1 are noninterference. The cutoff of 2.0 can be increased in case higher stringency of data is needed. But the above criteria generally work quite well.

3.5. Nucleotide Analogs: Incorporation and Properties

We and several others have successfully synthesized and tested phosphorothioate-tagged nucleotide analogs for various NAIM experiments. We will discuss incorporation properties within RNA of the published and a few unpublished analogs (see Table 1), and the information they provide on the chemical nature of RNA–RNA and RNA–ligand interactions. The ligands may include metal ions, small molecules, and proteins.

3.5.1. Adenosine Analogs

We will first discuss purine nucleotides starting with adenosine analogs, followed by pyrimidines, beginning with cytidine. So far, 15 adenosine analogs have been used for interference mapping (see Fig. 3a). Eleven among them are modified at various positions within the purine base. This subset includes purine riboside (PurαS), N-methyladenosine (m^{6}AαS), 8-aza-adenosine (n^{8}AαS), Formycin A (FormAαS), 7-deazaadenosine or tubercidin (7dAαS), 3-deazaadenosine (c^{3}AαS), 2-aminopurine riboside (2APαS), 7-fluoro-7-deazaadenosine (7F7dAαS), 2-fluoro-7-deazaadenosine (2F7dAαS), 2-fluoroadenosine (2FAαS), and diaminopurine riboside (DAPαS). The analogs modified at the ribose sugar include 2′-deoxyadenosine (dAαS) 2′-deoxy-2′-thioadenosine (^{SH}AαS), 2′-deoxy-2′-fluoroadenosine (^{F}AαS), and 2′-O-methyladenosine (^{OMe}AαS), all of which are modified at the 2′-OH position. Most of these analogs can be incorporated by the wild-type T7 RNA polymerase. The 7dAαS and m^{6}AαS require lower concentration than AαS for 5% level of incorporation, and DAPAαS incorporates more efficiently than AαS (35). PurαS is a difficult analog to incorporate and the intended 5% incorporation level is achieved by increasing the PurαS concentration and lowering the ATP concentration (35). PurαS can sometimes incorporate unevenly, especially at sites with contiguous adenosines. 2APαS is incorporated by the wild-type polymerase, but requires higher concentrations than the parent nucleotide. Both n^{8}AαS and FormAαS can be incorporated by the wild-type enzyme.

The minor groove modified A-analogs c^{3}AαS, dAαS, ^{SH}AαS, ^{F}AαS, and ^{OMe}AαS require the Y639F mutated form of T7 RNA polymerase for incorporation into RNA transcripts (35–37). By far the most difficult to incorporate of the adenosine analogs is c^{3}AαS. The reported best incorporation level is about 1%. Both dAαS and ^{F}AαS are incorporated efficiently, whereas ^{OMe}AαS incorporation is poor but generally even. It is difficult to obtain more than 2% incorporation level, although incorporation level in some smaller transcripts may reach about 5%.

a

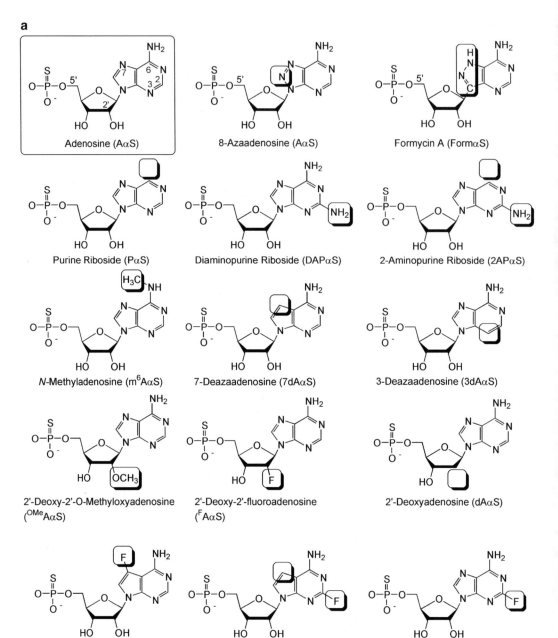

Fig. 3. Chemical structure of the modified nucleotide analogs. The site of chemical modification in individual analog is marked by a *box*. (**a**) Adenosine analogs. (**b**) Guanosine analogs. (**c**) Cytidine analogs. (**d**) Uridine analogs.

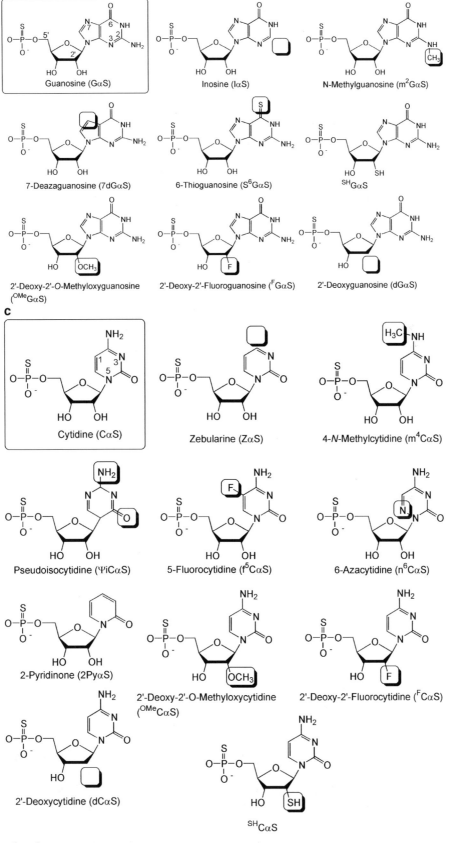

b

Guanosine (GαS)

Inosine (IαS)

N-Methylguanosine (m²GαS)

7-Deazaguanosine (7dGαS)

6-Thioguanosine (S⁶GαS)

ˢᴴGαS

2'-Deoxy-2'-O-Methyloxyguanosine (ᴼᴹᵉGαS)

2'-Deoxy-2'-Fluoroguanosine (ᶠGαS)

2'-Deoxyguanosine (dGαS)

c

Cytidine (CαS)

Zebularine (ZαS)

4-N-Methylcytidine (m⁴CαS)

Pseudoisocytidine (ΨiCαS)

5-Fluorocytidine (f⁵CαS)

6-Azacytidine (n⁶CαS)

2-Pyridinone (2PyαS)

2'-Deoxy-2'-O-Methyloxycytidine (ᴼᴹᵉCαS)

2'-Deoxy-2'-Fluorocytidine (ᶠCαS)

2'-Deoxycytidine (dCαS)

ˢᴴCαS

Fig. 3. (continued)

d

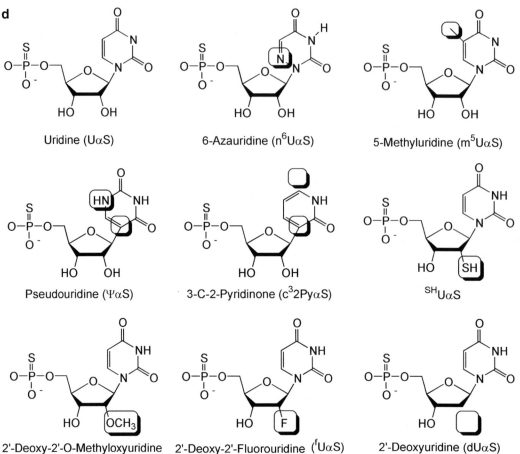

Fig. 3. (continued)

The set of A-analogs provide almost complete information on the chemical basis of adenosine function within RNAs (35). The PurαS, m⁶AαS, and 2APαS provide information on the N-6 exocyclic amine of adenosine by deleting the amine group as in the cases of PurαS and 2APαS, whereas a proton is replaced by a methyl group in m⁶AαS (7, 9, 10, 35). Interference at any site due to PurαS or 2APαS incorporation indicates importance of the amine at that site. The m⁶AαS interference indicates that either both protons are necessary for H-bonding or the methyl group creates steric hindrance in a potentially cramped space. The effect may also be due to modification of the electronic property of the purine ring, for example pK_as or π-stacking.

Interference with 7dAαS, in which the N-7 nitrogen is replaced by a C–H group indicates a critical major groove contact with the ring nitrogen (10, 35). Interference with PurαS, m⁶AαS, and 7dAαS strongly suggests Hoogsteen-type hydrogen bonding.

The DAPαS and 2APαS have an extra amine at the C-2 position, and general interference with both of these analog may indicate close minor groove packing (7, 10, 35, 38).

In c³AαS the N-3 imino group of adenosine is replaced by a C-3, which allows to probe involvement of N-3 group in RNA function (36). This analog is particularly useful in mapping A-minor motifs within RNA.

The fluorine substituted adenosine analogs changes the pK_a at N1 position and have been used to probe the effect of nucleobase protonation in functional RNA (39). The n⁸AαS and FormAαS are analogs with perturbed N1 pK_a, and are used to probe adenosine base ionization within ribozymes (13, 40).

3.5.2. Guanosine Analogs   So far, seven guanosine analogs have been used for NAIM (see Fig. 3b). Four among them have modified functional groups within the base (inosine, IαS; N^2-methylguanosine, m²GαS; 7-deazaguanosine, 7dGαS; and 6-thioguanosine, S⁶GαS) and the remaining three at the 2'-OH of the ribose moiety (2'-deoxyguanosine [dGαS],2'-O-methylguanosine (OMeGαS), and 2'-deoxy-2'-thioguanosine (SHGαS)).

Wild-type T7 RNA polymerase efficiently and uniformly incorporates both IαS and 7dGαS (8, 41). S⁶GαS is incorporated reasonably well by the wild-type enzyme, but only in the presence of Mn^{2+} in the transcription reaction (14). The S⁶GαS incorporated RNAs are highly unstable, which requires completion of all experiments within 48 h of transcription (see Note 6). OMeGαS incorporation by the wild-type enzyme has been reported using greatly increased ratio of analog to GTP (3). But in several of the constructs we tried, including the L-21 G414 RNA, resulted in the inability to incorporate the analog. m²GαS and dGαS are both minor groove modified analogs. dGαS and SHGαS require the Y639F mutant enzyme for efficient incorporation (27, 37).

IαS and m²GαS both probe the role of N-2 exocyclic amine of guanosine (8, 42). Due to the lack of the amine, inosine may cause reduced duplex stability or missing tertiary contact, whereas replacement of a proton with the methyl group on the amine generally results only in loss of tertiary bonding. 7dGαS lacks the N-7 ring nitrogen and interference with this analog is an indication of major groove contact at this position (41). Another major groove modified analog S⁶GαS has the O-6 replaced by a sulfur and interference with this analog may probe RNA–RNA contact as well as sites of divalent or monovalent metal ion binding site within RNA (14).

The dGαS is a minor groove ribose modified analog where the 2'-OH is replaced by a hydrogen atom. Any critical H-bonding with 2'-OH of a guanosine can be probed with this analog.

3.5.3. Cytidine Analogs   So far there are ten cytidine analogs synthesized and incorporated (see Fig. 3c). These include six analogs that are modified within the pyrimidine base: 6-azacytidine (n⁶CαS), 5-fluorocytidine (f⁵CαS), pseudoisocytidine (ΨiCαS), zebularine (ZαS), N-methylcytidine (m^{4}CαS), and 2-pyridinone (2PyαS); and four analogs are modified at the 2'-OH of the ribose sugar: 2'-deoxycytidine (dCαS), 2'-deoxy-2'-fluorocytidine (^{F}CαS), 2'-deoxy-2'-thiocytidine (^{SH}CαS), and 2'-O-methylcytidine (^{OMe}CαS). Three of these n^{6}CαS, f^{5}CαS, and ΨiCαS can be incorporated by the wild-type enzyme (14, 29). The other four of the base modified analogs of cytidine require Y639F mutant polymerase and Mn^{2+} for efficient incorporation (12, 21, 37, 43) (Oyelere A, Cardon J, Basu S, Strobel SA, unpublished results). dCαS, ^{SH}CαS, and ^{F}CαS require Y639F mutant enzyme for incorporation (11, 37) (Oyelere A, Kardon J, Basu S, Strobel, SA, unpublished results). Additionally, ^{F}CαS requires Mn^{2+} in the transcription buffer for incorporation into RNA transcripts. However, ^{OMe}CαS could be incorporated by the wild-type enzyme (3).

The n^{6}CαS has an extra ring nitrogen in place of C-6. f^{5}CαS has a fluoro functionality at C-5, which introduces bulkiness in the major groove. ΨiCαS is a C-linked nucleoside, in which the C-5 is also replaced by a nitrogen. In ZαS the N-4 exocyclic amine is missing, which can be used to probe its importance in major groove contacts. An interesting property of n^{6}CαS, f^{5}CαS, ΨiCαS, and ZαS is that they have different N-3 pK_as relative to cytidine, and this characteristic has been utilized to determine pK_a perturbation within RNA (12, 21). In m^{4}CαS a proton of the N-4 amine is replaced by a methyl group, whereas in 2PyαS the N-4 amine (the N-3 ring nitrogen is replaced by a C) is missing. They can be used to probe major groove hydrogen bonding contacts (Oyelere A, Kardon J, Basu S, Strobel SA, unpublished results). The additional bulk of the methyl may cause interference if present in the context of insufficient space. The nature of information obtained from the 2'-OH derivatives of cytidine are similar to those obtained from similarly modified adenosine analogs.

3.5.4. Uridine Analogs

A total of eight analogs of uridine have been synthesized and incorporated into RNA (see Fig. 3d). Four of them are base modified: 6-azauridine (n^{6}UαS), 3-C-2-pyridinone (c^3-2PyαS), pseudouridine ($\Psi\alpha$S), and 5-methyluridine (m^{5}UαS); and the other four are modified in the ribose sugar: 2'-deoxyuridine (dUαS), 2'-deoxy-2'-thiorouridine (^{SH}UαS), 2'-deoxy-2'-fluorouridine (^{F}UαS), and 2'-O-methyluridine (^{Ome}UαS).

Except $\Psi\alpha$S, the other base modified analogs n^{6}UαS and c^3-2PyαS can only be incorporated into a transcript by the Y639F mutant enzyme (Oyelere A, Cardon J, Basu S, Strobel SA, unpublished results). The ^{Ome}UαS (3) and the other three ribose sugar modified analogs also require the Y639F mutant enzyme (44). Additionally, ^{F}UαS needs Mn^{2+} for efficient incorporation.

ΨαS is a C-linked nucleoside and the C-5 is substituted by a nitrogen. This analog does not exhibit any interference within the L-21 G414 ribozyme (S. Basu and S. A. Strobel, unpublished results).

The 2′-OH modified analogs provide information similar in nature to identically modified adenosine analogs.

4. Notes

1. Many of the chemicals used for NAIM are hazardous and proper precaution must be taken while handling them.

2. Solubility of certain nucleosides can adversely affect the yield of NTPαS synthesis. To dissolve the nucleoside one should not use too much TEP. Excess TEP severely affects the reaction yield, so the volume should not exceed 1 mL. The majority of the causes, guanosine and its analogs are generally more difficult to dissolve than others. One can mildly heat the reaction vessel with air-gun with continuous stirring to facilitate the dissolving process.

3. There are several incorporation-related issues one must carefully take into consideration. In cases where transcription conditions require NTP concentration at some multiples of 1 mM, merely changing the analog concentration to that multiple may not achieve the desired level of incorporation. The conditions, especially the ratio of analog to NTP may need to be reoptimized for each such system. Similarly the concentration required for optimum incorporation, if pure S_p isomer is used, must be reoptimized; simply using half of the value used for the diastereomeric mixture may not work. This is true because the diastereomeric mixture may not contain equal amounts of the S_p and R_p isomers.

4. Sometimes, while trying to obtain optimum incorporation one may inadvertently create a condition for over-incorporation, exhibited by weaker bands compared to the parent analog. This may happen because an RNA over-incorporated with an analog may be nonselectively detrimental to the activity and result in overall weaker signal falsely suggesting poor incorporation. In case of a ribozyme ligation reaction this can be detected by the topmost band in a lane which represents the full-length uncleaved ligated product. An analog lane may have substantially reduced amount of such product indicating poor ligation reaction, which may be a tell-tale sign of over-incorporation. The analog concentration should be appropriately

adjusted to verify that the problem is over-incorporation and not under-incorporation.

5. Sometimes nonspecific cleavage may affect an experiment adversely, especially when the experiment requires cutting a band from a native gel to isolate the RNA. Under native conditions the nonspecifically cleaved RNAs may still form a folded structure which may create problems during the sequencing step, as the non-full-length RNAs will create nonspecific bands or bands of extra intensity. To solve this problem, the RNA is eluted into an SDS (1%) containing buffer to denature any extraneous nucleases, consequently limiting any nonspecific cleavage of the RNA. If SDS is used the eluate must be phenol/chloroform extracted to remove the SDS. Sometimes an additional denaturing gel purification before the sequencing step may be necessary to have a pool of full-length RNA.

6. We found that $S^6G\alpha S$ incorporated RNA transcripts are very unstable. All experiments must be concluded within 48 h of transcription. Storing the $S^6G\alpha S$ incorporated RNAs at –80°C does not increase the half-life.

Acknowledgments

We thank Dr. Scott Strobel for generous gift of some of the analogs and sharing with us unpublished results. This work was supported by funds to SB from the University of Pittsburgh and was conducted at University of Pittsburgh.

References

1. Conway, L., and Wickens, M. (1989) Modification interference analysis of reactions using RNA substrates. *Methods Enzymol.* **180**, 369–379.

2. Stern, S., Moazed, D., and Noller, H. F. (1988) Structural analysis of RNA using chemical and enzymatic probing monitored by primer extension. *Methods Enzymol.* **164**, 481–489.

3. Conrad, F., Hanne, A., Gaur, R. K., and Krupp, G. (1995) Enzymatic synthesis of 2′-modified nucleic acids: Identification of important phosphate and ribose moieties in RNase P substrates. *Nucleic Acids Res.* **23**, 1845–1853.

4. Gaur, R. K., and Krupp, G. (1993) Modification interference approach to detect ribose moieties important for the optimal activity of a ribozyme. *Nucleic Acids Res.* **21**, 21–26.

5. Gish, G., and Eckstein, F. (1988) DNA and RNA sequence determination based on phosphorothioate chemistry. *Science* **240**, 1520–1522.

6. Suydam, I. T., and Strobel, S. A. (2009) Nucleotide analog interference mapping, in *Methods Enzymol.* **468**, 3–30.

7. Strauss-Soukup, J. K., and Strobel, S. A. (2000) A chemical phylogeny of group I introns based upon interference mapping of a bacterial ribozyme. *J. Mol. Biol.* **302**, 339–358.

8. Strobel, S. A., and Shetty, K. (1997) Defining the chemical groups essential for *Tetrahymena* group I intron function by nucleotide analog interference mapping. *Proc. Natl. Acad. Sci. U. S. A.* **94**, 2903–2908.

9. Boudvillain, M., and Pyle, A. M. (1998) Defining functional groups, core structural features and inter-domain tertiary contacts

essential for group II intron self-splicing: a NAIM analysis. *EMBO J.* **17**, 7091–7104.

10. Siew, D., Zahler, N. H., Cassano, A. G., Strobel, S. A., and Harris, M. E. (1999) Identification of adenosine functional groups involved in substrate binding by the ribonuclease P ribozyme. *Biochemistry* **38**, 1873–1883.

11. Ryder, S. P., and Strobel, S. A. (1999) Nucleotide analog interference mapping of the hairpin ribozyme: implications for secondary and tertiary structure formation. *J. Mol. Biol.* **291**, 295–311.

12. Oyelere, A. K., Kardon, J. R., and Strobel, S. A. (2002) pK$_a$ perturbation in genomic Hepatitis Delta Virus ribozyme catalysis evidenced by nucleotide analogue interference mapping. *Biochemistry* **41**, 3667–3675.

13. Jones, F. D., and Strobel, S. A. (2003) Ionization of a critical adenosine residue in the *neurospora* varkud satellite ribozyme active site. *Biochemistry* **42**, 4265–4276.

14. Basu, S., Rambo, R. P., Strauss-Soukup, J., Cate, J. H., Ferre-D'Amare, A. R., Strobel, S. A., and Doudna, J. A. (1998) A specific monovalent metal ion integral to the AA platform of the RNA tetraloop receptor. *Nat. Struct. Biol.* **5**, 986–992.

15. Wrzesinski, J., and Jozwiakowski, S. K. (2008) Structural basis for recognition of Co^{2+} by RNA aptamers. *FEBS J.* **275**, 1651–1662.

16. Basu, S., and Strobel, S. A. (1999) Thiophilic metal ion rescue of phosphorothioate interference within the *Tetrahymena* ribozyme P4-P6 domain. *RNA* **5**, 1399–1407.

17. Cate, J. H., Hanna, R. L., and Doudna, J. A. (1997) A magnesium ion core at the heart of a ribozyme domain. *Nat. Struct. Biol.* **4**, 553–558.

18. Jansen, J. A., McCarthy, T. J., Soukup, G. A., and Soukup, J. K. (2006) Backbone and nucleobase contacts to glucosamine-6-phosphate in the *glm*S ribozyme. *Nat. Struct. Biol.* **13**, 517–523.

19. Ryder, S. P., Ortoleva-Donnelly, L., Kosek, A. B., and Strobel, S. A. (2000) Chemical probing of RNA by nucleotide analog interference mapping. *Methods Enzymol.* **317**, 92–109.

20. Arabshahi, A., and Frey, P. A. (1994) A simplified procedure for synthesizing nucleoside 1-thiotriphosphates: dATPαS, dGTPαS, UTPαS, and dTTPαS. *Biochem. Biophys. Res. Commun.* **204**, 150–155.

21. Oyelere, A. K., and Strobel, S. A. (2000) Biochemical detection of cytidine protonation within RNA. *J. Am. Chem. Soc.* **122**, 10259–10267.

22. Eckstein, F., and Goody, R. S. (1976) Synthesis and properties of diastereoisomers of adenosine 5'-(O-1-thiotriphosphate) and adenosine 5'-(O-2-thiotriphosphate). *Biochemistry* **15**, 1685–1691.

23. Chen, J. T., and Benkovic, S. J. (1983) Synthesis and separation of diastereomers of deoxynucleoside 5'-O-(1-thio)triphosphates. *Nucleic Acids Res.* **11**, 3737–3751.

24. Chamberlain, M., Kingston, R., Gilman, M., Wiggs, J., and de Vera, A. (1983) Isolation of bacterial and bacteriophage RNA polymerases and their use in synthesis of RNA in vitro. *Methods Enzymol.* **101**, 540–568.

25. Griffiths, A. D., Potter, B. V. L., and Eperon, I. C. (1987) Stereospecificity of nucleases towards phosphorothioate-substituted RNA: stereochemistry of transcription by T7 RNA polymerase. *Nucleic Acids Res.* **15**, 4145–4162.

26. Sousa, R. (2000) Use of T7 RNA polymerase and its mutants for incorporation of nucleoside analogs into RNA. *Methods Enzymol.* **317**, 65–74.

27. Sousa, R., and Padilla, R. (1995) A mutant T7 RNA polymerase as a DNA polymerase. *EMBO J.* **14**, 4609–4621.

28. Christian, E. L., and Yarus, M. (1992) Analysis of the role of phosphate oxygens in the group I intron from *Tetrahymena*. *J. Mol. Biol.* **228**, 743–758.

29. Milligan, J. F., and Uhlenbeck, O. C. (1989) Synthesis of small RNAs using T7 RNA polymerase. *Methods Enzymol.* **180**, 51–62.

30. Batey, R. T., Rambo, R. P., Lucast, L., Rha, B., and Doudna, J. A. (2000) Crystal structure of the ribonucleoprotein core of the signal recognition particle. *Science* **287**, 1232–1239.

31. Beaudry, A. A., and Joyce, G. F. (1992) Directed evolution of an RNA enzyme. *Science* **257**, 635–641.

32. Cech, T. R. (1990) Self-splicing of group I introns. *Annu. Rev. Biochem.* **59**, 543–568.

33. Mei, R., and Herschlag, D. (1996) Mechanistic investigations of a ribozyme derived from the *Tetrahymena* group I intron: insights into catalysis and the second step of self-splicing. *Biochemistry* **35**, 5796–5809.

34. Lingner, J., and Keller, W. (1993) 3'-End labeling of RNA with recombinant yeast poly(A) polymerase. *Nucleic Acids Res.* **21**, 2917–2920.

35. Ortoleva-Donnelly, L., Szewczak, A. A., Gutell, R. R., and Strobel, S. A. (1998) The chemical basis of adenosine conservation throughout the *Tetrahymena* ribozyme. *RNA* **4**, 498–519.

36. Soukup, J. K., Minakawa, N., Matsuda, A., and Strobel, S. A. (2002) Identification of A-Minor tertiary interactions within a bacterial group I intron active site by 3-deazaadenosine interference mapping. *Biochemistry* **41**, 10426–10438.

37. Schwans, J. P., Cortez, C. N., Olvera, J. M., and Piccirilli, J. A. (2003) 2'-Mercaptonucleotide interference reveals regions of close packing within folded RNA molecules. *J. Am. Chem. Soc.* **125**, 10012–10018.

38. Strobel, S. A., Ortoleva-Donnelly, L., Ryder, S. P., Cate, J. H., and Moncoeur, E. (1998) Complementary sets of noncanonical base pairs mediate RNA helix packing in the group I intron active site. *Nat. Struct. Biol.* **5**, 60–66.

39. Suydam, I. T., and Strobel, S. A. (2008) Fluorine substituted adenosines as probes of nucleobase protonation in functional RNAs. *J. Am. Chem. Soc.* **130**, 13639–13648.

40. Ryder, S. P., Oyelere, A. K., Padilla, J. L., Klostermeier, D., Millar, D. P., and Strobel, S. A. (2001) Investigation of adenosine base ionization in the hairpin ribozyme by nucleotide analog interference mapping. *RNA* **7**, 1454–1463.

41. Kazantstev, S. V., and Pace, N. R. (1998) Identification by modification-interference of purine N-7 and ribose 2'-OH groups critical for catalysis by bacterial ribonuclease P. *RNA* **4**, 937–947.

42. Ortoleva-Donnelly, L., Kronman, M., and Strobel, S. A. (1998) Identifying RNA minor groove tertiary contacts by nucleotide analog interference mapping with N2-methylguanosine. *Biochemistry* **37**, 12933–12942.

43. Szewczak, A. A., Ortoleva-Donnelly, L., Zivarts, M. V., Oyelere, A. K., Kazantsev, A. V., and Strobel, S. A. (1999) An important base triple anchors the substrate helix recognition surface within the Tetrahymena ribozyme active site. *Proc. Natl. Acad. Sci. U. S. A.* **96**, 11183–11188.

44. Szewczak, A. A., Ortoleva-Donnelly, L., Ryder, S. P., Moncoeur, E., and Strobel, S. A. (1998) A minor groove RNA triple helix within the catalytic core of a group I intron. *Nat. Struct. Biol.* **5**, 1037–1041.

Chapter 18

In Vitro Selection of Metal Ion-Selective DNAzymes

Hannah E. Ihms and Yi Lu

Abstract

The discovery of DNAzymes that can catalyze a wide range of reactions in the presence of metal ions is important on both fundamental and practical levels; it advances our understanding of metal–nucleic acid interactions and allows for the design of highly sensitive and selective metal ion sensors. A crucial factor in this success is a technique known as in vitro selection, which can rapidly select metal-specific RNA-cleaving DNAzymes. In vitro selection is an iterative process where a DNA pool containing a random region is incubated with the target metal ion. Those DNA sequences that catalyze the preferred reaction (the "winners") are amplified and carried on to the next step, where the selection is carried out under more stringent conditions. In this way, the selection pool becomes enriched with DNAzymes that exhibit desirable activity and selectivity. The method described can be applied to isolate DNAzymes selective to many different types of metal ions or different oxidation states of the same metal ion.

Key words: DNAzyme, In vitro selection, Functional DNA, Deoxyribozyme, Catalytic DNA, Metal ions, Bioinorganic chemistry

1. Introduction

Discovered by Breaker and Joyce (1), DNAzymes demonstrate that DNA is capable of much more than passive genetic information storage. DNAzymes, also known as deoxyribozymes or catalytic DNA, are strands of DNA that catalyze reactions such as RNA cleavage, porphyrin metallation, and DNA adenylation (2–8). Though DNAzymes can exhibit a wide range of structural diversity (9), DNA has fewer functional groups than protein or ribozymes. By making use of metal cofactors, however, DNAzymes can expand their activity. That anionic DNA could bind cationic metal ions was predictable; that it could do so selectively, and that these metal ions would enhance the DNA's catalytic activity was not predicted. In vitro selection has now been used to obtain DNAzymes with high selectivity toward Pb^{2+}, Mg^{2+}, Co^{2+}, and UO_2^{2+}. Since the mechanisms of these DNAzymes remains largely unknown, biochemical

Jörg S. Hartig (ed.), *Ribozymes: Methods and Protocols*, Methods in Molecular Biology, vol. 848,
DOI 10.1007/978-1-61779-545-9_18, © Springer Science+Business Media, LLC 2012

and biophysical studies are underway to increase the fundamental understanding of metal-to-nucleic acids interactions. Even without a detailed knowledge of their principle of action, DNAzymes with high selectivities have been converted into fluorescent (10), colorimetric (11), and electrochemical (12) sensors for metal ions such as Pb^{2+} (10, 13) and UO_2^{2+} (14). Since in vitro selection can isolate DNAzymes that bind a variety of metal ions, or even different oxidation states of the same metal ion, it is one of the most general approaches for developing metal sensors. As a further benefit, DNAzymes can withstand harsh conditions including multiple denaturing and annealing steps. Therefore, DNAzyme-based sensors can be tailored for a wide range of applications such as environmental monitoring and biomedical diagnostics.

Because DNAzymes are important on both fundamental and practical levels, it is desirable to obtain new DNAzymes for novel targets. In vitro selection is a powerful method to achieve the goal. During in vitro selection, a DNA pool of $\sim 10^{14}$ sequences is incubated with a solution of the selection target under application-relevant conditions (1, 15). In iterative rounds, the "winner" strands which catalyze the desired reaction in the presence of the target metal ion are amplified, enriching them in the pool. When the ensemble activity of the pool reaches a desirable level, the pool is sequenced, and the sequences are sorted into families based on their sequence similarities. The most active families, then the most active sequences in those families, are isolated. After characterizing this active sequence to determine its sensitivity and selectivity, it can be developed into a sensor for the target metal ion by functionalizing it with a fluorophore and quencher, nanoparticles, or electrochemical tags.

RNA-cleaving DNAzymes are one class of DNAzymes that are commonly selected for using in vitro selection. This type of DNAzyme consists of an enzyme strand that hybridizes to a substrate strand containing a ribonucleotide in its center. The phosphodiester bond in RNA is $\sim 10^5$ times more readily hydrolyzed than that in DNA, making it the most susceptible region of the substrate (16). The DNAzyme hybridizes to its substrate in such a way that the catalytic core of the enzyme strand can access the junction of the ribonucleotide and its 3'-adjacent base, and catalyze the cleavage of the phosphodiester bond. During the selection process, a single strand of DNA containing both the fixed substrate strand and a random region to give rise to the DNAzyme is generated through the polymerase chain reaction (PCR). The pool contains primer-binding regions for PCR amplification. During each selection round, the pool is incubated with the target analyte, then the cleaved product "winners" are isolated, amplified, and regenerated.

In the method described below, four different oligonucleotides are used to construct or regenerate the pool: template, and

Unfolded Pool

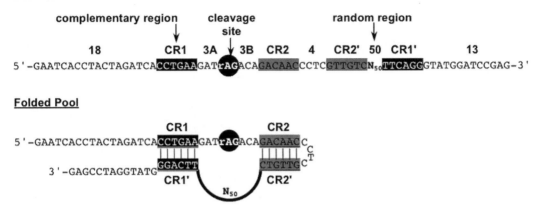

Fig. 1. During the selection process, the pool is a single-stranded DNA that folds back upon itself, as shown here. The pool has two sets of complementary regions (CR1, CR2, and their complements) that makes this folding possible. A cleavage site with an RNA nucleotide adjacent to a DNA nucleotide is marked here, as well as the 50-base random region. The pool should be designed so that its most energetically favorable state is the folded structure shown here, where the random region has full access to the cleavage site.

Fig. 2. Four oligonucleotides are used to construct or regenerate the pool. Primer A is a truncated version of Primer B, and both primers are complementary to the template. Primer C, on the other hand, is complementary to the pool, and contains a Stop Taq region (C_{18} spacer + AAC repeats) so its complementary product can be distinguished from the pool when PCR products are purified by PAGE.

Primers A–C (see Figs. 1 and 2). The template is a shortened and complementary version of the final random pool, and Primers A and B are forward primers of two different lengths that introduce the ribonucleotide. Primer C is a reverse primer containing a Stop Taq sequence that makes its PCR product longer than the desired pool; these two strands can thus be separated from one other during polyacrylamide gel electrophoresis.

There are many choices involved in designing an in vitro selection method. The oligonucleotides required to construct or regenerate the pool should be carefully designed to avoid homo- or heterodimers and undesirable secondary structure; the primers should have compatible melting temperatures. If the primer-binding regions contain extensive amounts of secondary structure, or if they can dimerize, it will make PCR less efficient. The number of nucleotides to include in the random region must be decided. The smaller the number of random nucleotides, the higher the probability that the entire sequence space will be probed. For example, if a 25 nucleotide random region was used, the entire sequence space could theoretically be covered if a selection began with a 2 nmol pool. On the other hand, it might not be necessary to probe all of sequence space to find a desired DNAzyme. Catalytic cores range in size, and with a longer random region, a desired sequence could appear multiple times in the random region (17), and a selection could easily become biased toward shorter motifs, because of their prevalence. This could explain the high incidence of a DNAzyme motif known as the "8–17." Many selections carried out by different labs have produced this motif. It can be a source of difficulty because it is often Pb^{2+}-sensitive, and it is often highly desirable to select a DNAzyme that is not sensitive to Pb^{2+}. Thus, various research labs have devised strategies to limit the incidence of the 8–17 motif. A final consideration in sequence design is that while all four nucleotides can be equally distributed in the random region, it can also be enriched with nucleotides of interest (18).

The selection buffer must contain the target metal ion in a stabilized form and available for the DNAzyme to bind in an application-relevant manner. Another crucial choice is the dinucleotide junction defining the cleavage site. RNA-cleaving DNAzymes can be selected to cleave any dinucleotide junction, but the global folding (19) and rate of cleavage is highly dependent on the identity of the two bases (20). In addition, the concentration of the selection buffer as well as the incubation time should be decreased as the selection continues; as the selection conditions become more stringent, the "winners" will be narrowed down to only include sequences with high catalytic rates.

In vitro selection provides not only a method of searching for a DNAzyme with preferred activity in the presence of a target analyte (positive selections), it also allows a pool to be biased against undesirable target analytes (negative selections) (21). These two types of selection rounds can be used in conjunction with selection pressures to select robust DNAzymes. A round can be made more stringent by decreasing the incubation time and/or the analyte concentration. When varying these characteristics, the results from previous selections should be taken into account. It is recommended the pool's activity be assessed multiple times throughout the selection, that the pool's activity be assessed at multiple points during the selection, and that the activities of major and minor sequences obtained from the selection be evaluated (22).

Single nucleotide mutations can substantially alter the selectivity and sensitivity of DNAzymes (23), and once an analyte-specific sequence has been discovered through in vitro selection, its sequence can be subsequently fine-tuned.

2. Materials

2.1. Design of DNAzyme Selection Pool

1. Random DNA Generator, http://www.faculty.ucr.edu/~mmaduro/random.htm (The Morris Maduro Lab, University of California at Riverside, Riverside, CA).

2. OligoAnalyzer, http://www.idtdna.com/analyzer/Applications/OligoAnalyzer (Integrated DNA Technologies, Coralville, IA).

3. Mfold http://mfold.ma.albany.edu/?q=mfold (The RNA Institute, University at Albany State University of New York).

2.2. Pool Generation (see Note 1)

1. 20-, 200-, and 1,000-μL barrier pipette tips.

2. 200-μL Seal-Rite thin-wall PCR tubes.

3. 1.7-mL microtubes.

4. 10× PCR buffer without $MgCl_2$ (see Note 2).

5. 2 mM dNTP solution (see Note 3).

6. 50 mM $MgCl_2$.

7. Primer B, Primer C, and Template (see Note 4).

8. [α ^{32}P]-deoxyadeonisine-5′-triphosphate.

9. Millipore water (see Note 5).

10. Platinum *Taq* DNA Polymerase.

11. C1000 PCR thermocycler (Bio-Rad, Hercules, CA).

12. 3 M sodium acetate (pH 5.2; made from sodium acetate dihydrate).

13. Ethanol.

14. Vacufage plus vacuum concentrator (Eppendorf, Hauppauge, NY).

2.3. 5′-γ ^{32}P Labeling of DNA Pool

1. DNA.

2. [γ^{32}P]-deoxyadeonisine-5′-triphosphate.

3. 10× forward reaction buffer (New England Biolabs, Ipswich, MA).

4. T4 Kinase.

5. Sep-Pak Plus® C18 cartridges (Waters Corporation, Milford, MA).

6. Acetonitrile.

7. Methanol.

8. 2 M ammonium acetate.

9. Vacufage plus vacuum concentrator (Eppendorf, Hauppauge, NY).

2.4. Polyacrylamide Gel Electrophoresis

1. Tris[hydroxymethyl]aminomethane.
2. Boric acid, crystalline powder, electrophoresis grade.
3. Ethylenediaminetetraacetic acid, disodium salt, dihydrate, crystalline powder, electrophoresis grade.
4. 10× TBE buffer (1.78 M Tris, 1.78 M boric acid, 0.05 M EDTA) (see Note 6).
5. 25% (w/v) ammonium persulfate (APS).
6. Tetramethylethylenediamine (TEMED).
7. 40% acrylamide/bisacrylamide (3.3% C).
8. Urea, high purity grade.
9. Gel electrophoresis apparatus.
10. Electrophoresis power supply.
11. Maximum Resolution (MR) film (Eastman KODAK Co., Rochester, NY).
12. FUTURA 200K™ Automatic X-ray film processor (Fischer Industries, Inc., Geneva, IL).
13. Soak solution (10 mM Tris, 1 mM EDTA, 300 mM NaCl, pH 7.5).
14. 3 M sodium acetate (pH 5.2).
15. Ethanol.
16. Vacufage plus vacuum concentrator (Eppendorf, Hauppauge, NY).

2.5. Positive In Vitro Selection

1. 2× selection solution (unique to each selection; see the details in Subheading 3).
2. Loading buffer (8 M urea, 50 mM EDTA, 1× TBE).
3. Bromophenol blue, electrophoresis purity.
4. Xylene cyanol FF, electrophoresis purity.
5. Loading buffer with dye (same as the loading buffer listed above, but with 0.05% (w/v) bromophenol blue and 0.05% (w/v) xylene cyanol).
6. 117 and 89 nt DNA Markers (Integrated DNA Technologies, Coralville, IA) (see Note 7).

2.6. Pool Regeneration

1. 200-μL Seal-Rite thin-wall PCR tubes.
2. 10× PCR buffer.
3. 2 mM dNTP solution.
4. 50 mM $MgCl_2$.
5. Primer A, Primer C, and template.
6. [α ^{32}P]-deoxyadeonisine-5′-triphosphate.
7. Platinum *Taq* DNA Polymerase (Invitrogen, Carlsbad, CA).

2.7. Activity Assays

1. 96-Well plate.
2. 96-Well adhesive plate cover.
3. 2× Selection solution (unique to each selection; see the details in Subheading 3).
4. Loading buffer with dye (8 M urea, 50 mM EDTA, 1× TBE, 0.3% (w/v) bromophenol blue and 0.3% (w/v) xylene cyanol).
5. Plastic transparency sheet.
6. Phosphorimager screen and cassette (Amersham Biosciences, Piscataway, NJ).
7. Storm 430 Phosphorimager (Molecular Dynamics, Inc., Sunnyvale, CA).
8. ImageQuant software (Molecular Dynamics Inc., Sunnyvale, CA).
9. OriginPro software (OriginLab, Northampton, MA).

3. Methods

3.1. Design of DNAzyme Selection Pool

1. Choose a length of random region, and decide whether or not to include secondary structure in it.
2. Choose a ribonucleotide–nucleotide cleavage junction (see Note 8).
3. Design the sequence of the random pool. As shown in Fig. 1, the random pool is 117 nucleotides long, and is designed so that as the six-base complementary regions anneal (CR1 binding to CR1′ and CR2 binding to CR2′), the pool folds back on itself, holding the cleavage site in place across from the random region that will give rise to the DNAzyme. Using the Random DNA Generator, choose 6-base complementary regions for the pool. Using Fig. 1 as a guide, design the rest of the pool (Note: the individual bases shown in Fig. 1 show one possible pool design, and this protocol describes how to choose a new pool design).
4. Use OligoAnalyzer and/or mfold to analyze the sequences generated (see Note 9). For parameters, use room temperature and the monovalent and divalent concentrations of metal ions present during PCR. See if the sequence's most stable predicted secondary structure shows the pool folding as shown in Fig. 1, and if the sequence forms homodimers. If it is prone to form unwanted secondary structures with high melting temperature or homodimers, identify the bases giving rise to the problem, use OligoAnalyzer to generate a new sequence, and try again (see Note 10).

Table 1
Oligonucleotides used to construct or regenerate the selection pool

DNA sequence	Purpose	Length (nt)	Complementary to	Note
Primer A	5'-GAATCACCTACTAGATCACCTGAAGATrAGACAGACAACCCTCG-3' Forward primer; introduce the ribonucleotide	28	Template	Ends with the ribonucleotide
Primer B	5'-GAATCACCTACTAGATCACCTGAAGATrA-3' Forward primer; introduce the ribonucleotide; elongate a truncated pool	43	Template	Is identical to Primer A, except it continues past the ribonucleotide
Primer C	5'-(AAC)$_{12}$-c$_{18}$spacer-CTCGGATCCATACCC-3' Reverse primer	51 + C$_{18}$ sequence	Random pool	Includes a C$_{18}$ Stop *Taq* sequence to distinguish the pool from its reverse complement during PAGE
Template	5'-CTCGGATCCATACCCTGAAN50GACAACGAGGGTTGTCTGTCTATCTTCAGGTGATCTAG-3' Provide a basis for the random pool	107	Random pool	The template is shorter than the random pool because the primers add length to it

5. Using the results from the previous step, and referencing Fig. 2 and Table 1, design the four oligonucleotides that will be used to construct or regenerate the random pool.

6. Use OligoAnalyzer to analyze these four oligonucleotides. Ensure that the primers have similar melting temperatures, and that they do not form homodimers, heterodimers, or unwanted secondary structures.

3.2. Pool Generation (see Note 11)

1. Make two batches of extension solution, and one batch of amplification solution, following Tables 2 and 3.

2. Aliquot the two batches of extension solution into 20 PCR tubes, each containing 99 μL of solution.

3. Add 1 μL of *Taq* to each tube.

4. Place the tubes in a PCR thermocycler and proceed with the protocol shown in Table 4 (see Note 12). At the eighth PCR step shown in Table 4, add 13.4 μL of amplification solution to each tube.

5. Divide the contents of all 20 PCR tubes into eight 1.5-mL centrifuge tubes, with each containing 284 μL of solution.

6. Ethanol precipitate the random pool by adding 28.4 μL of 3 M sodium acetate, pH 5.2 (10% of the DNA solution's volume) and 710 μL of 200 proof ethanol (250% of the DNA solution's volume). Pipette to mix, and place in a −80°C freezer for 1 h. Centrifuge at 17 krpm ($\approx$32,000 × g) for 30 min at −4°C.

7. Pipette off and discard the supernatant, being careful not to disturb the pellet DNA on the side of the tube.

8. Lyophilize in the Vacufage until dry (see Note 13).

3.3. 5′-γ^{32}P Labeling of DNA Pool

1. Make the solution described in Table 5.

2. Heat the solution at 37°C for 1.5 h in a PCR thermocycler.

3. Desalt the solution using a Sep-Pak Plus® C18 cartridge. Prepare the cartridge by washing it with 10 mL 95% (w/w) acetonitrile, 10 mL (1:1:1 acetonitrile:methanol:water), 20 mL water, and 10 mL 2 M ammonium acetate (see Note 14).

4. Add 200 μL of water to the DNA labeling solution, then load this solution on the Sep Park cartridge.

5. Wash the sample-loaded cartridge with 20 mL of water.

6. Elute the radiolabeled DNA in 2 mL of 1:1:1 acetonitrile:methanol:water.

7. Uncap the solution and immerse it in liquid nitrogen to flash freeze it.

8. Lyophilize it to dryness (see Note 15).

Table 2
Extension solution (make two)

10× PCR buffer	105 μL
2 mM dNTP solution	105 μL
50 mM MgCl$_2$	31.5 μL
100 μM N$_{50}$ template	1.05 μL
10 μM Primer B	21 μL
Water	776 μL

Table 3
Amplification solution (make one)

10× PCR buffer	26.4 μL
2 mM dNTP solution	26.4 μL
50 mM MgCl$_2$	7.92 μL
10 μM Primer B	57 μL
10 μM Primer C	154 μL
[α ^{32}P]-dATP	2 μL

Table 4
PCR protocol for pool generation

Step	Temperature (°C)	Time (min)	Note
1	95	3	
2	52	1.5	
3	72	1	
4	93	1	
5	52	1	
6	72	1	
7			Run PCR steps 4 through 6 (93°, 52°, 72°) six times
8	85	3	Add amplification solution
9	93	1	
10	52	1	
11	72	1	
12			Run PCR steps 9 through 11 (93°, 52°, 72°) six times
13	72	3	

Table 5
5′-γ ^{32}P DNA labeling solution

10 μM of the DNA to be labeled	2 μL
10× Forward reaction buffer	2 μL
γ ^{32}P dATP	3 μL
Water	10.5 μL
T4 kinase	2.5 μL

9. Dissolve the two samples in water and reconstitute by vortexing and spinning down.

3.4. Polyacrylamide Gel Electrophoresis

1. Prepare a 10% gel stock solution. Combine 250 mL of 40% acrylamide/bisacrylamide, 480 g urea, and 100 mL of 10× TBE. Add enough water to make 1 L and stir until dissolved. Filter through a 0.25-μm filter and store in an airtight container (see Note 16).

2. Dilute 10× TBE to make 1× TBE.

3. Place spacers between two glass plates, and tape the two sides and bottom with gel tape (see Note 17). Cast a 10% gel by adding 45 μL of TEMED and 45 μL of 25% APS to 35 mL of 10% PAGE gel stock in an Erlenmeyer flask. Swirl to dissolve, and pour between the two glass plates (see Note 18). Insert a well comb, and clamp the sides with binder clips (see Note 19). Allow the gel to polymerize for ~45 min, or until the gel is firm to the touch.

4. Cut the bottom tape from the gel, and carefully rinse the gel to remove any unpolymerized gel.

5. Place the gel on the gel rack, add an aluminum plate to evenly distribute the heat generated during electrophoresis, and pour 1× TBE into the top and bottom buffer reservoirs.

6. Pre-run the gel for 30 min at 25 W.

7. Meanwhile, dissolve the lyophilized DNA in water and combine the DNA from all the tubes together (see Note 20).

8. Add an equal volume of loading buffer to the reconstituted sample, and, after its pre-run is over, load it on the 10% gel.

9. Run radiolabeled markers alongside the pool (see Note 21).

10. Run the gel for 1.5 h at 25 W.

11. Remove the gel from the rack, remove the top plate, and wrap the gel and bottom plate in plastic wrap. Place a clean plate on top of this, and place in a metal cassette.

12. In a dark developing room, remove the top plate, and place an X-ray film on top of the plastic wrap to expose it to the radio-labeled DNA. With a razor blade, cut an "x" in two opposing corners so that the film and gel can be reoriented later. Replace the top plate, place the assembly back into the metal cassette, and wrap the cassette in a black sleeve.

13. After exposing the film, develop it.

14. Place the film on top of a light box, and place the gel/glass plate assembly on top of the film. Realign the film to the gel using the "x"s cut earlier.

15. With a fresh razor blade, cut out the lower band that runs alongside the upper, 117mer marker (see Note 22).

16. Place this gel slice in a 1.5-mL tube, add 1 mL of soak solution, and allow it to soak for 2 h.

17. After this time, pipette off the first soak batch and transfer it to another tube. Add another 1 mL of soak solution to the gel slice, and allow it to soak for another 2 h.

18. Pipette off this second solution, combine it with the first batch of soak solution, then ethanol precipitate and lyophilize it per the directions in Subheading 3.2.

3.5. Positive In Vitro Selection (see Note 23)

1. Choose the incubation solution pH based on the solubility, stability, and speciation of the metal, and the application the DNAzyme is intended for. Take into account the air sensitivity of the metal in solution, and other factors influencing its stability. If the metal is not stable in solution, make it fresh before use each time. If the metal solution is stable, make a concentrated solution, aliquot it, and freeze it. If the metal ion target is not very soluble, consider including a chelator such as citrate. Include monovalent cations to stabilize the random pool, and buffer the solution to the appropriate pH. Common solution concentrations are 100–500 mM NaCl, and 25–50 mM buffer.

2. When carrying out a selection round, choose a metal concentration higher than that intended for the final application, to increase the probability of the metal ion binding to the target analyte. Common starting concentrations for 2× metal solutions are 1–20 mM.

3. Choose an incubation time for the first selection. 1–5 h is a usual incubation time for the first selection round.

4. Dissolve the random pool in 2× selection buffer. Denature at 95°C for 3 min and anneal by allowing the DNA to cool to room temperature over a 30-min period.

5. Spin down the sample, then incubate it with an equal volume of 2× metal solution.

6. After the time for the selection has elapsed, add an equal volume of loading buffer to the reaction.

7. Gel purify the selection reaction on a 10% denaturing PAGE gel. Run this against radiolabeled markers, and cut out the lower band corresponding to the cleaved, 89 nt product (see Note 24).

8. Extract the cleaved pool from the gel by soaking it, then ethanol precipitate and lyophilize it as described in Subheading 3.4.

9. Dissolve the lyophilized, cleaved pool in water (~60 μL). Freeze half of it for future use and regenerate the rest to take on to the next round of selection.

3.6. Pool Regeneration   1. Prepare the solutions described in Tables 6 and 7.

Table 6
PCR1 solution: PCR extension of cleaved product

DNA product	30 μL
10× PCR buffer	10 μL
2 mM dNTP solution	10 μL
50 mM MgCl$_2$	3 μL
10 μM Primer A	4 μL
10 μM Primer C	4 μL
H$_2$O	38 μL
Taq polymerase	1 μL

Table 7
PCR2 solution: PCR amplification of cleaved product

PCR1 product	10 μL
10× PCR buffer	10 μL
2 mM dNTP solution	10 μL
50 mM MgCl$_2$	3 μL
10 μM Primer B	5 μL
10 μM Primer C	2.5 μL
H$_2$O	57.5 μL
Taq polymerase	1 μL

Table 8
PCR conditions to regenerate pool

Step	Temperature (°C)	Time (min)	Note
1	95	3	
2	52	0.5	
3	72	1	
4	93	0.5	
5	52	0.5	
6	72	1	
7			Go to step 4 nine times for PCR1 or 19 steps for PCR2
11	72	3	

2. Perform PCR1 with the solution described above, following the protocol outlined in Table 8.

3. Add 10 μL of the PCR1 product to the PCR2 solution. Following the protocol in Table 8, perform PCR2 with this solution. Freeze the remaining PCR1 volume at –80°C.

4. Purify the PCR2 product on a 10% PAGE gel as described in Subheading 3.4, but this time at step 14 cut out the regenerated pool that runs alongside the 89 nt marker. Extract, ethanol precipitate, and lyophilize the pool.

5. Use this DNA population for the next round of selection. Repeat the steps in Subheading 3.5, changing the selection pressures (reaction time and target analyte concentration) as desired.

3.7. Activity Assays
(see Note 25)

1. Following the protocol in Subheading 3.3, 5'-γ ^{32}P label 2 μL of a 10 μM solution (20 pmol) of Primer B. (This radiolabeled primer is now referred to as Primer B*) (see Note 26).

2. Dissolve the labeled primer in 20 μL water.

3. Label the DNA population from the round of interest by preparing the solution described in Table 9.

4. Run 20 cycles of PCR (PCR2 as described before).

5. Purify the labeled pool by 10% PAGE. Only one band—the strand incorporating P3*—will be visible; the complementary strand will not be visible. Cut out this band and soak as before to extract the DNA.

6. Sep-Pak purify the DNA and lyophilize.

Table 9
PCR2*: labeling of a DNA population for an activity assay

PCR1 product	10 μL
10× PCR buffer	10 μL
2 mM dNTP solution	3 μL
50 mM MgCl$_2$	3 μL
10 μM Primer B	1 μL
10 μM Primer C	0.5 μL
10 μM P3*	5 μL
Water	66.5 μL
Taq polymerase	1 μL

7. Dissolve the DNA population in 2× selection buffer. Allow for at least 30 μL of solution for each reaction, and ensure that you have at least two reactions: one with the metal solution used during selection rounds, and a control with water.

8. Denature at 95°C for 3 min and anneal by allowing the solution to cool to room temperature over a 30-min time period.

9. Prepare a 96-well plate with a separate well for each time-point to be taken. Pipette 10 μL of loading buffer with dyes into each time-point well, and pipette 60 μL of 2× metal solution into another well, as well as 60 μL of water into a final well. Finally, pipette 30 μL of the prepared pool solution for each reaction.

10. Initiate the reaction by adding 30 μL of the 2× metal solution to the 30 μL of labeled pool solution. Then initiate the control reaction by adding 30 μL of water to the second 30 μL aliquot of labeled pool solution.

11. At predetermined time-points, withdraw 5 μL of the metal reaction mixture and add it to 10 μL of loading buffer with dyes. Do the same with the control reaction.

12. Separate the uncleaved pool and product on a 20% PAGE gel (32 W, 2.5 h).

13. Remove the gel from the rack, transfer it from the glass plate to a plastic transparency sheet, and wrap the gel and plastic sheet in plastic wrap.

14. Expose the gel by placing it in a phosphorimager cassette (see Note 27).

15. Image using Molecular Dynamics Storm 430 Phosphorimager.

16. Analyze the fraction of pool cleavage using ImageQuant software. Place boxes of the same area around the uncleaved and cleaved pool bands, and a background position. In a spreadsheet, background subtract the intensity of each band, and compute the percentage of the pool cleaved at each time-point in minutes.

17. Plot kinetic curves using OriginPro and fit to the equation $\%Pt = \%P_0 + \%P\infty\,(1 - e{-}kt)$, where t is the reaction time in minutes, $\%Pt$ is the percent product at time t, $\%P_0$ is the initial percent product ($t = 0$), $\%P\infty$ is percent product at the end point of the reaction ($t = \infty$), and k is the observed rate of cleavage.

4. Notes

1. Any tube to be used for the long-term storage of any solution should be soaked in 10% nitric acid overnight and triple-rinsed with Millipore water to prevent metal ions from leaching out of the plastic into the solution. Use ultrapure materials for all solutions, and adjust their pH using a sterilized pH meter and ultrapure acid or base, such as that available from Alfa Aesar (Ward Hill, MA). After calibrating the pH meter, it can be sterilized by soaking it in 1 M HCl for 2 min, and then in 3 M NaOH for 2 min, rinsing it in Millipore water after each soak. Any solution that does not contain EDTA should be treated with Chelex 100 sodium form beads (Sigma-Aldrich, St. Louis, MO) overnight to remove any trace divalent cations. To treat a solution, add ~1 g of sodium Chelex beads per 100 mL of solution, and stir overnight. The solution should be stirred fast enough for the beads to circulate, but not so rapidly that the beads beat against the side of the container and are damaged. After treating the solution, remove the Chelex beads by filtering the solution. Since Chelex treatment can alter the pH of the solution, check the pH of the solution and readjust it, if necessary.

2. This solution is included with every *Taq* DNA polymerase purchase.

3. Dilute from the 10 mM solution provided by New England Biolabs.

4. DNA can be ordered at different purity levels, with the two most common levels being standard desalted (low purity) and HPLC-purified (high purity). All DNA should be purified before use, if not by HPLC, then by PAGE-purifying desalted DNA. The choice between these levels is individual, with convenience lying on the side of HPLC-purified DNA, and economy on the side of desalted DNA.

5. The water used for all experiments was purified using a Milli-Q system (Millipore, Billerica, MA).

6. The urea in this solution is prone to precipitate over time, so its shelf-life is only about a month. It is easiest to remove solid urea from the sample bottle by adding nitric acid, then rinsing.

7. These markers are the length of the intact pool (117 nt) and the length of the cleaved pool (89 nt). Choose random sequences of DNA that will not form homo- or heterodimers with the random pool.

8. It has been found that the 16 choices (rAA, rAC, rAG, rAT, etc.) vary in their cleavage rate, Pb^{2+} sensitivity, and susceptibility to producing the common 8–17 DNAzyme motif.

9. Use "N" for the random region, but include only five nucleotides in the random region, to separate the parts of the pool surrounding the random region, but not to generate a large number of side products from the random region binding to itself. The random region will interact with itself, but the point of this step is to ensure that there is minimal unwanted interactions between the primer-binding regions.

10. You can alter the %GC content for your own advantage during this process as well.

11. When working with radiolabeled DNA, standard radiation safety procedures should be followed, and all waste should be labeled and disposed of properly. Always use filtered pipettes for radioactive solutions.

12. These PCR conditions should be optimized for the DNA sequences used in each selection. An efficient way to do this is realtime PCR (RT-PCR). Two of the most important parameters to optimize are the annealing temperature and number of PCR cycles.

13. To protect a sample from dust, punch a hole in a 1.5-mL tube cap with the needle. Cut the cap off, and snap it onto the sample tube.

14. Squeeze the solution through the cartridge at a rate that is dropwise; do not push air through the cartridge or pull air back into the cartridge.

15. This typically is an overnight process, because of the large amount of liquid.

16. Do not refrigerate, because this would cause the urea to precipitate. Note: unpolymerized acrylamide/bisacrylamide is a carcinogen. Handle with care, and change your gloves after working with it.

17. Plates should be soaked in detergent (such as RBS 35® detergent (Pierce, Rockford, IL)) between uses, if contaminated with radioactive material.

18. Air bubbles should be carefully removed during this process. First, after pouring the gel solution between the plates, hold the gel vertically and tap it on the bench to remove air bubbles. Then, when inserting the well comb, use it to "scoop out" air bubbles as they rise to the top.

19. Binder clips, such as are available at office stores, are used. It should be noted that four clips are often used, with two on either side, directly across from one another. If the gel is unevenly clamped, it can cause the gel to polymerize crookedly, wreaking havoc on the gel as it runs.

20. Because the DNA can be spread around all sides of the tube, it is important to use a large enough volume of water when reconstituting the sample, and to vortex it thoroughly. On the other hand, it is also best to keep the sample as concentrated as possible at all steps. To accomplish this, add ~200 μL of water to the first tube, vortex it for 2 min, spin it down, and transfer the volume to the next tube. Continue this with all of the sample tubes, then lyophilize the sample to concentrate the sample to ~50 μL.

21. Adjust the amount of radiolabeled marker so it is of approximately the same activity as the sample. The two marker strands should be labeled separately, and Sep-paked together. The two fractions can be combined, and reconstituted in 100 μL of water. Typically, when fresh, 1 μL of this solution can be added to 10 μL of loading solution with dye, and run in a well alongside the sample. As the marker decays, increase the amount used, being sure that the solution loaded is at least 50% loading solution.

22. The upper band is the sequence complementary to the pool, which contains the C_{18} Stop *Taq* sequence.

23. This section describes a positive selection round. If a negative selection round is to be undertaken, change the metal solution to be incubated with to a metal that you do not desire your pool to be sensitive to. Then, instead of excising and purifying the cleaved pool, excise and purify the uncleaved pool.

24. Commonly, the cleaved product will not be visible for the first few rounds. Thus, the marker allows one to cut out the product.

25. When working with radiolabeled DNA, standard radiation safety procedures should be followed, and all waste should be labeled and disposed of properly. Always use filtered pipettes for radioactive solutions.

26. To assess the pool's activity at any given selection round, the DNA pool can be γ-labeled with ^{32}P. This is done instead of incorporating the α-labeled dATP into the strand PCR because a 5 ′-labeled oligonucleotide is uniformly labeled and can be used in quantitative evaluations while oligonucleotides labeled by the random incorporation of a radioactive nucleotide vary in their radioactivity and cannot be used quantitatively.

27. The exposure time depends on the radioactivity of the sample.

Acknowledgments

The authors would like to thank Andrea K. Brown, Debapriya Mazumdar, Nandini Nagraj, Tian Lan, and Seyed-Fakhreddin Torabi for developing and fine-tuning this method. This work was supported by the US National Institutes of Health (ES016865) and Department of Energy (DE-FG02-08ER64568).

References

1. Breaker R, Joyce G (1994) A DNA Enzyme that Cleaves RNA. Chem Biol 1:223-229.

2. Li Y, Sen D (1996) A Catalytic DNA for Porphyrin Metallation. Nat Struct Biol 3:743–747.

3. Lu Y (2002) New Transition Metal Ion-Dependent Catalytic DNA and Their Applications as Efficient RNA Nucleases and as Sensitive Metal Ion Sensors. *Chem Euro J* 8:4588–4596.

4. Li Y, Liu Y, Breaker R (2000) Capping DNA with DNA. Biochem 39:3106–3114.

5. Schlosser K, Li Y (2009) Biologically Inspired Synthetic Enzymes Made from DNA. Chem Biol 16:311–322.

6. Silverman S (2008) Catalytic DNA (Deoxyribozymes) for Synthetic Applications—Current Abilities and Future Prospects. Chem Commun 3467–3485.

7. Lu Y, Liu J (2006) Functional DNA Nanotechnology: Emerging Applications of DNAzymes and Aptamers. *Curr Opion Biotech* 17:580–588.

8. Franzen S (2010) Expanding the Catalytic Repertoire of Ribozymes and Deoxyribozymes Beyond RNA Substrates. Curr Opin Mol Ther 12:223–232.

9. McManus S, Li Y (2010) The Structural Diversity of Deoxyribozymes. Molecules 15:6269–6284.

10. Li J, Lu Y (2000) A Highly Sensitive and Selective Catalytic DNA Biosensor for Lead Ions. J Am Chem Soc 122:10466–10467.

11. Liu J, Lu Y (2003) A Colorimetric Lead Biosensor Using DNAzyme-Directed Assembly of Gold Nanoparticles. J Am Chem Soc 125:6642–6643.

12. Xiao Y, Rowe A, Plaxco K (2007) Electrochemical Detection of Parts-per-billion Lead via an Electrode-Bound DNAzyme Assembly. J Am Chem Soc 129:262–263.

13. Lan T, Furuya K, Lu Y (2010) A Highly Selective Lead Sensor Based on a Classic Lead DNAzyme. Chem Commun 46:3896–3898.

14. Liu J, Brown A, Meng X, Cropek D, Istok J, Watson D, Lu Y (2007) A Catalytic Beacon Sensor for Uranium with Parts-per-trillion Sensitivity and Millionfold Selectivity. P Natl Acad Sci USA 104:2056–2061.

15. Li J, Zheng W, Kwon A, Lu Y (2000) In Vitro Selection and Characterization of a Highly Efficient Zn(II)-dependent RNA-cleaving Deoxyribozyme. Nucleic Acids Res. 28:481–488.

16. Li Y, Breaker R (1999) Kinetics of RNA Degradation by Specific Base Catalysis of Transesterification Involving the 2′-Hydroxyl Group. J Am Chem Soc 121: 5364–5372.

17. Vant-Hull B, Gold L, Zichi D (2000) Theoretical Principles of In Vitro Selection using Combinatorial Nucleic Acid Libraries. In: Egli M, Herdewijn, P, Matusda, A, Sangyi Y (ed) Current Protocols in Nucleic Acid Chemistry. Wiley, New York.

18. Schlosser K, Li Y (2009) DNAzyme-mediated Catalysis with Only Guanosine and Cytidine Nucleotides. Nucleic Acids Res 37:413–420.

19. Lam J, Li Y (2010) Influence of Cleavage Site on Global Folding of an RNA-Cleaving DNAzyme. Chem Bio Chem 11:1710–1719.

20. Schlosser K, Li Y (2010) A Versatile Endoribonuclease Mimic Made of DNA: Characteristics and Applications of the 8–17 RNA-Cleaving DNAzyme. Chem Bio Chem 11:866–879.

21. Brueschoff PJ, Li J, Augustine III, AJ, Lu Y (2002) Improving Metal Ion Specificity During In Vitro Selection of Catalytic DNA. *Combinator Chem High Throughput Screening* 5:327–335.

22. Schlosser K, Lam J, Li Y (2009) A Genotype-to-Phenotype Map of *In Vitro* Selected RNA-cleaving DNAzymes: Implications for Accessing the Target. Nucleic Acids Res 37:3545–3557.

23. Lam J, Withers J, Li Y (2010) A Complex RNA-Cleaving DNAzyme That Can Efficiently Cleave a Pyrimidine–Pyrimidine Junction. J Mol Bio 400:689–701.

Chapter 19

Selecting Allosteric Ribozymes

Nicolas Piganeau

Abstract

Allosteric ribozymes can be designed to respond to virtually any molecule of choice. The resulting species may be used for example as synthetic regulators of gene expression or alternatively as biosensors. In vitro selection techniques allow the isolation of active molecules from libraries as large as 10^{15} different molecules. The present protocol describes an in vitro selection strategy for the de novo selection of allosteric self-cleaving ribozymes responding to virtually any drug of choice. We applied this method to select hammerhead ribozymes inhibited specifically by doxycycline or pefloxacin in the sub-micromolar range. The selected ribozymes can be converted into classical aptamers via insertion of a point mutation in the catalytic center of the ribozyme.

Key words: Allostery, Aptazyme, In vitro selection, Ribozyme

1. Introduction

The activity of catalytic RNAs can be regulated by small molecules. These so-called allosteric ribozymes or aptazymes can find applications in the field of basic biological research or applied biotechnology. For example they can be employed as molecular sensors detecting the presence of the effector molecule (1). Alternatively they can be inserted in genes and serve as synthetic switches for the control of gene expression (2) (Fig. 1).

The first allosteric ribozymes were generated via rational design by the fusion of a constitutive ribozyme to an RNA-aptamer (3). Later on in vitro selection methods were developed to optimize the "communication module" between the aptamer and the ribozyme (4). These methods depend on the preselection of an aptamer before the creation of the allosteric ribozyme. During the selection of aptamers, the small ligand must be immobilized to allow affinity chromatography. The immobilization can be difficult

Jörg S. Hartig (ed.), *Ribozymes: Methods and Protocols*, Methods in Molecular Biology, vol. 848,
DOI 10.1007/978-1-61779-545-9_19, © Springer Science+Business Media, LLC 2012

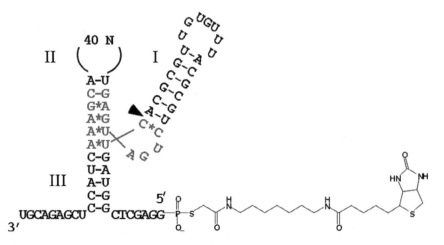

Fig. 1. Secondary structure of the transcripts from the initial pool. Helix II is shortened to two base pairs, and loop II is replaced with a 40 nt random region. *Gray*: nucleotides of the catalytic center of the hammerhead ribozyme (HHR). *Black arrow*: cleavage site. The chemical link between the biotin moiety and the RNA used during the selection is also shown.

to achieve and may mask one side of the molecule potentially important for interaction with RNA.

However, it is also possible to select de novo an allosteric ribozyme by linking a random sequence to the ribozyme and selecting for inhibition or activation of the catalytic activity via a small effector (5, 6). Using this method no immobilization procedure of the small ligand is required.

The method presented here allows the selection of allosteric hammerhead ribozymes inhibited by virtually any drug of choice. It can be easily adapted for the selection of ribozymes activated by the effector molecule.

2. Materials

2.1. Pool Synthesis

1. Primers.

   Pnp-rev: 5'-ACG TCT CGA GGT AGT TTC GT.

   Pnp-1: 5'-CGC GTT GTG TTT ACG CGT CTG ATG.

   Pnp-pool: 5'-CGC GTT GTG TTT ACG CGT CTG ATG AGT NNN NNN NNN NNN NNN NNN NNN NNN NNN NNN NNN NNN NNN NAC GAA ACT ACC TCG AGA CGT.

   Pnp-2: 5'-AGC TGG TAC CTA ATA CGA CTC ACT ATA GGA GCT CGG TAG TGA CGC GTT GTG TTT ACG CGT CTG ATG.

   Pnp-3: 5'-AGC TGG TAC CTA ATA CGA CTC ACT ATA GGA GCT CGG TAG TCA CGC GTT GTG TTT ACG CGT CTG ATG.

2. Double distillated water (ddH$_2$O). The water should be free of RNases. In our hand no further treatment was necessary. If needed add 0.1% diethyl pyrocarbonate (DEPC) to water, mix overnight, and autoclave 20 min to hydrolyze DEPC.

3. DAp Gold star DNA polymerase and Gold star Buffer, Eurogentec.

4. 25 mM MgCl$_2$, filtered through a 0.2-μm filter.

5. 4 mM dNTP (each). Store at –20°C.

6. Phenol/chloroform/isoamyl alcohol (25:24:1) saturated with TE (10 mM Tris pH 8.0, 1 mM EDTA). Store at 4°C protected from light.

7. 3 M sodium acetate pH 5.2 (adjust pH with glacial acetic acid).

8. Sephadex G50 medium (GE healthcare). Prepare 50% slurry according to manufacturer instructions.

2.2. Transcription

1. 5× T$_7$ reaction buffer: 200 mM Tris–HCl (pH 8.0), 40 mM MgCl$_2$, 250 mM NaCl, 10 mM spermidine, 150 mM DTT. Store at –20°C.

2. 25 mM NTP (each). Prepare aliquots and store at –20°C.

3. Alpha-P^{32} CTP: 10 μCi/μL, 3,000 Ci/mmol.

4. 100 mM GMPS (Guanosine-5′-O-monophosphorothioate), Biolog (see Fig. 2).

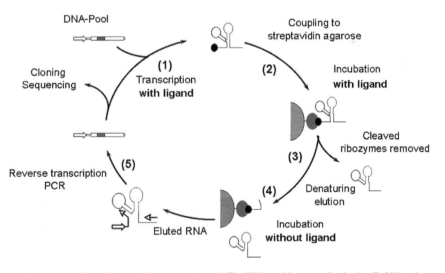

Fig. 2. Schematic representation of the selection procedure. (1) The DNA pool is transcribed using T$_7$ RNA polymerase in the presence of guanosine monophosphorothioate (GMPS) and 1.0 mM effector to avoid cleavage during transcription (2) To efficiently separate uncleaved from cleaved ribozymes, the whole RNA library is chemically biotinylated at the 5′-termini, immobilized on streptavidin agarose, and incubated with the effector. (3) Bound ligand and cleaved ribozyme products are removed by denaturing washing steps. (4) Uncleaved, immobilized RNAs are incubated for cleavage without effector. (5) Cleaved ribozymes are eluted, reverse transcribed, and amplified by PCR. The design of the PCR primers allows restoration of the 5′ cleaved fraction of the HHR and the T$_7$ promoter. The resulting DNA is used for the next selection cycle.

5. T$_7$ RNA polymerase.

6. 6 M Ammonium acetate pH 6.0 (adjust pH with glacial acetic acid).

7. DNase I (RNase-free).

2.3. Polyacrylamide Gel Electrophoresis

1. PAGE loading buffer: 9 M urea, 50 mM EDTA. For UV-shadowing the buffer should be free of dye. To follow electrophoresis add xylene cyanol and bromophenol blue (0.4% w/v each).

2. 40% acrylamide/bis solution (19:1) (this is a neurotoxin when unpolymerized and so care should be taken not to receive exposure) and N,N,N,N'-Tetramethyl-ethylenediamine (TEMED).

3. Ammonium persulfate: prepare 10% solution in water, store at 4°C for no more than 1–2 weeks.

4. Dichlordimethylsilane (5% in chloroform). Store and manipulate under a fume hood.

5. 10× TBE: 1.1 M Tris, 900 mM Borate, 25 mM EDTA, pH 8.3.

6. Thin layer chromatography plates F$_{254}$ (20 × 20 cm, Merck).

2.4. Selection

1. 10× Biotinylation buffer: 500 mM Tris, 50 mM EDTA, pH 8.3.

2. Iodoacetyl-LC biotin 4 mM in dimethyl formamide (DMF), Pierce. Prepare aliquots and store at –20°C.

3. Streptavidin agarose (Pierce) equilibrated in coupling buffer (PBS, 150 mM NaCl –50% slurry) according to manufacturer instructions.

4. WA: 25 mM HEPES pH 7.4, 1 M NaCl, 5 mM EDTA.

5. WB: 3 M urea, 5 mM EDTA.

6. 5× selection buffer: 200 mM Tris pH 8.0 (25°C), 250 mM NaCl, 10 mM spermidine. Store at –20°C. During the selection the 1× buffer should be supplemented with MgCl$_2$ (8 mM final concentration).

7. Glycogen (20 μg/μL).

8. 5× RT-PCR buffer: 250 mM bicine/KOH, pH 8.2 (25°C); 575 mM K-acetate; 40% glycerol (v/v).

9. *Tth* DNA polymerase.

10. 10× *Taq* reaction buffer: 100 mM Tris pH 8.3; 500 mM KCl; 0.01% Gelatin.

11. *Taq* polymerase.

2.5. Analysis of Selected Clones

1. Calf Intestine Alkaline Phosphatase, Fermentas.

2. RNasin, Promega.

3. Gamma-P^{32} ATP: 10 μCi/μL, 3,000 Ci/mmol.

4. T4 polynucleotide kinase.

3. Methods

3.1. Construction of Initial Pool

1. Prepare three water baths preheated at the following temperatures: 94°, 55°, and 72°C.

2. In a total volume of 80 mL mix the following components: Pnp-pool 5 nmol, primers (Pnp-1 and Pnp-rev) 80 nmol, MgCl$_2$ 1.5 mM, dNTP 0.2 mM, Gold star reaction buffer 1x, and DAp Gold star DNA polymerase 250 U. Aliquot PCR reaction into ten 15-mL tubes (8 mL each).

3. Perform five PCR cycles by transferring the tubes successively into the three water baths like following: 94°C 5 min, 55°C 5 min, and 72°C 7 min. Mix every 2 min by inversion. Take 5 μL aliquots after each cycle to follow amplification on a 2% agarose gel.

4. To purify the PCR reaction, add 7 mL phenol/chloroform/isoamyl alcohol to each tube, vortex strongly, and centrifuge for 10 min at 4,500×g. Transfer the aqueous phase to a new tube. Add 7 mL chloroform, vortex, and centrifuge as previously. Transfer aqueous phase to a new tube.

5. Pool PCR into six 50-mL tubes (13 mL each), add 1.3 mL 3 M sodium acetate and 30 mL 100% ethanol. Incubate for 30 min at –20°C and centrifuge for 30 min at 4,500×g at 4°C. Remove supernatant, add 10 mL 70% ethanol, and centrifuge for 10 min at 4,500×g at 4°C. Remove supernatant, let pellets dry, and resuspend the PCR product into 1 mL ddH$_2$O (total volume).

6. Apply the resuspended PCR product on a G50 column (0.7×20 cm) pre-equilibrated with ddH$_2$O and elute with ddH$_2$O. Collect 1 mL fractions and measure absorption at 260 nm (see Note 1). Analyze DNA containing fractions on a 2% agarose gel and pool fractions containing the PCR product. Typical recovery rates should be around 20–40 nmol.

7. Repeat the whole large-scale PCR procedure with 5 nmol of the amplified product using primers Pnp-3 and Pnp-rev.

3.2. In Vitro Selection

The following protocol describes a typical selection cycle (see Note 2). The reaction volumes and the concentration of the effector molecule should be adjusted during the selection to increase selection stringency. Conditions used during a successful selection are shown in Table 1.

3.2.1. Transcription

1. To produce GMPS-primed RNA mix following components on ice to a final volume of 100 μL: 5× T$_7$ reaction buffer, 20 μL; 25 mM NTP, 10 μL; alpha-P^{32} CTP, 3 μL; 100 mM GMPS, 20 μL; 200 pmol DNA template; 1 mM effector molecule; and 250 U T$_7$ polymerase. Start the reaction by addition of the polymerase. Incubate 4 h at 37°C (see Notes 3 and 4).

Table 1
Selection conditions

Cycle	Transcription volume	Biotinylation (RNA used) (nmol)	Streptavidin agarose (50% slurry)/RNA	Incubation volume	First incubation	Effector concentration	Second incubation (min)
1[a]	4 mL (6 nmol template)	40	5 mL/20 nmol	10 mL	2 h	1 mM	10
2	200 µL	4	500 µL/2 nmol	1 mL	4 h	1 mM	5
3–5	100 µL	2	250 µL/1 nmol	500 µL	4 h	1 mM	5
6 and 7	100 µL	2	250 µL/1 nmol	500 µL	4 h[b]	1 mM	5
8–10	100 µL	2	250 µL/1 nmol	500 µL	4 h[b]	100 µM	1
11[c]–13	100 µL	2	250 µL/500 pmol	500 µL	4 h[b]	10 µM	1
14[c]–16	50 µL	1	100 µL/100 pmol	250 µL	4 h[b]	1 µM	1

The modifications of the basic selection procedure used during a successful selection are shown (5). The actual conditions needed for a particular target may vary from the one presented here. If high ribozyme activity or high effector sensitivity is required a theoretical model can be employed to determine the optimal selection-parameters (6).

[a]Additional modification for this cycle: reverse transcription and PCR volumes are doubled

[b]The selection cycle is modified according to Subheading 3.3.2

[c]Before cycles 11 and 14 a mutagenic PCR is performed according to Subheading 3.3.1

2. The DNA template is then degraded by addition of 5 U DNAse I followed by 30 min incubation at 37°C. Stop enzymatic reaction by addition of 100 μL 0.25 M EDTA pH 8.0.

3. Add 100 μL 6 M Ammonium Acetate and 900 μL 100% ethanol. Vortex. Incubate 5 min at room temperature before centrifugation at $15,000 \times g$ for 15 min at 4°C. Remove supernatant, add 1 mL 70% ethanol, and centrifuge at $15,000 \times g$ for 5 min at 4°C. Remove supernatant, open test tube and dry pellet for a few minutes at 37°C. Resuspend pellet in 100 μL PAGE loading buffer.

3.2.2. Denaturing Polyacrylamide Gel Electrophoresis

1. Clean glass plates (16.5×22 cm), spacers (1.5 mm), and comb (ten wells 1 cm each) thoroughly with water and 70% ethanol. If necessary (usually every two to five gels) treat one of the glass plates with dichlordimethylsilane by gently wiping a few milliliters on the plate under a fume hood and letting the surface dry. Assemble gel plates and spacers, seal with adhesive tape.

2. Prepare 8% polyacrylamide solution by mixing 5 mL 10× TBE, 10 mL 40% polyacrylamide solution, and 25 g urea and ddH$_2$O to a final volume of 50 mL. After dissolution of urea start polymerization with addition of 250 μL 10% APS and 25 μL TEMED. Cast gel immediately. If necessary remove air bubbles by knocking gently on glass plates. Place comb and wait until gel is polymerized (30 min to 1 h). Remove tape and comb; wash slots with water to remove polyacrylamide rests.

3. Assemble gel on electrophoresis apparatus with aluminum plate for heat dispersion and fill reservoirs with 1× TBE (dilute 100 mL 10× TBE to 1 L with ddH$_2$O in a cylinder, seal with parafilm and mix by inverting a few times). If needed remove air bubbles in the wells and at the bottom of the gel using a 50-mL syringe.

4. Connect gel to power supply (minus electrode on top) and pre-run for 20 min at 300 V. Denature RNA probes by a short incubation at 95°C (1–2 min). Stop power supply and rinse the wells thoroughly with 1× TBE using a 50-mL syringe with a 21-gauge needle to remove urea. Immediately load RNA on gel (two wells). Load loading dye in an adjacent well to follow electrophoresis. Run gel for 1 h to 90 min at 300 V until bromophenol blue is out.

5. Remove gel from apparatus and carefully separate glass plates using one of the spacers as lever. The gel should remain on one plate. Place wrapping foil on the gel, invert and remove second glass plate. Place wrapping foil on the other side.

6. Place gel on a thin layer chromatography plate and illuminate with UV light (254 nm), the RNA should appear as a dark shadow. Two bands should be visible: intact ribozymes and

cleaved products. Cut out band corresponding to the full-length ribozyme. Place gel piece containing RNA into a 2-mL reaction tube and crush against the tube wall with a blue tip. Add 600 μL 0.3 M sodium acetate (pH 5.2) and incubate for 90 min at 65°C with strong shaking.

7. Insert glass wool into a syringe and use the piston to press the solution containing the gel pieces through the glass wool into a new 2-mL tube. Fill the tube with 100% ethanol, incubate at –20°C for 20 min, and centrifuge at 15,000×g for 15 min at 4°C. Remove supernatant and wash with 1 mL 70% ethanol.

8. Resuspend dried pellet into 50 μL ddH$_2$O. Use 5 μL of this solution in a total volume of 200 μL H$_2$O to determine optical density at 260 nm. Typical yield is 2–3 nmol RNA.

3.2.3. RNA Biotinylation

1. Mix the following component to a final volume of 1 mL: 100 μL 10× biotinylation buffer, 2 nmol GMPS-primed RNA, and 100 μL iodoacetyl-LC-biotin. Incubate for 90 min at room temperature protected from light with occasional shaking.

2. Precipitate the reaction products by addition of 100 μL 3 M sodium acetate and 2.5 mL 100% ethanol, incubation at –20°C for 20 min, and centrifugation at 15,000×g for 15 min at 4°C. Remove supernatant and wash with 1 mL 70% ethanol.

3. The dried pellet are resuspended in 50 μL PAGE loading buffer and purified with PAGE (in one slot) as before. The final amount of recovered RNA is quantified on a photometer at 260 nm.

3.2.4. Column Immobilization and Selection

1. 1 nmol biotinylated RNA is incubated for 30 min at room temperature on 250 μL streptavidin–agarose equilibrated in coupling buffer. The amount of RNA linked on the column typically ranges between 20 and 40%.

2. To eliminate unlinked species the column is washed thoroughly (6 times with alternatively 1 mL WA and 1 mL WB) and rinsed with water (5 times 500 μL). Collect flow-through and wash fraction for analysis.

3. For the first selection incubation, the column material is incubated in 500 μL selection buffer with the appropriate amount of effector molecule at 37°C with gentle shaking for the appropriate time (see Table 1). The incubation is initiated upon addition of MgCl$_2$.

4. Repeat steps 2 and 3 without effector molecule. Adjust incubation time according to Table 1.

5. Finally, the cleaved RNA is eluted with WB (2 times 500 μL). The different washing and eluting fractions are counted in a scintillation counter, and the amount of eluted RNA is determined.

6. The eluted RNA is purified with 3 phenol-chloroform-isoamyl alcohol extractions, one chloroform extraction and precipitated (sodium acetate) in the presence of glycogen (5 µg). The pellets is washed 2 additional times with 70% ethanol before resuspension in 20 µL ddH$_2$O with 200 pmol of Pnp-rev primer.

7. The RNA-oligonucleotide mix is denatured (1 min, 95°C) and mixed with 80 µL reverse transcription mix (5 µL dNTP 4 mM, 10 µL MnOAc 25 mM, 20 µL 5× RT-PCR buffer, 43 µL ddH$_2$O, 2 µL Tth DNA polymerase (2–10 U)). The total mix is incubated for 30 min at 72°C (at the same time a control without RNA is performed).

8. The reverse transcription mix is then diluted into a 500 µL PCR reaction under standard conditions with primers Pnp-3 and Pnp-rev (including a negative control). The number of cycles is calculated using the following rule:

$$n \geq \frac{1 + \ln\left(\dfrac{1,000}{x}\right)}{\ln(2)}$$

where x is the amount of RNA eluted in pmol.

9. The PCR products are analyzed on a 2% agarose gel, before phenol/chloroform extraction and precipitation (sodium acetate). 25% of the resulting DNA is used for the next selection cycle.

10. The whole selection cycle should be reproduced up to 16 times.

3.3. Optional Selection Components

New mutations can be inserted into the selected pool via mutagenic PCR (7). The following protocol should introduce on average one mutation per RNA molecule.

3.3.1. Mutagenic PCR

1. Prepare the following master-mix and make 20 83 µL aliquots in PCR tubes: 10× *Taq* reaction buffer, 210 µL; 10 mM dATP-dGTP, 42 µL; 10 mM dCTP-dTTP, 210 µL; 100 mM MgCl$_2$, 147 µL; 5 mM MnCl$_2$, 210 µL; 100 µM Pnp-3, 105 µL; 100 µM Pnp-rev, 105 µL; and ddH$_2$O, 714 µL.

2. Add 15 µL of PCR product from the last selection cycle to the first aliquot. Preheat mix to 55°C before adding 2 µL *Taq* polymerase (10 U). Perform three PCR cycles (94°C 50 s, 55°C 1 min, 72°C 1 min). Let the block cool down to 55°C.

3. After the three cycles, 15 µL of the reaction mixture is transferred into a new aliquot preheated at 55°C. Add *Taq* polymerase and perform PCR cycles as before. Repeat procedure 20 times (60 PCR cycles). Verify regularly the DNA levels (every 15 PCR cycles) on a 2% agarose gel.

4. Fix the mutations using a standard PCR protocol (without manganese) for four cycles using the last 100 μL as template in a total volume of 800 μL. Purify PCR product by phenol/chloroform extraction and precipitation. Use 15% of the final product for the next selection cycle.

3.3.2. Counter-Selection of Misfolded Ribozymes

The selection procedure described above can also enrich ribozymes folding into several different states, some active and some inactive. To avoid these ribozymes to overcome the population, a counter-selection can be performed. For this purpose replace the first incubation (with effector) during the selection (step 6) with the following procedure.

1. The column material is incubated in 500 μL selection buffer with the appropriate amount of effector molecule (see Table 1) at 37°C with gentle shaking for 15 min. The incubation is initiated upon addition of magnesium.

2. The column is washed twice with 1 mL WB and 3 times with 1 mL ddH$_2$O to allow denaturation and renaturation of the ribozymes.

3. Steps 1 and 2 are repeated 10 times.

4. Incubate column material in 500 μL selection buffer with the appropriate amount of effector molecule (see Table 1) at 37°C with gentle shaking for final 90 min.

5. Continue selection from step 7.

3.4. Analysis of Selected Allosteric Ribozymes

3.4.1. Cloning and Sequencing

1. Clone the selected DNA pool after the last selection cycle into a vector of choice using standard molecular biology methods. For cloning of the pool with the T$_7$ promoter use the restriction sites KpnI and XhoI. For cloning of the sole ribozyme use the sites SacI and XhoI.

2. Sequence the inserts and identify individual sequences.

3. Perform PCR from a clone of interest with primers Pnp3 and Pnp-rev. Use PCR product for a transcription reaction as described above (Subheading 3.2.1) in the absence of GMPS and radioactive nucleotide. Purify transcripts with PAGE (Subheading 3.2.2).

4. Dephosphorylate RNA with Calf Intestine Alkaline Phosphatase. 150 pmol RNA is incubated in 50 mM Tris (pH 8.5), 1 mM EDTA, Rnasin (20 U), and calf intestinal alkaline phosphatase (10 U) in 50 μL for 30 min at 37°C and 10 min at 75°C after addition of 0.5 μL 0.5 M EDTA pH 8.0 and vortexing. After the dephosphorylation, purify the RNA via a phenol/chloroform extraction and precipitate (with sodium acetate) in the presence of glycogen (5 μg). Finally, resuspend pellet in 20 μL ddH$_2$O.

5. 10 pmol of the dephosphorylated RNA (3 μL considering 50% loss during the purification) is then radioactively marked in a total volume of 20 μL in 70 mM Tris–HCl (pH 7.6), 10 mM MgCl$_2$, 5 mM dithiothreitol, with 10 U Rnasin, 30 μCi gamma-P^{32}-ATP (i.e., 10 pmol), and 20 U T4 polynucleotide kinase in the presence of 1 mM effector molecule. Incubate the reaction 30 min at 37°C before quenching with 1 μL 0.5 M EDTA pH 8.0.

6. Precipitate by addition of 10 μL 6 M ammonium acetate (pH 6.0) and 50 μL 100% ethanol. Incubate 5 min at room temperature and centrifuge for 20 min at 15,000×*g* at 4°C. Remove supernatant; add 50 μL 70% ethanol. Centrifuge under the same conditions for 5 min. After removing supernatant and shortly drying the pellet resuspend into 10 μL PAGE loading buffer.

7. Perform PAGE purification as before. Instead of UV-shadowing, the RNA band should be visualized using an X-ray film with 30 s exposure. Mark film position on the gel during exposure with waterproof marker. After film development replace the film under the gel at the position marked during exposure and cut out band corresponding to uncleaved ribozymes. After elution, resuspend RNA into 500 μL ddH$_2$O. Assuming a loss of 50% during purification the concentration of RNA should be 10 nM.

8. Incubate RNA (1 nM) in selection buffer with various amounts of effector molecule (typically between 5 nM and 5 μM). Start reaction upon addition of MgCl$_2$ and take 10 μL aliquots of the reaction every 20 s for the first 2 min of the reaction. Mix aliquots immediately with an equal volume of PAGE loading buffer on ice. Load 10 μL samples on an 8% polyacrylamide gel. For processing of large amounts of samples a sequencing-gel is recommended.

9. Separate plates and transfer gel on a Whatman paper. Cover with wrapping foil. Expose overnight with a PhosphorImager screen. Develop screen and quantify bands corresponding to cleaved and uncleaved ribozymes. Determine the percentage of uncleaved ribozyme for each sample. Determine cleavage rate k_{obs} by fitting each reaction with a simple exponential decay $(e^{(-k_{obs}t)})$. To determine the inhibition constant (K_i) fit the cleavage rates at different effector concentrations with the following formula (where k is the cleavage rate in the absence of effector):

$$k_{obs}([\text{Drug}]) = \frac{k}{1 + \dfrac{[\text{Drug}]}{K_i}}$$

3.4.2. Conversion
of Allosteric Ribozymes
into Aptamers

1. Perform PCR from a clone of interest with primers Pnp2 and Pnp-rev. Primer Pnp2 introduces a point mutation into the hammerhead ribozyme sequence abolishing self-cleaving activity.

2. Binding of the aptamer to the small molecule can be monitored via competition experiments with intact ribozymes.

4. Notes

1. Correspondence between absorbance at 260 nm (OD = 1) and concentration of nucleic acids: ssDNA: 33 µg/mL, dsDNA: 50 µg/mL, RNA: 40 µg/mL.

2. The protocol is set-up for the selection of aptazymes inhibited by the effector molecule. When selecting for activation instead of inhibition, remove the effector from all steps before the second incubation on the column and add it during this incubation.

3. When working with RNA special care should be taken to avoid contamination through RNAses. All solutions should be tested for the presence of RNases before use. For this incubate in the solution of interest a radioactively labeled RNA and check its integrity using polyacrylamide gel electrophoresis and autoradiography.

4. Radioactive labeling of the RNA is optional but is recommended to follow the molecules during selection procedure. The amount of RNA recovered at the end the initial cycles is too low to monitor via absorbance at 260 nm. It is advisable to perform every few cycles a control without effector to detect if the enrichment observed is due to the effector or to misfolded ribozymes. Counter-selection of misfolded ribozymes can be performed as described in Subheading 3.3.2.

References

1. Breaker, R. R. (2002) Engineered allosteric ribozymes as biosensor components. *Curr Opin Biotechnol* 13, 31–9.

2. Yen, L., Svendsen, J., Lee, J. S., Gray, J. T., Magnier, M., Baba, T., D'Amato, R. J., and Mulligan, R. C. (2004) Exogenous control of mammalian gene expression through modulation of RNA self-cleavage. *Nature* 431, 471–6.

3. Tang, J., and Breaker, R. R. (1997) Rational design of allosteric ribozymes. *Chem Biol* 4, 453–9.

4. Koizumi, M., Soukup, G. A., Kerr, J. N., and Breaker, R. R. (1999) Allosteric selection of ribozymes that respond to the second messengers cGMP and cAMP. *Nat Struct Biol* 6, 1062–71.

5. Piganeau, N., Jenne, A., Thuillier, V., and Famulok, M. (2001) An Allosteric Ribozyme Regulated by Doxycyline. *Angew Chem Int Ed Engl* 40, 3503.

6. Piganeau, N., Thuillier, V., and Famulok, M. (2001) In vitro selection of allosteric ribozymes: theory and experimental validation. *J Mol Biol* 312, 1177–90.

7. Cadwell, R. C., and Joyce, G. F. (1994) Mutagenic PCR. *PCR Methods Appl* 3, S136–40.

Chapter 20

Screening Effective Target Sites on mRNA: A Ribozyme Library Approach

Hoshang J. Unwalla and John J. Rossi

Abstract

Hammerhead ribozymes have been extensively used as RNA-inactivating agents for therapy as well as forward genomics. A ribozyme can be designed so as to specifically pair with virtually any target RNA, and cleave the phosphodiester backbone at a specified location, thereby functionally inactivating the RNA. Two major factors that determine whether ribozymes will be effective for posttranscriptional gene silencing are colocalization of the ribozyme and the target RNAs, and the choice of an appropriate target site on the mRNA. Complex secondary structures and the ability to bind to some of the cellular proteins mandate that some RNA sequences could stearically occlude binding of RNA-based antivirals like ribozymes to these sites. The use of ribozyme libraries in cell culture factors in these interactions to select for target sites on the RNA, which are more accessible to RNA-based antivirals like ribozymes or siRNA. This chapter provides a useful guide toward using ribozyme libraries to screen for effective target sites on mRNA.

Key words: Antiviral, Gene Expression, Hammerhead Ribozyme, mRNA cleavage, Target Site Selection, Therapeutics

1. Introduction

Ribozymes are small RNA molecules with distinct catalytic motifs that can be adapted for trans-target cleavage by modifying the hybridizing arms to bind to target RNAs through Watson–Crick base pairing. In the presence of magnesium, their intrinsic enzymatic activities cleave the phosphodiester backbone of targeted mRNAs (1). Specifically, any RNA can be cleaved as long as the ribozyme can pair with the target RNA and the target contains an NUC triplet where N = A, G, C, or U. This property, combined with the ability of ribozymes to undergo multiple turnover reactions, makes

Jörg S. Hartig (ed.), *Ribozymes: Methods and Protocols*, Methods in Molecular Biology, vol. 848,
DOI 10.1007/978-1-61779-545-9_20, © Springer Science+Business Media, LLC 2012

them attractive agents for modulating gene expression (1). The two most common forms of ribozymes employed for mRNA cleavage are the hairpin and the hammerhead motifs. For ribozymes, merely randomizing the hybridizing arms can generate a library of $4n$ distinct ribozyme sequences, depending on the length of the hybridizing arms (n). The application of ribozymes in forward genetic screens involves delivery of these libraries to target cells, either through transient transfection or transduction, followed by screening or selection of a particular phenotype. Importantly, the library approach selects only for target inhibition, and does not require 100% matched base pairing to the target (2). In general, therefore, this approach allows only the identification of ribozymes that achieve the desired phenotype, and not necessarily of ribozymes with the greatest cleavage activities. We have exploited the ability of small nucleolar RNA (snoRNA) U16 to localize to the nucleolus to compartmentalize our ribozyme and target (in our case HIV RNA) in the same cellular compartment (3). The nucleolar localization is accomplished by insertion of the hammerhead ribozyme into the structure of the U16snoRNA. Use of the U16 snoRNA also provides an added benefit of providing primer-binding regions for selection of ribozyme library members after each round by RT-PCR. There are two ways to introduce the ribozyme library into cells, by transient transfection in cells lines or stable transduction using viral vectors. Transient transfection has its advantages in that it is rapid and bypasses the steps of cloning and packaging the libraries in viral vectors each of which could lead to loss of library members and decrease representation. Stably transduced libraries often suffer from a potent disadvantage in that the site of vector integration can determine the level of expression of the ribozyme oftentimes leading to suboptimal levels of expression and loss of efficient library members. Moreover, transient transfection approach ensures maximal library representation allowing for multiple ribozyme sequences to be inserted within a single cell. These cells with even a single copy of an effective ribozyme would provide the desired phenotype leading to the entire pool within the cell being selected for the next round. Selecting a good end point is often a determining factor for success in isolating efficient ribozyme molecules. In our study, we have used cell survival as an end point to determine efficient ribozyme target sites by selecting against an infectious HIV proviral DNA containing the HSV-thymidine kinase (HSV-TK) gene within the infectious HIV proviral DNA pNL4-3. While alternate splicing, in case of HIV, facilitates the insertion of TK gene within the viral RNA, selection of ribozymes against other target genes based on cell survival would involve placing the HSV-TK ORF downstream of the target gene spaced by an IRES to generate a bicistronic transcription cassette such that the target RNA and HSV-TK are encoded by the same RNA. The pIRES vector (Clontech; catalog no. 631605) is ideally suited for this purpose as it provides an IRES element flanked by two multiple cloning sites. A cell survival end point assures that only the most potent of the ribozyme molecules make the cut in each and every round and this greatly reduces the number of rounds required for selection. Alternately, other end points that promote selection by FACS analysis where the readout is expression or shutdown of EGFP or other reporter genes could also be used. While the end point would differ with the gene or phenotype of interest this chapter essentially deals with the construction and screening by transient transfection of ribozyme library against HIV proviral DNA using cell survival as an end point.

2. Materials

Plasmids HIV-1 pNL4-3 and HIV-1 pNL-TK were obtained from the NIH AIDS Research and Reference reagent program (Rockville, MD).

All molecular biology enzymes e.g., Sal I, Xba I, *Taq* DNA polymerase, and T4 DNA ligase were obtained from New England Biolabs.

Ganciclovir was purchased from Sigma Aldrich-USA (Cat no. G2536).

High-efficiency competent cells for transforming ribozyme pools were purchased from New England Biolabs (Catalog no. C2987H).

p24 ELISA was done using a Beckman Coulter p24 ELISA kit (This kit has been discontinued. The Alliance HIV-1 P24 ANTIGEN ELISA Kit; Cat no. NEK050B001KT, Perkin Elmer—USA can be used in its place and has been found equally reliable). The p24 values were calculated using a Dynatech MR5000 enzyme-linked immunosorbent assay plate reader (Dynatech Lab, Chantilly, VA).

3. Methods

3.1. Construction of the Ribozyme Library

While designing a ribozyme library it is important to ensure that the library is embedded within two flanking disparate sequences at least 20 nucleotides long and which serve as primer-binding sites during amplification and selection. In our study, the use of U16 stem provides this primer-binding region. The Ribozyme library is prepared by an initial extension reaction using four overlapping PCR A, B, C, and D, under the conditions described earlier (4) and outlined in Fig. 1.

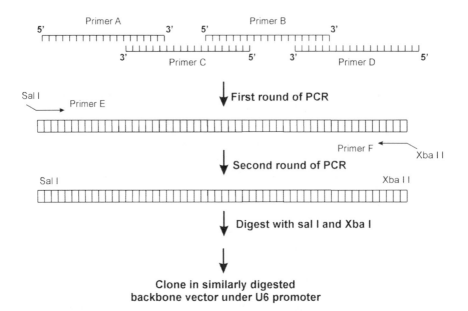

Fig. 1. Overlapping primers A, B C, and D are first amplified in the first round of PCR. During this PCR, smaller fragments corresponding to two or three primer extensions are also generated along with the full-length product of all four primers. The full-length product is selectively amplified in a second round of PCR using primers appended with restriction enzymes and designed to anneal to the ends of the full-length product. The PCR product is digested with the respective restriction enzymes and ligated to similarly digested backbone vector under the U6 promoter.

PCR-based generation of ribozyme library: The primers are allowed to hybridize and PCR amplified in a 50 µL reaction. Final concentrations for PCR Round 1 Reaction Mixture: Primer A (1 µM), Primer B (1 µM), Primer C (1 µM), Primer D (1 µM), 10× Magnesium free PCR Buffer (5.0 µL), MgCl2 (1.5 mM), dNTP mix (0.8 mM), and *Taq* DNA polymerase (2.5 U).

Amplification cycle:

Initial denaturation: 94°C/5 min

Cycling to 25 cycles

    Denaturation: 94°C/4 min

    Annealing: 50°C/3 min

    Extension: 72°C/2 min

Final extension: 72°C for 10 min.

10 µL of this amplification reaction is used as a template for a second round of amplification using primers (E and F) designed to hybridize the flanking sequences of the ribozyme library. These primers should be designed with flanking restriction sites to facilitate cloning the ribozyme library in the desired vector. Final concentrations for PCR Round 2 Reaction Mixture: Primer E (0.4 µM),

Primer F (0.4 µM), 10× Magnesium-free PCR Buffer (5.0 µL), $MgCl_2$ (1.5 mM), dNTP mix (0.8 mM), and *Taq* DNA polymerase (2.5 U).

Cycling parameters:

Initial denaturation: 94°C/5 min

Cycling to 27 cycles

   Denaturation: 94°C/1 min

   Annealing: 50°C/1 min

   Extension: 72°C/2 min

Final extension: 72°C for 10 min.

The PCR product from the second PCR is purified using PCR purification kit (Qiagen), digested with the appropriate restriction enzymes (Sal I and Xba I) and ligated to a similarly digested backbone vector under the promoter of interest in a 20 µL ligation reaction (vector: insert ratio of 1:5 w/w). In our study we used the Sal I and Xba I restriction sites to facilitate cloning in the vector pTzU6 + 1. This places the ribozyme library under the control of a U6 promoter and uses a set of six thymidine residues for transcription termination. While a U6 promoter expression system is desired due to ease of generating a transcription cassette and high-level expression, it is also possible to express the library from Pol II promoters using a minimal polyadenylation signal sequence reported by Xia et al. and us. (5, 6).

Transformation: A standard transformation protocol is used. Set up ten transformations using 2 µL of the ligation reaction each to use up the entire ligation mix. Incubate on ice for 30 min, Heat shock at 42°C/1 min. Chill on ice for 2 min and add 900 µL of SOC medium. Incubate with shaking at 37°C/1 h and plate on ten 10-cm LB antibiotic (ampicillin) plates. Isolate plasmid DNA from at least ten colonies (more is desirable) and sequence to confirm the randomness of the library. Once confirmed, add 1 mL LB broth to each plate and with a spreader mix all the colonies on each plate and pool into four different pools or mini-libraries. This approach maximizes the library representation.

3.2. Selection and Amplification

HEK 293 cells are maintained in Dulbecco's modified Eagle's medium (DMEM) 20% fetal bovine serum (FBS). Twenty-four hours before transfection, cells are replated in 10-cm dishes at ~10^6 cells per plate with fresh media, without antibiotics. HEK 293 cells are transiently transfected with selection plasmid (target gene-IRES-HSV-TK or pNL-TK in our studies) and the four mini-libraries (1:2 wt/wt ratio, respectively) taking care not to exceed 500 ng per plate of total DNA. Lipofectamine 2000 (Invitrogen) is used as transfection reagent according to the manufacturer's instructions. Twenty-four hours after transfection, ganciclovir is added at a final concentration of 5 nmol/L and the cells are incubated further. Forty-eight hours after ganciclovir addition, dead cells lift off the plate, see Note 1. Cells are washed to remove the detached cells and total RNA is extracted from adherent cells using RNA STAT-60 (TEL-TEST "B") according to manufacturer's instructions. The ribozyme sequences are rescued using primers E and F appended to the appropriate restriction enzymes using PCR conditions identical to that described above for PCR Round 2. The PCR products are analyzed on a gel, extracted using Gel extraction kit (Qiagen) and cleaved with the respective restriction enzymes (Sal I and Xba I) and recloned in similarly digested vector backbone carrying the U16 promoter (vector: insert ratio of 1:5 w/w). After three rounds of selection at least 50 colonies (more is desirable) should be randomly selected and sequenced to look for enrichment. Enrichment of sequences indicates that the selection is proceeding successfully. At this point one can either stop and look at the enriched sequences for their ability to demonstrate the desired phenotype or proceed with more rounds of selection, see Notes 2 and 3. In our study enrichment is observed within three rounds and the selected sequences were analyzed for their ability to inhibit HIV proviral DNA pNL4-3 either directly or via inhibiting a cellular factor during infection.

3.3. Screening Enriched Sequences for Their Ability to Inhibit the Cognate Target

The first step would be to test the ribozyme sequences with the selection process itself, see Note 3. HEK 293 cells are co-transfected with the pNL-TK and the enriched ribozymes under U6 promoter (1:2 wt/wt ratio, respectively taking care not to exceed 500 ng per plate of total DNA). Twenty-four hours after transfection, ganciclovir is added at a final concentration of 5 nmol/L and the cells are incubated further. After 48 h, cells are detached by trypsinization and viability is determined by trypan blue staining. The enriched sequences that demonstrate increased viability in presence of ganciclovir are analyzed further. In the next step, the selected ribozymes are tested against the target RNA itself. This step is essential to filter out sequences that are selected due to their ability to target the thymidine kinase gene. For this step, the HSV-TK in the selection plasmid is replaced with a reporter gene like GFP. The enriched ribozyme sequences in pTzU6 + 1 vector are tested in an acute challenge by transient transfection with the target gene (target gene-IRES- GFP) (4:1 wt/wt ratio, respectively). Three days post-transfection, inhibition of target RNA is monitored by microscopy or fluorometry. Since our target was HIV RNA, the progress of the infection is monitored over 3 days by collection of culture supernatants. Viral output from these cells is measured by measuring the p24 levels by ELISA.

3.4. Determining Putative Target Sites in RNA Based on Enriched Ribozyme Sequences

The first step would be to determine the possible target sequence from the enriched ribozyme sequence and try to align this sequence with the target RNA to get a perfect match (any of the commonly available alignment softwares could be used). Most of the times a perfect match would not be possible, as the rules governing ribozyme cleavage have been established under in vitro conditions which are more controlled and most of these studies have involved the catalytic motif or the cleavage site with scant attention to the effect loops, bulges, or mismatches that might have on cleavage. Also in vivo, a number of factors can influence ribozyme binding and cleavage. For someone to completely study rules governing ribozyme cleavage, different ribozyme would require studying all the possible combinations of mismatches and bulges, which would be a herculean task.

We modified the rules to allow us to consider mismatches, bulges of one or two nucleotides (nt), and wobble base pairing. Target sites containing an NU(A, C, U, or even G) as cleavage triplet flanked by at least five nucleotides complementary to helices one and three are selected.

3.5. Confirming the Ribozyme Target Sites

Once these target sites are determined, the target sequence is used to design variants of the enriched ribozymes that would match perfectly with the target RNA. The ribozyme sequence is cloned in the same transcription cassette and tested by transient transfection with the target RNA (4:1 wt/wt ratio, respectively) or cloned in a lentiviral vector (preferred) to test in a more gene therapy like setting. If the target RNA is cellular gene then one can determine inhibition by Quantitative RT-PCR or northern blot analysis. In our studies, CEM T-cells transduced by these ribozyme variants were challenged with infectious virus. Progress of the infection is monitored by measuring the p24 output in culture supernatants by ELISA.

4. Notes: Trouble Shooting and Tips

1. *Low cell death during selection*: It is expected that higher cell death will be observed in the initial rounds. Rounds 3 and later would demonstrate a lower cell death. Often this would be a good criterion to determine the point at which selection can be paused and the sequences analyzed for enrichment. If no cell death is observed during initial rounds it would be advisable to start over. Ganciclovir dose should be titrated to ensure cell killing in untransfected HEK 293 cells and this should be used for selection.

2. *No enrichment is observed after three to four rounds*: In our study, cell survival as an end point provides a high stringency of selection such that only the most potent molecules are selected. An inability to see any enrichment of ribozyme sequences is an indication that the selection process has not been optimal. This could be due to several reasons. The dose of ganciclovir for selection was not optimal resulting in noneffective ribozyme sequences being selected. Ganciclovir dose should be tittered to obtain maximal cell death in untransfected cells. Transfection efficiency during one of the rounds was not optimal resulting in a loss of selected sequences. Low passage HEK 293 cells should be used. Cells should be withdrawn from antibiotics 1 day before transfection. A control transfection with a reporter plasmid like GFP should be done to determine transfection efficiency.

3. *Excessive enrichment observed for one or two sequences*: While enrichment of sequences is desirable, excessive enrichment could be an indication of contamination of PCR reagents with one or two sequences. For instance if 20 out of 50 sequences are identical then this would be considered excessive enrichment and hence point to a possible contamination of the PCR reagents. The sequence/sequences should be tested against the target RNA to determine if the ribozyme sequence inhibits target gene expression. It is advisable to use good molecular biology practices and use fresh reagents. A good idea would be to sample a few sequences (ten or more) after each round to determine the randomness of the pool.

References

1. Rossi, J.J., *Ribozymes, genomics and therapeutics.* Chem Biol, 1999. **6**(2): p. R33–37.

2. Waninger, S., et al., *Identification of cellular cofactors for human immunodeficiency virus replication via a ribozyme-based genomics approach.* J Virol, 2004. **78**(23): p. 12829–37.

3. Unwalla, H.J., et al., *Use of a U16 snoRNA-containing ribozyme library to identify ribozyme targets in HIV-1.* Mol Ther, 2008. **16**(6): p. 1113–19.

4. Dillon, P.J. and C.A. Rosen, *A rapid method for the construction of synthetic genes using the polymerase chain reaction.* Biotechniques, 1990. **9**(3): p. 298, 300.

5. Unwalla, H.J., et al., *Negative feedback inhibition of HIV-1 by TAT-inducible expression of siRNA.* Nat Biotechnol, 2004. **22**(12): p. 1573–78.

6. Xia, H., et al., *siRNA-mediated gene silencing in vitro and in vivo.* Nat Biotechnol, 2002. **20**(10): p. 1006–10.

Chapter 21

A Computational Approach to Predict Suitable Target Sites for *trans*-Acting Minimal Hammerhead Ribozymes

Alberto Mercatanti, Caterina Lande, and Lorenzo Citti

Abstract

Trans-acting hammerhead ribozymes are challenging tools for diagnostic, therapeutic, and biosensoristic purposes, owing to their specificity, efficiency, and great flexibility of use. One of the main problems in their application is related to the difficulties in the design of active molecules and identification of suitable target sites.

The aim of this chapter is to describe ALADDIN, "SeArch computing tooL for hAmmerheaD ribozyme DesIgN," an *open-access* tool able to automatically identify suitable cleavage sites and provide a set of hammerhead ribozymes putatively active against the selected target.

ALADDIN is a fast, cheap, helpful, and accurate tool designed to overcome the problems in the design of *trans*-acting minimal hammerhead ribozymes.

Key words: Minimal hammerhead ribozyme, Ribozyme design, RNA computational analysis, RNA folding, Structural thermodynamics, Open-access tool

1. Introduction

The application of hammerhead ribozymes as "in-trans" active tools for diagnostic, therapeutic, or biosensoristic purposes has been slowed down by the difficulty in obtaining truly active catalytic molecules (1). The choice of the proper triplets to which the hammerhead catalytic motifs are to be addressed was the main problem to solve in the design of potentially active ribozymes. Ever since the discovery of their "in-trans" activity, several research groups have designed hammerhead ribozymes aimed at finding suitable cleavage sites by either an experimental or a theoretical approach.

Jörg S. Hartig (ed.), *Ribozymes: Methods and Protocols*, Methods in Molecular Biology, vol. 848,
DOI 10.1007/978-1-61779-545-9_21, © Springer Science+Business Media, LLC 2012

Experimental procedures have been applied using both "in vitro" molecular assays or intact biological systems accounting for the bio-environment and putative molecular interaction. These procedures can be divided into two categories: those based on a multitude of defined oligonucleotides and those based on combinatorial chemistry. The former approach aims to identify accessible cleavage sites by targeting each site with an appropriate antisense oligodeoxynucleotide (ODN) (2–5). The latter approach is based on combinatorial chemistry. This alternative strategy enables the isolation of a set of candidate ribozymes starting from a larger pool of oligonucleotide molecules (6–9). However, all the experimental methods are laborious, highly expensive, and time consuming, and although they provide effective ribozymes, they appear to be disadvantageous especially when investigations such as system biology studies request the contemporary analysis of different gene factors of a given phenotype (10).

On the other hand, theoretical approaches based on computational analyses can yield structures whose features meet the requirements of effective targeting by ribozymes. Core of the overall predictive investigation is the analysis of target RNA, which includes RNA folding. This is the most difficult issue to be solved, and implies the identification of a number of optimal and suboptimal structures. The first developed MFold algorithm (11) is based on the thermodynamics of RNA structural motifs, including base-paired intramolecular stems and unpaired loops. It provides the identification of the putative optimal, minimum free energy (MFE) structure and of the suboptimal folding enclosed in a selected divergence energy gap from MFE. A successive approach based on the equilibrium partition function was proposed by McCaskill (12). It allows the identification of all possible combinations of structural folding by searching the largest number of admissible base pairing. This approach escapes the imprecise energy evaluation involved in MFold elaboration, especially for loops, and allows to identify an extreme structure with maximal number of paired bases. Such structure conceptually corresponds to the former MFE as it is assumed to have minimal free energy. However, the extreme structure, as obtained by the partition function, may not correspond to the MFE obtained by the thermodynamic approach because the latter depends on which energy parameters have been used. In spite of their advancements toward the characterization of Boltzmann's ensemble of secondary structures, the equilibrium partition function and related base-pairing probabilities make it difficult to mathematically solve the exponentially growing number of suboptimal foldings. Therefore, a sampling algorithm has been described (13) that allows generation of a statistically representative sample of secondary structures which, in turn, may illustrate Boltzmann's ensemble of folded species. All the computational methods described have been successfully employed to predict accessible regions in RNA sequences. More recently, a "unified" method has been produced, "Unified Nucleic Acids Folding—UNAFold" (14). This is a software package where different RNA folding programs are provided in an integrated collection, which allows secondary structure prediction for single-stranded RNA or DNA combining free energy minimization, partition function calculations, and stochastic sampling.

We describe here an *open-access* tool, "SeArch computing tooL for hAmmerheaD ribozyme DesIgN—ALADDIN," specific for minimal hammerhead ribozyme design. This computing tool, an evolution of our formerly described computational method (15), starts by mapping all canonical NUH cleavage sites in the RNA sequence to be silenced, and then automatically identifies all the target sites including the flanking elements. Applying the "UNAFold" package, the tool predicts the accessibility of each target according to calculated secondary structures within Boltzmann's ensemble. Finally, it provides suitable hammerhead ribozyme sequences with optimal structural folding.

The proposed tool for ribozyme design is simple, cheap, and fast, and provides a small set of putatively effective ribozymes.

The obvious limit of the tool, shared with other theoretical approaches, concerns the hypothetical nature of structural studies so that it is impossible to guarantee for the actual "in vivo" situation. Consequently, the small collection of ribozymes obtained needs to be synthesized and experimentally tested in the appropriate biological systems.

2. Materials: Open-Access Computing Tool

Successful targeting of a defined RNA sequence by specific hammerhead ribozymes consists in realizing a series of crucial conditions necessary for the achievement of effective cleavage-depletion of the target. As a matter of fact, a ribozyme must be specifically tailored to its target and this implies significant predictive mapping of the most suitable cleavage sites contained in the sequence. Some of these critical points are conceptually addressed by this computing tool.

2.1. Mapping of NUH Cleavage Sites Along the Target Sequence

This step allows to identify the occurrences of a set of three nucleotides, NUH, where N = any nucleotide and H = A, C, or U according to IUB code.

$$S_{\text{pattern}} = \{N, U, H\} \text{ where } N \in \{A, C, G, U\} \text{ and } H \in \{A, C, U\}$$

The cleavage sites that have been found are then grouped as

$$\{AUA, AUC, AUU, CUA, CUC, CUU, GUA, GUC,$$
$$GUU, UUA, UUC, UUU\}$$

2.2. Definition of Binding Elements Flanking the NUH Site

This step is useful to fix the length of the binding regions flanking the NUH sites.

Let H be the base of the cleavage site, and let

$$N_i' \in \{A,C,G,U\}, i = 1,2,\ldots,n$$

$$N_j'' \in \{A,C,G,U\}, j = 1,2,\ldots,m$$

where n and m are respectively the length of the upstream and downstream regions of the cleavage site.

Then let

$$S_{\{N,U,H\}} = \{N_1', N_2',\ldots,N_n'\} \cup \{N,U,H\} \cup \{N_1'', N_2'',\ldots,N_m''\}$$

be the subsequence containing the searched pattern {N, U, H}. Consequently

$$N_{left} = \{N_1', N_2',\ldots,N_n'\} \cup \{N,U\}$$

$$N_{right} = \{N_1'', N_2'',\ldots,N_m''\}$$

are the flanking regions around the cleavage site.

2.3. Energetic Evaluation of Binding Elements Flanking the NUH Site

Energy evaluation of binding elements is necessary to establish which elements are suitable for proper binding with target mRNA.

The definition of flanking sequences is based on the computation of pairing energies ΔG_{left} and ΔG_{right}. Such computation is performed applying the Nearest Neighbor Nucleotide Algorithm (16) and its evolution, which uses the single contribution of free energy based on couples of adjacent nucleotides at standard conditions (25°C, 1 atm).

Thus, let

$$\Delta G_{Ni,i+1} = f_{\Delta G}(N_i, N_{i+1})$$

be the free energy contribution based on the positional law of nucleotides N_i and N_{i+1}.

Then

$$\Delta G_{left} = \Delta G_{N_{1,2}'} + \Delta G_{N_{2,3}'} + \ldots + \Delta G_{N_{n-1,n}'} + \Delta G_{N_{n,N}'} + \Delta G_{N,U}$$

$$= \ldots = \sum_{i=1}^{n-1} f\Delta G(N_i', N_{i+1}') + f\Delta G(N_n', N) + f\Delta G(N,U)$$

$$\Delta G_{\text{right}} = \Delta G_{N_{1,2}''} + \Delta G_{N_{2,3}''} + \ldots + \Delta G_{N_{m-1,m}''} = \sum_{i-1}^{m-1} f \Delta G(N_i'', N_{i+1}'')$$

is the free energy of the flanking sequences around the cleavage site.

The imposed limits for the choice of flanking region length are adjusted by the following rules:

$$-15\frac{\text{kcal}}{mol} \le \{\Delta G_{\text{left}}, \Delta G_{\text{right}}\} \le -10\frac{\text{kcal}}{mol}$$

$$\left|\Delta G_{\text{left}} - \Delta G_{\text{right}}\right| \le 1.2\frac{\text{kcal}}{mol}$$

The chosen values ensure proper energetic conditions in order to provide balanced binding.

2.4. Accessibility of Each Target Stretch

The next step concerns the choice of the target stretches. A target stretch defines the exact segment inside the target RNA sequence which includes both the binding elements and the NUH cleavage triplet.

Our proposed ALADDIN tool automatically accesses the predictive program UNAFold (14), a computing algorithm implemented to obtain a significant Boltzmann's ensemble of optimal and suboptimal secondary structures from which the accessibility score (AS) of each single target stretch can be deduced.

The data provided by the UNAFold program concern structures with almost the same probabilities. In our approach, these structural data are corrected by Boltzmann's distribution, which provides real meaning to the stored results for large amounts of samples.

The main data files include two classes of information. The former, the *ss-count file*, contains the "single strand" condition of each nucleotide for every folding, according to the following relation: let n_i be the generic nucleotide for $1 \le i \le S_L$ where S_L is structure length; then *ss-count(i)* is the number of times the n_i base is unpaired in the computed folding ensemble.

The latter class, the *ct file*, contains the data for all secondary predicted structures. For each one, the free energy of formation $\Delta G°$ is reported, and is expressed in kcal/mole. Furthermore, it gives information about which other n_j nucleotide is paired with the n_i nucleotide.

The complete collection of structures describes the dynamic interconversion folding of mRNA, where each geometry represents a single intermediate transition of the global process. The free formation energy of a given structure ($\Delta G°$) is proportional to the lifetime of this conformational geometry, the most stable being the most persistent. Generally speaking, the actual distribution of mRNA molecules among various structural forms is described by

Boltzmann's distribution, from which we can calculate the molar fraction χ_i of each structure according to the relationship

$$\chi_i = \frac{N_i}{N_0} = \frac{e^{\frac{-\Delta G_i}{kT}}}{Z}$$

where N_i is the number of molecules assuming the i structure, N_0 is the total number of molecules, and Z is Boltzmann's distribution function

$$Z = \sum_{i=1}^{S} e^{\frac{-\Delta G_i}{kT}}$$

The program analyses, nucleotide by nucleotide, the score of unpairing events among the different structures. For each n nucleotide, the score U_n is calculated according to: $U_n = \sum_{i=1}^{S} (\chi_i P_i)$, where P_i is the pairing condition of the nucleotide n inside the i secondary structure ($P_i = 1$ when the nucleotide is unpaired, $P_i = 0$ when paired), χ_i is the molar fraction calculated according to Boltzmann's distribution, and S is the total number of structures.

The mean of the scores over length L of the target stretch will give the AS, according to the following relationship:

$$AS = \frac{\sum_{n=1}^{L} U_n}{L} = \sum_{n=1}^{L} \left(\sum_{n=1}^{S} \left(P_i \frac{e^{-\Delta G_i / kT}}{Z} \right) \right)$$

where k is Boltzmann's constant, T is the absolute temperature of the system (°Kelvin), S is the number of secondary structures, and L is the length of the target sequence expressed as number of nucleotides.

Thus, let

$$AS_{i,i+l-1} = f_{AS}(N_{i,i+l-1})$$

be the function computing the AS of a generic base sequence

$$N_{i,i+l-1} = \{N_i, N_{i+1}, \ldots, N_{i+l-1}\} \text{ of length } l; \text{ then}$$

$$AS_{\text{left}} = f_{AS}(N_{\text{left}})$$

$$AS_{\text{right}} = f_{AS}(N_{\text{right}}) \text{ and}$$

$$AS_{S_{\{N,U,H\}}} = f_{AS}(S_{\{N,U,H\}})$$

are the AS values of the flanking regions and of the overall target sequence. $AS_{S_{\{N,U,H\}}}$ accounts for left + right regions including the cleavage site for the specified target pattern $\{N,U,H\}$.

The threshold values chosen for the evaluation of flanking sequences are

$$AS_{left} > 35.0 \text{ and } AS_{right} > 35.0 \text{ as first filter}$$

$$AS_{S_{\{N,U,H\}}} \geq 40.0 \text{ as second filter}$$

$$\left| AS_{left} - AS_{right} \right| \leq 50.0 \text{ as third filter.}$$

2.5. Uniqueness of Each Selected Target Stretch

The homology recurrences of the binding elements are searched within the RNA target molecule and are discarded when three or more events are found. A duplicate for each flanking region is allowed only if its AS conforms to the rules described later.

2.6. Deduction of Minimal Hammerhead Ribozyme Sequences Relative to Each Target Stretch

The ALADDIN tool automatically inserts the antisense version of binding elements at the end of the two arms of conserved canonical hammerhead sequence.

2.7. Folding Analysis of the Ribozyme Sequences Obtained

The final (fourth) filter is based on the structural characteristics of the ribozyme sequences obtained.

To overcome the last filter, two features deriving from the schematic representation of the intermediate catalytic complex between hammerhead ribozyme and target stretch (see Fig. 1) are necessary. First, the stem and loop of the structural domain (stem-loop II) must be correctly conformant and, second, at least four nucleotides at the ends of each binding domain must be devoid of any intramolecular constraint in order to allow binding to the target stretch.

The score from 0 to 1 is assigned according to the possibility of nucleotides other than those involved in stem II to be involved or not in the secondary structures. The score is 1 when no nucleotide except for stem II is base paired.

The described sequential operative list (see Fig. 2) addresses many of the crucial points involved in ribozyme design, but is unable to account for other topics. For instance, tertiary interactions, which are structural motifs involved in three-dimensional arrangement of RNA molecules, cannot be assessed by the described computational tools. However, because the secondary structure provides the scaffold for tertiary interactions and accounts for almost all the free energy of each molecular structure, the involvement of tertiary motifs in the predictive studies for ribozyme design may be neglected as first approximation. A second more important question concerns the actual situation of the RNA target molecule

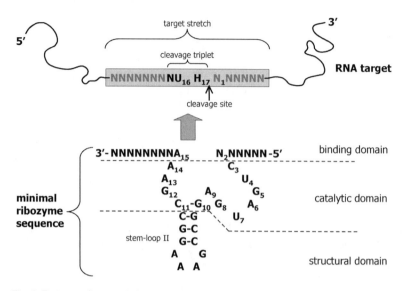

Fig. 1. Features of a generic hammerhead ribozyme binding (*large arrow*) its target sub-strate. Synthetic hammerhead moiety (*bottom*) is represented by highlighting its distinctive domains. The target molecule (*top*) is also shown and its target stretch schematically illustrated. NUH is the cleavage triplet on target sequence. The *small arrow* represents the cleavage site. The numbering system is conformant to Hertel et al. (20).

in living conditions. Indeed, mRNA could be involved in many interactions such as RNA–RNA and RNA–protein, and mask the selected target site directly or indirectly by preventing accessibility due to alternative folding of the molecule. Moreover, such interactions are site- and time-dependent in response to different types of metabolic signaling. Finally, RNAs may be stored in different cellular compartments and may require that the ribozymes are properly addressed. As concerns the bio-environmental problems, these cannot be treated by a single theoretical approach, thus representing the main limit of this strategy as compared to the experimental ones previously described. In spite of these drawbacks, the theoretical approach has proven to be very efficient in providing effective minimal hammerhead ribozymes (15) and therefore justifies this proposal as *open-access* tool.

3. Methods: Operative Options and Provided Outputs

The "SeArch computing tooL for hAmmerheaD ribozyme DesIgN—ALADDIN" is an *open-access* tool specific for minimal hammerhead ribozyme design.

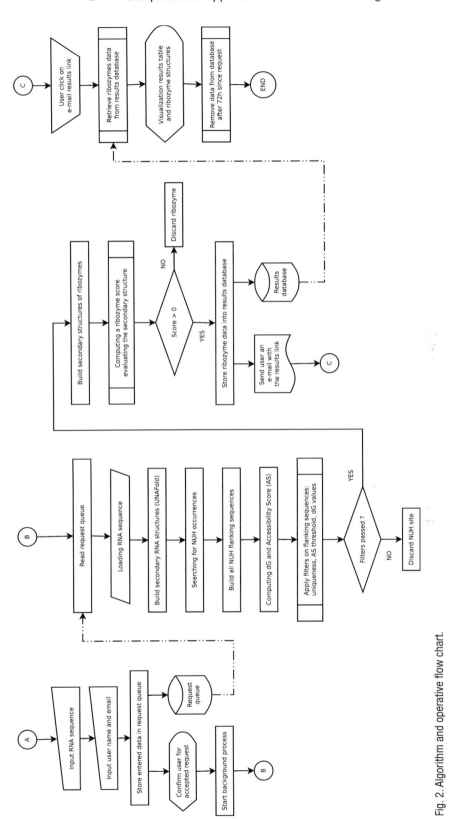

Fig. 2. Algorithm and operative flow chart.

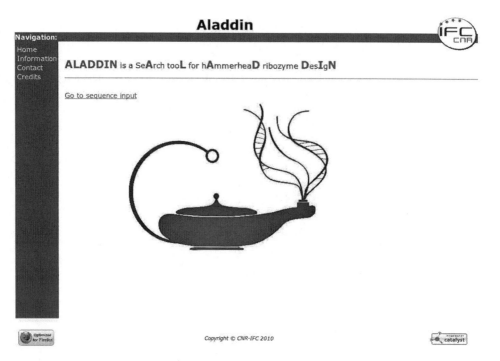

Aladdin

ALADDIN is a SeArch tooL for hAmmerheaD ribozyme DesIgN

Go to sequence input

Copyright © CNR-IFC 2010

Fig. 3. Screenshot of the home page of ALADDIN tool.

It allows the identification of a small number of putatively accessible NUH triplets right for ribozyme targeting and provides the sequences of related ribozymes.

The use of this tool is simple, fast, and intuitive. We report here the operating instructions of the system, showing the application of ALADDIN algorithms to human *Survivin* mRNA, a member of inhibitors of the apoptosis protein (IAP) family (17), as example of ribozyme design.

3.1. Tool Access and RNA Sequence Input

In order to start a new working session it is necessary to browse the web page (see Fig. 3) at the following link: http://aladdin.ifc.cnr.it/

You should then insert the sequence to be analyzed in the proper box by:

– Using the NCBI accession number.
– Typing or pasting a typed sequence.
– Loading a file in FASTA format.

You should then "Submit Sequence." If you make a mistake, you can "Reset Fields."

In this example the human *Survivin* sequence was loaded using the NCBI accession number U75285 (see Fig. 4).

3.2. Input of User's Data and Starting a New Work

After submission of the sequence input, you can visualize a summary of the information about your target RNA, including NCBI accession number, gene description, and complete nucleotide sequence (see Fig. 5).

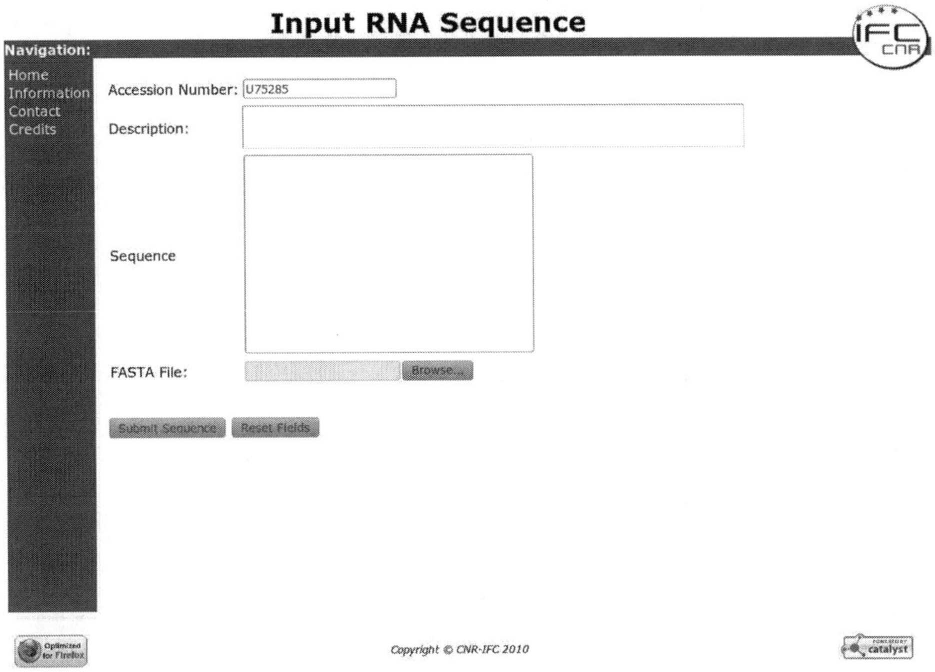

Fig. 4. Screenshot of input RNA sequence page. In the described example, the field "accession number" was filled with the NCBI accession number of human Survivin mRNA.

Fig. 5. Screenshot of the page with summary of the request and user's data.

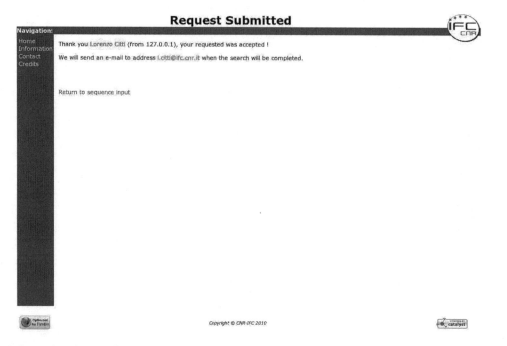

Fig. 6. Screenshot of the confirmation page: it reports the provided e-mail address necessary to receive the link to the results page.

To start the analysis, fill in the fields at the bottom of the page, typing your name and e-mail address before submitting the form.

Here we report the example for *Survivin*.

3.3. Results: Data Output

After your inquiry you will see a confirmation page (see Fig. 6) of your submission.

A few minutes later at the address requested you will receive an e-email containing a link by which you can visualize and download the page containing the produced results.

3.4. Data Rendering

The above link allows access to the results page (see Fig. 7), with a summary of your inquiry and some information about each target site that our computing tool identifies as suitable for ribozyme cleavage.

The table reports:

3.4.1. Sequence and Position of Each NUH Triplet Within the Submitted mRNA Sequence

The "*trans*-acting" hammerhead catalytic motif requests the very simple NUH consensus triplet at the cleavage site where, according to the IUB code, N = any nucleotide and H = A, C or U.

In the reported example, the program retrieves a number of 271 NUH triplets, but only 36 are selected on the basis of their suitability as target site.

The position number concerns the position of the "N" nucleotide of the target triplet along the RNA sequence.

Hammerhead Ribozyme List

Navigation:
Home
Information
Contact
Credits

User Name: Lorenzo Citti

e-mail: l.citti@ifc.cnr.it

NCBI Accession Number: U75285

Description:

Homo sapiens apoptosis inhibitor survivin gene, complete cds.

Sequence:

CCGCCAGATTTGAATCGCGGGACCCGTTGGCAGAGGTGGCGGCGGCGGCATGGGTGCCCCGACGTTGCC
CCCTGCCTGGCAGCCCTTTCTCAAGGACCACCGCATCTCTACATTCAAGAACTGGCCCTTCTTGGAGGGCT
GCGCCTGCACCCCGGAGCGGATGGCCGAGGCTGGCTTCATCCACTGCCCCACTGAGAACGAGCCAGACTT
GGCCCAGTGTTTCTTCTGCTTCAAGGAGCTGGAAGGCTGGGAGCCAGATGACGACCCCATAGAGGAACAT
AAAAAGCATTCGTCCGGTTGCGCTTTCCTTTCTGTCAAGAAGCAGTTTGAAGAATTAACCCTTGGTGAATTT
TTGAAACTGGACAGAGAAAGAGCCAAGAACAAAATTGCAAAGGAAACCAACAATAAGAAGAAAGAATTTG
AGGAAACTGCGAAGAAAGTGCGCCGTGCCATCGAGCAGCTGGCTGCCATGGATTGAGGCCTCTGGCCGG
AGCTGCCTGGTCCCAGAGTGGCTGCACCACTTCACAGGGTTATTCCCTGGTGCCACCAGCCTTCCTGTGGG
CCCCTTAGCAATGTCTTAGGAAAGGAGATCAACATTTTCAAATTAGATGTTTCAACTGTGCTCCTGTTTTGTC
TTGAAAGTGGCACCAGAGGTGCTTCTGCCTGTGCAGCGGGTGCTGCTGGTAACAGTGGCTGCTTCTCTCTC
TCTCTCTCTTTTTGGGGGCTCATTTTTGCTGTTTTGATTCCCGGGCTTACCAGGTGAGAAGTGAGGGAGGA

Search Results:

N°	Target Site NUH	Target Site Position	Binding & Energy (dG° kcal/mol) Left	Binding & Energy (dG° kcal/mol) Right	Accessibility Score Left	Accessibility Score Right	Uniqueness Left	Uniqueness Right	Ribozyme Sequence (dG° kcal/mol)	Score	Structure
1	AUC	14	GAUUUGAAU (-11.8)	GCGGGA (-14.8)	38.1%	41.1%			UCCCGCCUGAUGAGGCCGAAAGGCCGAAAUUCAAAUC (-12.4)	0.83	View
2	AUC	14	AGAUUUGAAU (-13.9)	GCGGGA (-14.8)	35.3%	41.1%			UCCCGCCUGAUGAGGCCGAAAGGCCGAAAUUCAAAUCU (-12.4)	0.85	View
3	CUA	108	CAUCUCU (-12.2)	CAUUCAAG (-11.6)	35.4%	39.3%			CUUGAAUGCUGAUGAGGCCGAAAGGCCGAAAGAGAUG (-11.3)	0.91	View
4	AUA	279	GGAACAU (-12)	AAAAGCAUU (-12.3)	36.1%	41.0%			AAUGCUUUUCUGAUGAGGCCGAAAGGCCGAAAUGUUCC (-11.3)	0.84	View
5	AUA	279	AGGAACAU (-14.1)	AAAAGCAUU (-12.3)	43.6%	41.0%			AAUGCUUUUCUGAUGAGGCCGAAAGGCCGAAAUGUUCCU (-11.3)	0.85	View
6	UUC	289	AAAAAGCAUU (-13.2)	GUCCG (-10.3)	38.4%	40.0%	1435 (0.2%)		CGGACCUGAUGAGGCCGAAAGGCCGAAAUGCUUUUU (-11.3)	1.00	View
7	UUC	289	AAAAAGCAUU (-13.2)	GUCCGG (-13.6)	38.4%	47.6%			CCGGACCUGAUGAGGCCGAAAGGCCGAAAUGCUUUUU (-11.3)	1.00	View
8	UUC	289	UAAAAAGCAUU (-14.5)	GUCCG (-10.3)	36.2%	40.0%	1435 (0.2%)		CGGACCUGAUGAGGCCGAAAGGCCGAAAUGCUUUUUA (-11.3)	1.00	View
9	UUC	289	UAAAAAGCAUU (-14.5)	GUCCGG (-13.6)	36.2%	47.6%			CCGGACCUGAUGAGGCCGAAAGGCCGAAAUGCUUUUUA (-11.3)	1.00	View
10	CUU	303	UGCGCU (-13.4)	UCCUUUC (-12)	60.6%	38.2%			GAAAGGACUGAUGAGGCCGAAAGGCCGAAAGCGCA (-11.3)	0.92	View
11	CUU	303	UUGCGCU (-14.3)	UCCUUUC (-12)	64.2%	38.2%			GAAAGGACUGAUGAGGCCGAAAGGCCGAAAGCGCAA (-11.3)	0.93	View
12	GUC	575	AGCAUGU (-13.9)	UUAGGAA (-10.9)	49.4%	55.5%			UUCCUAACUGAUGAGGCCGAAAGGCCGAAACAUUGCU (-11.3)	0.89	View
13	GUC	575	AGCAUGU (-13.9)	UUAGGAAA (-11.8)	49.4%	59.2%			UUUCCUAACUGAUGAGGCCGAAAGGCCGAAACAUUGCU (-11.3)	0.89	View
14	GUC	575	AGCAUGU (-13.9)	UUAGGAAAG (-13.9)	49.4%	52.6%			CUUUCCUAACUGAUGAGGCCGAAAGGCCGAAACAUUGCU (-11.3)	0.89	View
15	UUA	578	AAUGUCUU (-11.7)	GGAAAGG (-12.9)	62.1%	39.1%			CCUUUCCUGAUGAGGCCGAAAGGCCGAAAAGACAUU (-11.8)	0.87	View
16	UUA	578	CAAUGUCUU (-13.8)	GGAAAGG (-12.9)	55.6%	39.1%			CCUUUCCUGAUGAGGCCGAAAGGCCGAAAAGACAUUG (-11.8)	0.89	View
17	UUU	612	AUUAGAUGUU (-14.1)	CAACUGU (-11.6)	35.2%	42.7%			ACAGUUGCUGAUGAGGCCGAAAGGCCGAAAACAUCUAAU (-12.4)	0.79	View
18	CUA	1119	ACAAACU (-10.4)	CAAUUAAAAC (-11.2)	93.6%	91.1%			GUUUUAAUUGCUGAUGAGGCCGAAAGGCCGAAAGUUUGU (-12.4)	0.75	View
19	CUA	1119	ACAAACU (-10.4)	CAAUUAAAACU (-13.3)	93.6%	91.9%			AGUUUUUAAUUGCUGAUGAGGCCGAAAGGCCGAAAGUUUGU (-12.4)	0.75	View
20	UUA	1125	CUACAAUU (-10.6)	AAACUAAG (-10.4)	94.5%	82.0%			CUUAGUUUCUGAUGAGGCCGAAAGGCCGAAAAUUGUAG (-11.3)	0.92	View
21	UUA	1125	ACUACAAUU (-12.8)	AAACUAAG (-10.4)	95.1%	82.0%			CUUAGUUUUCUGAUGAGGCCGAAAGGCCGAAAAUUGUAGU (-11.3)	0.92	View
22	UUA	1125	AACUACAAUU (-13.7)	AAACUAAG (-10.4)	94.5%	82.0%			CUUAGUUUUCUGAUGAGGCCGAAAGGCCGAAAAUUGUAGUU (-11.3)	0.93	View
23	UUA	1125	AAACUACAAUU (-14.6)	AAACUAAG (-10.4)	93.9%	82.0%			CUUAGUUUCUGAUGAGGCCGAAAGGCCGAAAAUUGUAGUUU (-11.3)	0.93	View
24	CUA	1131	AAUUAAAACU (-11.2)	AGCACA (-11.9)	92.2%	51.9%	1111 (44.5%)		UGUGCUCUGAUGAGGCCGAAAGGCCGAAAGUUUUAAUU (-11.3)	0.94	View
25	CUA	1131	AAUUAAAACU (-11.2)	AGCACAA (-12.8)	92.2%	46.1%	1111 (50.8%)		UUGUGCUCUGAUGAGGCCGAAAGGCCGAAAGUUUUAAUU (-11.3)	0.94	View
26	CUA	1131	CAAUUAAAACU (-13.3)	AGCACA (-11.9)	91.9%	51.9%	1111 (44.5%)		UGUGCUCUGAUGAGGCCGAAAGGCCGAAAGUUUUAAUUG (-11.3)	0.95	View
27	CUA	1131	CAAUUAAAACU (-13.3)	AGCACAA (-12.8)	91.9%	46.1%	1111 (50.8%)		UUGUGCUCUGAUGAGGCCGAAAGGCCGAAAGUUUUAAUUG (-11.3)	0.95	View
28	CUA	1148	CCAUUCU (-11.9)	AGUCAUU (-10.8)	46.9%	40.4%			AAUGACUCUGAUGAGGCCGAAAGGCCGAAAGAAUGG (-12.8)	0.79	View
29	CUA	1148	CCAUUCU (-11.9)	AGUCAUUG (-12.9)	46.9%	35.7%			CAAUGACUCUGAUGAGGCCGAAAGGCCGAAAGAAUGG (-12.8)	0.80	View
30	AUU	1155	UAAGUCAU (-12.1)	GGGGAAA (-14.1)	58.7%	39.1%			UUUCCCCUGAUGAGGCCGAAAGGCCGAAAUGACUUA (-11.3)	0.97	View
31	AUU	1155	CUAAGUCAU (-14.2)	GGGGAAA (-14.1)	63.2%	39.1%			UUUCCCCUGAUGAGGCCGAAAGGCCGAAAUGACUUAG (-11.3)	0.97	View
32	UUU	1324	UAAAUCCUUUU (-14.7)	AAAUGACU (-11.7)	94.5%	97.9%			AGUCAUUUCUGAUGAGGCCGAAAGGCCGAAAAAAGGAUUUA (-13.2)	0.77	View
33	AUU	1506	GAUGGAU (-12.4)	UGAUUCG (-11.3)	71.3%	57.5%			CGAAUCACUGAUGAGGCCGAAAGGCCGAAAUCCAUC (-13.1)	0.76	View
34	AUU	1506	GAUGGAU (-12.4)	UGAUUCGC (-14.7)	71.3%	50.4%			GCGAAUCACUGAUGAGGCCGAAAGGCCGAAAUCCAUC (-13.1)	0.76	View
35	UUU	1507	GAUGGAUU (-13.3)	GAUUCGC (-12.6)	62.5%	57.5%			GCGAAUCCUGAUGAGGCCGAAAGGCCGAAAUCCAUC (-12.4)	0.80	View
36	AUA	1601	UGAGAAAU (-11.9)	AAAAGCC (-11.5)	57.2%	57.3%			GGCUUUUCUGAUGAGGCCGAAAGGCCGAAAUUUCUCA (-11.3)	0.83	View

Fig. 7. Screenshot of the results page: it reports the list with the ribozyme selected as suitable for target cleavage by ALADDIN.

3.4.2. Sequence of the Binding Domains and Binding Energy in kcal/mol for the Right and Left Region

After mapping the cleavage sites, the program selects two short stretches of flanking sequences, necessary for specific binding of the ribozyme molecule.

The stretches are selected on the basis of the best conditions acknowledged for hammerhead cleavage efficiency, taking into account specificity, binding stability, and structural compatibilities.

The selection of binding elements depends on the thermodynamics of their base-pairing energies: this aspect concerns the probable strength by which the ribozyme will bind its target. Excessively strong binding would affect the release of cleavage products, while weak binding would prevent cleavage from occurring. Suitable coupling free energies were estimated to range between −10 and −15 kcal/mol for each binding arm. Moreover, because the two binding ribozyme elements are independent, their binding energy should be balanced in order to ensure comparable kinetic behavior. Differences among the binding free energies of the left and right arm should not exceed 1.2 kcal/mol.

This selection represented the first filter of the program.

Looking at the *Survivin* example, the tool found four couples of binding stretches at position 289, UUC triplet, whose lengths and binding energies are slightly different but still between −10 and −15 kcal/mol (see Note 1).

3.4.3. Accessibility Score, for Both Left and Right Portion

The accessibility concept is the major critical point for target selection. RNA molecules spontaneously fold on themselves by pairing complementary stretches from different regions of their sequence. These intramolecular interactions may combine in several ways, producing a series of secondary structures with different stability and persistence, according to their free formation energy. Single-strand regions are directly exposed to ribozyme binding; in terms of overall energy balance they are better than double-stranded (partial or total) RNA, where considerable energy fraction is necessary to disrupt the structural constraint before binding. This aspect is conceivably related to cleavage kinetics and is expected to deeply affect ribozyme efficacy.

On the basis of the algorithm employed, the flanking sequences are evaluated and the target site selected if the following threshold values are satisfied:

$$AS_{left} > 35.0 \text{ and } AS_{right} > 35.0 \text{ (first filter)}$$

$$AS_{S_{[N,U,H]}} \geq 40.0 \text{ (second filter)}$$

$$\left| AS_{left} - As_{right} \right| \leq 50.0 \text{ (third filter)}$$

AS evaluation is automatically applied to the binding sites which have passed the first selection process (first filter) described above. A second filter discards all the regions displaying *AS* values below the imposed threshold ($AS_{S_{(N,U,H)}} = 40\%$). A refining analysis again discards all the target regions that, in spite of an *AS* value higher than that of the threshold, present highly unbalanced accessibility among up- and downstream segments ($\Delta AS > 50\%$ —third filter). An over 50% unbalance of ΔAS is conceivably correlated to a somewhat rigid structure.

Using again the example of *Survivin*, at position 289 the triplet AUC shows four possible results selected on the basis of four combinations of the flanking stretches. The result number 6 shows an AS of 38.4% for the left and 40.0% for the right arm which is more balanced than the result number 9, where the AS is 36.2% for the left and 47.6% for the right arm respectively (see Note 2).

3.4.4. Uniqueness of the Sequence

This is a crucial aspect affecting the specificity of the ribozyme. To this aim, analyses have to be performed over the whole RNA sequence, looking at recurrences (partial or total) of the binding elements.

The subsequences relative to the binding arms of the selected ribozyme (target) are probed along the entire RNA molecule and discarded when three or more occurrences are found. A duplicate for each flanking region would be accepted if only its AS conformed to the rules described later.

In the example of *Survivin*, the result 6 and 8 related to the previous UUC at position 289 both show another possible binding site in position 1,435, with very low probability (0.2%), so that it should be negligible. On the contrary, the selected triplet UUA at position 1,131 provides four results, 24, 25, 26, 27, all displaying the right binding region also recurring at position 1,111, with accessibility comparable to that of the wanted target (44.5–50.8% for the off-target element vs. 46.1–51.9% for the correct element) (see Note 3).

3.4.5. Nucleotide Sequence of the Ribozyme with Respective Folding Energy in kcal/mol

This a consequential step which allows to obtain a specifically related ribozyme from a given target stretch. Ribozyme sequence is automatically obtained by simply adding the binding elements in their antisense version, directly at the two nucleotide ends of the canonical hammerhead sequence.

For example, the structure of the ribozyme against the triplet CUA at position 108 of *Survivin* mRNA (result 3) is formed by the antisense version of the flanking sequences, the right CUUGAAUG and left AGAGAUG. It is linked to the conserved region of the canonical minimal hammerhead ribozyme composed by the structural stem-loop II domain GGCCGAAAGGCC and by the catalytic domains CUGAUGA and GAA.

3.4.6. Score Index

This index ranges from 0 to 1 and provides information about the expected folding of the ribozyme's secondary structure. The highest value 1 represents the expected final folding suitable for cleavage.

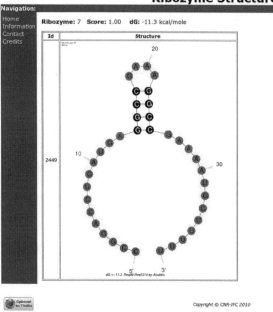

Fig. 8. Structure of ribozyme number 7 against triplet UUC at position 289: score 1 is consistent with the canonical structure where the binding arms are not involved in intramolecular interactions, and the catalytic core is structured correctly.

The computational study of probable ribozyme folding is performed applying the UNAfold algorithm. This analysis provides a final (fourth) filter which allows to select correctly folded molecules from others involved in molecular interactions responsible for catalytically inactive molecules.

Example of a putatively right structure endowed with score 1 appears in Fig. 8. This folded molecule, corresponding to the ribozyme against triplet UUC in position 289 (result number 7), displays free binding arms with no intramolecular interactions able, in principle, to freely bind the target mRNA. The catalytic core is not involved in tertiary interaction and the stem-II structural domain is arranged correctly.

The structure (see Fig. 9) corresponding to triplet CUA at position 1,148 (result number 28) shows a score of 0.79, due to an additional stem expected to be formed between catalytic core and binding region. This structural motif renders this molecule putatively less efficient to bind and cleave the target mRNA.

Another possibility (see Fig. 10) is to have multiple alternative conformational folding of the same molecule that shows different free energies. For example, the ribozyme targeting the triplet CUA at position 108 displays, for instance, a score of 0.91, and shows three different conformations (result number 3). In this case the energy values are very similar, so that each structure is likely to be equally represented at equilibrium. In spite of the presence of a

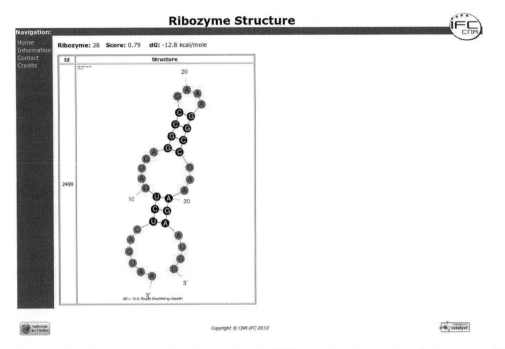

Fig. 9. Structure of the ribozyme corresponding to the result number 28. The structure shows an interaction between catalytic core and binding arm conferring a constraint which potentially reduces the activity of the molecule (score = 0.79).

very stable fully opened conformation, the overall score of this ribozyme is lower than 1, owing to the contribution of the alternative arrangements (see Notes 4–6).

4. Notes

1. It is generally better to choose a couple of binding regions with the closest similar energies.

2. It is usually better to choose a couple of binding regions with similar accessibility index. For example, the target position 289 (result 6) of *Survivin* is preferable to that of the result 9.

3. It is better to choose results with no alternative binding position. If necessary, choose alternative options with lower probability.

4. It is usually better to choose results with score 1. If impossible, it is better to choose structures with the highest possible score, namely those offering several possibilities, for example a structure with proper conformation is better than a structure with only one canonical conformation. In the example above structure

Ribozyme Structure

Navigation:

Home
Information
Contact
Credits

Ribozyme: 3 **Score:** 0.91 **dG:** -11.3 kcal/mole

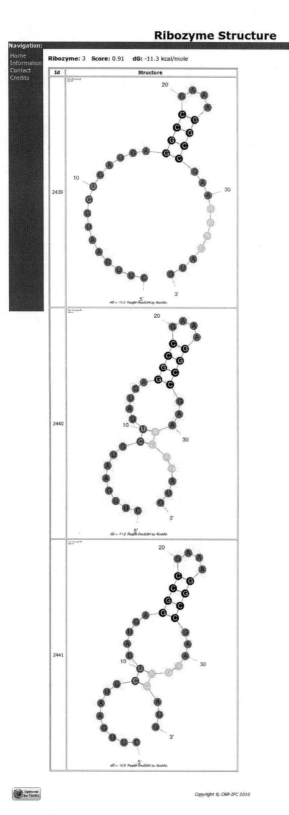

Fig. 10. Structures of the ribozyme corresponding to the result number 3. Three are the probable structures, almost all with the same energy. The most stable (*upper*) shows a canonical structure, while the least stable (*bottom*) shows intramolecular interactions lowering its score.

number 3 is better than number 28 which is supposed to have more rigid constraints.

5. Selectivity of the ribozyme is a crucial aspect to avoid any off-target effects, especially in system biology investigations and in marker validation studies. Therefore, homology analyses of the selected target stretches should be performed over the known transcriptome (Gene Bank known transcription repositories) relative to the given species looking at (partial or total) recurrences of binding elements in any unwanted RNA sequence. This aspect is not run automatically by the ALADDIN tool.

6. To experimentally validate the molecules, it is necessary to synthesize three or four different ribozymes and assess their kinetic properties "in vitro" (18, 19). Most reactive sequences should be selected and then need to be tested in viable biological systems, once they have been synthesized using stabilized monomers which provide better resistance against ubiquitous nucleases.

References

1. Citti, L., and Rainaldi, G. (2005) Synthetic hammerhead ribozymes as therapeutic tools to control disease genes, *Current Gene Therapy 5*, 11–24.

2. Usman N, Beigelman L, McSwiggen JA. (1996) Hammerhead ribozyme engineering. *Current Opinion in Structural Biology 4* 527–533.

3. Kronenwett, R., Haas, R., and Sczakiel, G. (1996) Kinetic selectivity of complementary nucleic acids: bcr-abl-directed antisense RNA and ribozymes, *Journal of Molecular Biology 259*, 632–644.

4. Jarvis, T. C., Wincott, F. E., Alby, L. J., McSwiggen, J. A., Beigelman, L., Gustofson, J., DiRenzo, A., Levy, K., Arthur, M., MatulicAdamic, J., Karpeisky, A., Gonzalez, C., Woolf, T. M., Usman, N., and Stinchcomb, D. T. (1996) Optimizing the cell efficacy of synthetic ribozymes - Site selection and chemical modifications of ribozymes targeting the proto-oncogene c-myb, *Journal of Biological Chemistry 271*, 29107–29112.

5. Scherr, M., and Rossi, J. J. (1998) Rapid determination and quantitation of the accessibility to native RNAs by antisense oligodeoxynucleotides in murine cell extracts, *Nucleic Acids Research 26*, 5079–5085.

6. Lieber, A., and Strauss, M. (1995) Selection of efficient cleavage sites in target RNAs by using a ribozyme expression library, *Molecular and Cellular Biology 15*, 540–551.

7. Mir, A. A., Lockett, T. J., and Hendry, P. (2001) Identifying ribozyme-accessible sites using NUH triplet-targeting gapmers, *Nucleic Acids Research 29*, 1906–1914.

8. Pan, W. H., Xin, P., Bui, V., and Clawson, G. A. (2003) Rapid identification of efficient target cleavage sites using a hammerhead ribozyme library in an iterative manner, *Molecular Therapy 7*, 129–139.

9. Ellington, A. D., Chen, X., Robertson, M., and Syrett, A. (2009) Evolutionary origins and directed evolution of RNA, *International Journal of Biochemistry & Cell Biology 41*, 254–265.

10. Tedeschi, L., Lande, C., Cecchettini, A., and Citti, L. (2009) Hammerhead ribozymes in therapeutic target discovery and validation, *Drug Discovery Today 14*, 776–783.

11. Zuker, M. (1989) On finding all suboptimal foldings of an RNA molecule , *Science 244*, 48–52.

12. McCaskill, J. S. (1990) The equilibrium partition function and base pair binding probabilities for RNA secondary structure, *Biopolymers 29*, 1105–1119.

13. Ding, Y., and Lawrence, C. E. (2003) A statistical sampling algorithm for RNA secondary structure prediction, *Nucleic Acids Research 31*, 7280–7301.

14. Markham, N. R., and Zuker, M. (2008) UNAFold - Software for nucleic acid folding and hybridization, *Methods in Molecular*

Biology:Volume II: Structure, function and applications, 3–31.

15. Mercatanti, A., Rainaldi, G., Mariani, L., Marangoni, R., and Citti, L. (2002) A method for prediction of accessible sites on an mRNA sequence for target selection of hammerhead ribozymes, *Journal of Computational Biology 9,* 641–653.

16. Mathews, D. H., Disney, M. D., Childs, J. L., Schroeder, S. J., Zuker, M., and Turner, D. H. (2004) Incorporating chemical modification constraints into a dynamic programming algorithm for prediction of RNA secondary structure, *Proceedings of the National Academy of Sciences of the United States of America 101,* 7287–7292.

17. Pennati, M., Binda, M., Colella, G., Zoppe, M., Folini, M., Vignati, S., Valentini, A., Citti, L., De Cesare, M., Pratesi, G., Giacca, M., Daidone, M. G., and Zaffaroni, N. (2004) Ribozyme-mediated inhibition of *Survivin* expression increases spontaneous and drug-induced apoptosis and decreases the tumorigenic potential of human prostate cancer cells, *Oncogene 23,* 386–394.

18. Citti L, Boldrini L, Nevischi S, Mariani L, Rainaldi G. (1997) Quantitation of in vitro activity of synthetic trans-acting ribozymes using HPLC, Biotechniques, 23, 898–903.

19. Poliseno L, Bianchi L, Citti L, Liberatori S, Mariani L, Salvetti A, Evangelista M, Bini L, Pallini V, Rainaldi G., (2004), Bcl2-low-expressing MCF7 cells undergo necrosis rather than apoptosis upon staurosporine treatment. *Biochemical Journal 1*;823–832.

20. Hertel, K.J. et al. (1992) Numbering system for the hammerhead ribozyme. *Nucleic Acids Res. 20,* 3252.

Chapter 22

Targeting mRNAs by Engineered Sequence-Specific RNase P Ribozymes

Yong Bai, Naresh Sunkara, and Fenyong Liu

Abstract

The methods of using engineered RNase P catalytic RNA (termed as M1GS RNA) for in vitro and in vivo in trans-cleavage of target viral mRNA are described in this chapter. Detailed information is focused on (1) mapping accessible regions of target viral mRNA in infected cells, (2) generation and in vitro cleavage assay of the customized M1GS ribozyme, (3) stable expression of M1GS RNAs and evaluation of its antiviral activity in cultured cells. Using these methods, we have constructed functional M1GS ribozyme that can cleave an overlapping region of the mRNAs coding for the human cytomegalovirus (HCMV) capsid scaffolding protein (CSP) and assemblin in vitro. Further study has demonstrated that, in cultured human cells expressing the functional M1GS ribozyme and infected with HCMV, more than 85% reduction in the expression of CSP and assemblin and a 4,000-fold reduction in viral growth were achieved. Our study provided the direct evidence that the customized M1GS ribozyme can be used as an effective gene-targeting agent for in trans-cleavage of viral genes and inhibition of viral growth in cultured cells.

Key words: Gene targeting, Gene therapy, Cytomegalovirus, Herpesvirus, Antiviral, Ribozyme, M1 RNA, RNase P

1. Introduction

Ribonuclease P (RNase P) is a ribonucleoprotein complex found in all organisms examined (1–3). This enzyme is responsible for the maturation of 5′ termini of all tRNAs by catalyzing a hydrolysis reaction to remove the 5′ leader sequence of precursor tRNA (ptRNA) (3–5) (see Fig. 1a). RNase P primarily recognizes the structure, rather than the sequence, of the substrate (6). Accordingly, any complex of two RNA molecules that resembles a tRNA molecule can be recognized and cleaved by RNase P (7–9). This unique property has allowed the engineering of RNase P ribozymes to

Jörg S. Hartig (ed.), *Ribozymes: Methods and Protocols*, Methods in Molecular Biology, vol. 848,
DOI 10.1007/978-1-61779-545-9_22, © Springer Science+Business Media, LLC 2012

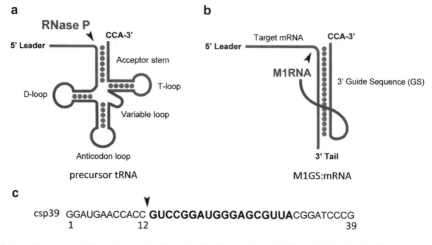

a

RNase P

CCA-3'

5' Leader

Acceptor stem

D-loop

T-loop

Variable loop

Anticodon loop

precursor tRNA

b

Target mRNA   CCA-3'

5' Leader

M1RNA

3' Guide Sequence (GS)

3' Tail

M1GS:mRNA

c

csp39   GGAUGAACCACC **GUCCGGAUGGGAGCGUUA**CGGATCCCG
          1   12   39

Fig. 1. (**a**) Schematic representation of natural substrates for Ribonuclease P (RNase P). (**b**) A hybridized complex of a target RNA (e.g., viral mRNA) and M1GS RNA. (**c**) Schematic representation of csp39 substrate used in this study with the targeted sequences that bind to the guide sequences of the ribozyme highlighted. The *arrowhead* shows the site of the cleavage by RNase P and M1 RNA.

cleave almost any target sequences other than naturally occurring tRNA substrates.

RNase P of *Escherichia coli* contains a catalytic RNA subunit (M1 RNA) that can be engineered to cleave tRNA-like substrates and other target RNAs, including specific mRNAs (4, 5, 7, 10). A sequence-specific ribozyme, M1GS, constructed by attaching to M1 RNA an additional small RNA (guide sequence [GS]) which contains a sequence complementary to a target mRNA and a 3' proximal CCA, is effective in blocking substrate mRNA expression (11, 12) (see Fig. 1b). Previous studies have demonstrated that customized M1GS RNA and RNase P are effective in cleaving both viral and cellular mRNAs and blocking their expression in cultured cells (12–17).

Human cytomegalovirus (HCMV) is a common opportunistic pathogen and causes significant morbidity and mortality in immunocompromised and immunologically immature individuals (18). Current antiviral therapy against HCMV infection showed lots of limitations, particularly, the emergence of drug resistance. Consequently, it is important that new treatment strategies utilizing novel mechanism against HCMV will be developed to combat this virus. RNase P-mediated inhibition of gene expression represents a novel and promising nucleic acid-based gene interference strategy for specific inhibition of target mRNA (11, 19). This chapter summarizes the protocol focusing on the generation and expression of M1GS catalytic RNA for in trans-cleavage of target mRNAs coding for HCMV essential capsid scaffolding protein (CSP) and assemblin in vitro and in cultured host cells. This study has demonstrated the feasibility of using RNase P as potential therapeutic agent against HCMV and other DNA/RNA viruses.

2. Materials

2.1. Reagents, Solutions, and Kits

1. Dimethyl sulfate (DMS).
2. 1 mM β-Mercaptoethanol.
3. 100 mM Dithiothreitol.
4. Chloroform.
5. Isopropanol.
6. Formaldehyde.
7. DMEM.
8. Fetal bovine serum (FBS).
9. Nonfat dry milk (NFDM).
10. Phosphate-buffered saline (PBS).
11. Neomycin.
12. [^{32}P]-labeled nucleotides.
13. Diethylpyrocarbonate (DEPC)-treated H$_2$O: double-distilled water is mixed with 0.1% DEPC and stirred overnight. The DEPC is inactivated by autoclaving for 20 min.
14. 30% Acrylamide/0.8% *N,N'* methylene bisacrylamide.
15. 1× Denhardt's solution: 0.1% bovine serum albumin (BSA), 0.1% polyvinylpyrrolidone, 0.1% Ficoll, 0.1% SDS, and 200 µg/mL of denatured salmon sperm.
16. 20× Standard saline citrate (SSC): 3 M NaCl, 0.3 M trisodium citrate.
17. 10× TBE: 0.89 M Tris-borate, 10 mM EDTA.
18. 2× RNA dye solution: 8 M urea, 20 mM EDTA, 0.25 mg/mL bromophenol blue, 0.25 mg/mL xylene cyanol FF.
19. Prehybridization buffer: 6× SSC, 0.05% sodium pyrophosphate, and 2× Denhardt's solution.
20. TNT: 10 mM Tris–HCl, pH 7.0, 150 mM NaCl, 0.05% Tween-20.
21. Cell lysis buffer: 150 mM NaCl, 10 mM Tris–HCl pH 7.4, 1.5 mM MgCl$_2$, 0.2% NP40.
22. Polymerase chain reaction (PCR) system including 10× PCR buffer, 25 mM MgCl$_2$, four 10 mM deoxynucleotide 5′ triphosphate (dNTP), *Taq* DNA polymerase (Promega).
23. 10× buffer A (cleavage assay buffer): 500 mM Tris–HCl, pH 7.5, 1 M NH$_4$Cl, 1 M MgCl$_2$.
24. T7 in vitro transcription system and 5× transcription buffer (Promega).
25. AMV RT and 5× RT buffer (Boehringer Mannheim).

26. TRIzol reagent (total RNA isolation reagent).

27. Mammalian transfection kit (GIBCO/BRL).

28. ECL Western blotting detection kit (GE Healthcare).

29. Random Primed Labeling Kit (Roche).

2.2. Virus, Cells, and Plasmids

1. HCMV (Strain AD169, ATCC).

2. Murine PA317 cells (amphotropic retrovirus packaging cell line).

3. Human foreskin fibroblasts (HFF).

4. Human U373MG cells (ATCC).

5. pFL117 (M1 RNA clone containing M1 DNA sequence driven by the T7 RNA polymerase promoter).

6. pLXSN (retroviral vector).

3. Methods

3.1. In Vivo Mapping of the Accessible Regions of Viral mRNA in HCMV Infected Cells

Using DMS methods, RNA secondary structure and protein association can be mapped in living cells (20–22). DMS can methylate N7 of guanine, N1 of adenine, and N3 of cytosine. The latter two modifications can further be detected by primer extension as it stops in the transcript one base before the modified base. Therefore, determining the site of modification by primer extension would reveal the regions that are accessible to DMS, and presumably to M1GS binding. Here, we employed this method to determine the accessibility of the region of the CSP mRNA in HCMV-infected cells. One of the most accessible regions to DMS modification is chosen as the in trans-cleavage site for M1GS RNA in future studies.

1. HFF cells which are maintained in T25 flask and in DMEM supplemented with 10% FBS are infected with HCMV at a multiplicity of infection (MOI) of 5 for 8–24 h prior to the treatment with DMS.

2. Cells are washed with fresh media once, and then incubated with 5 mL of fresh media that contain 1–2% of DMS for 5–10 min. After the incubation, the DMS media are immediately aspirated (see Note 1). Then cells are washed 3 times with cold PBS that contains 1 mM β-mercaptoethanol.

3. Cells are lysed by adding 0.5 mL of cold lysis buffer to the flask. The lysates are transferred to an Eppendorf tube and immediately spun in a microcentrifuge for 10 s at 4°C. The supernatant containing the cellular lysate is transferred to another tube.

4. Total RNAs are isolated by phenol-chloroform extraction of the supernatant cellular lysate 3 times, followed by ethanol-precipitation.

5. Oligonucleotides of 17–25 nt that are complementary to several regions of the targeted RNA are synthesized chemically in a 380B DNA synthesizer. These primers used for the primer extension are 5′-labeled by T4 polynucleotide kinase (PNK) in the presence of γ-[^{32}P]-ATP.

6. 10 μg of Total cellular RNA is mixed with 50,000 cpm of the primer in a volume of 8 μL, heated to 90°C for 2 min, and cooled down to allow them to be annealed to each other. Add the following sequentially: 4 μL 5× RT buffer, 2 μL 10 mM dATP, 2 μL 10 mM dTTP, 2 μL 10 mM dGTP, 2 μL 10 mM dCTP, 0.5 μL RNasin and 1 μL AMV reverse transcriptase. The primer extension reactions are preceded for 2 h at 42°C.

7. The reaction products are extracted with phenol chloroform, then precipitated by cold ethanol, and finally are separated in 8% denaturing gels. Autoradiograph of the gels will reveal the sites that block the primer extension reaction by reverse transcription. These sites are the DMS modification site (see Note 2).

3.2. Construction of M1-F Catalytic RNA and In Vitro Cleavage Assay for Substrate Csp39

Ribozyme M1-F was constructed by covalently linking the 3′ terminus of M1 RNA with a guide sequence which is complementary to the targeted CSP mRNA sequence. A substrate, csp39, which contained the targeted mRNA sequence of 39 nt, was used. M1-F catalytic RNA and csp39 substrates were synthesized in vitro, and then incubated for assaying ribozyme in trans-cleavage. Efficient cleavage by M1-F yielded two hydrolyzed products of 12 and 27 nt, respectively.

3.2.1. Construction of M1-F Ribozyme and Substrate Csp39

1. The DNA sequence of M1GS ribozyme (M1-F) targeting CSP mRNA is constructed by PCR amplification from plasmid pFL117 with the following oligonucleotides as 5′ and 3′ primers respectively: AF25 (5′-GGAATTCTAATACGACTCACTA TAG-3′) and M1AP3(5′-CCCGCTCGAGAAAAAATGGTGT CCGGATGGGAGCGTTATGTGGAATTGTG-3′). The following reagents are added sequentially to the PCR reaction: 10 μL 10× PCR buffer (Mg^{2+} free), 2 μL 10 mM dATP, 2 μL 10 mM dTTP, 2 μL 10 mM dGTP, 2 μL 10 mM dCTP, 8 μL MgCl$_2$, 10 pmol 5′ primer (AF25),100 pmol 3′ primer (M1AP3), 1 μg PvuII-digested pFL117 plasmid, 1 μL Taq DNA polymerase. Add DEPC-treated water to adjust the final reaction volume to 100 μL. The PCR reactions are performed using the following cycle: (1) denaturing for 2 min at 94°C; (2) 30 cycles of denaturing for 2 min at 94°C, annealing for 1 min at 47°C, extension for 1 min at 72°C; (3) final extension

for 10 min at 72°C. The PCR DNA products are then separated in 5% polyacrylamide gels under non-denaturing conditions, and are purified and used as the template for the in vitro transcription synthesis of the ribozymes.

2. M1-F ribozyme is synthesized by in vitro transcription which contains the following mixture: 4 μL (2 μg) M1-F DNA from PCR reaction, 8 μL 5× transcription buffer, 4 μL 100 mM DTT, 4 μL 10 mM ATP, 4 μL 10 mM GTP, 4 μL 10 mM CTP, 4 μL 10 mM UTP, 4 μL H_2O, 1 μL RNasin (5 U/μL), 2 μL T7 RNA polymerase. The reaction mixture is incubated at 37°C for 4 h to overnight. An equal volume of 2× RNA dye solution is added and the sample is loaded onto an 8% polyacrylamide-8-M urea gel. After electrophoresis, the gel is placed on a TLC plate (silica gel UV256). The RNA bands are visualized by briefly shadowing with a portable shortwave ultraviolet lamp. RNAs are extracted from the excised gel slice by the crush-soak method using DEPC-treated water.

3.2.2. Construction of M1-F Ribozyme and Substrate Csp39

1. The DNA sequence that encodes substrate CSP39 was constructed by PCR using pGEM3zf (+) as a template and oligonucleotides AF25 (5′-GGAATTCTAATACGACTCACTATAG-3′) and sAP3 (5′-CGGGATCCGTAACGCTCCCATCCGGACGGTGGT TCATCCTATAGTGAGTCGTATTA-3′) as 5′ and 3′ primers, respectively.

2. Labeled csp39 is synthesized by in vitro transcription as described above, except using 30 μCi α-[^{32}P]-GTP and 1 mM GTP instead of 10 mM GTP in the reaction.

3.2.3. In Vitro in trans-Cleavage Assay for Substrate Csp39

1. 10 nM of M1-F RNA and trace amount of α-[^{32}P]-labeled csp39 mRNA (10,000 cpm) are mixed together. 1 μL of 10× Buffer A is added and water is used to adjust the final volume to 10 μL. The cleavage reaction mixture is incubated at 37°C for 40 min and the reaction is terminated by adding 10 μL of 2× RNA dye solution.

2. The cleavage products are separated by 8% polyacrylamide-7-M urea gel. The gel is exposed to a phosphorscreen and quantitated with a STORM840 PhosphorImager (Molecular Dynamics).

3.3. Expression and Functional Analysis of M1-F Ribozyme in Cultured U373MG Cells

The DNA sequence coding for the M1-F ribozyme is cloned into retroviral vector pLXSN and is placed under the control of the small nuclear U6 RNA promoter. This promoter is transcribed by RNA polymerase III and has previously been shown to highly express M1GS RNA steadily in nucleus (8, 20, 23, 24). To construct cell lines that express M1-F ribozyme, amphotropic packaging cell line PA317 (25) is transfected with pLXSN-M1-F. After U373MG cells are infected with retroviral vectors containing supernatants, cells stably expressing the ribozyme are cloned. To determine the efficacy of the ribozymes, cells are challenged with HCMV AD169 strain and the expression pattern of CSP and assemblin is studied by Northern blot and Western analysis. The growth of HCMV in cultured cells is determined by plaque assay.

3.3.1. Stable Expression of M1-1 Ribozyme in Cultured Cells

1. PA317 cells are cultured onto 6-well plates 1 day before transfection. 12–16 h after cell plating, when the cells reach 80% confluent (see Note 3), the cells are transfected with 10–20 μg of plasmid LXSN-M1-F DNA by using the mammalian transfection kit (GIBCO/BRL) (see Note 4). Forty-eight hours after transfection, the culture supernatants are collected (about 3 mL). These supernatants are retroviral stock used to infect U373MG cultured cells.

2. One day before infection, U373MG cells are plated onto a 6-well plate. The cells should have 80% confluent the day of infection with the culture supernatant containing retroviruses. The media is aspirated out and 1.5 mL of the collected retroviral stock (from step 1) is added to the U373MG cell culture. The infected cells are incubated for 4–12 h with occasional shaking (see Note 5). The inoculums are replaced with fresh DMEM supplemented with 10% FBS (Collaborative Research).

3. 48–72 h After infection, the cells are incubated in cultured medium that contain 600 μg/mL neomycin. Neomycin-resistant cells are selected and cloned in the presence of neomycin for 2 weeks. Cells infected with retrovirus are subsequently split sparsely over ten cultured flasks and placed under neomycin to select for cloned ribozyme-expressing cell lines (see Note 6). The selected cells are aliquoted and frozen for long-term storage in liquid nitrogen or used for further studies.

3.3.2. Northern Analysis of Expression of M1-F in Cultured Cells

Total RNA Extraction from Cultured Cells

1. The total RNA from T75 flasks of cells (~10^7 infected cells) is isolated using standard TRIzol reagent RNA isolation protocol (see Note 7).

2. The RNA pellet is resuspended in RNase-free water and the concentration is tested by Bio-spectrophotometer (BioRad).

Northern Blot for Detection of M1-F in Cultured Cells (See Note 8)

1. 10 μg of Total RNA prepared above is loaded onto a 2% formaldehyde agarose gel and the RNAs are separated by running the gel at a constant voltage (see Note 9).

2. The gel is washed 3 times (10 min each) in deionized water and the RNAs are transferred from the agarose gel onto a nitrocellulose membrane.

3. The transfer sandwich is set up in a large glass dish filled with 1 L 20× SSC solution from bottom up: a long glass plate, a half

sheet of Whatman 3MM paper, three pieces of Whatman 3MM paper (larger than the gel), the gel, a saran wrap cut with a window to expose gel, a piece of nitrocellulose membrane wetted with RNase-free water (larger than the gel), five sheets of Whatman 3MM (the same size as the membrane), a stack of paper towels of about 4 cm in height, a glass, and a weight (1 kg). At each step, air bubbles are trapped by gently rolling a test tube over the last layer added (see Note 10). After overnight transferring, the membrane is rinsed 3 times with deionized water for 10 min. And then the membrane is placed on top of a piece of Whatman 3MM paper and baked in an oven at 80°C for 1.5 h.

4. The radioactive-labeled DNA probe used to detect M1GS RNAs is synthesized from plasmid pFL117 by using a random primed labeling kit (Roche). The nitrocellulose membrane is pre-hybridized for 4 h and hybridized for 16 h with labeled probe at 65°C in hybridization buffer (6× SSC, 0.05% sodium pyrophosphate, 2× Denhardt's solution, 0.1% SDS, and 200 μg/mL of salmon sperm).

5. At 42°C, the membrane is washed sequentially with 2×, 1×, and 0.5× SSC (containing 0.1% SDS). Each step lasts for about 15 min.

6. After dried, the membrane is exposed to a phosphorscreen and scanned with a STORM840 Phosphorimager.

3.3.3. Analysis of M1-F-Mediated Inhibition of Viral Gene Expression and Growth in Cultured Human U373MG Cells

Expression of CSP mRNA Is Analyzed by Northern Blot

1. Cells are infected with HCMV at a MOI 0.05–1. Total RNAs are isolated from the infected cells at 8–72 h post infection as described in Subheading 3.2.2.

2. The expression levels of viral CSP mRNA are determined by Northern blot as described in Subheading 3.2.2. The RNA fractions are separated in 1% agarose gels that contained formaldehyde, transferred to a nitrocellulose membrane, hybridized with the [^{32}P]-radiolabeled DNA probes that contained the HCMV DNA sequence, and analyzed with a STORM840 phosphorimager. The DNA probes used to detect CSP mRNA were synthesized from plasmids pM80.

Expression of CSP Protein Is Analyzed by Western Blot

1. Proteins are isolated from infected cells at 24–72 h post infection. The cells are rinsed twice with 5 mL of PBS and then pelleted by a 5-min spin with $3,000 \times g$ at 4°C. The cell pellet is resuspended with 50–100 μL of cold PBS carefully, and then the equal volume of 2× disruption buffer is added. The mixture is vortexed at maximum speed to break for 1 min and sonicated on ice for three 20–30-s sessions. The sample is boiled for 1 min before loading on SDS-polyacrylamide gel electrophoresis (PAGE) gel. The level of viral protein is determined by western blot as described below.

2. The total amount of 50 µg proteins is loaded onto 9% SDS-polyacrylamide gels crosslinked with *N, N″*-methylenebisacrylamide with a stacking layer of 4.5% acrylamide/bisacrylamide. The gels are run at a constant power setting.

3. The proteins are transferred from the polyacrylamide gel onto a nitrocellulose membrane, using an electrophoretic transfer apparatus with a constant current of 150 mA for 2 h.

4. The nitrocellulose membrane is pre-blocked by incubating with TNT buffer plus 2.5% NFDM for 1 h in an orbital shaker.

5. After blocking, the membrane is incubated for 1 h with primary antibody at a dilution of 1:500 in TNT supplemented with 1.25% of NFDM.

6. The membrane is washed with TNT buffer for three 5-min sessions. The membrane is then incubated with HRP-linked anti-species antibody (secondary antibody) diluted at 1:1,000 dilutions for 1 h with shaking.

7. The membrane is washed with TNT buffer for three 5-min sessions.

8. The membrane is incubated with ECL substrates for 1 min at room temperature.

9. The membrane is exposed to the film and the film is developed and analyzed.

Growth of HCMV Is Analyzed by Plaque Assay

The level of viral growth inhibition mediated by M1-F ribozyme was determined by assaying the viral titers in tissue culture cells.

1. 5×10^5 Ribozyme-expressing cells are infected with HCMV at a MOI of 1–5. The cells are incubated with DMEM at 37°C with 5% CO_2 for 2 h.

2. After incubation, the cells are washed with PBS, and then 0.5 mL fresh DMEM supplemented with 10% FBS is added and the cells are incubated for an interval of 1–7 days.

3. The cells are harvested at 1-day interval throughout 7 days after infection. Viral stocks are prepared by adding 0.5 mL of 10% FBS/10% skim milk solution. The cells are scraped and sonicated 3 times.

4. Plaque forming assay is performed as described below. First, tenfold serial dilution of the viral stock is prepared in 2 mL of DMEM for each dilution. Second, HFF cells are infected in 6-well tissue culture plates with 2 mL of viral dilutions. The cultured plates are incubated at 37°C with 5% CO_2 for 2 h. Third, the cells are washed with DMEM medium and then overlaid with fresh 2% agarose and DMEM containing 4% FBS in a 1:1 ratio. Fourth, the cells are incubated for 10–14 days post infection and the numbers of viral plaques are counted

under an inverted microscope. Plaque-forming unit (PFU) is determined by the highest viral dilution that yields plaque. The values obtained from plaque assay were the average of triplicate experiments.

4. Notes

1. Several concentrations of DMS and different period of incubation are suggested in order to obtain specific and reproducible modification patterns.

2. It is suggested to run a sequencing reaction on the same gel. This will reveal the sequence of the targeted region. This sequencing reaction can be carried out by using a plasmid as the sequencing template that contains the targeted region and the same primer for the primer extension as the sequencing primer.

3. It is suggested to check the cell condition before transfection. In order to achieve optimal transfection efficiency, the cells should be evenly distributed throughout each well.

4. The transfection works well when using the alternative transfection kit (Lipofectamine 2000, Invitrogen, San Diego, CA).

5. To increase the efficiency of retroviral infection to target cells, it is suggested to add polybrene (Sigma) to a final concentration of 8 μg/mL. The efficiency of infection can also be improved by carrying out multiple rounds of infection and using a high MOI.

6. It is recommended to test more clones at this stage. In our experiments, more than ten clones were selected and tested. Most of the cell lines were found to express a high level of ribozymes.

7. According to the manual provided by manufacture Invitrogen, the amount of Trizol Reagent added is based on the area of the culture dish (1 mL/10 cm^2). An insufficient amount of TRIZOL Reagent may result in contamination of the isolated RNA with DNA.

8. The M1GS RNA expressed by the U6 promoter is primarily localized in the nucleus. This can be tested with northern blot by using the RNAs isolated from the nuclear and cytoplasmic fractions separately.

9. Since mRNA is extremely sensitive to degradation, this step should be RNase-free. DEPC-treated H_2O is recommended for the entire gel-running process.

10. To achieve better transfer effect, it is important to remove any air bubbles between the gel and the membrane by simply rolling out air bubbles with a plastic pipette.

Acknowledgments

Gratitude goes to Dr. Phong Trang and Dr. Kihoon Kim. This research has been supported by NIH (AI041927 and DE014842).

References

1. Evans D, Marquez SM, & Pace NR (2006) RNase P: interface of the RNA and protein worlds. (Translated from eng) *Trends Biochem Sci* 31(6):333–341.

2. Gopalan V & Altman S eds (2007) *Ribonuclease P: structure and catalysis. The RNA World.* (Cold Spring Harbor Laboratory Press, New York).

3. Xiao S, Scott F, Fierke CA, & Engelke DR (2002) Eukaryotic ribonuclease P: a plurality of ribonucleoprotein enzymes. (Translated from eng) *Annu Rev Biochem* 71:165–189.

4. Frank DN & Pace NR (1998) Ribonuclease P: unity and diversity in a tRNA processing ribozyme. *Annu Rev Biochem* 67:153–180.

5. Altman S & Kirsebom LA (1999) *Ribonuclease P. In The RNA world. 2nd Ed* (Cold Spring Harbor Laboratory Press, Cold Spring Harbor, NY) 2nd Ed pp 351–380.

6. Altman S (2010) History of RNase P and overview of its catalytic activity. *Ribonuclease P*, eds Liu F & Altman S (Springer, New York).

7. Forster AC & Altman S (1990) External Guide Sequences for an RNA Enzyme. *Science* 249(4970):783–786.

8. Yuan Y, Hwang ES, & Altman S (1992) Targeted cleavage of mRNA by human RNase P. *Proc Natl Acad Sci USA* 89(17): 8006–8010.

9. Liu F (2010) Ribonuclease P as a tool. *Ribonuclease P*, eds Liu F & Altman S (Springer, New York).

10. Guerriertakada C, Li Y, & Altman S (1995) Artificial Regulation of Gene-Expression in *Escherichia coli* by Rnase-P. *Proceedings of the National Academy of Sciences of the United States of America* 92(24):11115–11119.

11. Raj SML & Liu F (2003) Engineering of RNase P ribozyme for gene-targeting applications. *Gene* 313:59–69.

12. Liu FY & Altman S (1995) Inhibition of Viral Gene-Expression by the Catalytic RNA Subunit of Rnase-P from *Escherichia coli. Genes & Development* 9(4):471–480.

13. Plehn-Dujowich D & Altman S (1998) Effective inhibition of influenza virus production in cultured cells by external guide sequences and ribonuclease P. *Proceedings of the National Academy of Sciences of the United States of America* 95(13):7327–7332.

14. Cobaleda C & Sanchez-Garcia I (2000) In vivo inhibition by a site-specific catalytic RNA subunit of RNase P designed against the BCR-ABL oncogenic products: a novel approach for cancer treatment. *Blood* 95(3):731–737.

15. Kilani AF, *et al.* (2000) RNase P ribozymes selected in vitro to cleave a viral mRNA effectively inhibit its expression in cell culture. *Journal of Biological Chemistry* 275(14): 10611–10622.

16. Trang P, *et al.* (2000) Effective inhibition of human cytomegalovirus gene expression and replication by a ribozyme derived from the catalytic RNA subunit of RNase P from *Escherichia coli.* (Translated from eng) *Proc Natl Acad Sci USA* 97(11):5812–5817.

17. Kim K, Trang P, Umamoto S, Hai R, & Liu F (2004) RNase P ribozyme inhibits cytomegalovirus replication by blocking the expression of viral capsid proteins. (Translated from eng) *Nucleic Acids Res* 32(11):3427–3434.

18. Mocarski ES, Shenk T, & Pass RF eds (2007) *in Fields Virology,* pp 2701–2772 (Lippincott, Philadelphia).

19. Gopalan V, Vioque A, & Altman S (2002) RNase P: variations and uses. *J Biol Chem* 277(9):6759–6762.

20. Liu F & Altman S (1995) Inhibition of viral gene expression by the catalytic RNA subunit of RNase P from *Escherichia coli*. (Translated from eng) *Genes Dev* 9(4):471–480.

21. Ares M, Jr. & Igel AH (1990) Lethal and temperature-sensitive mutations and their suppressors identify an essential structural element in U2 small nuclear RNA. (Translated from eng) *Genes Dev* 4(12A):2132–2145.

22. Zaug AJ & Cech TR (1995) Analysis of the structure of Tetrahymena nuclear RNAs in vivo: telomerase RNA, the self-splicing rRNA intron, and U2 snRNA. (Translated from eng) *RNA* 1(4):363–374.

23. Bertrand E, *et al.* (1997) The expression cassette determines the functional activity of ribozymes in mammalian cells by controlling their intracellular localization. (Translated from eng) *RNA* 3(1):75–88.

24. Das G, Henning D, Wright D, & Reddy R (1988) Upstream regulatory elements are necessary and sufficient for transcription of a U6 RNA gene by RNA polymerase III. (Translated from eng) *EMBO J* 7(2):503–512.

25. Miller AD & Rosman GJ (1989) Improved Retroviral Vectors for Gene-Transfer and Expression. *Biotechniques* 7(9):980.

Chapter 23

Target-Induced SOFA-HDV Ribozyme

Michel V. Lévesque and Jean-Pierre Perreault

Abstract

Small *cis*-acting ribozymes have been converted into trans-acting ribozymes possessing the ability to cleave RNA substrates. The Hepatitis Delta Virus (HDV) ribozyme is one of the rare examples of these that is derived from an RNA species that is found in human cells. Consequently, it possesses the natural ability to function in the presence of human proteins in addition to an outstanding stability in human cells, two significant advantages in its use. The development of an additional specific *on/off* adaptor (SOFA) has led to the production of a new generation of HDV ribozymes with improved specificities that provide a tool with significant potential for future development in the fields of both functional genomics and gene therapy. SOFA-HDV ribozyme-based gene inactivation systems have been reported in both prokaryotic and eukaryotic cells. Here, a step-by-step approach for the efficient design of highly specific SOFA-HDV ribozymes with a minimum investment of time and effort is described.

Key words: Ribozyme, Gene inactivation, Functional genomics, Gene therapy

1. Introduction

Ribozymes (RNA enzyme, Rz), are commonly found as self-cleaving RNA motifs that are essential in the life cycles of infectious RNA that replicate via a rolling circle mechanism. These self-cleaving RNA motifs include both the *hammerhead* and *hairpin* structures that are found in the viroid and viroid-like satellite RNAs infecting plants. A self-cleaving ribozyme is also retrieved in the hepatitis delta virus (HDV) that infects several eukaryotes including humans. All of these *cis*-acting RNA motifs have been converted into trans-acting ribozymes possessing the ability to specifically recognize and subsequently catalyze the cleavage of an RNA target. As a consequence, they have become attractive tools in the development of gene inactivation systems (for reviews see refs. (1–4)).

Jörg S. Hartig (ed.), *Ribozymes: Methods and Protocols*, Methods in Molecular Biology, vol. 848,
DOI 10.1007/978-1-61779-545-9_23, © Springer Science+Business Media, LLC 2012

For a long time, the only example of a ribozyme derived from an RNA species naturally found in human cells (i.e. infected hepatocytes) was the HDV ribozyme (5, 6). Its evolution in human cells confers to it several unique properties for its use as a potential tool, including the natural ability to function both in the presence of human proteins and at the physiological magnesium concentration (i.e. ~1 mM magnesium). Due to its outstanding stability in human cells, the HDV ribozyme is an interesting potential candidate for the development of a gene-knockdown system (7). However, the potential development of such a gene inactivation system based on the HDV ribozyme has been relatively neglected because it suffers from a lack of substrate specificity when used as a molecular tool (8). Specifically, its substrate specificity depends on the formation of stem I that includes only 7 bp (see Fig. 1, recognition domain, RD), while a total of 13–14 bp has been estimated to be required in order to ensure the targeting of a unique RNA species from the human transcriptome (9). In order to overcome this hurdle, a module named SOFA (Specific On/oFf Adaptor) was engineered for the HDV ribozyme (see Fig. 1) (10). The SOFA module switches the cleavage activity from the off to the on state solely in the presence of its cognate target. Initially, the recognition domain site forms a short duplex with an inserted sequence element (the blocker, Bl), thereby locking the ribozyme in an inactive conformation (off) by increasing the energetic barrier for non-specific base pairing interactions and thus reducing the potential for off-target cleavages. A second inserted sequence element (the biosensor, Bs) extends the base pairing with the target in order to favour the binding of the genuine target, and the formation of this duplex concomitantly results in the disruption of the short duplex involving the blocker and the recognition domain, switching the ribozyme into an active conformation (on). In other words, the blocker acts as a safety lock for the ribozyme in which the key is the recognition of the target RNA by the biosensor. The combined work of the biosensor and the blocker has been shown to increase the substrate specificity of the ribozyme's cleavage by several orders of magnitude when compared with the wild-type version (10, 11). Finally, the last component of the SOFA module is a stabilizer stem that brings both the 5′ and 3′ strand ends together. This additional stem has no effect on the cleavage activity of the SOFA-HDV ribozyme, but data suggests that it is important for its molecular stability in cellulo (7, 10). Importantly, experiments performed in *Lactococcus lactis* confirmed that SOFA-HDV ribozyme retained the property of enzyme turnover, meaning that one molecule of ribozyme can successively cleave several target molecules (12). To our knowledge, the SOFA-HDV ribozyme constitutes the first example of a ribozyme bearing a target-dependent module that is activated by its RNA target, an arrangement which greatly diminishes non-specific effects. This new approach provides a specific

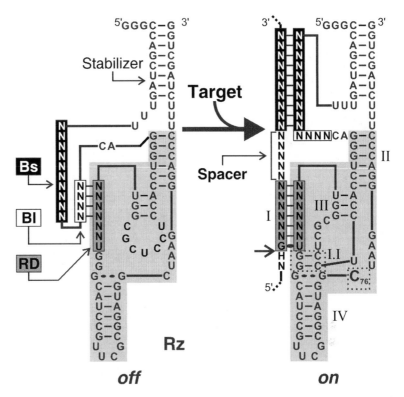

Fig. 1. Detailed representation of the SOFA-HDV ribozyme. The secondary structures of both the *off* (*left*) and *on* (*right*) conformations of the SOFA-HDV ribozyme are shown. The SOFA-HDV ribozyme (Rz) is illustrated in both cases, while the target is only shown for the *on* conformation. The hepatitis delta virus (HDV) ribozyme is highlighted in *grey* in each case. The recognition domain, the biosensor and the blocker are identified by RD (*black* on *grey*), Bs (*white* on *black*) and Bl (*black* on *white*), respectively. The stabilizer and the spacer (the region in the target that is localized between the recognition domain and the biosensor binding site) are also indicated. The roman numerals *I*, *I.I*, *II*, *III* and *IV* identify the corresponding stems or stem-loops in the HDV ribozyme. The variable nucleotides (i.e. those that can be A, C, G or U) of the ribozyme are indicated by the *boxed letter N* (*black* for Bs, *white* for Bl and *dark grey* for RD). All *unboxed* nucleotides are invariable and are constant from one ribozyme to the other, regardless of the target sequence. Positions where mutations can be inserted in order to produce a catalytically inactive SOFA-HDV ribozyme (I.I pseudoknot and the catalytic cytosine 76) are *boxed* with *dashed lines* on the *on* conformation. Finally, the *short arrow* indicates the cleavage site.

and improved tool with significant potential for application in the fields of both functional genomics and gene therapy. The development of several gene inactivation systems based on SOFA-HDV ribozyme in both prokaryotic and eukaryotic cells have been reported (10, 12–15). These studies have permitted the identification of the important features that must be considered in an optimized design process of SOFA-HDV ribozymes. Here, a simple method for designing SOFA-HDV ribozymes, highlighting the key points that need to be considered in order to improve the rate of success, is described.

2. Materials

2.1. Specificity Analysis

Ribosubstrate: http://www.riboclub.org/ribosubstrates.

2.2. Synthesis of DNA Template

1. SOFA-HDV RzX primer at a concentration of 100 μM: 5′-TAATACGACTCACTATAGGGCCAGCTAGTTT$(N)_{10Bs}$ $(N)_{4Bl}$CAGGGTCCACCTCCTCGCGGT$(N)_{6RD}$TGGGCAT CCGTTCGCGG-3′.

2. Universal SOFA reverse primer at a concentration of 100 μM: 5′-CCAGCTAGAAAGGGTCCCTTAGCCATCCGCGAAC GGATGCCC-3′.

3. 10× PCR Buffer: 200 mM Tris–HCl (pH 8.8), 100 mM KCl, 100 mM $(NH_4)_2SO_4$ and 1% Triton X-100.

4. 100 mM $MgSO_4$ stock solution.

5. dNTP stock solution: 10 mM of each dATP, dCTP, dGTP and dTTP.

6. *Pwo* DNA polymerase (Roche Diagnostics).

7. 3 M Sodium acetate (pH 5.2).

8. 100% and 70 Ice-cold ethanol.

2.3. In Vitro Transcription

1. NTP stock solution: 25 mM of each ATP, CTP, GTP and UTP.

2. 5× Transcription buffer: 400 mM HEPES-KOH (pH 7.5), 120 mM $MgCl_2$, 10 mM spermidine and 200 mM DTT.

3. Pyrophosphatase (1 U/μL, Roche Diagnostics) diluted 1:100 (V:V) in a dilution buffer containing 50% glycerol, 20 mM Tris–HCl (pH 8.0), 10 mM NaCl, 1 mM DTT, 1 mM EDTA, 100 μg/mL BSA and 0.03% NP40.

4. RNaseOUT (40 U/μL, Invitrogen) diluted 1:2 (V:V) in a buffer containing 50% glycerol, 20 mM Tris–HCl (pH 8.0), 50 mM KCl, 0.5 mM EDTA and 8 mM DTT.

5. Purified T7 RNA polymerase (5 μg/μL).

6. RQ1 RNase-free DNase (1 U/μL, Promega).

7. Phenol and chloroform.

8. Loading buffer: 98% formamide, 10 mM EDTA, 0.025% xylene cyanol and 0.025% bromophenol blue.

9. Elution buffer: 500 mM ammonium acetate, 1 mM EDTA and 0.1% SDS.

2.4. Target Labelling

1. 10× Antarctic Phosphatase Reaction Buffer (New England Biolabs).

2. Antarctic Phosphatase (5 U/μL, New England Biolabs).

3. 10× T4 polynucleotide kinase (PNK) (USB).

4. T4 PNK (30 U/μL, USB) diluted 1/10 times in T4 PNK dilution buffer (USB).

5. [γ-^{32}P]ATP (6,000 Ci/mmol, PerkinElmer).

6. Glycogen (20 mg/mL, Roche Diagnostics).

2.5. Cleavage Assays

1. 500 mM Tris–HCl (pH 7.5).

2. 100 mM $MgCl_2$.

2.6. Denaturing Polyacrylamide Gel Electrophoresis

1. Denaturing gels: 8 M urea and 19:1 ratio of acrylamide/bisacrylamide in 1× TBE buffer (89 mM Tris base, 89 mM boric acid and 2 mM EDTA).

2.7. Autoradiography and RNA Quantification

1. Hyperfilm MP films (GE Healthcare).

2. Phosphoscreen (GE Healthcare).

3. ImageQuant software (GE Healthcare).

3. Methods

The step-by-step protocol illustrated in Fig. 2 provides for a fast and simple selection of SOFA-HDV ribozymes directed against any target of choice. This protocol is composed of three main modules that are then subdivided into various steps. Briefly, the modules include, successively, the identification of the potential targeting sites, the design of the SOFA-HDV ribozymes and, finally, the selection of the most prominent ones for further development.

3.1. Target Module

The first step is solely related to the selected RNA target. It includes all of the prerequisite analyses and considerations that must be taken into account before moving on with the designing of SOFA-HDV ribozymes. This step can be subdivided into two parts: the documentation of the target and the identification of potential sites to be targeted.

3.1.1. Documentation of the Target

Initially, it is paramount to both gather all available information on the target and to establish all considerations that might influence the targeting of a given RNA species. For example, the targeting of a viral RNA with the aim of controlling the propagation of this pathogen may benefit from a sequence analysis in order to identify any stretches of highly conserved nucleotides that could then be used for ribozyme binding. This consideration increases the chances of targeting a wide range of viral genotypes. In addition, targeting

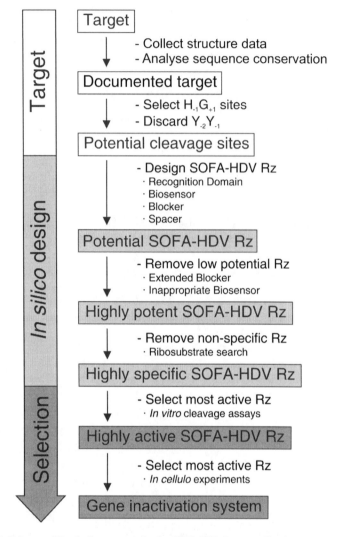

Fig. 2. Scheme of the design process for the SOFA-HDV ribozyme. The *large arrow* represent the three major, sequential, design steps. The target (*white box*), the in silico design (*light grey box*) and the selection (*dark grey box*) modules are divided into different analyses that are described in Subheading 3. The *horizontal boxes* indicate what is obtained after each procedure, while the text on the right side of each *small arrows* describes the key element in the process.

more conserved nucleotides should reduce the potential escape of any resistant mutants (16, 17). Conversely, the identification of sequences specific to a unique mRNA belonging to a family of relatively well-conserved proteins may lead to a reduction in the level of a single protein species even though several related ones are similarly expressed in a cell.

Effective cleavage at a specific site depends on its accessibility to the ribozyme. In principle, target sites located in single-stranded regions of an RNA species possess a higher potential because they should be more accessible for ribozyme binding than those located in double-stranded regions (18–20). Within the double-stranded regions, the ribozyme might compete unfavourably with intramolecular base pairing when trying to bind the target. Therefore, any available data on the secondary and tertiary structures of the target might be of great value. In the past, bioinformatic and biochemical procedures, as well as a combination of both, have been used (18, 20). However, most of these approaches remain exhausting and may not necessarily be relevant if they are performed in test tube, as many factors influence the structure of an RNA species in vivo. Therefore, when no structural data is available, many years of work has led us to suggest limiting these investigations to only bioinformatic analysis and instead compensating by designing and testing a larger number of ribozymes in the subsequent steps. Moreover, the biosensor has been demonstrated to act as a facilitator (i.e. unwinds the secondary structure in the neighbourhood of the target site after its binding), thereby reducing the importance of the accessibility of the target sites (10).

3.1.2. Identification of Potential Sites

The resulting sequences (i.e. conserved, specific and mostly accessible) are then further analyzed in order to identify those that fulfil the essential criteria for efficient HDV ribozyme cleavage. The minimum requirement for an RNA to be cleaved by an HDV-derived ribozyme is the presence of a $H_{-1}G_{+1}$ at the cleavage site, where H can be A, C or U (8). More specifically, the first nucleotide downstream of the cleavage site (position +1) must be a guanosine in order to allow for formation of the essential G–U Wobble base pair with the ribozyme. At position –1, the presence of a guanosine residue is detrimental to cleavage. A systematic analysis of the sequences in positions –1 to –4 of a collection of small targets revealed that each of these nucleotides contributes differently to the ability of a given target RNA to be cleaved (21). Further analysis using longer RNA targets indicates that the nucleotides located in positions –1 and –2 may significantly influence the cleavage level. Clearly, it is of interest to select the more favourable ones when identifying potential sites in order to improve the chances of success (14). Specifically, the presence of two consecutive pyrimidines in these positions (i.e. –1 and –2) must be avoided as it is detrimental for the cleavage. Based upon these analyses, a list of all potential target sites is generated, and the sequences surrounding the cleavage sites from positions –2 up to at least +27, which includes all of the features important for designing a ribozyme, are extracted and used in the next step.

3.2. Ribozyme Design Module

The second module is directed towards both the design and the pre-selection of ribozymes in silico.

3.2.1. Designing SOFA-
HDV Ribozymes

Over the years, the sequence of the SOFA-HDV ribozyme has been optimized in order to favour a high cleavage level (10, 13). Although the backbone of this RNA molecule is largely conserved, there are three domains that are considered to be variable in terms of the designing of ribozymes: the recognition domain, the biosensor domain and the blocker domain. The recognition domain is complementary to nucleotides +1 to +7 of the target and includes the uridine residue that forms a G–U Wobble base pair (see Fig. 1). In other words, the recognition domain is determined by the cleavage site ($H_{-1}G_{+1}$). Subsequently, the sequence composing the biosensor, which interacts with the target downstream of the recognition domain, must be determined (see Fig. 3). The regions of the target interacting with both the recognition domain and

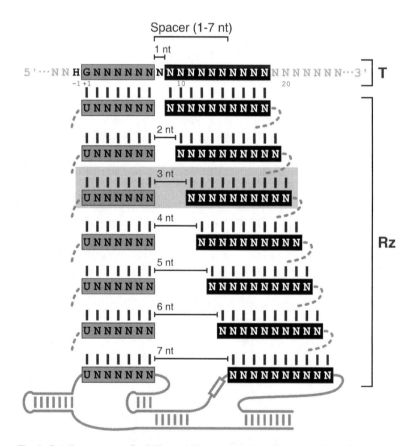

Fig. 3. Details on spacer flexibility and biosensor design. Representation of a potential cleavage site in a target (T), and of the corresponding SOFA-HDV ribozymes (Rz). The length in nucleotides (nt) of each spacer is indicated. The corresponding recognition and biosensor domains are shown below the target RNA. The optimal ribozyme with a spacer of 3 nt is highlighted in a *light grey box*. The sequences of the recognition domain, and of the biosensor binding sites on both the target and the ribozyme, are identified using the same colour code as in Fig. 1 (i.e. the recognition domain is in *black letters* within *grey boxes* and the biosensor is in *white letters* in *black boxes*).

biosensor are separated by a spacer sequence that was included in order to avoid the stacking of both domains, as they are in close proximity to one another. This stacking, should it occur, could be detrimental to the product release (see Figs. 1 and 3). Previous experiments demonstrated that it is preferable to have a spacer of at least 1 nt, but no longer than 7. The optimal length is 3 nt, but anything between 1 and 5 nt works well (see Fig. 3) (13). Therefore, a 3 nt spacer is generally used in the initial design; and consequently, the next 10 nt of the target can be considered as being the region bound by its complementary biosensor. A length of 10 nt for the biosensor has been shown to be optimal for efficient binding without preventing product release and thus conserving the turnover property of the ribozyme. Finally, the sequence of the blocker is determined. It consists of 4 nt that are complementary to the 5′ part of the ribozyme's recognition domain (see Fig. 1) (see Notes 1 and 2 for details).

3.2.2. Removal of Low-Potential SOFA-HDV Ribozymes

The various series of SOFA-HDV ribozymes developed in the past have led to the elucidation of guidelines for the pre-selection of the ribozymes possessing the greatest potential prior to the investment of any time in laboratory experiments.

The length of the blocker sequence that base pairs with the recognition domain has been shown to affect the cleavage activity ((13); unpublished data). The formation of more than 4 bp with the recognition domain may significantly reduce the cleavage activity by either preventing the switch from the *off* to the *on* conformation, or by generating a *cis*-acting ribozyme possessing the ability to self-cleave and therefore reduce the quantity of active SOFA-HDV ribozyme. The additional nucleotides that might be involved in these extra base pairs are in fact those located at the 3′ end of the ribozyme's biosensor. Therefore, it is important to analyze the biosensor's sequence in terms of the possibility of finding such base pairing. When it occurs, the simplest solution is to consider slightly displacing the biosensor's binding sequence by repeating the previous step. If no solution appears to be possible, the given SOFA-HDV ribozyme must be removed from the collection.

Previous experiments have shown that significant complementarity between either the recognition domain, or the biosensor, with the sequence composing stem-loop III may also impair the cleavage activity of a given ribozyme. Specifically, whenever the possibility exists of forming at least six consecutive base pairs, or of any stability equivalent (e.g. 7 bp including a bulge), it is preferable to either remove or redesign the SOFA-HDV ribozyme in question because it will most likely be inactive (unpublished data).

3.2.3. Specificity Analysis

When working towards the development of a gene inactivation system, a key question to ask is how to design target-specific ribozymes that do not produce any side effects. There are several potential sources of these side effects including, for example, the triggering of immunological responses. Moreover, in many cases, the side effects may result from the cleavage of non-desired mRNA species (i.e. off-target effects). In order to evaluate this possibility, the Ribosubstrates (http://www.riboclub.org/ribosubstrates) software was developed. This integrated software searches in selected cDNA databases for all potential targets of a given SOFA-HDV ribozyme (22). These potential targets include not only mRNAs having perfect matches with the catalytic RNA in question, but also for those interactions that include Wobble base pairs and/or mismatches. The results generated permit a rapid selection of the sequences suitable as targets for SOFA-HDV ribozymes. We suggest removing from the list, at a minimum, all ribozymes that would potentially recognize another target. Further selection criteria may also take into consideration the occurrence of any mismatches between the target and the ribozyme's binding domain. The presence of only one mismatch in the recognition domain is significantly more detrimental than the one occurring in the biosensor.

3.3. Ribozyme Screening Module

In this final module, an experimental design using in vitro cleavage assays for selecting the best candidate is presented. In addition, some important elements to be taken into consideration while using SOFA-HDV ribozyme in cellulo are discussed.

3.3.1. In Vitro Cleavage Assay

Preparations of DNA Templates

Most of the time, SOFA-HDV ribozymes are produced by in vitro transcription from PCR-generated DNA templates as described previously (14). Briefly, the PCR-based strategy includes two complementary and overlapping oligonucleotides. The sense primer, namely the SOFA-HDV RzX primer (where the X corresponds to the identification of the specific ribozyme, that is to say the cleavage site's position), is specific for each ribozyme and is composed of a T7 promoter followed by the SOFA-HDV ribozyme's sequence, stopping after loop IV (5'-TAATACGACTCACTATA GGGCCAGCTAGTTT$(N)_{10Bs}(N)_{4Bl}$CAGGGTCCACCTCCTCG CGGT$(N)_{6RD}$TGGGCATCCGTTCGCGG-3', where N represents A, C, G or T, and Bs, Bl and RD indicate the biosensor, the blocker and the recognition domain, respectively). The reverse primer, specifically the Universal SOFA reverse primer, is universal to all ribozymes (5'-CCAGCTAGAAAGGGTCCCTTAGCCATCCGC GAACGGATGCCC-3'). The underlined nucleotides represent the overlapping sequences between the primers. Those two DNA primers are then used in a filling PCR reaction in order to produce the template for the run-off transcription. The target's DNA template is obtained either from a digested plasmid, or through a PCR strategy as described below (see Note 3).

1. Mix 2 µL of both the SOFA-HDV RzX and the SOFA reverse primers (100 µM each) with 10 µL of 10× PCR buffer, 2 µL of

100 mM MgSO$_4$, 2 μL of 10 mM dNTP, 81.5 μL of sterilized deionised water and 0.5 μL of *Pwo* DNA polymerase (2.5 U) in a final reaction volume of 100 μL.

2. Run the PCR for 12 cycles (1 cycle = 45 s at 95°C, 45 s at 55°C and 45 s at 72°C).

3. Transfer the reaction to a tube containing 10 μL of sodium acetate. Add 2.5 volumes of ice-cold 100% ethanol and centrifuge at 16,200×*g* for 25 min. Remove the supernatant and wash the pellet by adding 150 μL of 70% ethanol and centrifuge at 16,200×*g* for 5 min.

4. After removing the supernatant, quickly dry the pellet and then dissolve it in 52 μL of RNase-free water.

Preparation of RNA Ribozymes and Targets

Both the targets and the SOFA-HDV ribozymes are produced by run-off transcription as described below.

1. To the 52 μL of DNA template from above, add 24 μL of 25 mM NTP, 20 μL of 5× transcription buffer, 1 μL of diluted pyrophosphatase (0.01 U), 1 μL of diluted RNaseOUT (20 U) and 2 μL of purified T7 RNA polymerase (10 μg) in a final reaction volume of 100 μL.

2. Incubate at 37°C for 2 h.

3. To each reaction, add 3 μL of DNase RQ1 (3 U) and incubate for 20 min at 37°C.

4. Add 0.5 volumes of both phenol and chloroform, vortex and centrifuge at 16,200×*g* for 10 min. Transfer the upper aqueous phase to a new tube containing 10 μL of sodium acetate (3 M, pH 5.2). Add 2.5 volumes of ice-cold 100% ethanol, vortex and centrifuge at 16,200×*g* for 25 min. Discard the supernatant and wash the pellet by adding 150 μL of 70% ethanol followed by centrifugation at 16,200×*g* for 5 min.

5. After discarding the supernatant, quickly dry the pellet.

6. Dissolve the pellet in 40 μL of RNase-free water. Add 80 μL of loading buffer and fractionate the RNA on a standard 8 M urea denaturing polyacrylamide gel (see Note 4). Visualize the RNA by UV shadowing (see Note 5) and excise the band. Transfer the band to a 1.5-mL tube and add 500 μL of elution buffer. Elute the RNA over night at 4°C on a rotating shaker.

7. Transfer the eluate to a fresh 2-mL tube containing 50 μL of sodium acetate (3 M, pH 5.2). Add 2.5 volumes of ice-cold 100% ethanol and centrifuge at 16,200×*g* for 25 min. Discard the supernatant and wash the pellet by adding 250 μL of 70% ethanol followed by centrifugation at 16,200×*g* for 5 min.

8. Discard the supernatant and quickly dry the pellet.

9. Dissolve the pellet in RNase-free water and quantify the RNA by absorbance at 260 nm.

RNA Target Labelling

Target RNA could either be transcribed in the presence of radiolabeled nucleotides ([α-^{32}P]UTP), or be labelled at either the 5′- or the 3′-end in the presence of either [γ-^{32}P]ATP or [α-^{32}P]Cp, respectively, and then be further purified by denaturing PAGE [10, 14, 23). Generally, the 5′-end labelling protocol described below is used (see also Note 6).

1. Mix 5 μL of 5 μM in vitro transcribed target (25 pmol) with 3 μL of RNase-free water, 1 μL of 10× Antarctic Phosphatase Reaction Buffer and 1 μL of Antarctic Phosphatase (5 U).

2. Incubate the 10 μL dephosphorylation reaction at 37°C for 30 min. To stop the reaction, incubate the reaction at 65°C for 7 min, then store it on ice for 5 min.

3. Transfer 2 μL (5 pmol of dephosphorylated RNA) of the previous reaction to a new tube containing 14 μL of RNase-free water and 2 μL of 10× T4 PNK Reaction Buffer. Add 1 μL of [γ-^{32}P]ATP and 1 μL of diluted T4 PNK (3 U).

4. Incubate the 20 μL reaction at 37°C for 1 h. Add 30 μL of loading buffer and gel purify the 5′-end radiolabeled RNA as described in subheading "Preparation of RNA Ribozymes and Targets". The RNA is visualized by autoradiography, cut out of the gel and eluted overnight with 500 μL of elution buffer at 4°C on a rotating shaker.

5. Add 1 μL of glycogen to the eluate followed by 50 μL of sodium acetate (3 M, pH 5.2) and 2.5 volumes of 100% ethanol. Centrifuge at 16,200×*g* for 25 min and discard the supernatant. Wash the pellet with 150 μL of 70% ethanol followed by centrifugation at 16,200×*g* for 5 min and finally dissolve it in 500 μL of RNase-free water for a final concentration of less than 10 nM.

Cleavage Assays

Usually the cleavage reactions are carried out under single turnover conditions in which the ribozyme concentration greatly exceeds that of the target ([SOFA-HDV Rz]»[S]) (10, 14). The following detailed procedure describes a 10 μL end-point cleavage reaction.

1. Mix 6 μL of RNase-free water with 1 μL of both 500 mM Tris–HCl (pH 7.5) and 100 mM MgCl$_2$ prior to the addition of 1 μL of radiolabeled target as mentioned in subheading "RNA Target Labelling" (see Note 7).

2. Start the reaction by adding 1 μL of 1 μM of SOFA-HDV ribozyme and then incubating at 37°C for 2 h.

3. Stop the reaction by adding 20 μL of loading buffer.

4. The completed reactions are fractionated on an 8 M urea denaturing PAGE gel of the appropriate concentration, and are then exposed to a Phosphoscreen.

5. The activity of each SOFA-HDV ribozyme is expressed as a percentage of the cleaved products over the total quantity of target $(100 \times [\text{cleaved}/\text{cleaved} + \text{uncleaved}])$ as determined using the ImageQuant (Molecular Dynamics) software.

Based on our experience, SOFA-HDV ribozymes that exhibit cleavage percentages over 75% under the conditions described above are considered as being active, and possess the greatest potential for further experiments. Conversely, SOFA-HDV ribozymes that exhibit cleavage activities below 20% are not suitable for targeting experiments. All of the SOFA-HDV ribozymes that exhibit median levels of cleavage activity could give good results in cellulo, and may therefore be considered depending on the desired number of ribozymes to test.

3.3.2. In Cellulo/In Vivo Assays

The next step in the establishment of a fully functional gene inactivation system is to test the selected SOFA-HDV ribozymes either in cellulo or in vivo. One important point to mention is that each case needs to be analyzed individually in order to design a suitable experiment. Depending on both the target and the cellular model, two major aspects should be considered: the appropriate controls and the expression system.

First, two types of controls are recommended in all experiments: a catalytically inactive ribozyme and an irrelevant, active ribozyme. The catalytically inactive ribozyme differentiates between any possible antisense effects and the catalytic cleavage of the target. Such a ribozyme can be produced by mutating either the catalytic cytosine to a guanosine (C76G), and/or the residues of the two GC base pairs that form the I.I pseudoknot to four uridines (see Fig. 1). The resulting SOFA-HDV ribozyme has the ability to bind the target, but does not exhibit cleavage. The other control is an active SOFA-HDV ribozyme that is irrelevant to the target, that is to say the sequence of the recognition domain and/or the biosensor are not complementary to the target. This control allows the measurement of the overall impact of the presence of a SOFA-HDV ribozyme in the cell without cleavage of the target. This control ribozyme should be analyzed with Ribosubstrate to confirm its irrelevancy to not only the target, but also to any important cellular genes. It is crucial that this ribozyme has a minimal impact on the overall life of the cell.

Finally, the selection of the expression system is crucial for the establishment of a gene inactivation system. According to the model of choice, either DNA or RNA transfection, as well as transduction with lentivirus encoding the ribozymes, could be used in in cellulo experiments. When expressing SOFA-HDV ribozymes from a DNA template, different options are available. First, the type of promoter has a significant impact on the results. A tissue-specific RNA polymerase II promoter will favour expression in a specific cell type, while an RNA polymerase III promoter should result in a more efficient transcription that yields a greater amount of SOFA-HDV ribozyme. The cellular localization of the target entails a specific transport of the ribozyme inside the cells. The direct use of any polymerase III promoter without any additional sequence is a good way to produce SOFA-HDV ribozymes that are restricted to the cell's nucleus. For cytoplasmic export of the ribozyme, a 5′ capped and 3′ polyadenylated mRNA-like expression driven by the RNA polymerase II is a good option. It is also possible to use hybrid ribozymes consisting of a fusion between SOFA-HDV ribozyme and an RNA motif that permits an active transport into the cytoplasm. A good example of this is the use of tRNA^{Val} as a leader sequence for SOFA-HDV ribozyme (14). Data from our laboratory suggest that hY RNA are also good candidates for driving the cellular localization of ribozymes (J. Perreault and J.P. Perreault, unpublished data) (see Note 8). Finally, a cocktail of ribozymes targeting two or three different sites may also be used in conjunction to knockdown of a gene during in cellulo assays.

In summary, no universal protocol governs the details of an in cellulo experiment with a ribozyme. Each unique context possesses its own constraints that define the optimal strategy.

4. Notes

1. The biosensor can be elongated up to 12 nt in order to increase the binding affinity without too significant effect on the ribozyme's turnover.

2. The stabilizer can be changed if undesired interactions can occur between the target and the stabilizer. As this part of the SOFA module only plays a structural role, the only limitation is to avoid base pairing with other parts of both the ribozyme and the target.

3. The target's DNA template must include a T7 RNA polymerase promoter upstream of the targeted sequence. If the template is a linearized plasmid, about 5 μg of DNA should be used in a 100 μL run-off transcription reaction.

4. For SOFA-HDV ribozyme purification, in general 8% PAGE is used and the electrophoresis is halted when the xylene cyanol reaches the bottom of the gel. For targets, the concentration of the gel needs to be adapted to their length.

5. In order to visualize RNA using UV shadowing, simply put the gel (covered with laboratory plastic wrap) on a TLC plate (silica gel, Whatman) and expose it to 254-nm UV light.

6. End labelling gives results that are easier to quantify as only two different products need to be quantified on the gel, and

the calculation is therefore more accurate. 5'-end labelling is usually favoured; however, when the target cannot be efficiently labelled in 5'-, 3'-end labelling involving the ligation of [α-^{32}P]Cp is performed. Alternatively, internal labelling by direct incorporation of [α-^{32}P]UTP during the transcription can be performed.

7. When testing a series of SOFA-HDV ribozymes targeting one RNA, a master mix of water, radiolabeled target, Tris–HCl and MgCl$_2$ is prepared in order to favour greater uniformity between the reactions.

8. The use of an RNA polymerase III promoter for the expression of SOFA-HDV ribozymes limits in the variety of sequences that can be used. All ribozymes that contain four or more consecutive uridines should be discarded because this number of uridines forms a transcription termination signal. The three uridines of the linker that are located between the stabilizer and the biosensor increase the possibility of creating a transcription terminator. In this particular case, the linker could be mutated in order to avoid having four consecutive uridines present.

Acknowledgments

We thank Jonathan Perreault and Gilles Boire for information about hY RNA. This work was supported by grants from Canadian Institute of Health Research (CIHR; grant numbers MOP-44002 and EOP-38322) to J.P.P. The RNA group is supported by grants both from CIHR and Université de Sherbrooke. M.V.L. was the recipient of a predoctoral fellowship from the Fonds de Recherche en Santé du Québec. J.P.P. holds the Canada Research Chairs in genomics and catalytic RNAs, and is a member of the Centre de Recherche Clinique Étienne Lebel.

References

1. Bagheri, S., and Kashani-Sabet, M. (2004) Ribozymes in the Age of Molecular Therapeutics. *Curr. Mol. Med.* **4**, 489–506.

2. Schubert, S., and Kurreck, J. (2004) Ribozyme- and Deoxyribozyme-Strategies for Medical Applications. *Curr. Drug Targets.* **5**, 667–681.

3. Asif-Ullah, M., Levesque, M., Robichaud, G., and Perreault, J. P. (2007) Development of Ribozyme-Based Gene-Inactivations; the Example of the Hepatitis Delta Virus Ribozyme. *Curr. Gene Ther.* **7**, 205–216.

4. Tedeschi, L., Lande, C., Cecchettini, A., and Citti, L. (2009) Hammerhead Ribozymes in Therapeutic Target Discovery and Validation. *Drug Discov. Today.* **14**, 776–783.

5. Teixeira, A., Tahiri-Alaoui, A., West, S., Thomas, B., Ramadass, A., Martianov, I., Dye, M., James, W., Proudfoot, N. J., and Akoulitchev, A. (2004) Autocatalytic RNA Cleavage in the Human Beta-Globin Pre-mRNA Promotes Transcription Termination. *Nature.* **432**, 526–530.

6. Salehi-Ashtiani, K., Luptak, A., Litovchick, A., and Szostak, J. W. (2006) A Genomewide Search for Ribozymes Reveals an HDV-Like Sequence in the Human CPEB3 Gene. *Science.* *313*, 1788–1792.

7. Levesque, D., Choufani, S., and Perreault, J. P. (2002) Delta Ribozyme Benefits from a Good Stability in Vitro that Becomes Outstanding in Vivo. *RNA. 8,* 464–477.

8. Bergeron, L. J., Ouellet, J., and Perreault, J. P. (2003) Ribozyme-Based Gene-Inactivation Systems Require a Fine Comprehension of their Substrate Specificities; the Case of Delta Ribozyme. *Curr. Med. Chem. 10*, 2589–2597.

9. Peracchi, A. (2004) Prospects for Antiviral Ribozymes and Deoxyribozymes. *Rev. Med. Virol. 14*, 47–64.

10. Bergeron, L. J., and Perreault, J. P. (2005) Target-Dependent on/off Switch Increases Ribozyme Fidelity. *Nucleic Acids Res. 33*, 1240–1248.

11. Bergeron, L. J., Reymond, C., and Perreault, J. P. (2005) Functional Characterization of the SOFA Delta Ribozyme. *RNA. 11*, 1858–1868.

12. Fiola, K., Perreault, J. P., and Cousineau, B. (2006) Gene Targeting in the Gram-Positive Bacterium *Lactococcus lactis*, using various Delta Ribozymes. *Appl. Environ. Microbiol. 72*, 869–879.

13. Bergeron, L. J., Reymond, C., and Perreault, J. P. (2005) Functional Characterization of the SOFA Delta Ribozyme. *RNA. 11*, 1858–1868.

14. Levesque, M. V., Levesque, D., Briere, F. P., and Perreault, J. P. (2010) Investigating a New Generation of Ribozymes in Order to Target HCV. *PLoS One. 5*, e9627.

15. Robichaud, G. A., Perreault, J. P., and Ouellette, R. J. (2008) Development of an Isoform-Specific Gene Suppression System: The Study of the Human Pax-5B Transcriptional Element. *Nucleic Acids Res. 36*, 4609–4620.

16. von Eije, K. J., ter Brake, O., and Berkhout, B. (2008) Human Immunodeficiency Virus Type 1 Escape is Restricted when Conserved Genome Sequences are Targeted by RNA Interference. *J. Virol. 82*, 2895–2903.

17. Haasnoot, J., Westerhout, E. M., and Berkhout, B. (2007) RNA Interference Against Viruses: Strike and Counterstrike. *Nat. Biotechnol. 25*, 1435–1443.

18. Amarzguioui, M., Brede, G., Babaie, E., Grotli, M., Sproat, B., and Prydz, H. (2000) Secondary Structure Prediction and in Vitro Accessibility of mRNA as Tools in the Selection of Target Sites for Ribozymes. *Nucleic Acids Res. 28*, 4113–4124.

19. Ryu, K. J., and Lee, S. W. (2004) Comparative Analysis of Intracellular Trans-Splicing Ribozyme Activity Against Hepatitis C Virus Internal Ribosome Entry Site. *J. Microbiol. 42*, 361–364.

20. Doran, G., and Sohail, M. (2006) Systematic Analysis of the Role of Target Site Accessibility in the Activity of DNA Enzymes. *J. RNAi Gene Silencing. 2*, 205–214.

21. Deschenes, P., Lafontaine, D. A., Charland, S., and Perreault, J. P. (2000) Nucleotides -1 to -4 of Hepatitis Delta Ribozyme Substrate Increase the Specificity of Ribozyme Cleavage. *Antisense Nucleic Acid Drug Dev. 10*, 53–61.

22. Lucier, J. F., Bergeron, L. J., Briere, F. P., Ouellette, R., Elela, S. A., and Perreault, J. P. (2006) RiboSubstrates: A Web Application Addressing the Cleavage Specificities of Ribozymes in Designated Genomes. *BMC Bioinformatics. 7*, 480.

23. Ouellet, J., and Perreault, J. P. (2004) Cross-Linking Experiments Reveal the Presence of Novel Structural Features between a Hepatitis Delta Virus Ribozyme and its Substrate. *RNA. 10*, 1059–1072.

Chapter 24

Ribozyme-Mediated Trans Insertion-Splicing into Target RNAs

P. Patrick Dotson II, Jonathan Hart, Christopher Noe, and Stephen M. Testa

Abstract

The trans insertion-splicing (TIS) reaction is a technique that can be used to site-specifically insert an RNA donor substrate into a separate RNA acceptor substrate. The TIS reaction, which is catalyzed by a group I intron-derived ribozyme from *Pneumocystis carinii*, is described with regards to system design, ribozyme preparation, and the overall protocol for conducting the TIS reaction.

Key words: Trans insertion-splicing, *Pneumocystis carinii*, Ribozyme, Group I intron

1. Introduction

Group I intron-derived ribozymes have previously been shown to be capable of catalyzing a wide array of enzymatic reactions using RNA substrates (1). One such reaction, termed trans insertion-splicing (TIS), has been developed for the site-specific insertion of a segment of RNA into a central region of a different RNA molecule (see Fig. 1) (2). The TIS reaction, which is catalyzed by a group I intron-derived ribozyme (see Fig. 2) from *Pneumocystis carinii* (see Note 1) has been shown to be capable of inserting modified RNA oligonucleotides in addition to non-modified substrates (2, 3). Modifications (3) have included substitutions to the phosphodiester backbone (phosphorothioate), to the base (2-aminopurine and 4-thiouridine), and to the sugar (deoxy and methoxy) (see Note 2).

The TIS reaction consists of three concerted chemical steps, schematically represented in Fig. 3 (2). Two substrates are utilized;

Jörg S. Hartig (ed.), *Ribozymes: Methods and Protocols*, Methods in Molecular Biology, vol. 848,
DOI 10.1007/978-1-61779-545-9_24, © Springer Science+Business Media, LLC 2012

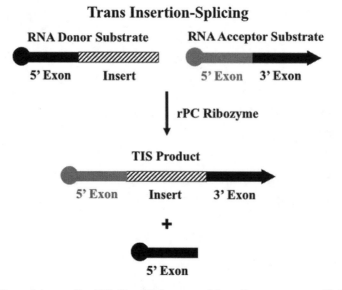

Fig. 1. The trans insertion-splicing reaction (TIS). The rPC ribozyme catalyzes the sequence-specific insertion of a portion of an RNA donor substrate into a predefined region of an RNA acceptor substrate.

one termed the RNA donor substrate (which contains the insert) and the other termed the RNA acceptor substrate (see Fig. 1). The two substrates bind to specific recognition elements (REs) within the *P. carinii* ribozyme (see Fig. 3). The sequence of the individual recognition elements are easily modified at the plasmid level by simple site-directed mutagenesis techniques, allowing the targeting of virtually any RNA substrate sequence (see Note 3).

In this chapter, we outline the steps for preparing the *P. carinii* rPC ribozyme, as well as the protocol for conducting the TIS reaction. Both non-modified and modified RNA oligonucleotides are shown (see Table 1; Fig. 4) to demonstrate the ability of the ribozyme to utilize modified substrates. In the future, we envision the TIS reaction as a potentially powerful biochemical tool for inserting RNA oligonucleotides into full-length RNA transcripts (see Note 4).

2. Materials

2.1. Oligonucleotide Synthesis and Preparation

1. RNA oligonucleotides (e.g., those shown in Table 1) (Dharmacon RNAi Technologies; Lafayette, CO).
2. Beckman DU 650 spectrophotometer (Beckman Coulter, Inc.; Fullerton, CA).

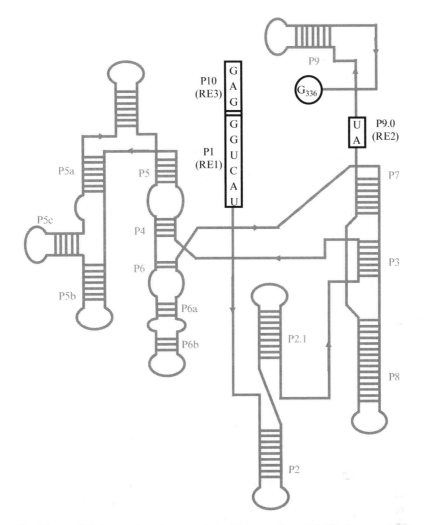

Fig. 2. Schematic of the predicted secondary structure of the *Pneumocystis carinii* rPC ribozyme. Positions within the ribozyme that constitute the recognition elements (RE1, RE2, and RE3) are *boxed*, and they are involved in P1, P10, and P9.0 helix formation. These are the sequences that can be changed, via site-directed mutagenesis, to alter the sequence specificity of the ribozyme. The terminal guanosine (G336), which is always required for the first reaction step, is *circled*.

2.2. Radiolabeling Oligonucleotide Substrates (4)

1. Adenosine 5′-triphosphate, [γ-^{32}P]-10 mM 250 μCi/mmol in 10 mM tricine buffer, pH 7.6 (PerkinElmer NEN®). Store at –20°C.

2. T4 Polynucleotide Kinase (New England Biolabs; Ipswich, MA). Supplied with 10× reaction buffer. 1× reaction buffer: 70 mM Tris–HCl (pH 7.6), 10 mM MgCl$_2$, and 5 mM dithiothreitol. Store at –20°C.

3. Glycerol.

4. 20% Acrylamide/bisacrylamide gel (29:1) (affects central and peripheral nervous systems and reproductive system if swallowed,

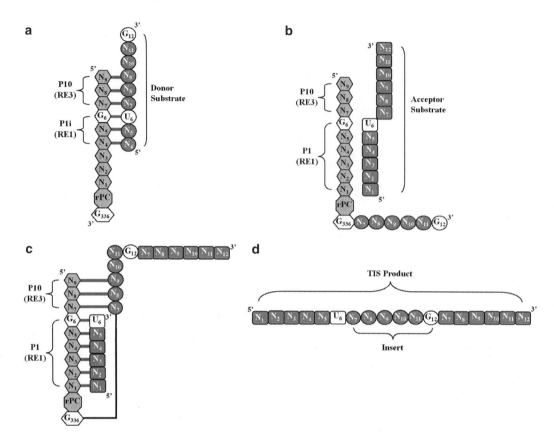

Fig. 3. Sequence requirements for the TIS reaction. (**a**) Binding of the RNA donor substrate (9-mer) via P1i (RE1) and P10 (RE3) helix formation prior to the first chemical step. (**b**) Binding of the RNA acceptor substrate via P1 (RE1) helix formation prior to the second chemical step. (**c**) Binding of the RNA donor and acceptor substrates via P1 (RE1) and P10 (RE3) helix formation prior to the third chemical step. (**d**) Sequence of the final TIS product. Specific nucleotide abbreviations (G and U) are used in positions where those specific nucleotides are required. For the ribozyme, only the nucleotides that encompass the REs and the terminal guanosine (G336) are shown. The RNA donor nucleotides are shown within *circles* and those in the acceptor are shown within *squares*. The designation N means that any base is suitable in the particular position, provided that it can base pair when necessary.

Table 1
Sequence of TIS substrates and products

RNA acceptor substrate (12-mer)	RNA donor substrate (9-mer)	TIS product (18-mer)
5′-AUGACUAAACAU-3′	5′-GCUCUCGUG-3′	5′-AUGACUCUCGUGAAACAU-3′
5′-AUGACUAAACAU-3′	5′-GCUCUC(dG)UG-3′	5′-AUGACUCUC(dG)UGAAACAU-3′

Sequences highlighted in *gray* represent the insert region of the donor substrate

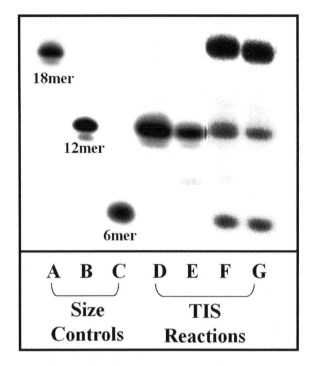

Fig. 4. Representative gel of the TIS reaction. *Lanes A–C* contain the 5′-end labeled size controls; 18-mer (TIS product, 5′-AUGACUCUCGUGAAACAU-3′), 12-mer (RNA acceptor substrate, 5′-AUGACUAAACAU-3′), and 6-mer (Intermediate, 5′-AUGACU-3′). *Lanes D, E* contain negative control reaction conditions consisting of either 0 nM ribozyme (*Lane D*) or HOMg (no magnesium, *Lane E*). *Lane F* contains the TIS reaction conducted with an unmodified donor substrate (5′-GCUCUCGUG-3′). *Lane G* contains a TIS reaction conducted with a modified donor substrate (5′-GCUCUCdGUG-3′, *dG* denotes a deoxyguanosine substitution).

inhaled, or absorbed through skin). Casting size for the polyacrylamide gel is 45 × 35 cm.

5. Running buffer (0.25× TBE). 1× Tris-Borate-EDTA buffer: 90 mM Tris (tris(hydroxymethyl)aminomethane), 90 mM boric acid, and 2 mM EDTA (ethylenediamineteraacetic acid), pH 8.4. Store at 20°C.

2.3. Ribozyme Synthesis

1. PC plasmid, which contains the sequence for the *P. carinii* rPC ribozyme (4).

2. *Xba*I (Invitrogen, Carlsbad, CA). Supplied with 10× REact® 2 buffer. 1× reaction buffer: 50 mM Tris–HCl (pH 8.0), 10 mM $MgCl_2$, and 50 mM NaCl. Store at –20°C.

3. rNTPs (Promega, Madison, WI). Each supplied at 10 mM.

4. T7 RNA polymerase (New England Biolabs). Supplied with 10× reaction buffer. 1× reaction buffer: 40 mM Tris–HCl (pH 7.9), 10 mM $MgCl_2$, 5 mM DTT, and 5 mM spermidine. Store at –20°C.

5. QIAquick PCR purification kit (Qiagen; Valencia, CA).

6. Qiagen Plasmid Midi Kit.

7. Beckman DU 650 spectrophotometer (Beckman Coulter, Inc).

2.4. Ribozyme Purification (4, 5)

1. QIAGEN-tip 100 anion-exchange columns (Qiagen).

2. Buffer QBT: 750 mM NaCl, 50 mM MOPS (pH 7.0), 15% ethanol, and 0.15% Triton X-100. Store at 4°C.

3. Buffer QC: 1.0 M NaCl, 50 mM MOPS (pH 7.0), and 15% ethanol. Store at 4°C.

4. 2-Propanol (isopropanol).

5. Ethanol.

6. Beckman DU 650 spectrophotometer (Beckman Coulter, Inc).

2.5. TIS Reactions

1. H10Mg buffer: 50 mM HEPES (25 mM Na$^+$), 135 mM KCl, and 10 mM MgCl$_2$, pH 7.5 (see Note 5).

2. 2× Stop buffer: 10 M urea, 3 mM EDTA, and 0.1× TBE.

3. 12% Polyacrylamide/8 M urea gel (casting size: 45×35 cm).

4. Phosphorimager (see Note 6).

3. Methods

3.1. Oligonucleotide Synthesis and Preparation

1. RNA oligonucleotides are deprotected using the manufacturer's recommended protocol (see below).

2. Nucleic acid concentrations are calculated (see Note 7) from UV-Vis absorption measurements using a Beckman DU 650 spectrophotometer.

3. Oligonucleotides are 5′-end radiolabeled (see Note 8) with [γ-^{32}P] ATP (see below).

3.2. RNA Deprotection Protocol

1. Briefly centrifuge the tubes provided.

2. Add 400 μL of 2′-Deprotection Buffer to each tube of RNA.

3. Completely dissolve RNA pellet by pipetting up and down.

4. Vortex for 10 s and centrifuge for 10 s.

5. Incubate at 60°C for 30 min.

6. Lyophilize or SpeedVac to dryness before use.

7. Dissolve the RNA in water or an appropriate buffer (see Note 9).

3.3. 5′-Radiolabeling of Oligonucleotides

1. Mix 6.4 pmol of the RNA to be radiolabeled with 27 pmol of [γ-^{32}P] ATP in 50 mM Tris (pH 7.5), 10 mM MgCl$_2$, 5 mM dithiothreitol, 0.1 mM spermidine, and 0.1 mM Na$_2$EDTA. Incubate the resulting solution with 20 units of T4 polynucleotide kinase in a reaction volume of 20 μL for 30 min at 37°C.

2. Separate products on a 20% native non-denaturing polyacrylamide gel (see Note 10).

3. First run the 20% polyacrylamide gel at 1,500 V for 2 min to ensure products enter the polyacrylamide matrix quickly.

4. Continue running the polyacrylamide gel at 1,000 V for 2 h or until the orange-G dye reaches the middle of the gel.

5. Remove the top (or bottom) glass plate from the gel sandwich and cover the gel with saran wrap.

6. Expose the polyacrylamide gel to photographic film in a dark room (see Note 11).

7. Develop the film, preferably with an automated film developer.

8. Excise the 5′-end labeled product from the polyacrylamide gel using a microscope glass cover slide. Place the excised gel slice in a 1.5-mL microcentrifuge tube.

9. Add 1 mL of RNAse-Free H$_2$O to the tube and let sit overnight at –70°C. Centrifuge the gel pieces to the bottom of the tube and siphon off the solution on top (which contains most of the radiolabeled product). Put the solution in a new tube (see Note 12).

10. Evaporate to a final oligonucleotide concentration of 8 nM (see Note 13).

3.4. Ribozyme Synthesis and Preparation

1. Linearize the PC plasmid in a 50 μL reaction mixture consisting of 16 μg plasmid (4), 5 μL of 10× REACT 2 buffer, and 50 units of XbaI at 37°C for 2 h.

2. Purify the linearized plasmid using a QIAquick PCR purification kit using the manufacturer's recommended protocols.

3. Transcribe the linearized plasmid by incubating 1 μg of plasmid, 50 units of T7 RNA polymerase, and 1 mM rNTP in 1× reaction buffer for a total volume of 50 μL for 2 h at 37°C.

4. Purify the transcribed RNA (called rPC) using a single QIAGEN-tip 100 anion-exchange column.

5. The column is first equilibrated with 4.0 mL of Buffer QBT.

6. The transcription reaction is then loaded onto the column, and the column washed with 7.0 mL of Buffer QBT.

7. The transcript is eluted using 4.0 mL of Buffer QC.

8. Following a 2-propanol precipitation and then an ethanol precipitation, the ribozyme sample is dried under vacuum and then redissolved in water (typically 20 µL ddH$_2$O).

9. The resultant ribozyme concentration is then quantified using a Beckman UV-Vis DU 650 spectrophotometer using the extinction coefficient of 3.5×10^6/M/cm for the rPC ribozyme.

10. Store ribozyme stocks at –20°C (see Note 14).

3.5. TIS Reactions

1. TIS reactions are conducted using the rPC ribozyme under previously optimized reaction conditions (2).

2. 240 nM of the Ribozyme is pre-annealed in 1× H10Mg buffer (a total reaction volume of 5.0 µL) at 60°C for 5 min. The reaction is then slow cooled to 44°C for 2 min (see Note 15).

3. In a separate tube, preincubate 6 nM (5′-end radiolabeled) acceptor substrate with 30 µM donor substrate in 1× H10Mg at 44°C.

4. The TIS reaction is then initiated by the addition of 1.0 µL of the substrate solution to the 5.0 µL ribozyme solution. Final nucleic acid concentrations are 200 nM rPC ribozyme, 1.0 µM donor substrate, and 1.0 nM acceptor substrate.

5. Reactions incubate for 2 h at 44°C (see Note 16).

6. Terminate the reaction by the addition of an equal volume of 2× stop buffer (see Note 17).

7. The reaction components are then denatured for 1 min at 90°C (see Note 18) and separated on a 12% polyacrylamide/8 M urea gel.

8. First run the 12% polyacrylamide gel at 1,500 V for 2 min.

9. Continue running the polyacrylamide gel for 2 h at 1,000 V or until the orange-G dye reaches 5 cm from the bottom of the gel.

10. Transfer the gel to chromatography paper and dry under vacuum.

11. Visualize and quantify the radioactive bands on a Phosphorimager (see Fig. 4 for a typical example).

4. Notes

1. It is expected that other group I intron-derived ribozymes, other than the *P. carinii* rPC ribozyme, are capable of catalyzing the TIS reaction. Each individual group I intron-derived ribozyme would have to be constructed to ensure that the ribozyme has

a 3′ terminal guanosine. The TIS reaction parameters would have to be optimized, including reaction time, magnesium concentration, RNA donor substrate concentration, temperature, pH, and ribozyme concentration.

2. We expect that many other types of modifications are amenable to the TIS reaction.

3. For problematic TIS reactions, the recognition elements of the ribozyme can also be lengthened to strengthen the binding of the substrates to the ribozyme.

4. This could be useful for real time in vivo tagging and/or site-directed mutagenesis (via insertion mutations) of cellular transcripts, which is possible because the ribozyme is inducible.

5. This buffer can be made as a 10× stock. It can be stored for about 6 months at –20°C.

6. We use a Molecular Dynamics Storm 860 phosphorimager, but any phosphorimager will work.

7. Nucleic Acid concentrations are calculated based on absorbance values using an in-house program called STread.

8. The RNA acceptor substrate can be radiolabeled at either the 5′ or 3′ end of the molecule (6), which is useful for determining TIS product yields and tracking the progress of the individual reaction steps. Of course, the TIS reaction can be conducted without a radiolabel.

9. If conducting experiments utilizing varying buffer components, it is helpful to dissolve the RNA in RNAse-free water. However, the ribozyme will lose activity within 3–6 months. Therefore, it is important to test the activity of the ribozyme regularly.

10. Glycerol should be added to the sample immediately before loading so that it sinks in the well. The final concentration of glycerol can be as low as 5% v/v.

11. The gel can be pushed into a corner such that it is immovable. A hole can be punched in the film for orientation, and then placed on the gel for exposure. After developing the film, the film can be placed back over the gel, in the same orientation, such that the positions of the bands in the gel are known.

12. Radiation is hazardous material, and all safety precautions dictated by your governing entity should be meticulously followed.

13. The concentration can be estimated based on the level of radioactivity.

14. Ribozyme stocks should not be kept over 6 months due to loss of activity. It is best to thaw ribozyme stocks as infrequently as possible.

15. Many annealing protocols were analyzed, and slow cooling of the ribozyme to 44°C after preincubation at 60°C is optimal for obtaining properly folded catalytically active ribozyme.

16. The small volumes in these reactions are prone to evaporation at 44°C. Therefore, the microcentrifuge tubes are sunk in a water bath by putting a flask weight on a floating microcentrifuge rack.

17. The stop buffer is dense enough for the sample to sink to the bottom of the well in PAGE gels.

18. Denaturation of the TIS reaction mixture at 90°C is critical for release of all bound RNA species from the ribozyme. Otherwise, the radiolabel gets stuck in the loading well with the ribozyme.

Acknowledgments

This work was supported by grants from the Kentucky Lung Cancer Research Program and the Muscular Dystrophy Association.

References

1. Dotson II PP, Testa SM (2006) Group I intron-derived ribozyme recombination reactions. *Recent Dev Nucleic Acids Res* **2**, 307–324.

2. Johnson AK, Sinha J, Testa SM (2005) *Trans* insertion-splicing: Ribozyme-catalyzed insertion of targeted sequences into RNAs. *Biochemistry* **44**, 10702–10710.

3. Dotson PP II, Frommeyer KN, Testa SM (2008) Ribozyme mediated *trans* insertion-splicing of modified oligonucleotides into RNA. *Archives of Biochemistry and Biophysics* **478**, 81–84.

4. Testa SM, Haidaris CG, Gigliotti F, Turner DH (1997) A *Pneumocystis carinii* group I intron ribozyme that does not require 2' OH groups in its 5' exon mimic for binding to the catalytic core. Biochemistry **36**,15303–15314.

5. Bell MA, Johnson AK, Testa SM (2002) Ribozyme-catalyzed excision of targeted sequences from within RNA. *Biochemistry* **41**,15327–15333.

6. Dotson PP, Sinha, J., and Testa, S. M. (2008) A *Pneumocystis carinii* Group I Intron-Derived Ribozyme Utilizes an Endogenous Guanosine as the First Reaction Step Nucleophile in the *Trans* Excision-Splicing Reaction. *Biochemistry* **47**, 4780–4787.

Chapter 25

Developing Fluorogenic RNA-Cleaving DNAzymes for Biosensing Applications

M. Monsur Ali, Sergio D. Aguirre, Wendy W.K. Mok, and Yingfu Li

Abstract

Deoxyribozymes (or DNAzymes) are single-stranded DNA molecules that have the ability to catalyze a chemical reaction. Currently, DNAzymes have to be isolated from random-sequence DNA libraries by a process known as in vitro selection (IVS) because no naturally occurring DNAzyme has been discovered. Several IVS studies have led to the isolation of many RNA-cleaving DNAzymes (RNase DNAzymes), which catalyze the transesterification of a phosphodiester linkage in an RNA substrate, resulting in its cleavage. An RNase DNAzyme and its substrate can be modified with a pair of donor and acceptor fluorophores (or a fluorophore and quencher pair) to create a fluorescence-signaling system (a signaling DNAzyme) where the RNA-cleaving activity of the DNAzyme is reported through the generation of a fluorescent signal. A signaling DNAzyme can be further coupled with an aptamer (a target-binding nucleic acid sequence) to generate a fluorogenic aptazyme in which the aptamer–target interaction confers an allosteric control of the coupled RNA-cleaving and fluorescence-signaling activity of the DNAzyme. Fluorogenic aptazymes can be exploited as valuable molecular tools for biosensing applications. In this chapter, we provide both a detailed description of methods for isolation of signaling DNAzymes by IVS and general approaches for rational engineering of fluorogenic aptazymes for target detection.

Key words: DNAzyme, In vitro selection, Fluorescence signaling, Biosensing, Aptazyme, Aptamer, ATP

1. Introduction

Enzymes are highly efficient and selective catalysts that play a critical role in biological systems. They have also been extensively explored as molecular tools for wide-ranging applications. Although most enzymes are protein-based, there exists a small subset of enzymes that consist solely of nucleic acids. The discovery of naturally occurring catalytic RNAs (ribozymes) (1, 2) has prompted the conceptualization and discovery of synthetic enzymes made from DNA (3–7). These catalytic DNA molecules have been coined

Jörg S. Hartig (ed.), *Ribozymes: Methods and Protocols*, Methods in Molecular Biology, vol. 848,
DOI 10.1007/978-1-61779-545-9_25, © Springer Science+Business Media, LLC 2012

deoxyribozymes or simply DNAzymes. DNAzymes can be obtained in a test tube by subjecting a synthetic random-sequence DNA library to an iterative process known as in vitro selection (IVS) (3, 8–10). Since DNA is a more stable polymer than RNA (11) and protein (12), and can be conveniently produced by automated solid-phase chemical synthesis, DNAzymes have been coveted for many potential applications (13–16).

To date, a large number of DNAzymes have been isolated and many of them catalyze chemical reactions involving nucleic acid substrates, such as cleavage of RNA and DNA phosphorylation (13, 17, 18). Among these DNAzymes, RNA-cleaving DNAzymes (denoted "RNase DNAzymes" in this chapter) represent a very diverse class. Several RNase DNAzymes have been extensively characterized (18, 19).

One area of DNAzyme research is the exploitation of DNAzymes for biosensing applications (14, 15). To engineer a DNAzyme-based biosensor, a reporting mechanism needs to be coupled to the catalytic action of the DNAzyme. Optical detection by fluorescence spectroscopy is a widely used technique due to its detection sensitivity, the availability of many fluorescent dyes and quenchers, and the capability for multiplex detection. Because DNA can be easily modified with fluorophores and quenchers, engineering fluorogenic DNAzymes for biosensing is of significant interest. Our group and others have been conducting research toward developing fluorogenic DNAzymes with RNA-cleaving activities (20–27).

The choice of fluorophores and quenchers, their modification locations, and metal-ion cofactors can all contribute to the catalytic and signaling performance of a fluorescent DNAzyme (26, 28). Our laboratory has developed a fluorescent DNAzyme system where a DNAzyme cleaves a single RNA linkage embedded in a DNA chain and the RNA site is immediately flanked by a fluorophore-containing nucleotide and a quencher-labeled nucleotide. By this design, the RNA-cleaving activity of the DNAzyme is synchronized with the generation of a fluorescent signal. For the convenience of description, in this chapter we will refer to such DNAzymes as "signaling DNAzymes". To date, we have isolated and characterized a large group of signaling DNAzymes with various catalytic efficiencies, metal-ion and pH dependencies, and structural properties (25, 26, 29–36). We have also shown that signaling DNAzymes can be further engineered into signaling aptazymes for target detection (26, 33, 37, 38).

In this chapter, we will describe a detailed method for isolating signaling DNAzymes by IVS. Moreover, we will discuss general approaches for designing signaling aptazymes for target detection.

2. Materials and Equipment

2.1. Synthetic DNA Oligonucleotides

1. *Random DNA library*—A DNA library to be used for IVS of signaling DNAzymes may contain 40–70 random nucleotides (nt) in the middle of the sequence, flanked by two primer-binding sites used for polymerase chain reaction (PCR)—an essential technique in an IVS experiment. A 98-nt DNA library used in some of our IVS experiments, denoted here as "DL1," has the sequence of 5′-CACGGAT CCTGACAAG-(N_{70})-CAGCTCCGTCCG-3′ (N = an equimolar mixture of A, C, G, and T). This library was purchased from Keck Oligo Synthesis Facilities (Yale University, New Haven, CT, USA), purified in house by 10% denaturing polyacrylamide gel electrophoresis (10% dPAGE; the protocols for purification are provided in Subheading 3), dissolved in double-deionized water (ddH_2O), quantified by UV spectroscopy, and stored at –20°C.

2. *DNA primers*—For DL1, the following three primers are routinely used by us for the amplification step in each selection cycle: forward primer 1 (FP1), 5′-CACGGATCCTGACAAG-3′; reverse primer 1 (RP1), 5′-CGGACGGAGCTG-3′; and RP2, 5′-A_{20}-S9-CGGACGGAGCTG-3′ (S9 is a triethylene glycol linker, which can be incorporated into a DNA oligonucleotide using "Spacer 9" amidite from Glen Research, Cat. No.10-1909). These oligonucleotides were purchased from Integrated DNA Technologies (IDT, Coralville, IA, USA), purified by 10% dPAGE, dissolved in ddH_2O, quantified, and stored at –20°C.

3. *Fluorogenic substrate*—The featured fluorogenic substrate, denoted here as "FS1," contains 23 nt and has a sequence of 5′-GATGTGTCCGTGC-F-R-Q-GGTTCGA-3′ (F, R, and Q represent a fluorescein-labeled dT, an adenine ribonucleotide, and a DABCYL-labeled dT, respectively). It can be purchased from Keck Oligo Facilities. F, R, and Q are incorporated into FS1 using 5′-dimethoxytrityloxy-5-(N-((3′,6′-dipivaloylfluoresceinyl)-aminohexyl)-3-acrylimido)-2′-deoxyuridine-3′-((2-cyanoethyl)-(N,N-diisopropyl))-phosphoramidite (from Glen Research, catalogue number: 10-1056), 5′-dimethoxytrityl-N-benzoyl-adenosine 2′-O-TBDMS-3′-((2-cyanoethyl)-(N,N-diisopropyl))-phosphoramidite (Glen Research catalogue number: 10-3003), and 5′-dimethoxytrityloxy-5-((N-4′-carboxy-4-(dimethylamino)-azobenzene)-aminohexyl-3-acrylimido)-2′-deoxyuridine-3′-((2-cyanoethyl)-(N,N-diisopropyl))-phosphoramidite (Glen Research catalogue number: 10-1058). Its purification is described in Subheading 3.

4. *Ligation template*—DLT1 is a synthetic DNA oligonucleotide designed to facilitate the ligation of FS1 to DL1. Its sequence is 5'-GTCAGGATCCGTGTCGAACCATAGC-3'. It was obtained from IDT, purified by 10% dPAGE, dissolved in ddH$_2$O, quantified, and stored at –20°C.

2.2. Chemical and Biochemical Reagents

1. 100 mM ATP solution. Store at –20°C.

2. *(γ-^{32}P)ATP*—Store at 4°C in a secured beta box. (Caution! Radioisotopes are highly hazardous and can cause radiation burn. They should be used carefully behind a beta shield. Lab coat and gloves should be worn.)

3. *Acrylamide/bis-acrylamide (29:1) mixture*—Purchased as 40% solution. Store at 4°C.

4. *Ammonium persulfate (APS)*—10% stock solution is prepared by dissolving 5 g of power in 50 mL of ddH$_2$O. Store at 4°C.

5. *Boric acid* (Biotechnology grade)—Store at room temperature.

6. *Bromophenol blue* (ACS grade)—Store at room temperature.

7. *Cadmium chloride* (ACS grade)—1M stock solution is prepared, autoclaved, and stored at room temperature.

8. *Cobalt chloride* (ACS grade)—1M stock solution is prepared, autoclaved, and stored at room temperature.

9. *(α-^{32}P)dGTP*—Store at 4°C in a secured beta box. (Caution! See item 2 above)

10. *dNTP stock solution*—The purchased set consists of 10 mM of each of dATP, dCTP, dTTP, and dGTP (see Note 1).

11. *EDTA* (Biotechnology grade)—EDTA stands for ethylenediaminetetraacetic acid. 0.5M stock solution (pH 8.0) is prepared as follows: Add 186.1 g of EDTA in 800 mL ddH$_2$O. Adjust the pH of the solution to 8.0 using NaOH pellets (Caution! NaOH is highly corrosive and should be handled with care). Bring the final volume to 1 L with ddH$_2$O. Autoclave and store at 4°C.

12. *HEPES* (Biotechnology grade)—1M stock solution is prepared as follows: Dissolve 23.83 g of HEPES in 60 mL of ddH$_2$O and adjust the solution pH to 7.5 using 1M NaOH. Make up the volume to 100 mL with ddH$_2$O, autoclave, and store at 4°C.

13. *InSTAclone™ PCR Cloning Kit* (MBI Fermentas)—The kit contains a TA cloning vector (pTZ57R/T), 5× ligation buffer, T4 DNA ligase (5 units/μL), and TransformAid bacterial transformation system. Any strain of *Escherichia coli* can be utilized since the kit provides chemical solutions for the preparation of chemically competent *E. coli* cells. We typically use *E. coli* XL1-Blue cells.

14. *LB (Luria Bertani) Medium*—Dissolve 12.5 g of LB powder in 500 mL of ddH$_2$O. Autoclave and store at room temperature.

15. *Magnesium chloride* (ACS grade)—1M stock solution is prepared, autoclaved, and stored at room temperature.

16. *Manganese chloride* (ACS grade)—1M stock solution is prepared, autoclaved, and stored at room temperature.

17. *N,N,N,N'-tetramethyl-ethylenediamine* (TEMED)—Store at room temperature. (Caution! This chemical is harmful if inhaled and should be handled in a fume hood.)

18. *Nickel chloride* (ACS grade)—1M stock solution is prepared, autoclaved, and stored at room temperature.

19. *Sodium chloride* (ACS grade)—5M stock solution is prepared, autoclaved, and stored at room temperature.

20. *Sodium dodecyl sulfate* (Electrophoresis grade)—Purchased as 10% solution. Store at 4°C.

21. *Sucrose* (Ultrapure grade)—Store at room temperature.

22. *T4 DNA ligase* (MBI Fermentas)—5 units/μL. The corresponding 10× reaction buffer is also supplied by the vendor. At 1× concentration, the buffer contains 40 mM Tris–HCl, 10 mM MgCl$_2$, 10 mM DTT, and 0.5 mM ATP (pH 7.8). Store at –20°C.

23. *T4 polynucleotide kinase* (PNK) (MBI Fermentas)—10 units/μL. The corresponding 10× reaction buffer A is also supplied by the vendor. At 1× concentration, the buffer contains 50 mM Tris–HCl (pH 7.6), 10 mM MgCl$_2$, 5 mM DTT, and 0.1 mM spermidine. Store at –20°C.

24. *Taq DNA polymerase* (Biotool B&M Labs, Madrid, Spain)—5 units/μL. The corresponding 10× reaction buffer is also supplied by the vendor. At 1× concentration, the buffer contains 75 mM of Tris–HCl (pH 9.0), 2 mM of MgCl$_2$, 50 mM of KCl, and 20 mM of (NH$_4$)$_2$SO$_4$.

25. *Tetra-butylammonium fluoride* (TBAF)—Purchased as 1.0M solution in THF. Store at 4°C.

26. *Tris base* (Ultrapure grade)—1M Tris–HCl (pH 7.5) is prepared as follows: Weigh 12.11 g of Tris base in a 200-mL glass beaker. Dissolve in 60 mL of ddH$_2$O by stirring with a magnetic stirrer. When completely dissolved, adjust the pH to 7.5 using 1M HCl. Make up the volume to 100 mL with ddH$_2$O, transfer to a glass bottle, and autoclave. Store at 4°C

27. *Urea* (Ultrapure grade)—Store at room temperature.

28. *Xylene cyanol FF*—Store at room temperature.

2.3. Specially Prepared Reagents

1. *10× TBE solution* (89 mM Tris, 89 mM boric acid, 2 mM EDTA, pH 7.5)—Mix 432 g of Tris base, 220 g of boric acid, 80 mL of 0.5M EDTA (pH 8.0), and add ddH$_2$O to a final volume of 4 L. Thoroughly mix the solution with a magnetic stirrer bar. Autoclave and store at 4°C.

2. *10% denaturing polyacrylamide gel stock*—Mix 1681.7 g of urea, 400 mL of 10× TBE, and 1 L of 40% acrylamide/bis-acrylamide (29:1) mixture. Add ddH$_2$O to 4 L and dissolve with mild heating and stirring. Store in amber bottles at 4°C. (Caution! Acrylamide is a neurotoxin prior to polymerization and thus, care should be taken to avoid contact with the body—gloves, mask, goggles, and lab coat should be worn when handling this chemical.)

3. *2× Gel loading buffer (2× GLB)*—Mix 44 g of urea, 8 g of sucrose, 10 mg of bromophenol blue, 10 mg of xylene cyanol FF, 400 µL of 10% sodium dodecyl sulfate, and 4 mL of 10× TBE. Adjust the final volume to 40 mL with ddH$_2$O and dissolve with mild heating and stirring. Store at 4°C.

4. *2× Selection Buffer (2× SB)*—2× SB contains 100 mM HEPES (pH 7.5), 400 mM NaCl, and 10–20 mM of divalent metal ion M(II), where M(II) may represent Mg(II), Mn(II), Co(II), Ni(II), and Cd(II) or a mixture of these divalent metal ions.

5. *DNA Elution buffer*—Mix 8 mL of 5M NaCl, 2 mL of 1M Tris–HCl (pH 7.5), 0.4 mL of 0.5M EDTA (pH 8.0), and add ddH$_2$O to a final volume of 200 mL. Autoclave and store at 4°C.

6. *LB agar plates:* Mix 1.5 g of agar (Bioshop) with 100 mL of LB media and sterilize by autoclaving. The autoclaved media may be stored at room temperature. To prepare LB agar plates, melt the solidified LB-agar media using a standard microwave and cool to 40–60°C. Add antibiotic that is relevant to the cloning vector. Since pTZ57R/T contains an ampicillin resistant gene, in this case, we add ampicillin (Sigma-Aldrich) to a final concentration of 50 µg/mL. Pour media into Petri dishes under sterile conditions and allow it to solidify at room temperature. The solidified plates may be stored at 4°C for 3–4 weeks.

2.4. Plasticware and Equipment

1. 0.5- and 1.5-mL polypropylene microfuge tubes.

2. 2-, 10-, 100-, 200-, and 1,000-µL pipettors.

3. Pipette tips and gel loading tips—10, 100, 200 and 1,000 µL.

4. 0.2-mL thin-wall Axygen PCR tubes with domed caps.

5. Low-retention filter-barrier micropipette tips.

6. 20×20-cm silica 60 F$_{254}$ gel plate.

7. 3K molecular size spin columns.

8. Stratagene Robocycler Gradient 96 PCR machine (Agilent Technologies Canada Inc., Mississauga, ON, Canada).

9. Typhoon 9200 variable mode imager (GE healthcare, Baie d'Urfe, PQ, Canada) with ImageQuant software (Molecular Dynamics, Sunnyvale, CA, USA).

10. Refrigerated centrifuge.

11. Bench-top vortexer.

12. Mini-centrifuge.

13. Fluorescence spectrophotometer, such as the CARY Eclipse (Varian Inc. Mississauga, Canada).

14. UV–vis absorbance spectrophotometer, such as the CARY 100 (Varian Inc. Mississauga, Canada).

15. Tube rocker.

16. Temperature adjustable oven.

17. DNA dryer or concentrator, such as Savant DNA120 SpeedVac Concentrator (Fisher Scientific).

3. Methods

IVS is an iterative process that can be used to enrich a subset of functional sequences from a vast random-sequence DNA library. IVS begins with a library of seemingly inactive molecules, which is subjected to repetitive rounds of selection and amplification using a well-thought-out experimental plan. For example, we have developed the IVS scheme shown in Fig. 1 for the isolation of signaling DNAzymes. The cyclic selection-amplification step is repeated as many times as needed until the evolving population exhibits a significant level of catalysis. Cloning and sequencing techniques are then applied to reveal the sequences of individual DNAzymes in the enriched pool. These DNAzymes are chemically synthesized and assessed for catalytic performance. The catalytically efficient DNAzymes are chosen for further characterization in order to obtain the minimal sequence and a secondary structure model. A fully characterized DNAzyme can be used for the engineering of an aptazyme with a specific aptamer.

3.1. Purification of the DNA Library DL1

1. The instruction here is specifically established for ASG-400, an adjustable gel electrophoresis apparatus (CBS Scientific, Del Mar, CA, USA). Two 16×28-cm glass plates, one consists of notches on the upper edge while the other does not, are used with 1.0-mm spacers, a 16×16-cm metal plate, and a power supply (model EC3000-90 from E-C Apparatus Corp., Milford, MA, USA). Before casting the gel, thoroughly cleanse both glass plates with ddH$_2$O and 95% ethanol. Dry the plate with Kimwipes. Place the notched plate on top of the rectangular

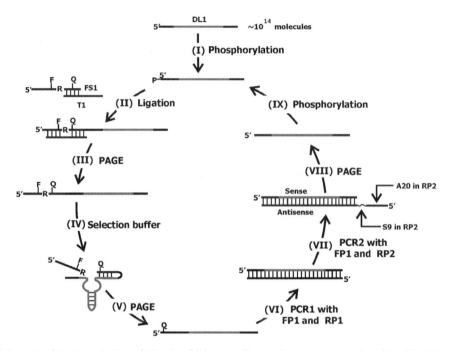

Fig. 1. Schematic of in vitro selection of signaling DNAzymes. The random-sequence region of the DNA library DL1 is depicted in *gray*. *F*, *R*, and *Q* in the substrate FS1 denote a fluorescein-labeled dT, adenine ribonucleotide, and a DABCYL-labeled dT, respectively.

plate separated by the spacers and secure them with clamps on both edges of the plates. Lay the clamped glass plate set on a plastic box placed evenly on a flat surface. Wash a 12-well comb (0.8 mm thick) with ddH$_2$O and 95% ethanol.

2. To cast 10% dPAGE gel, mix 70 mL of 10% denaturing poly-acrylamide gel stock with 70 µL of TEMED and 700 µL of 10% APS in a beaker. Promptly pour the mixture between the glass plates, insert the comb, and leave the solution to polymerize for about 30 min. Full polymerization can be confirmed by checking the residual gel mixture in the beaker, which should have completely solidified.

3. Once the gel is cast, remove the comb gently. Rinse the wells immediately with ddH$_2$O in a wash bottle to remove residual gel pieces lodged in the wells.

4. Mount the plates onto the gel electrophoresis apparatus with the rectangular plate facing out and place a metal plate behind the notched plate for cooling.

5. Add 1× TBE to the upper and the lower chambers of the apparatus such that the wells are filled with buffer and the lower edge of the gel is immersed.

6. Pre-run the gel by applying 40 mA (or 550 V) of current until the plates have warmed up to approximately 45°C (it may take 20 min).

7. Meanwhile, dissolve DL1 in 100 μL of 1× GLB. Vortex the mixture and heat at 90°C for 1 min. Vortex the sample again and perform a quick centrifugation using a bench-top mini-centrifuge.

8. Following the pre-run, rinse the wells of the gel with 1× TBE using a syringe and a needle.

9. Load the DL1 sample into two wells (50 μL/well) using a pipettor with a gel-loading tip. (Caution! Dispense each sample into the well slowly to ensure that it does not flow out of the well.)

10. Run the gel at 40 mA (550 V) until the bottom dye (bromophenol blue) is approximately 5 cm above the bottom edge of the gel.

11. Turn off the power supply, remove electrodes, and dismount the gel from the apparatus. Carefully separate the glass plates and wrap the gel using two sheets of plastic films (Saran Wrap).

12. Place the wrapped gel on top of a silica 60 F_{254} TLC plate and visualize the DNA band using a handheld UV lamp at a wavelength of 260 nm. For DL1 (98 nt long), a band should be observed above the top dye (xylene cyanol). Trace the DNA band with a fine permanent marker.

13. Excise the DNA band with a sterile razor blade. Cut the large gel piece into small pieces, and transfer them into a 15-mL conical tube. Add 2.5 mL of DNA elution buffer. Shake at 37°C overnight.

14. Following the elution, leave the tube undisturbed for 10 min to allow the gel pieces to settle. Transfer 400 μL of the supernatant into a 1.5-mL microfuge tube (four tubes are used). To each tube, add 40 μL of 3M sodium acetate (pH 7.0) and 1 mL of cold ethanol. Mix each sample thoroughly. Leave the tubes at −20°C for at least 2 h.

15. Centrifuge the tubes at $12,000 \times g$ for 25 min at 4°C. Carefully remove the supernatant and add 200 μL of cold 70% ethanol. Mix using a vortex and recentrifuge at $12,000 \times g$ for 10 min at 4°C. Remove the supernatant carefully, and dry the DNA pellet for 10 min in a speedvac.

16. Dissolve the purified DNA in 100 μL of ddH$_2$O. Determine the DNA concentration by measuring absorbance at 260 nm (see Note 2). Store the DNA sample at −20°C.

3.2. Purification of the DNA Primers and the Ligation Template

The DNA primers FP1, RP1, and RP2 and the ligation template DLT1 are purified similarly as described above for the DL1 purification. It is recommended that the primers be purified separately from the DNA library to avoid the possible cross-contamination (see Note 3).

3.3. Preparation of the Substrate

The fluorogenic substrate FS1 is typically synthesized at the scale of 0.2 μmol and received as a dry pellet. The 2′-hydroxyl group at the adenine ribonucleotide in FS1 is protected by a *tert*-butyldimethyl silyl group. It is necessary to remove this protective group in order to create a cleavable substrate. The deprotection and subsequent purification can be performed according to the instructions below.

1. Dissolve the dry pellet with 200 μL of ddH$_2$O and mix by vortexing. Transfer 50 μL to a 1.5-mL microfuge tube (the rest of the sample is archived and stored at –20°C). Add 150 μL of ddH$_2$O and 20 μL of 3M NaOAc (pH 7.0). Mix by vortexing. Perform a quick centrifugation using a bench-top mini-centrifuge.

2. Add 550 μL of cold ethanol. Mix thoroughly and leave at –20°C for 2 h.

3. Follow step 15 in Subheading 3.1 for DNA recovery by ethanol precipitation. It is necessary to make the substrate completely anhydrous, and therefore at the end of this step, dry the pellet in a speedvac for 1 h at room temperature.

4. Add 500 μL of TBAF to the pellet using a 1.0-mL Hamilton syringe. (Caution! TBAF is harmful if inhaled. It is extremely destructive to the tissues of mucosal membrane and the upper respiratory tract. It also causes skin and eye burns. Therefore, it should be handled in a fume hood; gloves, mask, goggles, and lab coat should be worn.)

5. Cap the tube and seal it tightly with paraffin film to minimize the evaporation of THF. Wrap it with aluminum foil (to protect fluorescein and DABCYL from damage by light). Place the sample on a tube rocker located within an oven that is pre-equilibrated at 55°C. Inspect the temperature of the oven to ensure that it remains at this temperature after 20 min. Leave the sample rocking in the oven for 12–14 h.

6. Remove the tube from the oven and cool it to room temperature. The mixture should appear as a homogeneous pale yellow solution with no residual pellet. The total volume may have been reduced to 200–300 μL since THF is volatile with a boiling point of 66°C.

7. Add ddH$_2$O to bring the volume of the solution to 500 μL. Add 25 μL of 1M Tris–HCl (pH 7.5) and mix by vortexing. Centrifuge the solution using a mini-centrifuge for 1 min to pellet any solids. Transfer the supernatant to a new tube.

8. Transfer 250 μL of the above solution into two 3K spin columns and centrifuge at 2,000×g at room temperature until 90% of the solution has passed through the filter in each column. The colored substrate should remain on top of the filter, while the flow-through should appear colorless. If the

flow-through appears yellow, the membrane has been damaged and the solution must be transferred to a new spin column and centrifuged once again.

9. Resuspend the concentrated substrate on the membrane by adding 20 μL of ddH$_2$O. Transfer the solution from both columns into a single tube. Adjust the final volume to 50 μL with ddH$_2$O. Add an equal volume of 2× GLB, mix, and keep on ice.

10. Prepare a 10% dPAGE gel following the steps 1–6 in Subheading 3.1.

11. Load the substrate sample from step 9 into two wells and run the gel until the bottom dye migrates past the midpoint of the gel. It is recommended to cover the plates during gel electrophoresis with aluminum foil to protect the chromophores from light-induced damages.

12. Follow steps 11 and 12 in Subheading 3.1 to disassemble the glass plates, visualize the gel, and excise the DNA band in each lane. Place the excised gel in two 1.5-mL microfuge tubes (one gel band in each tube). Since FS1 contains fluorescein, the DNA band on the gel can be seen by naked eyes.

13. Crush the gel into a paste with a pipette tip carefully. Add 500 μL of the DNA elution buffer to each tube and vortex for 10 min. Make sure that the tubes are protected from light by wrapping them with aluminum foil. Centrifuge the tubes at 12,000×g at room temperature for 2 min to separate the supernatant from the gel pieces. Transfer the supernatant to a 1.5-mL microfuge tube. Add 200 μL of the DNA elution buffer to the gel-containing tubes and vortex again for 10 min. Transfer the supernatant to the same tube or a new tube. If needed, adjust all the volumes to 400 μL with ddH$_2$O. Add 40 μL of 3M NaOAc (pH 7.0) to each tube and 1.0 mL of cold ethanol. Mix thoroughly and incubate at −20°C for at least 2 h.

14. Recover FS1 by centrifugation as described in step 15 in Subheading 3.1. Dissolve the purified FS1 in 100 μL of ddH$_2$O. Determine its concentration by measuring the absorbance at 260 nm. Store at −20°C until further use.

3.4. Preparation of the DNA Pool for the First Round of Selection (Steps I–III of Fig. 1)

1. Phosphorylate the DNA library as follows: Sequentially mix 7.1 μL of 70.0 μM (500 pmol) DL1, 5.0 μL of 10× PNK buffer A, 1 μL of (γ-^{32}P)ATP, 35.9 μL of ddH$_2$O, and 1 μL of PNK (10 units) in a 1.5-mL microfuge tube. Gently mix the components and incubate at 37°C for 15 min. (Caution! Isotopes should be handled behind a beta shield with extreme care to minimize exposure to radiation.)

2. Add 1.0 μL of nonradioactive ATP (from a 100 mM stock) to the reaction mixture, mix by gentle pipetting, and incubate at 37°C for another 20 min (see Note 4). Quench the reaction by

heating at 90°C for 5 min. Allow the sample to cool at room temperature for 10 min.

3. Add 6.7 μL (75 μM) of FS1 and 4.3 μL (120 μM) of DL1T to the reaction mixture prepared in step 2. Vortex the mixture and heat at 90°C for 40 s to denature the DNA. Leave the sample at room temperature for 20 min. Add 26 μL of ddH$_2$O to this reaction mixture, followed by 10 μL of 10× T4 DNA ligase buffer and 2 μL (10 units) of T4 DNA ligase. Incubate the mixture at room temperature for 2 h.

4. Recover DNA following the ethanol precipitation steps outlined in Subheading 3.1. (Caution! The supernatant is highly radioactive due to the presence of free (γ-^{32}P)ATP. Therefore, collect the supernatant in a falcon tube and store within a closed beta box until the radioisotope has completely decayed. Follow institutional guidelines for safe radioisotope handling.)

5. Dry the DNA pellet using the speedvac at room temperature. Dissolve the pellet with 20 μL of 1× GLB. The sample should be kept behind a beta shield and covered with aluminum foil until it is ready for gel loading.

6. Prepare a DNA marker by phosphorylating 1 μL of the unligated DNA library with (γ-^{32}P)ATP as described in step 1 of this section. Mix this mixture with 20 μL of 1× GLB and leave behind a beta shield until loading into the gel.

7. Purify the ligated product by 10% dPAGE using the instructions provided in steps 1–6 of Subheading 3.1 with three minor modifications: (1) 0.75-mm spacers and comb are used; (2) the gel is made from 40 mL of 10% dPAGE gel stock, 40 μL of TEMED, and 400 μL of 10% APS; (3) the gel is run at 40 mA and 900 V. (Caution! Following gel electrophoresis, the solution in the bottom chamber contains radioactive ATP. Therefore, it is recommended to place a beta shield in front of the gel apparatus when handling the gel to protect from radiation exposure. Carefully transfer the radioactive solution to a container placed in a beta box using a syringe and a needle. Store the buffer in the box until the radioactivity has completely decayed.)

8. Follow steps 11 and 12 in Subheading 3.1 to disassemble the glass plates, visualize the gel, and excise the DNA band. Since the ligated DNA contains fluorescein, the DNA band on the gel can be seen by naked eyes. Place the excised gel in a 1.5-mL microfuge tube.

9. Crush the gel into a paste with a pipette tip carefully. Add 300 μL of elution buffer to the crushed gel and vortex for 10 min at room temperature. Spin down the sample and carefully transfer ~200 μL of the supernatant to a fresh tube. Add

200 μL of fresh elution buffer to the gel remaining in the tube and vortex again for a second elution. Combine the supernatants together and make the final volume 400 μL. Add 40 μL of 3M NaOAc (pH 7.0) and 1.0 mL of cold 100% ethanol, mix thoroughly and incubate at –20°C for 2 h.

10. Recover the ligated DNA by centrifugation as described in step 15 in Subheading 3.1. Dissolve the purified DNA in 110 μL of ddH$_2$O (the DNA concentration is estimated to be ~1.0 μM). Store at –20°C until further use.

3.5. The First Round of Selection and Amplification (Steps IV–IX of Fig. 1)

1. Transfer 100 μL of the ligated DNA library to a fresh tube (the remaining DNA library will be used for preparing a marker to track the cleavage product in the selection), heat at 90°C for 30 s, and cool to room temperature for 20 min. Add 100 μL of 2× SB, mix the reaction mixture, and incubate at room temperature for 5 h while covering the reaction mixture with aluminum foil.

2. In the meantime, prepare the DNA size marker for electrophoresis using the remaining 10 μL of ligated DNA library. Add 80 μL of 0.25M NaOH to the 10 μL of ligated DNA library, vortex, spin down, and place in a 90°C heating block for 15 min. Remove the tube from the heating block and place behind a beta shield for 10 min to cool to room temperature. Spin down and add 10 μL of 3M NaOAc (pH 5.2). Vortex the tube, spin down again, and add 250 μL of cold ethanol. Mix thoroughly and incubate at –20°C for 2 h. Precipitate the DNA as described in step 15 in Subheading 3.1.

3. After 5 h of incubation, the cleavage reaction is quenched with 12 μL of 0.5M EDTA (pH 8.0). Recover the DNA by ethanol precipitation following the procedures described in Subheading 3.1.

4. Dry the DNA pellets obtained in steps 2 and 3 by speedvac for 10 min at room temperature and place behind a beta shield.

5. Meanwhile, prepare a 10% dPAGE gel and pre-run the gel following the instructions in step 7 in Subheading 3.4.

6. Dissolve the DNA pellets obtained from step 4 with 20 μL of 1× GLB, vortex, and spin down.

7. Load the samples into the gel (leave one free lane between the marker lane and the sample lane). Run the gel until the lower dye reaches the bottom edge of the plate.

8. Following gel electrophoresis, wrap gel using plastic wrap and expose it to a phosphorimage plate for 5 min. Scan the gel using Typhoon 9200 to obtain either a phosphorimage or a fluorimage.

9. In order to excise the cleaved DNA band that contains the cleaved DNA molecules, expose the gel on an X-ray film for

10 min in the dark. It is necessary to align the gel with the film in order to mark the exact position of the DNA afterward. This can be done by poking a few holes into the aligned film and gel with a pin. Develop the film using an X-ray developer. Align the holes on the X-ray film to those on the gel and view the alignment of bands (see Note 5). Mark the location in the sample lane of the gel that is of the same size as the marker band in the marker lane.

10. Excise the band, crush the gel, and add 500 μL of elution buffer. Vortex or leave the tube rocking for 2–3 h to elute the DNA from the gel. Spin down the tube, transfer 400 μL of the supernatant to a fresh tube and precipitate the DNA as described in steps 15 and 16 in Subheading 3.1. Dissolve the DNA pellet in 50 μL of ddH$_2$O. Amplify the cleaved DNA sequences in two rounds of PCR in the next two steps.

11. *PCR1*: Prepare a 50 μL reaction with primers FP1 and RP1. Mix 25 μL of the cleaved library (half of the total recovered sample), 1.25 μL FP1 (20 μM), 1.25 μL RP1 (20 μM), 17.0 μL ddH$_2$O, 5 μL 10× PCR reaction buffer, 5 μL of the dNTP mix (see Note 1), and 0.5 μL of DNA polymerase (2.5 units). Use the following conditions for PCR: denaturation at 94°C for 1 min, followed by 15 cycles of denaturation at 94°C for 30 s, annealing at 50°C for 45 s and extension at 72°C for 45 s, and finally a single extension cycle at 72°C for 2 min. Apply 5 μL of the PCR1 products in a 2% agarose gel supplemented with 0.005% (v/v) SyberSafe (Invitrogen Inc. Burlington, Canada) to confirm successful amplification. Visualize the gel using a fluorimager.

12. *PCR2*: Dilute 1 μL of PCR1 product into 20 μL with ddH$_2$O for using in PCR2. Take 1 μL of this diluted PCR1 product as the template for PCR2. Similar to PCR1, prepare a 50 μL of a reaction mixture using primers FP1 and RP2. The PCR conditions employed here are the same as those indicated for PCR1. Analyze the PCR product on a 2% agarose gel as indicated for PCR1 to ensure proper amplification. If needed, adjust the number of PCR cycles to obtain a higher yield of specific PCR products (see Note 6). Typically 2–4 individual PCR reactions should be conducted in order to obtain sufficient DNA for the next round of selection. PCR products from these reactions are combined, and the DNA molecules in the sample are precipitated following steps 14 and 15 in Subheading 3.1.

13. Prepare a 10% dPAGE gel as described in step 7, Subheading 3.4.

14. Dissolve the pellet with 20 μL of 1× GLB and load in the dPAGE. Run the gel until the lower dye reaches the bottom edge of the plate. Wrap the gel with Saran wrap and visualize by a handheld UV lamp. UV shadowing should produce two

bands in the gel. The upper band corresponds to the antisense strand since it contains a 20-nt overhang at the 5'-end separated by a triethylene glycol linker; the bottom band represents the desired sense strand. Excise this band, elute the DNA from the gel, and precipitate as described in step 9 of Subheading 3.4.

3.6. The Second Round of Selection and Amplification

1. Based on the amount of primers used in PCR2, the amount of DNA population recovered after gel purification following the first round of selection is approximately 50 pmol. Phosphorylate this purified PCR product following the instructions in steps 1 and 2 of Subheading 3.4.

2. Prepare fresh 60 μM stocks of FS1 and DLT1. These stocks are used in the second round and all subsequent rounds of selection. Add 1 μL of FS1, 1 μL of DLT1, and 35 μL of ddH$_2$O to the phosphorylated DNA population, vortex, and spin down. Heat the mixture at 90°C for 40 s and cool to room temperature for 20 min. Then, add 10 μL of 10× T4 DNA ligase buffer and 2 μL of T4 DNA ligase (10 units). Gently mix by pipetting and incubate the reaction mixture at room temperature for 2–3 h.

3. Following ligation, purify the DNA following the steps 7–10 outlined in Subheading 3.4.

4. The second round and all the subsequent rounds of selection are conducted in 100 μL reaction volumes. Dissolve the purified FS1-DL1 from the step above in 60 μL of ddH$_2$O. Transfer 50 μL of the library to a new 1.5-mL microfuge tube and initiate the cleavage reaction by the addition of 50 μL of 2× SB. Incubate the reaction mixture for 5 h at room temperature. Use the remaining pool (10 μL) of DNA to prepare the DNA marker as indicated in step 2 of Subheading 3.5.

5. Quench the reaction by adding 6 μL of 0.5M EDTA. Then, add 270 μL of cold ethanol, mix and incubate at −20°C for 2 h.

6. Repeat steps 4–14 in Subheading 3.5.

7. Conduct additional rounds of selection using the same protocol outlined in steps 1–6 above (see Note 7). Progress of the selection can be monitored by analyzing the percentage of cleaved product in each round using the ImageQuant software.

3.7. Cloning and Sequencing

1. *PCR for cloning*: The DNA population from the final round of selection is subjected to cloning and sequencing. For molecular cloning, both PCR1 and PCR2 are conducted with the same set of primers FP1 and RP1. Conduct PCR1 following the same protocol as described for the selection rounds. Dilute 1 μL of PCR1 to 20 μL with ddH$_2$O. Use 1 μL from this diluted sample for the PCR2 using the following

conditions: 94°C for 1 min 1 cycle; 94°C for 30 s, 50°C for 45 s, and 72°C for 45 s for 10–12 cycles. Assess the quality of the PCR product by 2% agarose gel containing 0.005% Sybersafe; adjust the number of cycles to obtain a clean product band on agarose gel. After this, incubate the PCR mixture at 72°C for 10 min; this step ensures that PCR products contain an A overhang at each 3′ termini.

2. *Ligation to the vector*: Dilute 5 μL of the PCR product to 50 μL with ddH$_2$O before cloning. Mix 5 μL of vector (pTZ57R/T), 15 μL of the diluted PCR products, 10 μL of 5× ligation buffer, 19 μL of ddH$_2$O, and 1 μL of T4 DNA ligase. Incubate the reaction mixture overnight at room temperature. Store at 4°C until further use.

3. To prepare chemically competent cells, inoculate *E. coli* XL1-Blue cells in 2 mL of C-medium and grow overnight at 37°C with shaking. To ensure that the media is not contaminated, incubate a 2 mL fraction of the media as a negative control.

4. Following overnight growth, take 150 μL of the cultured bacteria and transfer it to a fresh tube containing 1.5 mL of C-medium. Incubate this suspension for 20 min at 37°C and place on ice. Transfer 1 mL of this culture to a fresh tube and pellet cells using a mini-centrifuge for 1 min. Discard the supernatant and keep the cell pellet on ice.

5. Prepare fresh T-mix solution by mixing 200 μL each of solutions A and B supplied with the InsTAclone kit. Resuspend the pellet using 200 μL of T-mix. Chill the suspended cells on ice for at least 5 min before pelleting the cells again using the mini-centrifuge for 1 min.

6. Discard the supernatant and resuspend pellet with another 80 μL of T-mix. Transfer 70 μL of this cell suspension to a fresh 1.5-mL microfuge tube. Add 3.5 μL of the ligation mixture to the cells, mix gently by pipetting, and keep the cells on ice for 5 min.

7. Transfer 15 μL of the cell suspension to several LB agar plates. Spread the cells homogeneously using a sterile spreader. Incubate the plates at 37°C for 10–12 h until colonies appear on the plates (see Note 8). Prepare a positive and a negative control by transforming cells with the supplier provided positive control vector or with water in lieu of vector, respectively.

8. Following overnight incubation, the numbers of colonies on the experimental plate are compared to those on the control plates.

9. Pick colonies from the sample plates and inoculate them individually in culture tubes containing 2 mL of LB supplemented with ampicillin (50 μg/mL). Carefully select ~100 colonies for culturing. Incubate the cultures for 16–20 h at 37°C with

shaking. To ensure that the media is not contaminated, incubate a 2 mL fraction of the media as a negative control.

10. Transfer 1.4 mL of each culture into a single well of a 96-deep-well plate. After transferring the cultures into every well, seal the plate with 3M tape and centrifuge for 5 min at $1,900 \times g$ at 4°C. Remove tape and invert the plate gently to discard the supernatant. Seal the plate again with aluminum foil tape and store at –80°C until it is used for sequencing. We routinely send the samples to Functional Bioscience Inc. (Madison, WI, USA, http://www.functionalbio.com/) for direct sequencing. The M13 reverse or M13 forward primers are used.

11. After obtaining the sequencing results, sort out the sequences that correspond to the DNA library and delete the sequences belonging to the vector using the Chromas software (available from http://www.technelysium.com.au/chromas.html). Align the sequences with BioEdit sequence alignment editor (available through http://www.mbio.ncsu.edu/BioEdit/bioedit.html). Finally, group the sequences into classes based on sequence homology.

12. Purchase synthetic oligonucleotides for the identified DNAzyme classes, and test their cleavage activities. In the first step, the cleavage efficiency is usually tested in the *cis*-acting form (a single piece of molecule prepared by ligating the substrate FS1 to the purified synthetic DNAzyme). The cleavage efficiency can be assessed by quantifying the amount of cleavage products by phosphorimaging or fluorimaging. Alternatively, the cleavage performance can be assessed using a fluorescence spectrometer. An example of the cleavage activity of a signaling DNAzyme, SD1, assessed by gel and by a fluorimeter is shown in Fig. 2.

13. Once the catalytic performance of individual sequences is tested, the most efficient sequence is chosen for further characterization. An effort is made to convert the *cis*-acting DNAzyme into a *trans* version where the DNAzyme and the substrate are separated into individual entities and mixed together for reaction. This is also called a bimolecular system. This *trans*-acting derivative may be produced by deleting the primer-binding arms from the DNAzyme sequence. Thereafter, sequential deletions from the 5′ and 3′ ends or internal deletions can be introduced into this sequence. For example, in sequential deletion experiments, 5 nt are sequentially removed from each terminus of the sequence. This sequence truncation approach produces a set of sequences with each being 5 nt shorter than previous one. The systematic testing of each sequence provides information regarding the catalytic importance of nucleotides at a specific location. Once the sequence length is minimized, the possible secondary structures of the DNAzyme bound to

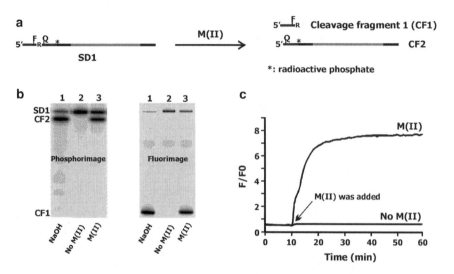

Fig. 2. Illustration of a cleavage assay by a signaling DNAzyme. (a) A radioactively labeled substrate-DNAzyme construct (indicated as "SD1" in the figure) is allowed to cleave in the presence of divalent metal ions, M(II). The cleavage reaction generates two cleavage fragments (CFs): a nonradioactive, fluorophore-containing CF1 and a radioactive CF2. (b) A phosphorimage (*left*) and a fluoroimage (*right*) of the same gel. Lane 1 contains a DNA marker prepared by treating SD1 with NaOH, lane 2 represents a negative control reaction in which no M(II) was added, and lane 3 is a test sample in which SD1 was allowed to cleave in the presence of M(II). (c) Monitoring signal generation of SD1 using a fluorimeter.

the substrate can be obtained using mfold, which is available at the following web site: http://www.bioinfo.rpi.edu/applications/mfold. The predicted secondary structure can be verified by covariation studies in which the identity of a base pair within predicted stem regions is changed from one Watson-Crick pair to another, followed by the activity test.

3.8. Applying DNAzymes for Biosensing

Applying a signaling DNAzyme for biosensing involves conjugation of an aptamer with a suitable signaling DNAzyme to create a signaling aptazyme. Often the sequence of the aptamer is integrated with the sequence of the DNAzyme in a special way such that unbound aptamer domain can mask the catalytic function of the DNAzyme. Ligand binding, however, promotes the formation of the catalytically active conformation of the aptazyme. Secondary structure foreknowledge, of both the DNAzyme and aptamer, greatly facilitates the construction of an aptazyme. Nevertheless, it is also possible to engineer aptazymes with a relatively less characterized DNAzymes by employing various approaches. Our group has rationally designed a few aptazymes for ATP detection using different signaling DNAzymes and different designing strategies (26, 33, 37, 38), which are discussed below.

An aptazyme using DEC22-18 DNAzyme for ATP sensing (see Fig. 3a): DEC22-18 is a DNAzyme that exhibits optimal cleavage of the substrate FS1 at pH 6.8 with 1 mM Co(II) (26). The proposed secondary structure of this signaling DNAzyme is depicted in

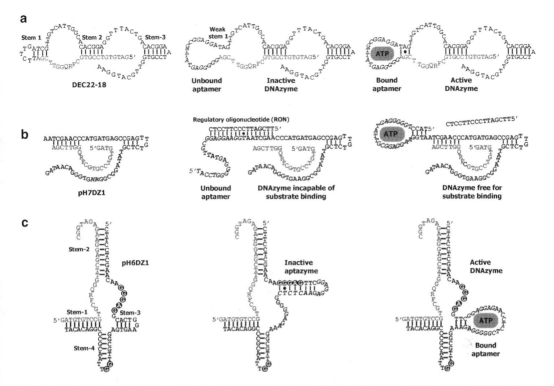

Fig. 3. Examples of designing ATP-dependent aptazymes using three different DNAzymes: DEC22-18 (**a**), pH7DZ1 (**b**), and pH6DZ1 (**c**). Each substrate sequence is depicted by nucleotides in *gray* and each DNAzyme is depicted by nucleotides in *black*. The ATP aptamer domains in the aptazyme constructs are represented by *italicized letters*.

Fig. 3a (left structure). It was found that stem 1 and its loop could be altered by incorporating additional nucleotides. Hence, the ATP aptamer was appended to this stem loop such that the stem becomes relatively unstable and cannot support the cleavage reaction (see Fig. 3a, middle structure). Note that a G:A mismatch was introduced at stem 1 to create a weak stem bridging the aptamer and the DNAzyme. Upon the addition of ATP, the aptamer and ATP form a complex, strengthening the stem 1 and promoting the catalytic activity (see the right structure in Fig. 3a).

An aptazyme with pH7DZ1 DNAzyme for ATP sensing: pH7DZ1 was isolated in our laboratory by IVS to cleave FS1 (25). Following selection and characterization, a *trans*-acting form of this DNAzyme was obtained. This DNAzyme was found to exhibit an optimal activity in the presence of 50 mM HEPES (pH 7.5), 200 mM NaCl, 100 mM KCl, and 15 mM $MnCl_2$ at room temperature. A predicted secondary structure is shown in Fig. 3b (left structure). Although this DNAzyme was not fully characterized, it was found that it binds to only one arm of the substrate (7 nt at the 3'-end of FS1). Therefore, it was assumed that it would be possible to block substrate binding by employing a regulatory oligonucleotide (RON). In order to confer allostery, the ATP aptamer sequence was ligated at the 5'-end of pH7DZ1. The RON was designed to simultaneously hybridize to both the aptamer and the DNAzyme as indicated in Fig. 3 (the middle structure). A single nucleotide mismatch was also introduced into the sequence of RON in order to reduce the duplex stability, allowing for efficient structure switching. In the absence of ATP, the aptazyme forms a duplex with RON, making the DNAzyme domain unavailable for substrate binding. As a result, no catalysis occurs. However, when ATP binds to the aptamer domain, the RON dissociates from the DNAzyme and consequently, it is able to interact with the substrate to perform catalysis (see the right structure in Fig. 3b) (38).

An aptazyme with DNAzyme pH6DZ1 for ATP sensing: pH6DZ1 is another signaling DNAzyme created in our laboratory. This enzyme exhibits an optimal activity at pH 6.0 in presence of both Ni(II) and Mn(II). The pH6DZ1 sequence was thoroughly investigated to identify the catalytically important nucleotides with a series of sequence truncation and covariation studies; a *trans*-acting derivative of pH6DZ1, coined pH6-ET1, was obtained (34, 37). Finally, a predicted secondary structure was deduced as shown in Fig. 3c (left structure). It was hypothesized that stem 3 and its loop might be appropriate for the incorporation of the ATP aptamer because none of the catalytically essential nucleotides reside in this region. Therefore, the aptamer sequence was used to replace this stem loop. Part of the aptamer sequence was specially altered such that, in the absence of ATP, this part of the aptamer forms a short duplex with some of the catalytically essential nucleotides of the DNAzyme domain (see Fig. 3, middle structure), rendering it inactive. When ATP binds to the aptamer domain, the catalytically active conformation of the DNAzyme is revived (see Fig. 3c, right structure).

4. Notes

1. dNTP mixture (10 mM each of dATP, dCTP, dGTP, and dTTP) can be purchased from MBI Fermentas. This dNTP mixture is diluted to make a 2 mM stock to be used in PCR. We routinely aliquot the dNTP stock into small (e.g., 50 μL) fractions in order to prevent cross-contamination and extensive freezing and thawing of the stock. In many cases, it is also beneficial to conduct PCR in the presence of radioactive dGTP (α-^{32}P(dGTP)). This is usually done for PCR2. For these radioactive reactions, a radioactive dNTP mixture can be prepared. A dNTP set in which each dNTP is available in individual 100 mM stocks can be obtained through MBI Fermentas. The radioactive mixture can be then prepared by incorporating 2 mM of each of dATP, dCTP, and dTTP and 0.2 mM of dGTP. Immediately prior to PCR, 1 or 2 μL of α-^{32}P(dGTP) is added to the reaction mixture.

2. (a) Because a DNA library is produced by chemical synthesis and its sequence is quite long, the amount of DNA obtained after purification may be small. However, we have found that a DNA synthesis at both 40-nmole and 200-nmole scales can produce a sufficient quantity of purified DNA to be used for IVS (100–1,000 pmol of purified DNA are usually used in our experiments). (b) During gel purification a smeared DNA band may be observed with some DNA libraries, particularly if the sequences are long (80–120 nt). In this case, it is advisable to only excise the thicker portion of the band located at the top. (c) After the completion of the purification step, we recommend checking the quality of the library (the length of DNA and amplification compatibility) by PCR.

3. Contaminating reagents and equipment with a trace amount of the DNA library is always a key concern for IVS. The success of IVS largely depends on avoiding such contaminations. Specially, DNA primers, PCR buffer, dNTP mixture, selection buffer, and pipettors should be handled with great care to avoid such a contamination. It is a good practice to purify the primers and the DNA library separately (using separate PAGE gels or on different days). Aliquoting the PCR primers and buffers into small volumes is also a good practice.

4. In preparing the DNA library pool for the first round of selection, we usually perform two different phosphorylation reactions for the library, one with radioactive ATP and another with nonradioactive ATP. These samples can then be combined for ligation in a single reaction.

5. In the first few rounds of selection it is common that the amount of cleaved substrate is too insignificant to be detected. However, a piece of the gel at the expected position of the cleaved sequence, as indicated by the marker DNA band, should be excised following PAGE purification. Although a DNA band is not visible, a population of DNA sequences of the expected size should be produced following PCR. The cleaved product from the gel can be eluted for a longer time, especially for the first few rounds of selection, to maximize the DNA recovery from the gel.

6. During PCR amplification of the enriched pool, it is important to optimize the number of amplification cycles and the concentration of the template in order to prevent overamplification, which can result in the overrepresentation of certain sequence classes and reduce the quality and quantity of the amplicons. The selected DNA pool can be dived into several aliquots, one of which can be used to establish the needed number of cycles for full amplification (but not overamplification). To check the depth of amplification, the agarose gel electrophoresis can be used. If it appears that overamplification

has occurred, the number of cycles to be used for the remaining aliquots can be decreased. Throughout the course of selection, the amount of the template (the cleaved product from the cleavage reaction in each round) can be progressively diluted before PCR1. For example, 1/3 may be used in round 4 and 1/5 may be in round 6.

7. The stringency of IVS can be progressively increased in successive rounds in order to isolate highly efficient DNAzymes. For example, after observation of a strong signal (such as >10% cleavage), the cleavage reaction time may be reduced to 10 min from 5 h. After a few more rounds, the reaction time may be further reduced to 1 min or even 5 s. Moreover, the reaction condition can be altered after a few rounds of selection to produce diversified DNAzymes that are active under different conditions. For example, in one of our previous studies, a DNAzyme population established after a few rounds of selective amplification at pH 4 was divided for parallel selection experiments conducted at different pH settings. This resulted in the isolation of a collection of DNAzymes with intriguing pH dependencies (25).

8. It is advised not to incubate the *E. coli* transformants on the LB agar for more than 16 h. Otherwise, formation of satellite colonies resulting from cells that have not taken up the plasmids will be observed around the transformants due to degradation of ampicillin by secreted β-lactamase. Satellite colonies make it difficult to identify the authentic transformants. Sequencing a good number of recombinant plasmids (~50) is recommended. A good indication of a successful selection experiment is the presence of conserved sequence domains.

Acknowledgments

The DNAzyme and aptamer work in the Li Lab has been supported by the Canadian Institutes of Health Research (CIHR), Natural Science and Engineering Research Council of Canada (NSERC), and SENTINEL Bioactive Paper Network. YL holds a Canada research chair.

References

1. Kruger, K., Grabowski, P. J., Zaug, A. J., Sands, J., Gottschling, D. E., and Cech, T. R. (1982) Self-splicing RNA: Autoexcision and autocyclization of the ribosomal RNA intervening sequence of tetrahymena, *Cell 31*, 147–157.

2. Guerrier-Takada, C., Gardiner, K., Marsh, T., Pace, N., and Altman, S. (1983) The RNA moiety of ribonuclease P is the catalytic subunit of the enzyme, *Cell 35*, 849–857.

3. Breaker, R. R., and Joyce, G. F. (1994) A DNA enzyme that cleaves RNA, *Chem Biol 1*, 223–229.

4. Cuenoud, B., and Szostak, J. W. (1995) A DNA metalloenzyme with DNA ligase activity, *Nature 375*, 611–614.

5. Li, Y., and Sen, D. (1996) A catalytic DNA for porphyrin metallation, *Nat Struct Biol 3*, 743–747.

6. Breaker, R. R., and Joyce, G. F. (1995) A DNA enzyme with Mg(2+)-dependent RNA phosphoesterase activity, *Chem Biol 2*, 655–660.

7. Carmi, N., Shultz, L. A., and Breaker, R. R. (1996) In vitro selection of self-cleaving DNAs, *Chem Biol 3*, 1039–1046.

8. Tuerk, C., and Gold, L. (1990) Systematic evolution of ligands by exponential enrichment: RNA ligands to bacteriophage T4 DNA polymerase, *Science 249*, 505–510.

9. Robertson, D. L., and Joyce, G. F. (1990) Selection in vitro of an RNA enzyme that specifically cleaves single-stranded DNA, *Nature 344*, 467–468.

10. Ellington, A. D., and Szostak, J. W. (1990) In vitro selection of RNA molecules that bind specific ligands, *Nature 346*, 818–822.

11. Li, Y., and Breaker, R. R. (1999) Kinetics of RNA Degradation by Specific Base Catalysis of Transesterification Involving the 2′-Hydroxyl Group, *J Am Chem Soc 121*, 5364–5372.

12. Smith, R. M., and Hansen, D. E. (1998) The pH-Rate Profile for the Hydrolysis of a Peptide Bond, *J Am Chem Soc 120*, 8910–8913.

13. Achenbach, J. C., Chiuman, W., Cruz, R. P. G., and Li, Y. (2004) DNAzymes: From Creation In Vitro to Application In Vivo, *Curr Pharm Biotech 5*, 321–336.

14. Navani, N. K., and Li, Y. (2006) Nucleic acid aptamers and enzymes as sensors, *Curr Opin Chem Biol 10*, 272–281.

15. Liu, J., Cao, Z., and Lu, Y. (2009) Functional nucleic acid sensors, *Chem Rev 109*, 1948–1998.

16. Silverman, S. K. (2008) Catalytic DNA (deoxyribozymes) for synthetic applications-current abilities and future prospects, *Chem Commun* 3467–3485.

17. Baum, D., and Silverman, S. (2008) Deoxyribozymes: useful DNA catalysts in vitro and in vivo, *Cell Mol Life Sci 65*, 2156–2174.

18. Schlosser, K., and Li, Y. (2009) Biologically Inspired Synthetic Enzymes Made from DNA, *Chem Biol 16*, 311–322.

19. Silverman, S. K. (2005) In vitro selection, characterization, and application of deoxyribozymes that cleave RNA, *Nucleic Acids Res 33*, 6151–6163.

20. Liu, J., Brown, A. K., Meng, X., Cropek, D. M., Istok, J. D., Watson, D. B., and Lu, Y. (2007) A catalytic beacon sensor for uranium with parts-per-trillion sensitivity and millionfold selectivity, *Proc Natl Acad Sci USA 104*, 2056–2061.

21. Liu, J., and Lu, Y. (2003) Improving fluorescent DNAzyme biosensors by combining inter- and intramolecular quenchers, *Anal Chem 75*, 6666–6672.

22. Liu, J., and Lu, Y. (2006) Fluorescent DNAzyme biosensors for metal ions based on catalytic molecular beacons, *Methods Mol Biol 335*, 275–288.

23. Liu, J., and Lu, Y. (2007) A DNAzyme catalytic beacon sensor for paramagnetic Cu2+ ions in aqueous solution with high sensitivity and selectivity, *J Am Chem Soc 129*, 9838–9839.

24. Liu, J., and Lu, Y. (2007) Rational design of "turn-on" allosteric DNAzyme catalytic beacons for aqueous mercury ions with ultrahigh sensitivity and selectivity, *Angew Chem Int Ed 46*, 7587–7590.

25. Liu, Z., Mei, S. H., Brennan, J. D., and Li, Y. (2003) Assemblage of signaling DNA enzymes with intriguing metal-ion specificities and pH dependences, *J Am Chem Soc 125*, 7539–7545.

26. Mei, S. H., Liu, Z., Brennan, J. D., and Li, Y. (2003) An efficient RNA-cleaving DNA enzyme that synchronizes catalysis with fluorescence signaling, *J Am Chem Soc 125*, 412–420.

27. Chiuman, W., and Li, Y. (2007) Efficient signaling platforms built from a small catalytic DNA and doubly labeled fluorogenic substrates, *Nucleic Acids Res 35*, 401–405.

28. Rupcich, N., Chiuman, W., Nutiu, R., Mei, S., Flora, K. K., Li, Y., and Brennan, J. D. (2006) Quenching of fluorophore-labeled DNA oligonucleotides by divalent metal ions: implications for selection, design, and applications of signaling aptamers and signaling deoxyribozymes, *J Am Chem Soc 128*, 780–790.

29. Kandadai, S. A., and Li, Y. (2005) Characterization of a catalytically efficient acidic RNA-cleaving deoxyribozyme, *Nucleic Acids Res 33*, 7164–7175.

30. Chiuman, W., and Li, Y. (2006) Revitalization of six abandoned catalytic DNA species reveals a common three-way junction framework and diverse catalytic cores, *J Mol Biol 357*, 748–754.

31. Chiuman, W., and Li, Y. (2006) Evolution of high-branching deoxyribozymes from a catalytic DNA with a three-way junction, *Chem Biol 13*, 1061–1069.

32. Ali, M. M., Kandadai, S. A., and Li, Y. (2007) Characterization of pH3DZ1-an RNA cleaving deoxyribozyme with optimal activity at pH 3, *Can J Chem 85*, 261–273.

33. Chiuman, W., and Li, Y. (2007) Simple fluorescent sensors engineered with catalytic DNA 'MgZ' based on a non-classic allosteric design, *PLoS One 2*, e1224.

34. Shen, Y., Brennan, J. D., and Li, Y. (2005) Characterizing the secondary structure and identifying functionally essential nucleotides of pH6DZ1, a fluorescence-signaling and RNA-cleaving deoxyribozyme, *Biochemistry 44*, 12066–12076.

35. Kandadai, S. A., Mok, W. W., Ali, M. M., and Li, Y. (2009) Characterization of an RNA-cleaving deoxyribozyme with optimal activity at pH 5, *Biochemistry 48*, 7383–7391.

36. Kandadai, S. A., Chiuman, W., and Li, Y. (2006) Phosphoester-transfer mechanism of an RNA-cleaving acidic deoxyribozyme revealed by radioactivity tracking and enzymatic digestion, *Chem Commun*, 2359–2361.

37. Shen, Y., Chiuman, W., Brennan, J. D., and Li, Y. (2006) Catalysis and rational engineering of trans-acting pH6DZ1, an RNA-cleaving and fluorescence-signaling deoxyribozyme with a four-way junction structure, *Chembiochem 7*, 1343–1348.

38. Achenbach, J. C., Nutiu, R., and Li, Y. (2005) Structure-switching allosteric deoxyribozymes, *Analytica Chimica Acta 534*, 41–51.

Chapter 26

Development of Trainable Deoxyribozyme-Based Game Playing Automaton

Renjun Pei, Joanne Macdonald, and Milan N. Stojanovic

Abstract

Molecular automata are self-operating machines serially exchanging information with their environment while changing their configurations. Molecular protoautomata are devices trained in a series of sessions with an operator to become molecular automata. Reconfigurable deoxyribozyme-based logic gates can be used to build multipurpose protoautomata, reprogrammable devices that go beyond single-purpose "hard-wired" molecular automata. Molecular array of YES and AND gates (MAYA)-III is such a protoautomaton that can be taught by example to play a game. Training of MAYA-III is a process that does not require the operator to be familiar with the underlying molecular programming. Herein we provide the instructions to construct this protoautomaton, with particular focus on the optimization of computing components.

Key words: Logic gates, Deoxyribozyme, Game playing, Molecular automaton, Trainable, Optimization

---

1. Introduction

Much progress has been made in the field of molecular computing, as witnessed by the reports of programmable molecular automata and various complex molecular networks (1–5). Over the past several years, we constructed a full set of molecular logic gates using deoxyribozymes. These logic gates analyze sets of oligonucleotides as inputs and produce cleaved oligonucleotides as outputs (6). In the past, they enabled us to establish molecular circuits such as half- and full-adders for performing basic arithmetical operations (7, 8). As an unbiased test of new approaches to molecular computing, we also constructed molecular automata that play games (9–11), resulting in a series of automata known as *MAYAs* (abbreviated initially from *M*olecular *A*rray of *Y*ES and *A*ND gates). To date,

Jörg S. Hartig (ed.), *Ribozymes: Methods and Protocols*, Methods in Molecular Biology, vol. 848,
DOI 10.1007/978-1-61779-545-9_26, © Springer Science+Business Media, LLC 2012

we have constructed three generations of *MAYAs* (9–11). Our first two automata, *MAYA-I* and *MAYA-II*, were large molecular circuits hardwired to play versions of the game tic-tac-toe. The focus of the current protocol, *MAYA-III*, is, in contrast, a programmable protoautomaton that can be trained to perform any strategy of an invented retributive game. In all automata, we introduce human moves of the game to the automaton through input oligonucleotides, the "language" through which molecular circuits communicate. We form a sequence of inputs, and we obtain a sequence of responses that a human interprets as moves of a successful game play by the automata.

MAYA-III introduces two new concepts into the molecular automata arena, field programmable (reconfigurable) molecular logic arrays and "teaching by example" (11). It also introduces one new game, *tit-for-tat* (see Fig. 1a). The game is an example of a trivial two-player two-move game played on a board split into four fields, invented explicitly for the purpose of studying the ability to train/program molecular automata. These two-player two-move games are indeed trivial, but have one great advantage: after some restrictions in the way we observe moves (focusing only on the remaining fields for the second move), their complete action space can be represented with a field array consisting of four single input *YES* gates, responding to the inputs keyed to the first moves, and 12 dual input *AND* gates, responding to all legal combinations of inputs keyed to the first and second moves. This arrangement allows us to select any function within this game space, all with only 16 gates.

The automaton's goal in the game is to match each human move into one field with a move to another field (thus, *tit-for-tat*). The game has 81 winning strategies, defined as comprehensive sets of responses to all possible moves by the human (cf. Fig. 1b) leading to the automaton fulfilling its goal (each strategy has 4×2 possible game plays). Besides oligonucleotides that represent human moves (total of eight, four for the first moves, and four for the second moves),

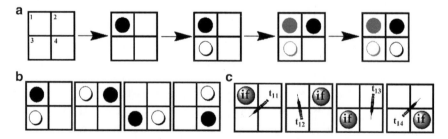

Fig. 1. (a) An example of a tit-for-tat game-play, with human moves shown in *filled circles*, and automaton moves in *unfilled*. Past moves are shown in *gray*, current moves in *black*. (b) An example of the strategy (a set of possible responses to all human first moves; in winning strategies the automaton does have any choices in the second moves). (c) A part of the training session (first move), leading playing strategy shown in (b). Training consists of injecting training inputs (t_{mn}) in an intuitive way, mimicking the actual game-play. The complete training consists of covering also second moves.

we need to introduce so-called training inputs. The training inputs allow us to activate the required gates in individual wells of the automaton in training sessions. Individual training sessions resemble playing individual games, and result in the automaton learning how to play all possible game plays within a single strategy (Fig. 1c shows a training session for all possible responses to first moves). Each training session turns a fully symmetric distribution of gates in four wells into *MAYA-I*-like situations with the automaton playing only one strategy, chosen by training. The human player/trainer does not need to understand any molecular logic in order to select a strategy; the procedure requires one only to have a key for using the training inputs.

The procedures described in this chapter permit the building and optimization of different types of molecular logic gates to combine into complex trainable molecular automata. The importance of *MAYA-III* is in the mechanism through which the device is programmed or selected, by exposure to instructional inputs in a very intuitive way. Eventually, we want to pursue complex and adaptive autonomous networks that will serve as standalone molecular devices in a two-way information exchange with the environment.

2. Materials

2.1. Instrument

1. Fluorescent plate reader (e.g., PerkinElmer Wallac Victor II 1420 Multilabel Reader) with TAMRA filters (530-nm excitation filter and a 580-nm emission filter).

2. Corning black 384-well microplates.

3. Amersham Biosciences Ultrospec 3300 pro UV/visible spectrophotometer.

4. Eppendorf L50 centrifuge.

5. Fisher Scientific Vortex mixer.

2.2. Chemicals

1. Fisher Scientific sterile DEPC-treated and nuclease-free water (see Note 1).

2. Reaction buffer: 50 mM Hepes, pH 7.0, 1 M NaCl, 1 mM $ZnCl_2$ (see Note 2).

2.3. Oligonucleotides (see Note 1)

1. Double-labeled Substrate, double RP-HPLC purified, and stored in aliquots as 300 μM stock solutions in nuclease-free water at −20°C; sequence: 5′-TAMRA-TCA CTA TrAG GAA GAG-BH_2, where preceding "r" indicates a ribonucleotide, TAMRA represents tetramethylrhodamine, and BH_2 represents Black Hole 2 quencher (see Note 3).

2. Molecular logic gate oligonucleotides (ANDNOT gates and ANDANDNOT gates; see Table 1), purified by denaturing

Table 1
The oligonucleotides used for the development of molecular array of YES and AND gates (MAYA)-III protoautomaton[a]

Game-playing input oligonucleotides (i_a and i_b)

i_{11}: GCTAGGCTATCGCGT; i_{12}: GGATCACTTACGTAT; i_{13}: AACGACTGCACCACG
i_{14}: TGCGTACTTTGGGTC; i_{21}: CTCAGGCTGTGTATT; i_{22}: CTCTCCCTGTACCCA
i_{23}: CACTATCTCGAATCA; i_{24}: ATCGCTCTCCATGCA

Molecular logic gate oligonucleotides[b]

i_{11}ANDNOTinh$_{11}$: CTGAAGAGAACGGCGATAGCCTAGCCTCTTCAGCGATGACTGACACGTAGCATCTGACAGTCCACCATGTTAGTGA

i_{12}ANDNOTinh$_{12}$: CTGAAGAGATACGTAAGTGATCCCTCTTCAGCGATGACTGACTGATGCATGAAAGTTCCCAGTCCACCATGTTAGTGA

i_{12}ANDNOTinh$_{12}$-a: GCTGAAGAGATACGTAAGTGATCCCTCTTCAGCGATGACTGATGCATGAAAGTTCCCAGTCCACCATGTTAGTGA

i_{13}ANDNOTinh$_{13}$: CTGAAGAGGCGTGGTGCAGTGCGTTCTCTTCAGCGATGACTGATGGATAAGACAACAGTCCACCATGTTAGTGA

i_{14}ANDNOTinh$_{14}$: CTGAAGAGGACCCAAAGTACGCACTCTTCAGCGATGACTGCCCGGTCAGGTATTACAGTCCACCATGTTAGTGA

i_{12}ANDi$_{21}$ANDNOTinh$_{21}$: CTGAAGAGATACGTAAGTGATCCCTCTTCAGCGATGACTGGAACAGTCCACCCATGTTAGTGAAATACAC AGCCTGAGTCACTAAC

i_{12}ANDi$_{21}$ANDNOTinh$_{21}$-a: GCTGAAGAGATACGTAAGTGATCCCTCTTCAGCGATGACTGGAACAGTCCACCCATGTTAGTGAAATAC ACAGCCTGAGTCACTAACA

i_{13}ANDi$_{21}$ANDNOTinh$_{21}$: CTGAAGAGGCGTGGTGCAGTGCGTTCTCTTCAGCGATGACTGGAACAGTCCACCCATGTTAGTGAAATACAC AGCCTGAGTCACTAAC

i_{13}ANDi$_{21}$ANDNOTinh$_{21}$-a: CTGAAGAGGCGTGGTGCAGTGCGTTCTCTTCAGCGATGACTGGAACAGTCCACCCATGTTAGTGAAATACAC ACAGCCTGAGTCACTAACA

i_{14}ANDi$_{21}$ANDNOTinh$_{21}$: CTGAAGAGGACCCAAAGTACGCACTCTTCAGCGATGACTGGAACAGTCCACCCATGTTAGTGAAATACAC AGCCTGAGTCACTAAC

i_{11}ANDi$_{22}$ANDNOTinh$_{22}$: CTGAAGAGACGGCGATAGCCTAGCCTCTTCAGCGATGACTGGAACAGTCCACCCATGTTAGTGATGGGTACA GGGAGAGTCACTAAC

i_{13}ANDi$_{22}$ANDNOTinh$_{22}$: CTGAAGAGGCGTGGTGCAGTGCGTTCTCTTCAGCGATGACTGGAACAGTCCACCCATGTTAGTGATGGGTAC AGGGAGAGTCACTAAC

i_{14}ANDi$_{22}$ANDNOTinh$_{22}$-a: CTGAAGAGGACCCAAAGTACGCACTCTTCAGCGATGACTGGAACAGTCCACCCATGTTAGTGATGGGT ACAGGGAGAGTCACTAACA

i_{14}ANDi$_{22}$ANDNOTinh$_{22}$: CTGAAGAGGACCCAAAGTACGCACTCTTCAGCGATGACTGGAACAGTCCACCCATGTTAGTGATGGGTA CAGGGAGAGTCACTAAC

i_{11}ANDi$_{23}$ANDNOTinh$_{23}$: CTGAAGAGACGGCGATAGCCTAGCCTCTTCAGCGATGACTGGAACAGTCCACCCATGTTAGTGATGATTCGA GATAGTGTCACTAAC

i_{11}ANDi_{23}ANDNOTinh$_{23}$-a: CTGAAGAGAGACGCGGATAGCCTAGCCTCTTCAGCGATGACTGGAACAGTCCACCCATGTTAGTGATGATTCG *AGATAGTGTCACTAACA*

i_{12}ANDi_{23}ANDNOTinh$_{23}$: CTGAAGAGAGATACGTAAGTGATCCCTCTTCAGCGATGACTGGAACAGTCCACCCATGTTAGTGATGATTCG *AGATAGTGTCACTAAC*

i_{12}ANDi_{23}ANDNOTinh$_{23}$-a: CTGAAGAGAGATACGTAAGTGATCCCTCTTCAGCGATGACTGGAACAGTCCACCCATGTTAGTGATGATTCG *AGATAGTGTCACTAACA*

i_{14}ANDi_{23}ANDNOTinh$_{23}$: CTGAAGAGAGGACCCAAAGTACGCACTCTTCAGCGATGACTGGAACAGTCCACCCATGTTAGTGATGATTCG *AGATAGTGTCACTAAC*

i_{11}ANDi_{24}ANDNOTinh$_{24}$: CTGAAGAGACGCGGATAGCCTAGCCTCTTCAGCGATGACTGGAACAGTCCACCCATGTTAGTGATGCATGG *AGAGCGATTCACTAAC*

i_{12}ANDi_{24}ANDNOTinh$_{24}$: CTGAAGAGAGATACGTAAGTGATCCCTCTTCAGCGATGACTGGAACAGTCCACCCATGTTAGTGATGCATGG *AGAGCGATTCACTAAC*

i_{13}ANDi_{24}ANDNOTinh$_{24}$: CTGAAGAGCGGTGGTGCAGTCGTTCTCTTCAGCGATGACTGGAACAGTCCACCCATGTTAGTGATGCATGG *AGAGCGATTCACTAAC*

Inhibitor oligonucleotides (inh$_a$ and inh$_b$)[c]

inh$_{11}$: **TGTCAGATGCTACGTGTATTATA**

inh$_{11}$-a: **TCAGATGCTACGTGTATTATA**

inh$_{11}$-b: **TGTCAGATGCTACGTGTATTATTA**

inh$_{11}$-c: **TGTCAGATGCTACGTGTATTAT**

inh$_{12}$: **TGGGAACTTTCATGCAATAAAT**

inh$_{12}$-a: **CTGGGAACTTTCATGCATAATAAT**

inh$_{12}$-b: **GGGAACTTTCATGCATCTTGGTTG**

inh$_{13}$: **TGTGTTGTCTTATCCATTATATT**

inh$_{13}$-a: **GTGTTGTCTTATCCATCTATATTA**

inh$_{14}$: **TGTAATACCTGAGCGGGGTTAATA**

inh$_{14}$-a: **GTAATACCTGAGCGGGGCCTTCTTG**

inh$_{21}$: **ACTCAGGCTGTGTATTTGCGATTATAACATGGG**

inh$_{21}$-a: **GACTCAGGCTGTGTATTCGATTATAACATGG**

inh$_{21}$-b: **TGACTCAGGCTGTGTATTCGATTATAACATGG**

inh$_{21}$-c: **GACTCAGGCTGTGTATTTGGATTATAACATGG**

inh$_{21}$-d: **GACTCAGGCTGTGTATTTGCGATTATAACATGG**

inh$_{22}$: **GACTCTCCCTGTACCCAAAAAATAACATGG**

inh$_{22}$-a: **CTCTCCCTGTACCCACAAAAAAATAACATGG**

(continued)

Table 1
(continued)

inh$_{22}$-b: *GACTCTCCCTGTACCCATAAAAATAACATGG*

inh$_{22}$-c: *ACTCTCCCTGTACCCATAAATAAATAACATGGG*

inh$_{22}$-d: *GACTCTCCCTGTACCCATAAATAAATAACATGGG*

inh$_{23}$: *GACACTATCTCGAATCAAATAATGTAACATGGG*

inh$_{23}$-a: *GACACTATCTCGAATCAATAATGTAACATGG*

inh$_{23}$-b: *TGACACTATCTCGAATCAATAATGTAACATGG*

inh$_{23}$-c: *GACACTATCTCGAATCATCATATAATGTAACATGG*

inh$_{23}$-d: *GACACTATCTCGAATCAATAATGTAACATGGG*

inh$_{23}$-e: *GACACTATCTCGAATCAATAATGCTAACATGG*

inh$_{24}$: *GAATCGCTCTCCATGCATCGTAATTTAACATGG*

inh$_{24}$-a: *GAATCGCTCTCCATGCATGCAGTAATTTAACATGG*

inh$_{24}$-b: *TGAATCGCTCTCCATGCATGCAGTAATTTAACATGG*

inh$_{24}$-c: *GAATCGCTCTCCATGCATGTAATTTAACATGG*

inh$_{24}$-d: *TGAATCGCTCTCCATGCATGTAATTTAACATGG*

inh$_{24}$-e: *TGAATCGCTCTCCATGCATGCGTAATTTAACATGG*

inh$_{24}$-f: *GAATCGCTCTCCATGCATGTTAATTTAACATGGG*

inh$_{24}$-g: *GAATCGCTCTCCATGCATCGTTAATTTAACATGG*

Training oligonucleotides (t_a and t_b)[d]

t_{11}: TATAATACACGTAGCATCTGACA

t_{11}-a: TATAATACACGTAGCATCTGA

t_{11}-b: TAATAATACACGTAGCATCTGACA

t_{12}: ATTATTATTGCATGAAAGTTCCCA

t_{12}-a: ATTATTATTGCATGAAAGTTCCCAG

t_{12}-b: CAACCAAGATGCATGAAAGTTCCC

t_{13}: AATATAATGGATAAGACAACACA

t_{13}-a: TAATATAGATGGATAAGACAACAC

t_{14}: TATTAACCCGCTCAGGTATTACA

t_{14}-a: CAAGAAGGCCCGCTCAGGTATTAC

t_{21}: CCATGTTATAATCGCAAATACACAGCCTGAGT
t_{21}-a: CCATGTTATAATCGAATACACAGCCTGAGTC
t_{21}-b: CCATGTTATAATCGAATACACAGCCTGAGTCA
t_{21}-c: CCATGTTATAATCCAAATACACAGCCTGAGTC
t_{21}-d: CCATGTTATAATCGCAAATACACAGCCTGAGTC
t_{22}: CCATGTTATTTTTTGGGTACAGGGAGAGTC
t_{22}-a: CCATGTTATTTTTTGGGTACAGGGAGAG
t_{22}-b: CCATGTTATTTTTTATGGGTACAGGGAGAGTC
t_{22}-c: CCCATGTTATTTATTTATGGGTACAGGGAGAGT
t_{22}-d: CCCATGTTATTTATTTATGGGTACAGGGAGAGTC
t_{23}: CCATGTTACATTATTTGATTCGAGATAGTGTC
t_{23}-a: CCATGTTACATTATTCGAGATAGTGTC
t_{23}-b: CCATGTTACATTATTGATTCGAGATAGTGTCA
t_{23}-c: CCATGTTACATTATATGATTCGAGATAGTGTC
t_{23}-d: CCCATGTTACATTATTGATTCGAGATAGTGTC
t_{23}-e: CCATGTTAGCATTATTGATTCGAGATAGTGTC
t_{24}: CCATGTTAAAATTACGCATGGAGAGCGATTC
t_{24}-a: CCATGTTAAAATTACTGCATGGAGAGCGATTC
t_{24}-b: CCATGTTAAAATTACTGCATGGAGAGCGATTCA
t_{24}-c: CCATGTTAAAATTACATGCATGGAGAGCGATTC
t_{24}-d: CCATGTTAAAATTACATGCATGGAGAGCGATTCA
t_{24}-e: CCATGTTAAAATTACGATGCATGGAGAGCGATTCA
t_{24}-f: CCCATGTTAAAATTAACATGCATGGAGAGCGATTC
t_{24}-g: CCATGTTAAATTAACGATGCATGGAGAGCGATTC

[a] All sequences are 5′–3′. The optimized sequences are named without suffix and used to construct MAYA-III (e.g., inh_{23} is the optimized sequence, and inh_{23}-a, inh_{23}-b, inh_{23}-c, inh_{23}-d, and inh_{23}-e are not optimal and are sequences studied in the optimization of inhibitor). The different bases (comparing with the optimized sequence) are shown in *italics* and *underlined* for the logic gates with suffix-names (i_{11}ANDi_{23}ANDNOTinh$_{23}$, etc.)

[b] The first input binding regions are shown in **bold**, the inhibitor binding regions are shown in ***bold and italics***, and for ANDANDNOT gates the second input binding region (which has crossover with the inhibiting binding region) is shown in gray

[c] For inh$_1$n, the loop binding region is shown in ***bold and italics***, and the overhang region is shown in *italics*. For inh$_2$n, the loop binding region is shown in ***bold and italics***, the catalytic core binding region is shown in **bold**, and the nonbinding linker bases are shown in *italics*

[d] Each training oligonucleotide is complementary to its inhibitor sequence and is named to match its inhibitor name, e.g., t_{23} for inh$_{23}$, t_{23}-c for inh$_{23}$-c

polyacrylamide gel electrophoresis (PAGE) and stored as 100 μM stock solutions in nuclease-free water at –20°C.

3. Inhibitor oligonucleotides (see Table 1), purified by HPLC and stored as 100 μM stock solutions in nuclease-free water at –20°C.

4. Training input oligonucleotides (see Table 1), purified by HPLC and stored as 100 μM stock solutions in nuclease-free water at –20°C.

5. Game-playing input oligonucleotides (see Table 1), purified by desalting and stored as 500 μM stock solutions.

2.4. Computer Program

Gates and oligonucleotides are tested for folding before custom synthesis using this web site: http://www.idtdna.com/analyzer/Applications/OligoAnalyzer/.

3. Methods

3.1. Design of Reconfigurable Deoxyribozyme-Based Logic Gates

Deoxyribozyme-based molecular logic gates are constructed from a modular design strategy. Two modules are combined: a catalytic module that consists of a deoxyribozyme (12), and a recognition module, usually a stem-loop oligonucleotide inspired by molecular beacons (13). Stem opening can change the conformation and allosterically control the activity of the deoxyribozyme (see Note 4). A YES gate is established by using the phosphodiesterase-based deoxyribozyme E6 (12) and positive allosteric regulation by a stem-loop attached at the 5′ or 3′ end, which blocks the substrate-recognition region of the deoxyribozyme (see Fig. 2a). Upon binding of the complementary oligonucleotide to the loop, the stem will open and the deoxyribozyme is activated, cleaving a double-labeled substrate (see Note 5). A NOT gate is established by using E6 and negative allosteric regulation by a stem loop which replaces the nonconserved loop of the E6 catalytic core (see Fig. 2b). The binding of a complementary oligonucleotide to the loop opens the stem and distorts the shape of the catalytic core, inhibiting the catalytic cleavage (6).

Four $YESi_a$ gates are required for construction of the MAYA-III protoautomaton, and are specifically designed such that they can be selected to be active through the addition of a training input (t_a) to a particular well. By combining a $YESi_a$ gate and a $NOTinh_a$ gate, an $ANDi_aNOTinh_a$ gate can be established, which is active in the presence of one input as long as a second inhibitory input inh_a is not present ((7); see Fig. 3a). For the purposes of developing trainable automata, such as MAYA-III, the inhibitory input (inh_a) that is complementary to the training input can be added prior to training, effectively creating a modified i_aANDt_a gate. During the

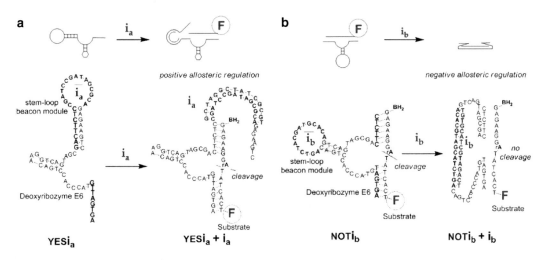

Fig. 2. (**a**) Structure of a YES gate (YESi$_a$). This gate is constructed by attaching a stem-loop beacon module to one of the substrate-recognition regions of an E6 deoxyribozyme module. Upon the addition of an input (i$_a$) complementary to the loop, the stem is opened and the gate is turned into its active form. (**b**) Structure of a NOT gate (NOTi$_b$). This gate is constructed by replacing the nonconserved loop of the E6 catalytic core with a beacon module. Opening the stem by an input (i$_b$) will distort the catalytic core and inhibit the activity of E6.

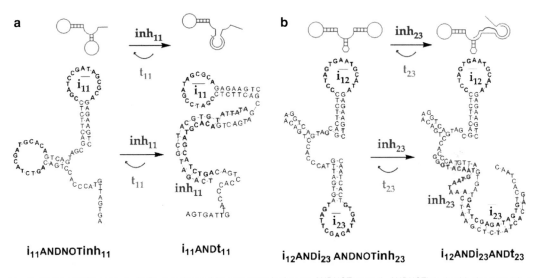

Fig. 3. (**a**) An i$_a$ANDt$_a$ gate (i$_{11}$ANDt$_{11}$) is derived by pre-complexing an ANDNOT gate (i$_{11}$ANDNOTinh$_{11}$) with the matching inhibitor oligonucleotide (inh$_{11}$) that inhibits the enzymatic activity of the deoxyribozyme; addition of t$_{11}$ (*gray*), which is complementary to inh$_{11}$, reconfigures the gate back to its original state. (**b**) An i$_a$ANDi$_b$ANDt$_b$ gate (i$_{12}$ANDi$_{23}$ANDt$_{23}$) is derived by pre-complexing an ANDANDNOT gate (i$_{12}$ANDi$_{23}$ANDNOTinh$_{23}$) with the matching inhibitor oligonucleotide (inh$_{23}$); addition of t$_{23}$ (*gray*), which is complementary to inh$_{23}$, reconfigures the gate back to its original state.

training session, addition of a training input (t$_a$) complementary to inh$_a$ will remove the inhibitory input, reconfiguring the gate back to its original YESi$_a$ function (see Fig. 3a).

MAYA-III also requires 12 i$_a$ANDi$_b$ gates that can be selected from i$_a$ANDi$_b$ANDt$_b$ gates by adding t$_b$. As above, an i$_a$ANDi$_b$ANDNOTinh$_b$

gate (9) can be established by placing first two stem-loop controlling regions on the both ends of E6, creating i_aANDi$_b$. Then, instead of a complement of t_b (inh$_b$) using a third stem loop, we used the catalytic core of the gate and the existing stem-loops recognizing i_b (see Fig. 3b). In this way, we reduced the size of oligonucleotides used from what was reported previously. Again, the inhibitory input can be added prior to training, effectively creating an i_aANDi$_b$ANDt$_b$ gate, which can be reconfigured during a training session by addition of a complementary training input.

MAYA-III incorporates 4 i_aANDt$_a$ and 12 i_aANDi$_b$ANDt$_b$ molecular logic gates, created by pre-complexing inhibitors inh$_a$ and inh$_b$ with i_aANDNOTinh$_a$ gates and i_aANDi$_b$ANDNOTinh$_b$ gates, respectively. Training of the protoautomaton is achieved through the use of eight training inputs (t_a and t_b), and game-playing is enacted by the addition of eight game-playing inputs (i_a and i_b). The process for developing a protoautomata such as MAYA-III is to first choose game playing input sequences (i_a and i_b), and design both i_aANDNOTinh$_a$ and i_aANDi$_b$ANDNOTinh$_b$ molecular logic gates. Inhibitor and training input pairs are then designed for each gate, which enables reconfiguration into both i_aANDt$_a$ and i_aANDi$_b$ANDt$_b$ molecular logic gates, respectively. The sequences used in the development of the MAYA-III protoautomaton are provided in Table 1. In the table we use i_{mn}, inh$_{mn}$ and t_{mn} notation, with m denoting first or second move, and n one of the four wells, e.g., i_{12} refers to the oligonucleotide input keyed to the first move to the second well.

We next explain construction and optimization of these logic elements.

3.1.1. Design of Logic Gates and Game-Playing Inputs

1. Build ANDNOT gates by attaching a stem-loop at the 5′ end of E6 and replacing the nonconserved loop of the E6 catalytic core by the second stem-loop (see Note 6; Fig. 3a and Table 1).

2. Build ANDANDNOT gates by attaching stem-loops at both ends of E6 (see Note 6; Fig. 3b and Table 1).

3. Choose 15-mer oligonucleotides as game-playing inputs. These inputs are complementary oligonucleotides to the loop domain (input-binding region) of the stem-loop at the 5′ and 3′ ends of E6 (see Note 7).

4. Test the designed gates and inputs using the DNA folding program.

5. Computer folding should produce the expected stem-loop structures (see Fig. 3).

6. For 15-mer loops inserted into E6 (see Fig. 3a, b), if the designed gates did not give canonical gate structures through computer program, choose other 15-mer oligonucleotides as loops (see Notes 6 and 7).

7. For all loops used for 16 logic gates, the sequences should not have a strong interaction with each other.

8. In the event that a specific sequence must be used but folding has indicated the gate will not fold appropriately, several manipulations can be performed that may help stabilize the gate structure (please refer to ref. (14)).

3.1.2. Design of Inhibitors and Training Inputs (see Table 1)

1. Inhibitors for ANDNOT gates are oligonucleotides which contain an overhang portion and a NOT loop binding portion complementary to the NOT loop region of the second stem-loop. The overhang portion should not bind with the enzyme.

2. Inhibitors for ANDANDNOT gates are constructed from oligonucleotides complementary to the input binding region of the logic gate, and by extending the oligonucleotide to also bind the catalytic core of gates. There are several nonbinding bases between two binding domains of the inhibitor.

3. Training inputs are complementary oligonucleotides of the inhibitors (see Note 7).

4. Test these sequences by using the DNA folding program.

5. These sequences should not have strong second structures.

3.2. Optimization of Deoxyribozyme-Based Logic Gates

We first test the individual logic gates with their game-playing inputs, and maximize their response strength. Then we choose a pair of inhibitor and training inputs for each gate by the procedure of optimization, in which we need the inhibitor to almost completely suppress the logic gate activity, and the training input to completely recover it.

3.2.1. Handling Oligonucleotides

1. Wear gloves and use RNase- and DNase-free water for all procedures (see Note 1).

2. Centrifuge the vial with custom oligonucleotides by minicentrifuge for 1 min before opening the cap.

3. Confirm the concentration of the received oligonucleotides by taking the OD measurements by UV/visible spectrophotometer.

4. Distribute the stock solution of substrate into multiple tubes as aliquots prior to storage at $-20°C$ in the dark.

5. After taking oligonucleotides from the freezer, wait till complete defrost, then vortex, and centrifuge.

3.2.2. Testing of Logic Gates

1. Prepare 1 μM gate solutions, 5 μM game-playing input solutions, and 2 μM substrate solution by diluting the stock solutions by the reaction buffer (see Note 1).

2. Calculate the adding amount for each component for preparing 80 μL samples with final concentrations as 50 nM gate, 500 nM substrate, 150 or 250 nM game-playing inputs (see Note 8).

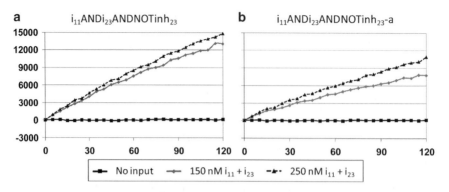

Fig. 4. Optimization of an ANDANDNOT gate. Logic gate (50 nM) i_{11}ANDi_{23}ANDNOTinh$_{23}$ (**a**) or i_{11}ANDi_{23}ANDNOTinh$_{23}$-a (**b**) was mixed with 500 nM substrate in the presence or absence of 150- or 250-nM game-playing inputs. Spectra show experimental results with time (min) on the *X*-axis and fluorescence response on the *Y*-axis. The additional nucleotide present at the 3′ end of i_{11}ANDi_{23}ANDNOTinh$_{23}$-a (*see* Table 1) reduced the rate of fluorescence increase compared to the original gate (i_{11}ANDi_{23}ANDNOTinh$_{23}$). The optimized logic gate (e.g., i_{11}ANDi_{23}ANDNOTinh$_{23}$; shown in (**a**)) should be in maximum activation in the absence of game-playing inputs, and minimum activation in the presence of game-playing inputs.

3. First add the reactive buffer and gate, then add substrate and inputs.

4. Transfer 75 µL samples into the wells of a black 384-well plate.

5. Start a fluorescence measurement with the fluorescence reader containing TAMRA filters. Follow the fluorescence increase in all wells over 120 min with measurements every 5 min (see Fig. 4).

3.2.3. Optimization of Logic Gate Structures

1. For increasing the reactive activity of the gate, remove one or two bases from the 5′- or 3′-end of the gate to shorten the stem of the stem-loop (see Fig. 4).

2. Some gates can be semi-activated in the absence of inputs or in the presence of only one input (for ANDANDNOT gates). For reducing the activity of these semi-active gates, add more bases to the 5′- or 3′-end of the gate to extend the stem of the stem-loop.

3.2.4. Testing of Inhibitors and Training Inputs

1. Prepare 1 µM gate solutions, 10 µM game-playing input solutions, 2 µM substrate solution, 1 µM inhibitor solutions, 1 µM training input solutions by diluting the stock solutions by the reaction buffer (see Note 1).

2. Calculate the adding amount for each component for preparing 80 µL samples with final concentrations as 50 nM gate, 500 nM substrate, 250 nM game-playing inputs, 75 nM inhibitor, and 100 nM training input.

3. First add the reactive buffer, gate, and inhibitor and incubate for 5 min, then add training input and incubate for another 5 min, finally add substrate and inputs.

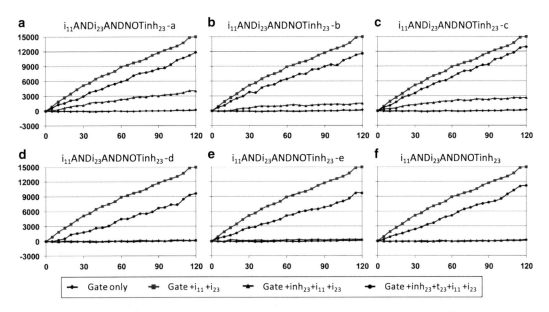

Fig. 5. Results for $i_{11}ANDi_{23}ANDNOTinh_{23}$ on optimization of inhibitors and training inputs. ANDANDNOT gate (50 nM) was premixed with 75 nM inhibitor, then mixed with 500 nM substrate and 250 nM game-playing inputs in the reaction buffer in the presence (*maroon circle*) or absence (*blue triangle*) of 100 nM training oligonucleotide. A control experiment was performed using 50 nM ANDANDNOT gate with 500 nM substrate in the presence (*red square*) or absence (*black diamond*) of 250 nM game-playing inputs. Fluorescence activity was monitored every 5 min for 120 min. Spectra show experimental results with time (min) on the X-axis and fluorescence response on the Y-axis. The results for five inhibitor-training input pairs are shown in (**a**) (inh_{23}-a, t_{23}-a), (**b**) (inh_{23}-b, t_{23}-b), (**c**) (inh_{23}-c, t_{23}-c), (**d**) (inh_{23}-d, t_{23}-d), (**e**) (inh_{23}-e, t_{23}-e), (**f**) (inh_{23}, t_{23}). The optimized inhibitor should inhibit the gate completely, and its training input should restore the activity to be close to the original activity of the gate. The optimized inhibitor-training input pair for $i_{11}ANDi_{23}ANDNOTinh_{23}$ is (inh_{23}, t_{23}; (**f**)).

4. Different inhibitor-training input pairs may need to be tested (see Subheadings 3.2.5 and 3.2.6; Table 1 and Figs. 5–7).

5. Transfer 75 μL samples into the wells of a black 384-well plate.

6. Start a fluorescence measurement with a fluorescence reader containing TAMRA filters. Follow the fluorescence increase in all wells over 120 min with measurements every 5 min (see Figs. 5–7).

3.2.5. Optimization of Inhibitors for ANDNOT Gates

1. The NOT loop binding portion is complementary to the NOT loop region. To increase the inhibition effect to prevent any "leaking" behavior, the NOT loop binding portion can be extended to bind more bases on the stem region of the stem-loop (see Table 1).

2. Changing the number of bases in the overhang portion (5 to 7-mer) will adjust the reactive rate and efficiency of the training input in subsequently removing the inhibitor (similar to Figs. 5–7).

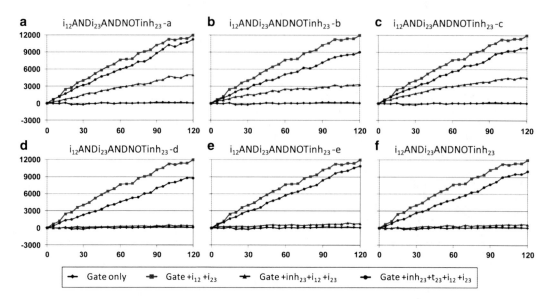

Fig. 6. Results for $i_{12}ANDi_{23}ANDNOTinh_{23}$ on optimization of inhibitors and training inputs. The experimental description and conditions are as shown in Fig. 5. The results for five inhibitor-training input pairs are shown in (**a**) (inh_{23}-a, t_{23}-a), (**b**) (inh_{23}-b, t_{23}-b), (**c**) (inh_{23}-c, t_{23}-c), (**d**) (inh_{23}-d, t_{23}-d), (**e**) (inh_{23}-e, t_{23}-e), (**f**) (inh_{23}, t_{23}). There are two optimized inhibitor-training input pairs for $i_{12}ANDi_{23}ANDNOTinh_{23}$: ($inh_{23}$-e, t_{23}-e; (**e**)) and (inh_{23}, t_{23}; (**f**)).

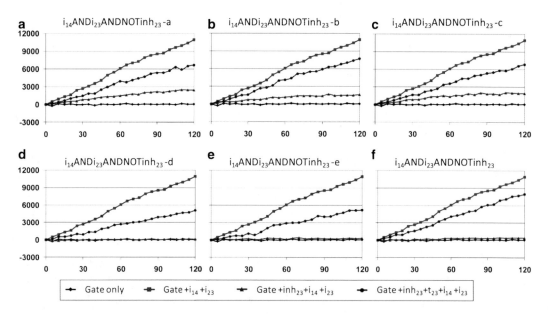

Fig. 7. Results for $i_{14}ANDi_{23}ANDNOTinh_{23}$ on optimization of inhibitors and training inputs. The experimental description and conditions are as shown in Fig. 5. The results for five inhibitor-training input pairs are shown in (**a**) (inh_{23}-a, t_{23}-a), (**b**) (inh_{23}-b, t_{23}-b), (**c**) (inh_{23}-c, t_{23}-c), (**d**) (inh_{23}-d, t_{23}-d), (**e**) (inh_{23}-e, t_{23}-e), (**f**) (inh_{23}, t_{23}). The optimized inhibitor-training input pair for $i_{14}ANDi_{23}ANDNOTinh_{23}$ is (inh_{23}, t_{23}; (**f**)), and is thus the optimized inhibitor-training input pair for the set of all three inh_{23} gates ($i_{11}ANDi_{23}ANDNOTinh_{23}$, $i_{12}ANDi_{23}ANDNOTinh_{23}$, and $i_{14}ANDi_{23}ANDNOTinh_{23}$).

1. The inhibitor binding domain for the loop is complementary to one of the logic gate stem-loops. To increase the inhibition effect, extend this region in the inhibitor to bind more bases on the stem region of the logic gate stem-loop.

2. The binding domain for the catalytic core is an 8 or 9-mer, complementary to the catalytic core of the gate. The inhibition effect can be adjusted by changing the base number and the bound position of the catalytic core.

3. The middle nonbinding portion of the inhibitor is a 6 or 7-mer. Changing the base number can adjust the reactive rate and efficiency of the training input in removing the inhibitor.

4. For MAYA-III ANDANDNOT gates, there are three gates in a set that share the same inhibitor and training input. These inhibitor-training input pairs should be tested for all three ANDANDNOT gates (see Figs. 5–7).

3.3. Training and Play-Back of the Automaton

We now describe how to assemble the protoautomaton and train it into an automaton that plays one strategy.

The tit-for-tat game is played on a board split into four fields, therefore experimentally in four wells of 384-well plate. A premade mixture of 4 AND and 12 ANDAND logic gates, covering all possible first inputs, and combinations of first and second inputs (listed above and optimized as described), is distributed into each of four wells ahead of training (see Fig. 8a). Prior to playing the game against the automaton, we must teach it how to play. We do so by example, and we train (by adding training inputs) for the complete set of games in a chosen strategy once and in that way show the moves to the automaton. To do this, a human (trainer) points to the well, and says (for the first move training) "if on the first move I play in this well, you will play *here*," simultaneously injecting a training oligonucleotide keyed to the human's move into the *here* well (see Fig. 1c). For example, if the human wants automaton to respond in well 1 after his/her first move into well 2, human will inject t_{12} into well 1.

For the second move, the procedure is the same, except we use a second move training inputs (t_{2n}), stating "if my second move is in this well, you play *here*." The procedure is repeated for all moves in the strategy. At the end of the procedure, eight training inputs are distributed into the assigned individual wells based on the chosen strategy. This training procedure results in a pruning or partial evaluation of the generic Boolean formulae into individual Boolean formulae specific for each of the wells; thereafter, the particular well is primed to accept a human move input (see Fig. 8b). After training, the automaton is ready and will play back the game according to the trained strategy (see Fig. 9).

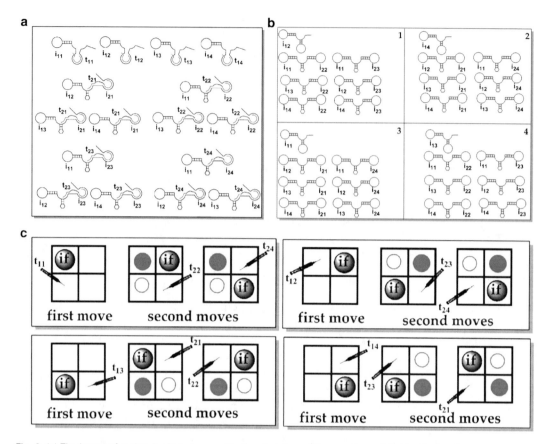

Fig. 8. (**a**) The layout of molecular logic in a well before training; all four wells contain identical molecular logic units. (**b**) The layout of molecular logic in the four wells of the automaton trained according to a "counterclockwise" strategy as shown in (**c**) (only the trimmed gates are shown, for simplicity of presentation). (**c**) An example of a complete training session using the "counterclockwise" strategy. In board representation, the human's potential new moves are represented by *black filled circles* with a word "if" inside, previous moves by *gray filled circles*, while the automaton's potential first moves are represented by *gray unfilled circle*. At the end of the training procedure, the training inputs t_{12}, t_{22}, t_{23} were added into well 1, t_{14}, t_{21}, t_{24} into well 2, t_{11}, t_{21}, t_{24} into well 3, and t_{13}, t_{22}, t_{23} into well 4.

3.3.1. Experiments for the Training and Game-Play of the Automaton

A so-called "counterclockwise" strategy will be used as an example here (see Fig. 1), all the other strategies (11) follow the similar procedure.

1. Prepare 5 μM gate solutions, 10 μM game-playing input solutions, 20 μM substrate solution, 5 μM inhibitor solutions, 5 μM training input solutions by diluting the stock solutions by the reaction buffer (see Note 1).

2. Calculate the adding amount for each component for preparing 80 μL samples with final concentrations as 50 nM for each gate (16 gates), 1,000 nM substrate, 75 nM inhibitor $inh_1 n$, 200 nM inhibitor $inh_2 n$, 100 nM training input $t_1 n$, 250 nM training input $t_2 n$ ($n = 1, 2, 3, 4$), 250 nM game-playing inputs.

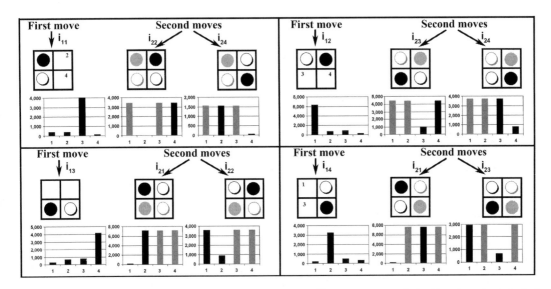

Fig. 9. A representative game play for the "counterclockwise" strategy. The automaton was trained by a "counterclockwise" strategy as shown in Fig. 1c. In board representation, the human's new moves are represented by *black filled circles*, previous moves by *gray filled circles*, while the automaton's new moves are represented by *black unfilled circles* and previous moves by *gray unfilled circles*. Inputs triggering these moves are also indicated. In bar graph representation, the experimental results indicating the automaton's move 30 min after the addition of inputs into all wells are shown with well numbers on the *horizontal* axis and fluorescence change (ΔF) on the *vertical* axis. Readouts for new moves are given in *black*; in the second-move displays, old (first) move readouts are indicated by maximized *bars* of gray color (see Note 9).

3. Mix the optimized 16 gates (i_{11}ANDNOTinh$_{11}$, i_{12}ANDNOTinh$_{12}$, i_{13}ANDNOTinh$_{13}$, i_{14}ANDNOTinh$_{14}$, i_{12}ANDi$_{21}$ANDNOTinh$_{21}$, i_{13}ANDi$_{21}$ANDNOTinh$_{21}$, i_{14}ANDi$_{21}$ANDNOTinh$_{21}$, i_{11}ANDi$_{22}$ANDNOTinh$_{22}$, i_{13}ANDi$_{22}$ANDNOTinh$_{22}$, i_{14}ANDi$_{22}$ANDNOTinh$_{22}$, i_{11}ANDi$_{23}$ANDNOTinh$_{23}$, i_{12}ANDi$_{23}$ANDNOTinh$_{23}$, i_{14}ANDi$_{23}$ANDNOTinh$_{23}$, i_{11}ANDi$_{24}$ANDNOTinh$_{24}$, i_{12}ANDi$_{24}$ANDNOTinh$_{24}$, i_{13}ANDi$_{24}$ANDNOTinh$_{24}$) and the optimized eight inhibitors (inh$_{11}$, inh$_{12}$, inh$_{13}$, inh$_{14}$, inh$_{21}$, inh$_{22}$, inh$_{23}$, inh$_{24}$) with the reactive buffer and incubate for 5 min. Prepare four identical solutions representing game wells 1, 2, 3, 4.

4. For "counterclockwise" strategy, through the training procedure, the final addition of the training inputs in each well as following: t_{12}, t_{22}, t_{23} into well 1, t_{14}, t_{21}, t_{24} into well 2, t_{11}, t_{21}, t_{24} into well 3, t_{13}, t_{22}, t_{23} into well 4 (see Fig. 8c). Incubate for another 5 min.

5. Add substrate into four wells, then inject game-playing input i_{11} into all four wells.

6. Transfer 75 µL samples into the four wells of a black 384-well plate.

7. Start a fluorescence measurement with a fluorescence reader containing TAMRA filters. Follow the fluorescence increase in all wells over 120 min with measurements every 5 min (record the 30-min response; see Fig. 9).

8. Repeat steps 3 and 4 above to prepare four samples, add substrate into each sample, then inject i_{11} into each sample for 30-min incubation. Finally, inject i_{22} into each sample.

9. Transfer 75 μL samples into the four wells of a black 384-well plate. Monitor the fluorescence response every 5 min for 120 min (taking 30-min response for Fig. 9).

10. There are eight game-plays for each strategy. Repeat this procedure for the other seven game-plays.

4. Notes

1. Always wear gloves and use RNase- and DNase-free water for all procedures. Although RNase-free water is not required for DNA oligonucleotides, the substrates contain embedded ribonucleotides.

2. To prevent possible precipitation, $ZnCl_2$ (from a 100 mM acidified stock solution) is added after adjusting the Hepes buffer to pH 7 by the addition of NaOH.

3. Dye-labeled substrates are light sensitive and should be stored in the dark to prevent photo-bleaching of the fluorescent labels over time.

4. Gates are designed to be fully modular by combining a catalytic module E6 and a recognition module stem-loop. With some limitations imposed by strong alternative secondary structures, gates could potentially be constructed to sense almost any input oligonucleotide.

5. Substrate is labeled with a fluorophore and a quencher on both ends. Upon cleavage by E6, there is an increase in fluorescence due to the separation of the fluorophore and the quencher.

6. The general rule is that introduced loops should not interact with one another, and should not interact with the enzyme portion of the gates.

7. There are certain limitations on the structures of inputs that can be accepted by the gates. Some inputs have strong secondary structures, and cannot open up stems efficiently, while some loops favor alternative conformations with E6.

8. Multiple concentrations of the game-playing inputs should be tested to confirm digital behavior of gates, and to titrate for maximum activation of gates.

9. To simplify the training procedure and to minimize the number and size of molecular logic gates in solution, for the automaton's second move, we observe the output only in the remaining fields and ignore any secondary responses in the two fields already played on the first move.

Acknowledgments

The authors were supported by the National Science Foundation. M.N.S. is a Lymphoma and Leukemia Society Fellow. We also thank our collaborator Darko Stefanovic and other coauthors on the initial publication.

References

1. Benenson, Y., Gil, B., Ben-Dor, U., Adar, R. , and Shapiro, E. (2004) An autonomous molecular computer for logical control of gene expression. *Nature* **429**, 423–429.

2. Ran, T., Kaplan, S. , and Shapiro, E. (2009) Molecular implementation of simple logic programs. *Nature Nanotechnol.* **4**, 642–648.

3. Seelig, G., Soloveichik, D., Zhang, D. Y. , and Winfree, E. (2006) Enzyme-free nucleic acid logic circuits. *Science* **314**, 1585–1588.

4. Zhang, D. Y., Turberfield, A. J., Yurke, B. , and Winfree, E. (2007) Engineering entropy-driven reactions and networks catalyzed by DNA. *Science* **318**, 1121–1125.

5. Wilner, O. I. et al. (2009) Enzyme cascades activated on topologically programmed DNA scaffolds. *Nature Nanotechnol.* **4**, 249–254.

6. Stojanovic, M. N., Mitchell, T. E., and Stefanovic, D. (2002) Deoxyribozyme-based logic gates. *J. Am. Chem. Soc.* **124**, 3555–3561.

7. Stojanovic, M. N., and Stefanovic, D. (2003) Deoxyribozyme-based half-adder. *J. Am. Chem. Soc.* **125**, 6673–6676.

8. Lederman, H., Macdonald, J., Stefanovic, D. , and Stojanovic, M. N. (2006) Deoxyribozyme-based three-input logic gates and construction of a molecular full adder. *Biochem.* **45**, 1194–1199.

9. Stojanovic, M. N., and Stefanovic, D. (2003) Deoxyribozyme-based automaton. *Nature Biotechnol.* **21**, 1069–1074.

10. Macdonald, J. et al. (2006) Medium scale integration of molecular logic gates in an automaton. *Nano. Lett.* **6**, 2598–2603.

11. Pei, R., Matamoros, E., Liu, M., Stefanovic, D., and Stojanovic, M.N. (2010) Training a molecular automaton to play a game. *Nature Nanotech.* **5**, 773–777.

12. Breaker, R. R., and Joyce, G. F. (1995) A DNA enzyme with Mg^{2+}-dependent RNA phosphoesterase activity. *Chem. Biol.* **2**, 655–660.

13. Tyagi, S., and Krammer, F. R. (1996) Molecular beacons: Probes that fluoresce upon hybridization. *Nature Biotechnol.* **14**, 303–309.

14. Macdonald, J., Stefanovic, D., and Stojanovic, M.N. (2006) Solution-phase molecular-scale computation with deoxyribozyme-based logic gates and fluorescent readouts. *Methods in Molecular Biology*, **335**, 343–363.

Chapter 27

Rational Design and Tuning of Ribozyme-Based Devices

Joe C. Liang and Christina D. Smolke

Abstract

A synthetic gene-regulatory device platform was described by modularly assembling three RNA components encoding distinct functions of sensing, transmission, and actuation. The molecular binding at the sensor component is translated by the transmitter component through a strand-displacement event to modulate activity of the actuator component, which then interacts with cellular transcriptional machinery to affect gene expression levels. Here, we provide some general guidelines on linking RNA components to construct gene-regulatory devices and strategies to tune device regulatory activities through an RNA folding program for specific cellular applications.

Key words: Rational design, Ribozyme, RNA folding, RNA switch, Synthetic biology

1. Introduction

Significant research efforts have been directed toward developing synthetic genetic devices that are capable of integrating diverse biological signals to control cellular function (1–5). An RNA-based gene-regulatory device platform was recently described to support the modular assembly of ribozyme-based switches responsive to diverse molecular inputs (2, 6). A ribozyme-based device is composed of three functional components: an actuator component, comprising a hammerhead ribozyme; a sensor component, comprising an RNA aptamer; and a transmitter component, comprising a sequence that transmits information between the actuator and sensor components through a strand-displacement mechanism (see Fig. 1). The device platform specifies a standard method for linking these functional components and utilizing the resulting RNA device to modulate gene expression.

The gene-regulatory activities of the ribozyme-based switches are largely determined by the relative partitioning between the two primary functional conformations. For example, in the case of

Jörg S. Hartig (ed.), *Ribozymes: Methods and Protocols*, Methods in Molecular Biology, vol. 848,
DOI 10.1007/978-1-61779-545-9_27, © Springer Science+Business Media, LLC 2012

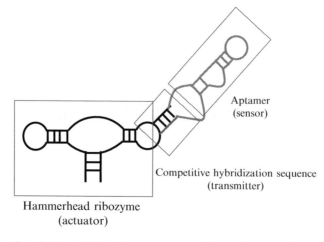

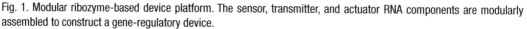

Fig. 1. Modular ribozyme-based device platform. The sensor, transmitter, and actuator RNA components are modularly assembled to construct a gene-regulatory device.

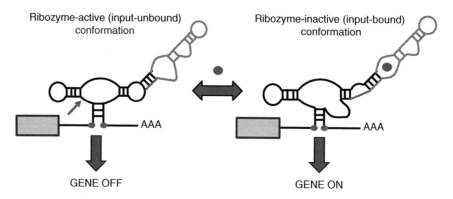

Fig. 2. Ribozyme-based ON switch. Two primary functional conformations are adopted by the ON switch. The binding of ligand to the aptamer component in the ribozyme-inactive (input-bound) conformation shifts the distribution of the two conformations, thereby modulating expression levels of the coupled gene.

an ON switch, the device can adopt either a ribozyme-active (input-unbound) conformation or ribozyme-inactive (input-bound) conformation (see Fig. 2). In the ribozyme-active conformation, the ribozyme component in the device undergoes self-cleavage and results in an unprotected transcript that is subject to rapid degradation by cellular ribonucleases, thereby lowering gene expression. The binding of the molecular input to the aptamer component in the device lowers the free energy of the input-bound (ribozyme-inactive) conformation, and thus shifts the distribution of structures to favor the formation of the input-bound conformation, resulting in increased gene expression. The level of gene expression from the RNA device can therefore be modulated through the concentration of molecular ligand present in the system to shift the distribution of conformations between ribozyme-active and -inactive states.

The gene-regulatory activities of a synthetic RNA device are characterized by a number of quantitative performance descriptors associated with the device response curve (see Fig. 3). In particular, the most common performance descriptors for RNA switches are basal expression level, ligand-saturating expression level (theoretical and experimental), dynamic range (difference and ratio), and EC_{50} value. In practice, these device regulatory properties must be fine-tuned when a device is integrated into a particular biological system in order to match activities of the RNA device with those of other system components, such as promoter strength or enzyme activity. The modular device framework enables the tailoring of individual component activities without significant device redesign. As an example, the transmitter component linking the sensor and actuator components is composed of two strands that compete for binding to a common hybridization region. The relative energetic contribution of the strand hybridization event determines the partitioning between the two primary functional conformations and thus the resulting regulatory properties. RNA folding programs can be used to predict the energetic differences between the alternative RNA hybridization events. In this section, we provide some general guidelines in the design of ribozyme-based switches and tuning of their basal and ligand-saturating expression levels. We also discuss the impacts of such tuning on the resulting dynamic ranges of the devices.

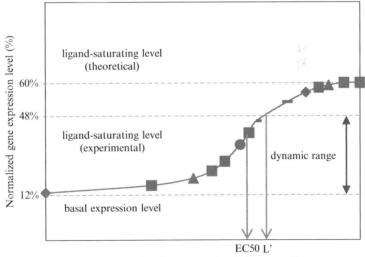

Fig. 3. Dose–response curve for an RNA switch. Here, we have annotated common performance descriptors for RNA switches based on a case where the maximum ligand concentration that can be experimentally added to the system (L') does not allow for the full gene expression ON state of the device to be attained. The switch has a dynamic range of four-fold (ratio) or 36% (difference) and has a theoretical ligand-saturating level of 60%.

2. Materials

2.1. RNA Sequences

1. RNA component sequences. Here, we have included sequences of the actuator component (sTRSV hammerhead ribozyme), sensor component (theophylline aptamer), and insulator regions, which isolate the device from interacting with neighboring sequences or structures (see Table 1). Several sample sequences of theophylline-responsive ribozyme ON switches (i.e., increased gene expression with increasing theophylline concentrations) have also been included for demonstration purposes (see Table 2). Each component in the device sequence has been formatted according to its function: actuator component (italicized), transmitter component (underlined), and sensor component (bold-faced).

2. RNA folding program. There are many RNA folding programs available and each uses slightly different algorithms in computing potential structures (7–9). We have found that *RNAstructure* in particular predicts the desired structures and is simple to browse structures with different energetics without any sophisticated programming skills (7). Here, we are using *RNAstructure v5.2* for demonstration purposes and will detail steps to analyze ribozyme-based devices. This software package can be downloaded for free from the following website: http://rna.urmc.rochester.edu/RNAstructure.html.

2.2. Device Performance Characterization

1. Characterization plasmid that harbors an appropriate reporter protein for measuring the quantitative performance of the devices. Here, we choose yeast as our cell host and use a yeast-enhanced version of green fluorescent protein (yEGFP) on a low-copy plasmid under control of the wild-type TEF1 promoter. The details of the cloning strategy, yeast strain, cell media and growth conditions, and fluorescence reporter assay have been previously described (2).

Table 1
Basic component sequences for ribozyme-based devices

Component	Sequence
sTRSV	GCUGUCACCGGAUGUGCUUUCCGGUCU GAUGAGUCCGUGAGGACGAAACAGC
Theophylline aptamer	AUACCAGCAUCGUCUUGAUGCCC UUGGCAG
5′ Insulator region	AAACAAACAAA
3′ Insulator region	AAAAAGAAAAAUAAAAA

Table 2
Sequences for ribozyme-based devices with different transmitter components

Device	Sequence
L2bulge1	*GCUGUCACCGGAUGUGCUUUCCGGUCUGAUGAGUCCGU*<u>*GUCC*</u> **AUACCAGCAUCGUCUUGAUGCCCUUGGCAG**<u>**GGACGGGAC**</u> *GAGGACGAAACAGC*
L2bulge9	*GCUGUCACCGGAUGUGCUUUCCGGUCUGAUGAGUCCGU*<u>*GUCC*</u> **AAUACCAGCAUCGUCUUGAUGCCCUUGGCAG**<u>**UGGACGGGAC**</u> *GAGGACGAAACAGC*
L2bulge10	*GCUGUCACCGGAUGUGCUUUCCGGUCUGAUGAGUCCGU*<u>*GUCCG*</u> **AUACCAGCAUCGUCUUGAUGCCCUUGGCAG**<u>**CGGACGGGAC**</u> *GAGGACGAAACAGC*
L2bulge11	*GCUGUCACCGGAUGUGCUUUCCGGUCUGAUGAGUCCGUU*<u>*UGUCC*</u> **AUACCAGCAUCGUCUUGAUGCCCUUGGCAG**<u>**GGACGGGACA**</u> *GAGGACGAAACAGC*
L2bulge12	*GCUGUCACCGGAUGUGCUUUCCGGUCUGAUGAGUCCGUU*<u>*UGUCC*</u> **AUACCAGCAUCGUCUUGAUGCCCUUGGCAG**<u>**GGACGGGACG**</u> *GAGGACGAAACAGC*

3. Methods

The design of ribozyme-based switches involves three major steps: (1) assembly of RNA components (2) integration of ribozyme-based device in the genetic constructs (3) tuning of device regulatory activities to meet application-specific performance requirements. In this section, we begin by providing some basic directions on using the RNA folding program (*RNAstructure*) to analyze RNA switches and discuss some general guidelines on how to use a combined directed and iterative (trial-and-error) process to design ribozyme-based switches for specific applications. For simplicity, we focus on the design of ribozyme-based ON switches, but the strategy is generalizable to OFF switches.

3.1. Basics on RNAstructure

1. Upon starting *RNAstructure*, under *File*, a new or an existing sequence file with a *seq extension can be created/opened.

2. After the file is created/opened, a *RNA Fold* window will appear and the following parameters are shown:

   - *Max% Energy Difference* is set to 10% by default. Since there are a large number of possible secondary structures for a given sequence, this parameter sets a range that is determined by the energetics of the minimal free energy (MFE) structure, such that only structures that have energetic values within 10% of the MFE structure energetic value will be computed. For example, if the MFE structure has an energetic value of −30 kcal/mol. Then, only structures that have energetic values between −27 and −30 kcal/mol will be computed.

- *Max Number of Structures* is set to 20 by default. This parameter sets the maximum number of structures that are computed within the *Max% Energy Difference* range, whichever constraint is met first.

- *Window Size* is set to 5 by default. This parameter indicates the degree of similarity in the base-pairing patterns between suboptimal structures. A smaller window will result in a larger set of structures that are similar. A larger window will result in fewer structures that are more different in predicted pairs. We typically set *Window Size* to 0, which is the lowest value possible, in order to observe all possible functional conformations for ribozyme-based switches.

3. Click *Start* to generate structures for the given sequence. To browse different structures, go to *Draw* and then *Go to Structure Number*.

3.2. Assemble RNA Components to Construct a Ribozyme-Based Device

The composition framework of the ribozyme-based device supports the modular assembly of RNA components. As an example, a tetracycline-responsive ribozyme-based switch was previously constructed by directly replacing the theophylline aptamer component of a switch with a tetracycline aptamer (2). Aptamers selected from in vitro strategies typically are composed of a base stem and a binding core, where the base stem is often not involved in molecular binding, and thus typically has less sequence restriction than nucleotides in the binding core.

As a first step to construct a new switch, we suggest removing the base stem sequence from the aptamer of interest and inserting the remaining sequence into an existing functional switch (Note 1 and 2). The sequence of the full RNA device is analyzed by *RNAstructure*. In the case of ribozyme-based ON switches, there are two primary conformations of interest: ribozyme-active (input-unbound) and ribozyme-inactive (input-bound) conformations. The ribozyme-active conformation is associated with the correct formation of the ribozyme component and incorrect formation of the aptamer component, while the ribozyme-inactive conformation is associated with the incorrect formation of the ribozyme component and correct formation of the aptamer component. We use L2bulge1 here as an example (see Tables 2 and 3). Two structures are generated from an analysis of L2bulge1 (see Fig. 4). The MFE structure indicates the correct formation of the ribozyme component and therefore represents the ribozyme-active (ligand-unbound) conformation. The second lowest free energy structure indicates the correct formation of the aptamer component and therefore represents the ribozyme-inactive (input-bound)

Table 3
Gene-regulatory activities of theophylline-responsive ribozyme-based ON switches

Device	Ribozyme-active	Ribozyme-inactive	$\Delta\Delta G$ (kcal/mol)	0 mM (%)	5 mM (%)
L2bulge1	−36.4	−35.5	−0.9	40	75
L2bulge9	−37.0	−36.9	−0.1	70	80
L2bulge10	−36.7	−37.9	1.2	80	84
L2bulge11	−37.7	−33.9	−3.8	8	12
L2bulge12	−36.9	−34.8	−2.1	9	37

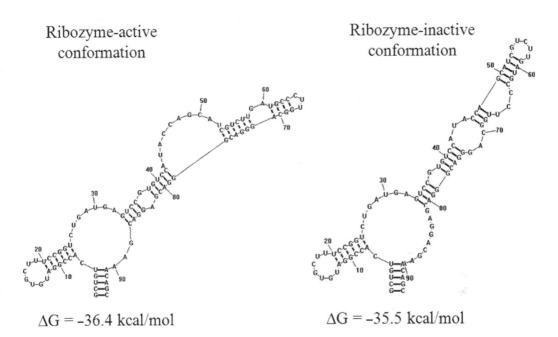

Ribozyme-active conformation

$\Delta G = -36.4$ kcal/mol

Ribozyme-inactive conformation

$\Delta G = -35.5$ kcal/mol

Fig. 4. *RNAstructure* analysis of L2bulge1. The predicted energetic difference between the two primary conformations is −0.9 kcal/mol.

conformation. The energetic difference between the two structures is predicted to be 0.9 kcal/mol.

In general, ribozyme-based devices may only exhibit regulatory activities within a certain range of energetic difference between the two primary conformations. Upon linking of a new aptamer component to the device platform, further strengthening or weakening the hybridization energies of the transmitter component may be necessary for the device to be functional in the cell host. Since RNA folding programs rely on energetic values obtained from in vitro measurement, the predicted energetics may not reflect exactly the values in the cellular environment. Therefore, users need to design switches with a range of energetic differences between the two primary conformations to pinpoint a functional range. Details of such tuning strategies are discussed in Subheading 3.5.

3.3. Integrate Ribozyme-Based Device into Genetic Construct

The RNA devices are flanked with insulation sequences (see Table 1) and inserted in the 3′ untranslated region (UTR) of the target transcript. The insulation sequences prevent the device from interacting with neighboring sequences that may disrupt its structure. The 3′ UTR is a more flexible sequence space for harboring regulatory elements, where the insertion of the device is anticipated to have minimal impact on gene expression. When constructing a new device, RNA folding program should be used to confirm that the insulation sequences do not interfere with the proper formation of the functional device conformations.

3.4. Characterize Ribozyme-Based Device

To compare regulatory activities across different ribozyme-based devices, it is essential to have a standard characterization method to determine the performance descriptors for the switches, including basal expression level, ligand-saturating expression level, dynamic range, and EC_{50} value. The gene-regulatory activities of switches can be measured in expression constructs harboring fluorescent proteins (2, 3) or reporter enzymes (4). Fluorescent reporter assays provide a linear relationship between the fluorescence measurement and device gene-regulatory activities, whereas enzyme activity assays may introduce nonlinear effects into the resulting activity measurement. We suggest using fluorescent proteins to quantify gene-regulatory activities of RNA devices, because these measures are more reliable predictors of the quantitative gene-regulatory effects when the device is coupled to a different gene.

Before we provide definitions for the switch performance descriptors, we define a maximum gene expression level to which all device activities are normalized. The maximum gene expression level for a given system is determined by cloning an sTRSV ribozyme with a scrambled core into the 3′ UTR of *yegfp* gene and is set to 100%. The expression level from an inactive device (here, scrambled-core ribozyme) should be compared to the expression level from a construct harboring no device, to determine the nonspecific effects of the device structure on expression levels from the characterization construct. For example, inactive device controls have indicated that the nonspecific effects of device structure result in an ~20% decrease in *yegfp* gene expression in yeast.

The device regulatory activities are characterized in the presence of a range of ligand concentrations, which are selected to appropriately define the dose–response curve. An appropriate negative control (e.g., cells that harbor a plasmid without the fluorescent reporter protein) is analyzed to determine the background fluorescence level. We define a basal expression level from a device to be

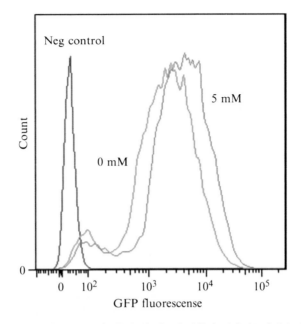

Fig. 5. Fluorescence histograms of cells that harbor the L2bulge1 device. Cells harboring the L2bulge1 device are characterized at 0 mM and 5 mM theophylline. A histogram of the negative control (cells harboring a plasmid construct lacking the *yegfp* gene) is also shown.

the mean fluorescence levels of cells harboring the device characterization plasmid in the absence of the appropriate input ligand. Fluorescence histograms of cells harboring a functional device should show a noticeable shift in cell population upon addition of appropriate ligand (see Fig. 5). Background fluorescence levels (represented by cells not harboring a construct) should be gated out prior to analyzing the mean fluorescence levels of cells harboring a functional device. In cases of devices having low basal expression levels, the fluorescence histograms of cells harboring a functional device may have a significant overlap with the histogram of the negative control, such that exclusion of the background fluorescence level is not appropriate. Then, it is important to report both the mean fluorescence levels of the negative control and cells harboring the functional device. The inactive device is also characterized in the same range of ligand concentrations, and the gene expression levels from the inactive device at the respective ligand concentration are used to normalize device regulatory activities to account for any nonspecific effects of the ligand on fluorescence levels in the cell host. Device characterization should not be conducted at ligand concentrations that significantly affect cell viability due to potential nonspecific effects of cell stress and health on gene expression or cellular fluorescence. Also, ligand solubility and cell membrane permeability may affect the maximum intracellular ligand concentration. In many cases, the maximum ligand concentration that can be added to the cells is lower than the concentration of ligand required to saturate the dose–response curve. In this situation, one refers to the experimental and theoretical ligand-saturating expression levels, where the latter can be determined by fitting parameters to the dose–response curve (see Eq. 1). It is important to note that the maximum ligand concentration for a given system is specific to the cell host and ligand, as ligands may exhibit varying toxicities to different cell hosts and permeability across the cell membrane.

$$\Upsilon = b + \frac{(M - b) * L}{EC_{50} + L},\qquad(1)$$

where Υ = normalized gene expression level at L, L = ligand concentration, b = basal expression level, M = maximum theoretical ligand concentration.

The dynamic range of a switch is set by the normalized ligand-saturating expression level (experimental) and the normalized basal expression level. The dynamic range can either be reported as the difference between these two measures or as the ratio of these two measures. Dynamic ranges reported as ratios will tend to skew toward lower basal expression level devices. Dynamic ranges reported as differences will skew toward maximal differences between the on and off states of the device. The EC_{50} value of a switch is set by the ligand concentration at which a half-maximal expression level is achieved. If theoretical ligand-saturating expression levels cannot be achieved due to properties of the ligand, the EC_{50} value is derived from fitting parameters to the dose–response curve (see Eq. 1).

3.5. Tune Regulatory Activities of the Ribozyme-Based Device

The regulatory activities of a ribozyme-based device depend on the partitioning between the two primary functional conformations (Note 3). An RNA folding program is used to compute the predicted energetic differences between the primary conformations. In the case of ribozyme-based ON switches, if the ribozyme-active (input-unbound) conformation is more thermodynamically favorable than the ribozyme-inactive (input-bound) conformation, then the basal activity of the device will generally be low. On the other hand, if the ribozyme-inactive (input-bound) conformation is more thermodynamically favorable than the ribozyme-active (input-unbound) conformation, the basal activity of the device will generally be high (Note 4). Here, we use *RNAstructure* to analyze a set of theophylline-responsive switches with varying energetic differences between the primary conformations and present some general observations on tuning switch regulatory activities (see Table 3).

First, we want to identify the boundary of energetic differences in which the switch exhibits regulatory activities. We designed L2bulge10 and L2bulge11 as two extreme cases. For L2bulge10, the ribozyme-inactive (input-bound) conformation is much

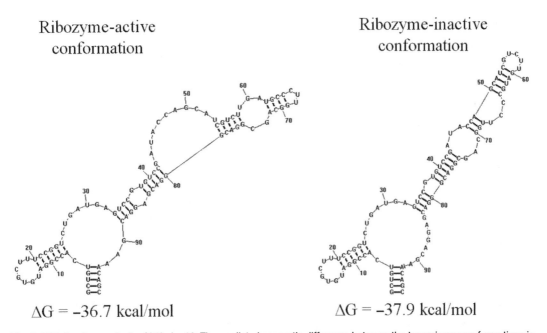

Fig. 6. *RNAstructure* analysis of L2bulge10. The predicted energetic difference between the two primary conformations is 1.2 kcal/mol.

more thermodynamically favorable than the ribozyme-active (input-unbound) conformation (see Fig. 6). We postulate that the majority of the RNA folds into the input-bound conformation even in the absence of ligand, and therefore the basal expression level is high (80%) and there is little response to ligand addition. For L2bulge11, the ribozyme-active (input-unbound) conformation is much more thermodynamically favorable than the ribozyme-inactive (input-bound) conformation (see Fig. 7). We postulate that since most of the RNA folds into the ribozyme-active conformation, the conformation switching between the two primary conformations may not be rapid enough to compete with the ribozyme cleavage, and therefore the basal expression level is low (8%) and there is little response to ligand addition. From these two cases, we conclude that for the device to exhibit regulatory activities, the predicted energetic differences should be somewhere in between those of L2bulge10 and L2bulge11.

Based on the observation on L2bulge10 and L2bulge11, three additional switches (L2bulge1, L2bulge9, and L2bulge12) with moderate energetic differences between the two functional conformations are designed. We have designed L2bulge1 to have an energetic difference that still thermodynamically favors the ribozyme-active (input-unbound) conformation but to a lesser extent than L2bulge11 (see Fig. 4). The resultant L2bulge1 device has a basal expression level of 40% and has an expression level of 75% at 5 mM theophylline. By further increasing the energetic

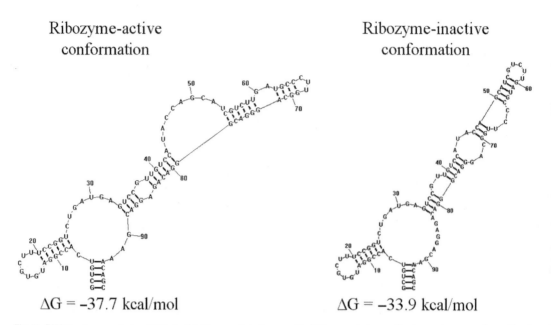

Fig. 7. *RNAstructure* analysis of L2bulge11. The predicted energetic difference between the two primary conformation is −3.8 kcal/mol.

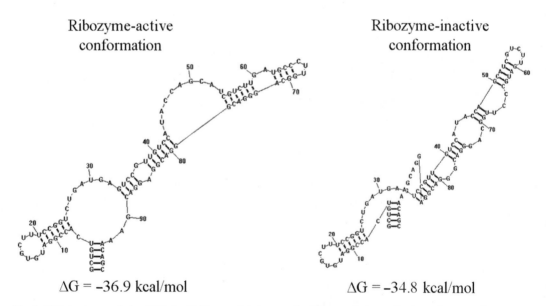

Fig. 8. *RNAstructure* analysis of L2bulge12. The predicted energetic difference between the two primary conformation is −2.1 kcal/mol.

difference between the two functional conformations, but still to an extent less than L2bulge11, we have designed L2bulge12, which has a lower basal expression level (9%) compared to L2bulge1, but also exhibits a lower expression level at 5 mM theophylline (37%) (see Fig. 8). As the majority of the L2bulge12 RNA folds

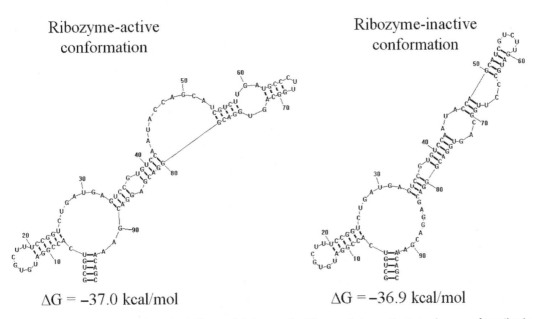

Fig. 9. *RNAstructure* analysis of L2bulge9. The predicted energetic difference between the two primary conformation is −0.1 kcal/mol.

initially into the ribozyme-active conformation and cleaves quickly, we postulate that the rates of ligand binding and/or conformational switching may not be fast enough to stabilize the ribozyme-inactive (input-bound) conformation. In addition, the practical limitations on theophylline levels that can be added to the cell do not allow for the dose–response curve to saturate. These two factors contribute to compromising the dynamic range (difference) of this switch. We have also designed L2bulge9 by decreasing the energetic difference between the two functional conformations, which leads to a nearly equal partitioning between the two states (see Fig. 9). L2bulge9 has a high basal expression (70%), which is close to that of L2bulge10. Since some of the L2bulge9 RNA has already adopted a ribozyme-inactive (input-unbound) conformation in the absence of ligand, we postulate that the slow ribozyme cleavage and/or conformational switching results in a high basal expression level. On the other hand, since there is sufficient amount of L2bulge9 RNA that adopts the ribozyme-inactive (input-bound) conformation, addition of high concentration of theophylline shifts almost all RNA to the input-bound conformation, thereby resulting in high gene expression levels (79%). It has to be noted that there are only one to two nucleotide differences among L2bulge1, L2bulge9, and L2bulge12, but the resultant regulatory activities are substantially different (see Fig. 10).

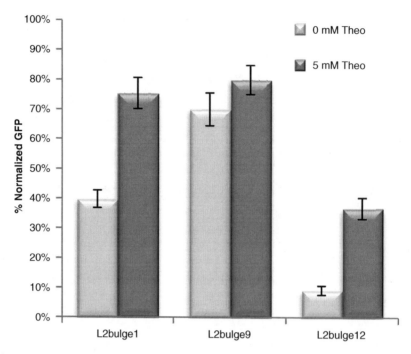

Fig. 10. Regulatory activities of L2bulge1, L2bulge9, and L2bulge12 in 0 mM and 5 mM theophylline. Despite their strong sequence similarities, all three switches exhibit substantially different regulatory activities.

4. Notes

1. When integrating a new aptamer into the device platform, the base stem of the aptamer often needs to be modified to ensure compatibility with the transmitter components. It is important to confirm that upon modification of this region of the aptamer sequence, the binding properties of the aptamer are not affected.

2. Aptamers selected and characterized in vitro may not be functional in cellular environments due to differences in folding, intracellular ligand concentrations, or incompatibility with other device components. However, it may be challenging to confirm the in vivo functionality of new aptamers, particularly to small molecule ligands, as independent components and often this verification is done by integrating the aptamer into a device platform.

3. Regulatory activities of ribozyme-based devices may depend on the rates of several processes, including conformational switching of the RNA device between the functional conformations, ribozyme cleavage, and ligand binding. The rates of these events in cellular environments are challenging to measure.

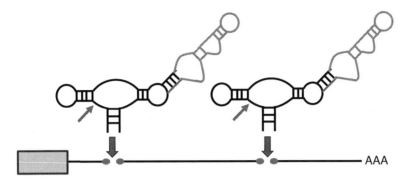

Fig. 11. Multicopy assembly of ribozyme-based devices. Multiple single-input ribozyme-based switches can be integrated into the 3′ UTR of a target transcript. This assembly strategy is supported by the modular framework of the device itself. Devices composed of multiple copies of identical or different switches can be constructed through this strategy (Note 5).

Therefore, predicted energetic values computed by the RNA folding program may not fully capture the regulatory mechanism of the device.

4. RNA folding programs may sometime generate several structures that share the same functional conformation. As an example, *RNAstructure* predicts the L2bulge12 sequence to have the two lowest free energy structures as the ribozyme-active (input-unbound) conformations, and the ribozyme-inactive (input-bound) conformation is predicted as the third lowest energy structure. Having multiple ribozyme-active conformations as lower free energy structures may result in a lower basal gene expression level.

5. The framework of the ribozyme-based device platform also supports extension to multi-input genetic devices by integrating multiple single-input ribozyme-based switches into the 3′ UTR of a target transcript (see Fig. 11). This design strategy has resulted in the construction of two-input logic gates (AND, NOR gates) by integrating single-input devices to different molecular inputs and filters (signal and bandpass) by integrating single-input devices to the same molecular input (6). In one example, the integration of multiple copies of the same single-input device was used to tailor the regulatory stringency of a T-cell signaling circuit for in vivo control of T-cell proliferation and survival (5).

Acknowledgments

The Smolke Laboratory is supported by funds from the NIH, NSF, and the Alfred P. Sloan Foundation.

References

1. Isaacs, F. J., Dwyer, D. J., and Collins, J. J. (2006) RNA synthetic biology., *Nature Biotechnology*. Nature Publishing Group *24*, 545–554.

2. Win, M. N., and Smolke, C. D. (2007) A modular and extensible RNA-based gene-regulatory platform for engineering cellular function., *Proceedings of the National Academy of Sciences of the United States of America*. National Academy of Sciences *104*, 14283–14288.

3. Wieland, M., Benz, A., Klauser, B., and Hartig, J. S. (2009) Artificial ribozyme switches containing natural riboswitch aptamer domains., *Angewandte Chemie International Edition 48*, 2715–2718.

4. Sinha, J., Reyes, S. J., and Gallivan, J. P. (2010) Reprogramming bacteria to seek and destroy an herbicide., *Nature Chemical Biology*. Nature Publishing Group *6*, 464–470.

5. Chen, Y. Y., Jensen, M. C., and Smolke, C. D. (2010) Genetic control of mammalian T-cell proliferation with synthetic RNA regulatory systems., *Proceedings of the National Academy of Sciences of the United States of America 107*, 8531–8536.

6. Win, M. N., and Smolke, C. D. (2008) Higher-order cellular information processing with synthetic RNA devices., *Science*. AAAS *322*, 456–460.

7. Reuter, J. S., and Mathews, D. H. (2010) RNAstructure: software for RNA secondary structure prediction and analysis., *BMC Bioinformatics 11*, 129.

8. Zadeh, J. N., Steenberg, C. D., Bois, J. S., Wolfe, B. R., Pierce, M. B., Khan, A. R., Dirks, R. M., and Pierce, N. A. (2010) NUPACK: Analysis and design of nucleic acid systems, *Journal of Computational Chemistry 31*.

9. Zuker, M. (2003) Mfold web server for nucleic acid folding and hybridization prediction., *Nucleic Acids Research*. Oxford University Press *31*, 3406–3415.

Chapter 28

In Vivo Screening of Ligand-Dependent Hammerhead Ribozymes

Athanasios Saragliadis, Benedikt Klauser, and Jörg S. Hartig

Abstract

The development of artificial switches of gene expression is of high importance for future applications in biotechnology and synthetic biology. We have developed a powerful RNA-based system which allows for the ligand-dependent and reprogrammable control of gene expression in *Escherichia coli*. Our system makes use of the hammerhead ribozyme (HHR) which acts as molecular scaffold for the sequestration of the ribosome binding site (RBS), mimicking expression platforms in naturally occurring riboswitches. Aptamer domains can be attached to the ribozyme as exchangeable ligand-sensing modules. Addition of ligands to the bacterial growth medium changes the activity of the ligand-dependent self-cleaving ribozyme which in turn switches gene expression. In this chapter, we describe the in vivo screening procedure allowing for reprogramming the ligand-specificity of our system.

Key words: Allosteric, Aptamer, Aptazyme, Artificial, Functional RNA, HHR, Ligand, Riboswitch, Ribozyme, Screening

1. Introduction

Riboswitches are a widespread class of naturally occurring functional RNAs acting as genetic regulatory elements. Riboswitches are usually found intramolecularly (in *cis*) within mRNAs, sensing a variety of chemical (and sometimes physical) stimuli that affect transcription elongation, mRNA stability, and translation initiation. Their modular architecture includes specific domains, such as aptamers binding to metabolites with high specificity, and expression platforms which are responsible for mediating changes in gene expression, see also Chapter 8 and (1).

In recent years, scientists have made use of the modular nature of RNA architecture for the construction of artificial, RNA-based

Jörg S. Hartig (ed.), *Ribozymes: Methods and Protocols*, Methods in Molecular Biology, vol. 848,
DOI 10.1007/978-1-61779-545-9_28, © Springer Science+Business Media, LLC 2012

switches of gene expression (2). In a pioneering study Tang and Breaker attached an ATP-binding aptamer to stem II of the minimal HHR motif resulting in the first example of an allosterically controlled ribozyme, also termed aptazyme (3). An aptazyme undergoes a conformational change upon ligand-binding, resulting in an altered phosphodiester cleavage activity. In order to act as a genetic switch, aptazymes are incorporated into 5′-untranslated regions of mRNAs. For efficient, site-directed RNA strand cleavage in vivo, fast-cleaving ribozymes that show enhanced activity even at low magnesium ion concentrations are necessary. An example of such a HHR is the sequence derived from *Schistosoma mansoni*, whose tertiary interactions between stems I and II lock the catalytic core in the active conformation (4), resulting in the mentioned rate enhancement and sufficient activity in vivo (5) (see Fig. 1). In order to render the ribozyme reaction ligand-dependent, aptamer sequences can be connected to the ribozyme. However, in order to identify ligand-regulated aptazymes, the connection

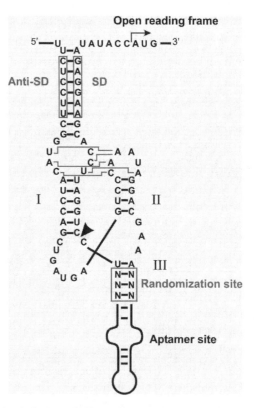

Fig. 1. Hammerhead ribozyme (HHR) used as artificial switch of gene expression. The ribozyme is inserted into the 5′-untranslated region of an mRNA. Boxed sequences are the ribosome binding site (SD) masked by extending the 5′-end of the ribozyme as well as the randomized positions in order to screen for optimized connection sequences. The cleavage site is marked by the *arrowhead*.

sequence linking the aptamer and HHR domain needs to be optimized so that it transmits the binding event in the aptamer domain to structural changes in the catalytic ribozyme domain.

Recent studies by our group and others demonstrated that aptazymes can be utilized as molecular toolbox for the construction of artificial, reprogrammable, and ligand-dependent switches of gene expression (6–12). We have developed general principles which allow for the aptazyme-mediated regulation of gene expression in *Escherichia coli*. In a first study we showed that the bacterial translation of an mRNA can be controlled if the ribosome binding site (RBS) is incorporated into stem I of the hammerhead ribozyme (HHR) (7).The sequestered RBS is only liberated upon self-cleavage of the HHR which leads to efficient initiation of translation. Importantly, by attaching a theophylline aptamer to stem III of the HHR scaffold, an artificial theophylline-dependent riboswitch was obtained. The essential experimental step for the development of such gene regulatory elements is an in vivo screening procedure of a randomized pool of connection sequences. In a second study, we utilized the same screening method for reprogramming the aptazyme-based switch by rendering them thiamine pyrophosphate (TPP)-dependent (8). We attached a naturally occurring TPP aptamer from the thiM riboswitch via a randomized connection sequence to the HHR scaffold. With the described screening protocol ON- and OFF-switches were obtained. We have recently been able to transfer the aptazyme-based concept for controlling mRNA translation to other classes of functional RNAs in *E. coli*, namely switching tRNA maturation (13)and the integrity of the 16S ribosomal subunit (14). In this chapter, we will focus on the experimental details enabling the in vivo screening of ligand-dependent HHRs for controlling mRNA translation in *E. coli* but the general procedure can as well be applied to changing ligand specificity in tRNA or rRNA switches.

2. Materials

2.1. PCR

1. 3 M Na-acetate, pH 5.2.

2. 6× agarose gel loading buffer and suited DNA ready-to-use ladder.

3. Agarose gels. For 0.8% (wt/vol) dissolve 0.8 g agarose in 100 mL 0.5× TBE buffer by boiling in a microwave. Dissolved solutions can be stored in an incubator at >65°C for several days (see Note 1).

4. DNA oligonucleotides as primers for PCR.

5. *Dpn*I and NEB buffer 4 (NEB).

6. Ethanol.

7. Ethidium bromide solution. Dissolve 200 μg in 400 mL ddH$_2$O. Can be kept at room temperature and reused (see Note 2).

8. PCR cycler.

9. PCR template: eGFP expression vector (pet16b_eGFP, AG Hartig, University of Konstanz).

10. Phusion Hot Start DNA polymerase (NEB), HF buffer, and DMSO.

11. Quick Ligation Kit (NEB).

12. Razor blade.

13. Tabletop centrifuge.

14. TBE buffer. Prepare 5× stock with 59 g Trizma base, 27.5 g boric acid, and 4.7 g EDTA in 1 L of ddH$_2$O. Adjust pH to 8.3 if needed with use of concentrated HCl. Dilute 100 mL with 900 mL ddH$_2$O for use.

15. UV light table.

16. DNA Gel Recovery Kit (e.g., Zymo Research).

2.2. Cell Culture and Screening

1. 96 deep-well plate (Nunc) and 96-well plate incubator (Heidolph Incubator 1000 and Titramax 1000) (see Note 3).

2. Carbenicillin stock solution. Prepare 1,000× stock solution by dissolving 100 mg/mL carbenicillin in 50% (vol/vol) ethanol. Can be stored at −20°C for several weeks.

3. Electrocompetent *E. coli* (*E. Coli* BL21 (DE3) gold, Stratagene) (see Note 4).

4. Electroporator (Eppendorf Elektroporator 2510) and electroporation cuvette.

5. Fluorescence plate reader (e.g., TECAN M200).

6. LB-medium. Dissolve 20 g LB medium in 1 L H$_2$O and autoclave. Add 1 mL 1,000× carbenicillin stock solution and stir (see Note 5). Store at 4°C for no longer than 1 week.

7. LB-agar plates. Dissolve 20 g LB medium and 10 g agar in 1 L H$_2$O and autoclave. Supplement with 1 mg/L carbenicillin (see Note 5). Stir briefly and pour the plates in 15-mm petri dishes. The plates can be stored at 4°C for up to 6 weeks.

8. Miniprep kit for plasmid isolation.

9. SOC medium. Dissolve in 1 L of water, 20 g tryptone, 5 g yeast extract and to a final concentration 10 mM NaCl, 2.5 mM KCl, 10 mM MgCl$_2$, 10 mM MgSO$_4$, 20 mM glucose. Autoclave and store at −20°C for several months.

10. Toothpicks sterilized by autoclaving.

3. Methods

3.1. PCR Conditions

1. Mix the necessary materials as follows and split the mixture into six PCR tubes:

Starting concentration	Material	µL to add	Final concentration
5×	HF buffer	30	1×
2 mM	dNTP mix	15	200 µM
100 µM	Forward primer	0.9	600 nM
100 µM	Reverse primer	0.9	600 nM
20 ng/µL	Template	1.5	30 ng
100% (vol/vol)	DMSO	4.5	3% (vol/vol)
2 U/µL	Phusion hot start DNA polymerase	1.5	3 U
	MilliQ H_2O	95.7	
	Total	150	

2. Program the PCR-cycler with the following setup (see Note 6):

Step	Stage	Temperature (°C)	Time	Go to step	Repeat
1	Initial denaturation	98	30 s		
2	Denaturation	98	10 s		
3	Annealing	57	30 s		
4	Extension	72	15–30 s/kb	2	24
5	Final extension	72	7 min		
6	Pause	4	–		

3. Pool the PCR reaction mix into one 1.5-mL Eppendorf tube. Add 16 µL of 3 M Na-acetate, pH 5.2, 480 µL ethanol and mix by inverting.

4. Place samples at –80°C for 20 min or –20°C for more than 2 h.

5. Centrifuge samples on tabletop centrifuge at 13.4 k rpm for 15 min at room temperature. Discard supernatant and resuspend pellet in 44 µL H_2O. Add 5 µL NEB buffer 4 and 1 µL *Dpn*I and mix thoroughly. Incubate sample at 37°C for 50 min followed by 10 min incubation at 80°C, in order to heat inactivate the enzyme.

6. Add 10 μL of 6× NEB agarose loading buffer to the sample and mix. In the meantime, pour an 0.8% (wt/vol) TBE-agarose gel and let it solidify. Once the gel is prepared, load sample in a long well, and run the agarose gel at 10 V/cm (distance between electrodes) for 75 min. In a smaller lane, use 2.5 μL GeneRuler DNA ladder.

7. After the 75-min run, place the agarose gel in a tank with EtBr for 10 min and visualize the DNA bands over a UV light table.

8. Excise the appropriate band corresponding to the correct size, and purify the DNA fragment using Zymoclean DNA Gel DNA Recovery Kit according to manufacturer's protocol.

9. After measuring the sample concentration, combine 30 ng of purified PCR product in 9 μL of water with 10 μL Quick Ligase buffer and add 1 μL Quick Ligase, mix well and briefly centrifuge samples (see Note 7). Incubate at 25°C for 15 min.

10. Purify the DNA sample by using Zymo DNA Clean and Concentrator according to manufacturer's protocol.

3.2. Transformation, Screening, Characterizing

1. Thaw electro-competent *E. Coli* BL21 (DE3) gold (80 μL) on ice for 15 min. In the meanwhile, preheat 1 mL SOC medium on 37°C, place LB-agar plates with the correct antibiotic on 37°C, a fresh 1.5-mL eppendorf tube on 37°C, and chill one electroporation cuvette on ice along with the DNA purified ligated sample from the previous step.

2. Add 1.5 μL of the purified sample to 80 μL competent cells and mix carefully. Transfer the cells into the chilled electroporation cuvette carefully, not to introduce any air bubbles. Place cuvette into the electroporator and transform cells. Transfer the cells using the prewarmed 1 mL SOC medium, into the fresh 1.5-mL eppendorf tube and incubate for 1 h at 37°C. Plate the transformed cells on the prewarmed LB-agar plates and incubate plates overnight at 37°C.

3. Prepare as many 96 deep-well plates with 1 mL LB-medium supplemented with carbenicillin as required in order to achieve sufficient pool coverage of the randomized area (see Note 8). Save first and last placeholders of the 96 deep-well plate for the control culture (see Note 9). Using sterilized toothpicks, pick single colonies. Incubate plates at 37°C overnight in the 96-well plate incubator.

4. For each master plate of the previous day (step 3), prepare two additional destination plates. The medium added to every second column should be supplemented with the appropriate ligand as shown in Fig. 1. Incubate the destination plates overnight in the 96-well plate incubator.

5. Transfer 100 μL of culture into a black 96-well plate and measure expression levels by determining eGFP fluorescence (see Note 10). By comparing the fluorescence for each screened clone in absence and presence of the ligand, potential hits are identified.

6. Isolate the clones of interest by picking them from the master plate and inoculating fresh cultures of ~10 mL LB-carbenicillin medium. Incubate by shaking vigorously at 37°C for at least 6 h.

7. Isolate plasmid from liquid culture using any Miniprep Kit and identify the screened clone by sequencing.

8. Additionally, prepare new 96-well plates containing 1 mL of LB medium with carbenicillin and a concentration gradient of the ligand from 0 to maximum concentration for the columns from 1 to 12. Inoculate the 96-well plate in triplicates for each clone isolated as potential hit from the initial screening process (see Fig. 1). Incubate at 37°C overnight and measure eGFP to confirm functional, concentration-dependent riboswitch.

4. Notes

1. For a good separation or resolution of 5–10 kb PCR products, an 0.8% (wt/vol) agarose gel is sufficient. Higher percentage (e.g., 2% (wt/vol)) can offer better resolution for smaller (0.2–1 kb) products.

2. Ethidium bromide is toxic and mutagenic. Wear nitrile gloves while working with it.

3. Alternatively, for larger screening purposes 384 deep-well plates can be used for cell culture and fluorescence measurements.

4. For the preparation of electrocompetent *E. coli*, a detailed protocol can be found at www.eppendorf.com.

5. Let the medium cool down to ~50°C before adding the antibiotic stock solution.

6. Alternatively to Phusion Hot Start polymerase, other polymerases can be used, and the protocol needs to be adapted to the appropriate buffer and conditions. Although an annealing temperature of 57°C is usually sufficient for primers with a binding region of around 20 nucleotides, we suggest that initially a gradient PCR should be carried out with a temperature range in order to determine the optimal annealing temperature.

7. The dilution of the buffer in the correct way is crucial for the efficiency of the ligation. Additionally, one should be careful not to introduce air bubbles while mixing the sample.

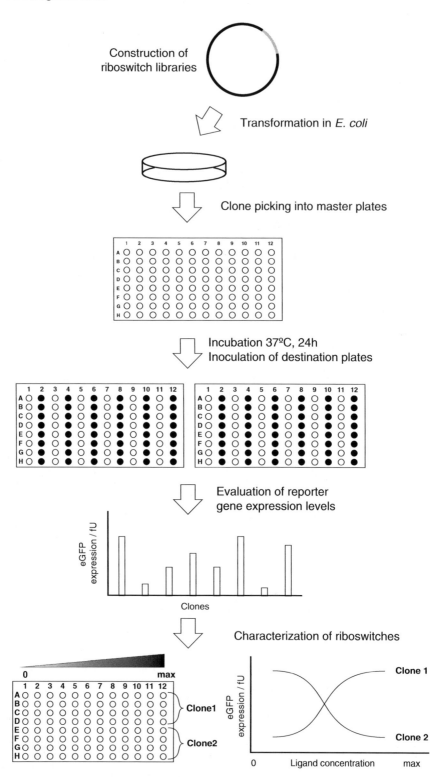

Fig. 2. Schematic illustration of the workflow for riboswitch clonal screening. In this example the process for a 96-well plate format is depicted but it can be used for 384-well plates as well.

8. Pool coverage can be calculated by obtaining the oversampling factor (O_f) which is calculated as

$$O_f = T \,/\, V = -\ln(1 - P_i)$$

where T is the number of clones actually screened, V the maximum number of different clones of a randomized sequence(sequence space), and P_i the probability that a particular sequence occurs in the library (15). Due to the exponential relationship, high pool coverage is only achieved when the theoretical sequence space is oversampled multiple times.

9. As control clones, one can use a plasmid expressing eGFP without any insert affecting the gene expression (positive control) and a negative control that bears an inactivated HHR variant by an A to G point mutant in the catalytic core, masking the RBS permanently.

10. Wait until eGFP expression has reached a plateau in outgrown cultures (at least 16 h). Measure eGFP expression levels by measuring fluorescence of cultures with excitation wavelength $\lambda_{ex} = 488$ nm and emission wavelength $\lambda_{em} = 535$ nm.

References

1. Roth, A., and Breaker, R. R. (2009) The structural and functional diversity of metabolite-binding riboswitches, *Annu Rev Biochem 78*, 305–334.

2. Wieland, M., and Hartig, J. S. (2008) Artificial riboswitches: synthetic mRNA-based regulators of gene expression, *Chembiochem 9*, 1873–1878.

3. Tang, J., and Breaker, R. R. (1997) Rational design of allosteric ribozymes, *Chem Biol 4*, 453–459.

4. Khvorova, A., Lescoute, A., Westhof, E., and Jayasena, S. D. (2003) Sequence elements outside the hammerhead ribozyme catalytic core enable intracellular activity, *Nat Struct Biol 10*, 708–712.

5. Yen, L., Svendsen, J., Lee, J. S., Gray, J. T., Magnier, M., Baba, T., D'Amato, R. J., and Mulligan, R. C. (2004) Exogenous control of mammalian gene expression through modulation of RNA self-cleavage, *Nature 431*, 471–476.

6. Win, M. N., and Smolke, C. D. (2007) A modular and extensible RNA-based gene-regulatory platform for engineering cellular function, *Proc Natl Acad Sci USA 104*, 14283–14288.

7. Wieland, M., and Hartig, J. S. (2008) Improved aptazyme design and in vivo screening enable riboswitching in bacteria, *Angew Chem Int Ed Engl 47*, 2604–2607.

8. Wieland, M., Benz, A., Klauser, B., and Hartig, J. S. (2009) Artificial ribozyme switches containing natural riboswitch aptamer domains, *Angew Chem Int Ed Engl 48*, 2715–2718.

9. Kumar, D., An, C. I., and Yokobayashi, Y. (2009) Conditional RNA interference mediated by allosteric ribozyme, *J Am Chem Soc 131*, 13906–13907.

10. Ogawa, A., and Maeda, M. (2008) An artificial aptazyme-based riboswitch and its cascading system in E. coli, *Chembiochem 9*, 206–209.

11. Auslander, S., Ketzer, P., and Hartig, J. S. (2010) A ligand-dependent hammerhead ribozyme switch for controlling mammalian gene expression, *Mol Biosyst 6*, 807–814.

12. Wieland, M., Gfell, M., and Hartig, J. S. (2009) Expanded hammerhead ribozymes containing addressable three-way junctions, *RNA 15*, 968–976.

13. Berschneider, B., Wieland, M., Rubini, M., and Hartig, J. S. (2009) Small-molecule-dependent regulation of transfer RNA in bacteria, *Angew Chem Int Ed Engl 48*, 7564–7567.

14. Wieland, M., Berschneider, B., Erlacher, M. D., and Hartig, J. S. (2010) Aptazyme-mediated regulation of 16S ribosomal RNA, *Chem Biol 17*, 236–242.

15. Reetz, M. T., Kahakeaw, D., and Lohmer, R. (2008) Addressing the numbers problem in directed evolution, *Chembiochem 9*, 1797–1804.

Chapter 29

Flexizymes as a tRNA Acylation Tool Facilitating Genetic Code Reprogramming

Yuki Goto and Hiroaki Suga

Abstract

Genetic code reprogramming is a method for the reassignment of arbitrary codons from proteinogenic amino acids to non-proteinogenic ones, and thus specific sequences of nonstandard peptides can be ribosomally expressed according to their mRNA templates. We here describe a protocol that facilitates the genetic code reprogramming using flexizymes integrated with a custom-made in vitro translation apparatus, referred to as the flexible in vitro translation (FIT) system. Flexizymes are flexible tRNA acylation ribozymes that enable the preparation of a diverse array of non-proteinogenic acyl-tRNAs. These acyl-tRNAs read vacant codons created in the FIT system, yielding the desired nonstandard peptides with diverse exotic structures, such as *N*-acyl groups, D-amino acids, *N*-methyl amino acids, and physiologically stable macrocyclic scaffolds. Facility of the protocol allows for a wide variety of applications in the synthesis of new classes of nonstandard peptides with biological functions.

Key words: Aminoacylation, FIT system, Flexizyme, Genetic code, Ribosome, Ribozyme, Translation, tRNA

1. Introduction

The translation apparatus polymerizes amino acids in accord with the sequence information encoded in mRNA templates with remarkably high fidelity. In spite of the governance of the genetic code restricting the use of 20 proteinogenic amino acids in translation, it is possible to alter this rule by appropriate manipulations of the translation system where some translation components are excluded and mischarged acyl-tRNAs with desired non-proteinogenic amino acids (Xaa-tRNAs) are added in order to reassign the codons from proteinogenic to non-proteinogenic (or artificial) amino acids. Such a custom-reconstituted in vitro translation system could enable us to "reprogram" the governance of the genetic code and thereby

Jörg S. Hartig (ed.), *Ribozymes: Methods and Protocols*, Methods in Molecular Biology, vol. 848,
DOI 10.1007/978-1-61779-545-9_29, © Springer Science+Business Media, LLC 2012

the translation apparatus could turn into machinery for the synthesis of nonstandard polypeptides consisting of non-proteinogenic amino acids (1–3).

To facilitate the preparation of a wide variety of Xaa-tRNAs, we have developed flexizyme (flexible tRNA acylation ribozyme) system, consisting of three artificial ribozymes, referred to as dFx, eFx, and aFx (3, 4). The complete set of flexizymes accepts virtually any amino acid as acyl-donor substrates by simply matching the flexizyme with the activating ester group in the acyl-donor substrate (see below and Table 1). Indeed, the flexizyme system has been applied to the synthesis of a wide variety of Xaa-tRNAs where Xaa represents ordinary proteinogenic amino acids (3) as well as non-proteinogenic amino acids such as those with artificial side chain (3, 5), D-configuration (3, 6), N-acyl (7), and N-alkyl modifications (8, 9) (see Table 2). Moreover, it is also capable of charging hydroxyacyl (10) and peptidyl groups involving diverse non-proteinogenic amino acids (11) onto tRNAs. In addition to the versatility to acid substrates, the flexizymes are able to accept a

Table 1
Summary of appropriate combination of active groups and flexizymes for various acid substrates

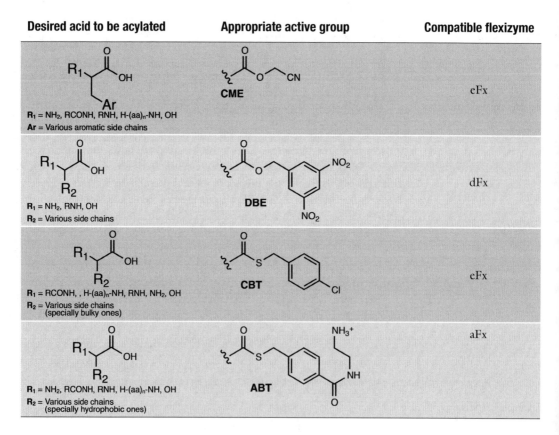

Desired acid to be acylated	Appropriate active group	Compatible flexizyme
$R_1 = NH_2, RCONH, RNH, H-(aa)_n-NH, OH$ Ar = Various aromatic side chains	**CME**	eFx
$R_1 = NH_2, RNH, OH$ R_2 = Various side chains	**DBE**	dFx
$R_1 = RCONH, , H-(aa)_n-NH, RNH, NH_2, OH$ R_2 = Various side chains (specially bulky ones)	**CBT**	eFx
$R_1 = NH_2, RCONH, RNH, H-(aa)_n-NH, OH$ R_2 = Various side chains (specially hydrophobic ones)	**ABT**	aFx

Table 2
Recommended acylation conditions for representative acid substrates and their reference yields of acylated microhelix RNAs

Amino acid	Active group/flexizyme	Reaction time (h)	Yield (%)
CIAcPhe	CME/eFx	2	89
MeTyr	CME/eFx	10a	40
DE-DK-DE-DK-F	CME/eFx	2	62
HOYme	CME/eFx	2b	91
MeAla	DBE/dFx	2	60
McGly	DBE/dFx	2	64
PrGly	DBE/dFx	6	73
BuGly	DBE/dFx	6	70
DAsp	DBE/dFx	2	47
β-Ala	DBE/dFx	22c	75
AcArg	CBT/eFx	72	51
Ile	CBT/eFx	6	17
Opa	ABT/aFx	2	48

a Final concentration of acid substrate is increased to 40 mM
b Reaction was carried out in HEPES-KOH buffer (pH 8.0)
c Reaction was carried out in Bicine-KOH buffer (pH 9.0)
Note: CIAcPhe N-chloroacetylphenylalanine; MeTyr N-methyltyrosine; DE-DK-DE-DK-F D-glutamyl-D-lysinyl-D-glutamyl-D-lysinyl-phenylalanine; HOYme p-methoxyphenyllactic acid; MeAla N-methylalanine; McGly N-methylglycine; PrGly N-propylglycine; BuGly N-butylglycine; D-Asp D-Aspartic acid; β-Ala β-Alanine; AcArg N-acetylarginine; Ile isoleucine; Opa 2-amino-3-(octanamido) propanoic acid

wide array of tRNAs without the dependence on body and anti-codon sequences, unlike ARSs, via simple base-pair recognition of the 3′ end of tRNA (12). Since the largest benefit of the flexizymes is the technical simplicity and facility, i.e., mixing the appropriate flexizyme with the desired tRNA and acid substrate followed by incubation on ice for a few hours yields the desired Xaa-tRNAs, the flexizyme system eliminates the previous technical barrier in the preparation of Xaa-tRNA for genetic code manipulation.

We here describe protocols for genetic code reprogramming using flexizymes integrated into a customized reconstituted translation system (3, 7–9), referred to as FIT (Flexible In Vitro Translation) system (see Fig. 1). In this FIT system, unnecessary translation components such as ARS(s), amino acid(s), methionyl-tRNA formyl transferase, its formyl donor substrate, and/or release factor(s) can be excluded depending on the desired codons to be reprogrammed.

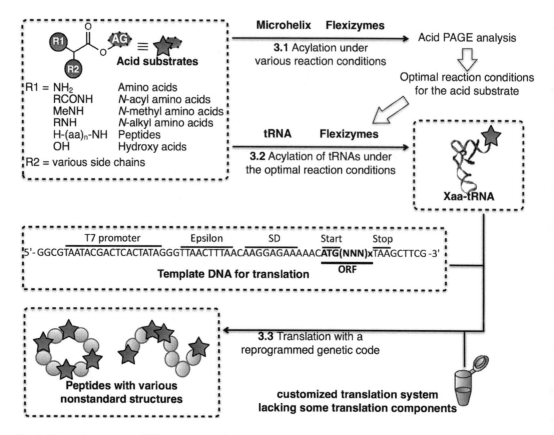

Fig. 1. Schematic overview of FIT system; genetic code reprogramming using flexizymes and expression of nonstandard peptides.

The minimal components of the translation apparatus are customized to match any desired codon(s) usage, while retaining maximum peptide expression levels and decoding fidelity. Each customized translation apparatus is then supplemented with Xaa-tRNAs prepared by the flexizyme technology to occupy the vacant codons, yielding in high purity the desired peptide product. By means of the genetic code reprogramming along with unique adopting chemistry, the translation system turns into machinery for the synthesis of natural product-like nonstandard peptides consisting of N-methyl peptide backbone and macrocyclic scaffolds (7–9, 13).

2. Materials

2.1. Acylation of Microhelix RNA and Acid PAGE Analysis

1. Microhelix RNA: Synthesize it by conventional in vitro transcription method (14) according to procedures described previously (3). See Table 3 for its sequence. Prepare 250 μM stock solution in RNase-free water. Store it in aliquots at −20°C (stable at least for a year).

Table 3
Sequences of flexizymes (eFx, dFx, and aFx), microhelix RNA, and tRNAs (tRNA$_{NNN}^{AsnE2}$ and tRNA$_{CAU}^{fMetE}$) used in this method

Name	RNA sequence
eFx	5′-GGAUC GAAAG AUUUC CGCGG CCCCG AAAGG GGAUU AGCGU UAGGU-3′
dFx	5′-GGAUC GAAAG AUUUC CGCAU CCCCG AAAGG GUACA UGGCG UUAGG U-3′
aFx	5′-GGAUC GAAAG AUUUC CGCAC CCCCG AAAGG GGUAA GUGGC GUUAG GU-3′
Microhelix RNA	5′-GGCUC UGUUC GCAGA GCCGC CA-3′
tRNA$_{NNN}^{AsnE2}$	5′-GGCUC UGUAG UUCAG UCGGU AGAAC GGCGG ACU**NN NAA**UC CGUAU GUCAC UGGUU CGAGU CCAGU CAGAG CCGCC A-3′[a]
tRNA$_{CAU}^{fMetE}$	5′-GGCGG GGUGG AGCAG CCUGG UAGCU CGUCG GGCUC AUAAC CCGAA GAUCG UCGGU UCAAA UCCGG CCCCC GCAAC CA-3′

[a]"NNN" shown in bold indicates anticodon triplet. This triplet sequence varies depending on the objective reprogrammed codon sequence

2. Flexizymes: Synthesize flexizyme RNAs by conventional in vitro transcription method according to procedures described previously (3, 4) (see Note 1). See Table 3 for their sequences. Prepare 250 μM stock solution in RNase-free water. Store it in aliquots at −20°C (stable at least for a year).

3. 10× Flexizyme reaction buffer: 500 mM HEPES-KOH buffer, pH 7.5 (see Note 2). Store it in aliquots at −20°C (stable for years).

4. Acid substrate: Synthesize acid substrate activated with an appropriate active group (see below and Table 1) for flexizyme reaction in two steps from properly protected amino acids according to the previously reported method (3, 4, 7, 15). To a solid of acid substrate add DMSO-d$_6$ to make a 200 mM stock solution. Keep it at −20°C (generally stable for a year). To make 25 mM working solution, mix 10 μL of 200 mM stock solution with 70 μL of DMSO-d$_6$. 25 mM working solution can be stored in aliquots at −20°C (generally stable for a year) (see Note 3).

5. Flexizyme quenching buffer: 0.3 M NaOAc, pH 5.2. Can be stored at room temperature for at least 6 months.

6. Acid PAGE loading buffer: 150 mM NaOAc, pH 5.2, 10 mM EDTA, 0.02% BPB, and 93% formamide. Can be stored at room temperature for at least a year.

7. Acid-acrylamide gel solution (20%): Mix 1.8 g of urea, 83 μL of 3 M NaOAc (pH 5.2), and 2.5 mL of 40% acrylamide/ bisacrylamide (19/1) solution, and add RNase-free water up

to 5 mL. Mix gently until urea dissolves completely. Add 50 μL of 10% APS and 4 μL of TEMED to the solution right before pouring to a mini-gel equipment.

8. Acid-polyacrylamide gel electrophoresis: Assemble gel plates (Bio-rad, Mini-PROTEAN 3) and pour acid-acrylamide gel solution (20%) into the gel plates. Stand it at room temperature until the gel solidifies. Use 50 mM NaOAc (pH 5.2) as a running buffer. Make gels freshly for each experiment.

9. EtBr solution for gel staining: Mix 10 mL of 1× TBE and 0.5 μL of 10 mg/mL EtBr. Prepare freshly upon use. Can be stored at 4°C for at least a year.

10. Fluorescent gel scanner (FLA-5100, Fuji).

2.2. Acylation of tRNAs by Flexizyme to Prepare Xaa-tRNA

1. Reagents/buffers for flexizyme reaction: See materials 2–5 in the Subheading 2.1.

2. tRNAs used for flexizyme reactions: Prepare required tRNAs by conventional in vitro transcription method according to procedures described previously (7, 10) (see Note 4). See Table 3 for their sequences. Store it in aliquots at –20°C (stable at least for a year).

3. tRNA pellet washing solution I: 70% ethanol containing 0.1 M NaOAc, pH 5.2. Can be stored at room temperature for at least 6 months.

4. tRNA pellet washing solution II: 70% ethanol. Can be stored at room temperature for at least 6 months.

2.3. Translation of Peptides Using Reprogrammed Genetic Code

1. 10× TS solution: 500 mM HEPES-KOH (pH 7.6), 1 M potassium acetate, 120 mM magnesium acetate, 20 mM ATP, 20 mM GTP, 10 mM CTP, 10 mM UTP, 200 mM creatine phosphate, 1 mM 10-formyl-5,6,7,8-tetrahydrofolic acid, 20 mM spermidine, 10 mM DTT, 15 mg/mL E. coli total tRNA. Use and store it in low binding tubes. Store it in aliquots at –80°C for at least 6 months.

2. 10× RP solution: 3 mM magnesium acetate, 12 μM E. coli Ribosome, 6 μM MTF, 27 μM IF1, 4 μM IF2, 15 μM IF3, 2.6 μM EF-G, 100 μM EF-Tu, 6.6 μM EF-Ts, 2.5 μM RF2, 1.7 μM RF3, 5 μM RRF, 40 μg/mL creatine kinase, 30 μg/mL myokinase, 1 μM inorganic pyrophosphatase, 1 μM nucleotide diphosphate kinase, and 10 μM T7 RNA polymerase. Use and store it in low binding tubes. Store it in aliquots at –80°C for at least 6 months.

3. 10× ARS solution: Mix the necessary ARSs for the objective genetic code reprogramming with the following concentrations; 7.3 μM AlaRS, 0.3 μM ArgRS, 3.8 μM AsnRS, 1.3 μM AspRS, 0.2 μM CysRS, 0.6 μM GlnRS, 2.3 μM GluRS, 0.9 μM GlyRS, 0.2 μM HisRS, 4.0 μM IleRS, 0.4 μM LeuRS, 1.1 μM

LysRS, 0.3 μM MetRS, 6.8 μM PheRS, 1.6 μM ProRS, 0.4 μM SerRS, 0.9 μM ThrRS, 0.3 μM TrpRS, 0.2 μM TyrRS, and 0.2 μM ValRS. Use and store it in low binding tubes. Store it in aliquots at –80°C for at least 6 months.

4. 5 mM Each amino acids solution: Mix proteinogenic amino acids, except for ones to be replaced with non-proteinogenic amino acids in the objective reprogrammed genetic code, to make 5 mM stock solution. Can be stored at –20°C for at least 6 months.

5. DNA template for translation: Prepare a DNA duplex coding a desirable peptide as described previously (3). The template DNA for translation requires the following elements in the 5′ to 3′ order; a T7 promoter, GGG triplet (enhancer of transcription efficiency), epsilon sequence (AU-rich bacterial translation enhancer element which is originally found in non-translated region of T7 phage gene) (16), Shine–Dalgarno sequence (purine-rich sequence that functions as ribosomal binding site), start codon (ATG), peptide coding sequence, stop codon (TAA or TGA), and additional 6–10 bp sequence at the downstream of stop codon. Can be stored at –20°C for at least a year.

3. Methods

We must choose an appropriate pair of flexizyme and leaving group depending upon the structure of objective acyl-donor substrates (see Table 1). For those containing aromatic side chains, e.g., Phe, Tyr, Trp, and their analogs, eFx is the choice to pair with those activated by a cyanomethyl ester (CME). For those bearing non-aromatic side chains, dFx is the primary choice pairing with the 3,5-dinitrobenzyl ester (DBE) leaving group. Although this pair is the most versatile and generic choice for tRNA acylation, acyl-donor substrates with certain structures (e.g., α-N-acyl, β-branched or significantly bulky side chain) occasionally give sluggish reaction rates and poor charging yields. In such a case, an alternative choice is eFx paired with 4-chlorobenzyl thioester (CBT) leaving group, which often gives significant improvement. The last pair is aFx and acyl-donors bearing the leaving group of 4-[(2-aminoethyl) carbamoyl]benzyl thioester (ABT). This pair has been developed for a special case where the activation of acid with any of above leaving group results in poor solubility in the reaction buffer. All flexizymes are able to accept tRNAs bearing any choice of anticodon and body sequence.

Optimal conditions for the flexizyme reactions vary depending on the acyl-donor substrates. The empirically determined optimal conditions for representative substrates are summarized in Table 2. Although the generic conditions, where the mixture of tRNA and flexizyme with 25 mM acyl-donor substrate is incubated for 2 h at pH 7.5 in the presence of 600 mM MgCl$_2$, would work in most cases, it is occasionally necessary to optimize conditions depending upon substrate structures. The recommended method for optimizing the reaction conditions when using a novel acid substrate is to use a microhelix RNA instead of tRNA (see Subheading 3.1). The convenience of using this small RNA is that the acylation yield can be directly determined by the band intensity of Xaa-microhelix RNA separated from that of free RNA on acid polyacrylamide gel electrophoresis (PAGE) (3, 17). The acylation reaction mixture containing the microhelix RNA, flexizyme, and acyl-donor substrate is incubated on ice and sampled at several time points, with the resulting RNA analyzed by acid PAGE to determine the optimal conditions (see Note 5; Fig. 2). The optimized conditions by acid PAGE analysis with microhelix RNA can be just applied for the flexizyme reaction to synthesize Xaa-tRNA by performing flexizyme reaction using the objective tRNA (see Note 4) instead of microhelix RNA. The resulting Xaa-tRNA is prepared just by simple ethanol precipitation and no further purification is required.

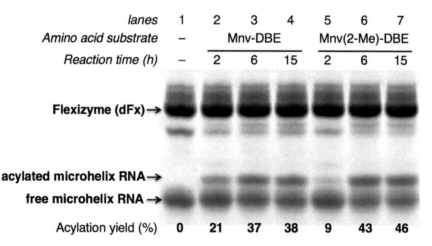

Fig. 2. An example for acid PAGE analysis flexizyme reaction products with microhelix RNA under various reaction conditions. Microhelix RNA (25 μM) was incubated with dFx (25 μM) and acid substrate (5 mM) in HEPES-KOH buffer (50 mM, pH 7.5) containing MgCl$_2$ (600 mM), and then the resulting reaction mixture was analyzed by 20% acid PAGE. Lane 1, a negative control without any acid substrate; lanes 2–4 reactions in the presence of activated 5-mercapto-norvaline (Mnv-DBE) with various reaction times (2, 6, and 15 h); lanes 5–7 reactions in the presence of activated 5-[(2-hydroxyethyl) dithio]-norvaline [Mnv(2-Me)-DBE] with various reaction times (2, 6, and 15 h). Bands corresponding to flexizyme, acylated microhelix RNA, and free microhelix RNA were labeled in the figure. Acylation yields were calculated based on the band intensity of acylated microhelix RNA (A) and free microhelix RNA (F), presented as (A)/[(A) + (F)]. This experiment demonstrates that the optimal reaction time for these acid substrates is 6 h.

Translation system used in the FIT system consists of the essential components (18, 19) used for translation and transcription (e.g., ribosome, initiation factors, elongation factors, release factors, tRNAs, amino acids, ARSs, T7 RNA polymerase, NTPs, and template DNA). Expression of the desired nonstandard peptide containing non-proteinogenic amino acids is executed using a translation system where the designated amino acids and/or ARSs are simply excluded and desired Xaa-tRNAs are supplied (see Note 6). Peptides expressed using the FIT system can be analyzed in various ways, such as quantitative analysis by tricine-SDS PAGE (20), qualitative identification of expressed peptide by MALDI-TOF mass spectrometry, and bioactivity analysis by proper biochemical or cell-based assays. In Fig. 3, we show genetic code reprogramming with four artificial amino acids and expression of a cyclic *N*-methyl peptide as a demonstration of FIT system.

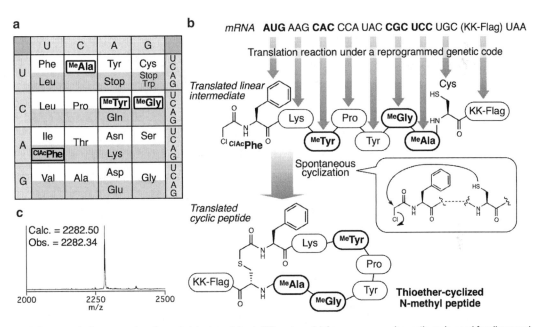

Fig. 3. An example for expression of nonstandard peptides in FIT system. (a) A reprogrammed genetic code used for ribosomal synthesis of cyclic *N*-methyl peptide. In this reprogrammed table, *N*-chloroacetylphenylalanine (ClAcPhe) is assigned to AUG start codon, and *N*-methylalanine (MeAla), *N*-methyltyrosine (MeTyr), and *N*-methylglycine (MeGly) are assigned to UCC, CAC, and CGC elongation codons, respectively. (b) Ribosomal synthesis of cyclic *N*-methyl peptide. Sequences of mRNA template is shown in the figure. The KK-Flag in parenthesis indicates the RNA sequence encoding a KK-Flag peptide (KK-DYKDDDDK). Reprogrammed translation apparatus affords a linear intermediate peptide containing a chloroacetyl group at the N-terminus, three *N*-methylated amino acids, and a Cys residue at the C-terminal region. Intramolecular attack of the thiol group in Cys to the N-terminal chloroacetyl group then spontaneously occurs to yield a thioether-cyclized *N*-methyl peptide. Note that this intramolecular cyclization occurs without the need for any additional external reagents in a high yield (in most cases in a nearly quantitative manner) in situ of the translation mixture. Schematic structure of the cyclic *N*-methyl peptide is shown in the figure. The cyclic structure closed by a thioether bond exhibits remarkable resistance against peptidases under reducing conditions unlike disulfide-closed cyclic ones. (c) MALDI-TOF mass spectrum of the cyclic *N*-methyl peptide synthesized by FIT system via spontaneous posttranslational cyclization. The calculated mass (Calc.) and observed mass (Obs.) for proton-adduct ions ([M + H]⁺) are shown in the spectrum.

3.1. Optimization of Conditions for Flexizyme Reaction

1. Mix 1 μL of 500 mM HEPES-KOH buffer (pH 7.5) (see Note 2), 1 μL of 250 μM flexizyme of choice (see Note 1), and 1 μL of 250 μM microhelix RNA with 3 μL of RNase-free water (see Note 7).

2. Heat the sample at 95°C for 2 min, then slowly cool it at room temperature over 5 min.

3. Add 2 μL of 3 M MgCl$_2$ into the sample, then incubate it at room temperature for 5 min followed by on ice for 3 min.

4. Add 2 μL of 25 mM acid substrate in DMSO-d$_6$ into the sample and mix well.

5. Incubate the acylation reaction mixtures on ice for several hours (see Note 8).

6. Add 40 μL of 0.3 M NaOAc (pH 5.2) and 100 μL of ethanol into the reaction mixtures to quench the acylation reaction.

7. Centrifuge the samples at $15,000 \times g$ for 15 min at 25°C. Then, remove the supernatant completely.

8. Add 4 μL of 10 mM NaOAc (pH 5.2) to the tubes and dissolve the RNA pellets quickly (see Note 9).

9. Add 12 μL of acid PAGE loading buffer to the samples, and mix well.

10. Apply 2 μL each of the resulting sample onto an acid-polyacrylamide gel and run it with 120 V for 2.5 h.

11. Wash the gel with 50 mL of 1× TBE by gently shaking for 10 min.

12. Stain the gel with 10 mL of EtBr gel-staining solution by gently shaking for 10 min.

13. Wash the gel briefly with 50 mL of RNase-free water.

14. Wash the gel with 50 mL of 1× TBE by gently shaking for 5 min.

15. Scan the gel image by a fluorescent gel scanner (see Note 10).

16. Quantify the bands corresponding to free and acylated microhelix RNA to determine the yield of acylation (see Fig. 2 for an example of acid PAGE analysis) (see Note 11).

3.2. Acylation of tRNAs by Flexizyme to Prepare Xaa-tRNA

1. Mix 2 μL of 500 mM HEPES-KOH buffer (pH 7.5), 2 μL of 250 μM flexizyme of choice (see Note 1), and 2 μL of 250 μM tRNA of choice (see Note 4) with 6 μL of RNase-free water.

2. Repeat Steps 2–4 in the Subheading 3.1 using double the volumes specified in those steps in all cases.

3. Incubate the acylation reaction mixtures on ice for several hours (see Note 12).

4. Repeat Steps 6 and 7 in the Subheading 3.1 using double the volumes specified in those steps in all cases (see Note 13).

5. Add 50 μL of 70% ethanol containing 0.1 M NaOAc (pH 5.2) to the tube and vortex the tube well to break the RNA pellet into pieces (see Note 14).

6. Centrifuge the sample at $15,000 \times g$ for 5 min at 25°C. Then, remove the supernatant completely.

7. Repeat Steps 5–6 one more time.

8. Add 50 μL of 70% ethanol to the tube.

9. Centrifuge the sample at $15,000 \times g$ for 3 min at 25°C. Then, remove the supernatant completely.

10. Open the tube lid and cover it with tissues, then dry the RNA at room temperature for 5 min (see Note 15).

3.3. Translation of Peptides Using Reprogrammed Genetic Code

1. Dissolve the Xaa-RNA pellet quickly into 1 μL of 1 mM NaOAc (pH 5.2) (see Note 16).

2. Set up the translation reaction (5 μL) as follows: Mix 0.5 μL of 10× TS Solution, 0.5 μL of 10× RP Solution, 0.5 μL of 10× ARS Solution, 0.5 μL of 5 mM each amino acids solution (see Note 17), 0.5 μL of 0.4 μM DNA template (see Note 18), 1 μL of Xaa-tRNA (from Step 1) (see Note 6), 1.5 μL of RNase-free water.

3. Incubate the reaction mixture at 37°C for 1 h.

4. Analyze the translated peptide by an appropriate method such as tricine-SDS PAGE, a SPE column desalting followed by MALDI mass spectrometry, and bioactivity assays.

4. Notes

1. Appropriate type of flexizymes should be used depending on the side chain and active group of acid substrates (see above and Table 1).

2. Use of buffer at higher pH such as HEPES-KOH (pH 8.0) and Bicine-KOH (pH 9.0) may improve the yield of acylation in some cases.

3. In case that the quality of acid substrate stock solution needs to be checked after long storage, measure its NMR spectrum.

4. An appropriate body sequence and anticodon of tRNA should be chosen for reprogramming codon. Transfer RNAs used for genetic code reprogramming should be inert against endogenous ARSs, so-called "orthogonal" tRNAs. We recommend to use tRNA$_{NNN}$^{AsnE2} (NNN denotes anticodon triplet) (10) and tRNA$_{CAU}$^{fMetE}, whose body sequences have been artificially engineered, for the reprogramming of elongation codons and start codon, respectively.

5. Conditions giving over a 10% yield in this analysis can be applied to the flexizyme-catalyzed tRNA acylation, and thus prepared Xaa-tRNAs are generally sufficient for the experiments of genetic code reprogramming.

6. The optimal concentration of Xaa-tRNAs in the translation mixture varies depending on the type of Xaa, but 50–100 μM of each Xaa-tRNA is generally used in our laboratory. We have also confirmed that a total concentration up to 600 μM Xaa-tRNAs does not disturb the translation when aa-tRNAs are well desalted by the procedure described in this protocol.

7. All the following steps should be performed in an RNase-free manner. Use RNase-free tubes, pipettes, pipette tips, and water. Wear gloves at all times.

8. Proper reaction time strongly varies depending on acid substrates. Perform several reactions with various reaction times to find the optimal reaction time (see above and Table 2).

9. Acyl-RNAs may be unstable, so dissolve RNA pellet right before applying to an acid-polyacrylamide gel. After adding 10 mM NaOAc (pH 5.2), perform the following steps promptly.

10. In case you cannot observe clear bands corresponding to flexizyme, microhelix RNA, and acylated microhelix RNA, but detect smear, the reagents and/or reaction mixture may be contaminated by RNase. Wear gloves and use RNase-free apparatus when performing experiments. Prepare all reagents freshly when RNase contamination is suspected.

11. If the yield of acylation is extremely low, you can try the following conditions, which may improve the acylation efficiency: Increase the pH of flexizyme reaction from 7.5 up to 9.5. The concentration of acid substrate can be also increased from genetic concentration (5 mM) to 25 or 40 mM. In some acid substrates, particularly those with α-N-acyl group or bulky side chain, long incubation time (up to 72 h) may be required. If precipitation of acid substrate is observed in the reaction mixture, you can increase the final concentration of DMSO up to 40% to help dissolution of substrate. Alternatively, change the leaving group of substrate to ABT in combination with aFx to enhance the water solubility of substrate.

12. Proper reaction time strongly varies depending on acid substrates. The reaction conditions optimized by using microhelix RNA and acid PAGE analysis in the Subheading 3.1 should be used for the synthesis of Xaa-tRNA.

13. Ethanol precipitation and the following washing steps (Steps 5–9) should be carried out at around room temperature not at lower temperature to avoid undesirable precipitation of salts.

14. The washing steps (Steps 5–9) are critical for the following translation reaction. Carry-over of magnesium and sodium ions would decrease the efficiency of translation reaction.

15. Dried Xaa-tRNA pellets can be stored at –80°C for 1 day. Some of them may be stored over weeks depending on the structure of amino acids.

16. Since Xaa-tRNAs may be unstable (especially under alkaline conditions), dissolve RNA pellet right before preparing the translation reaction mixture. Keep the resulting Xaa-tRNA solution on ice and proceed to Step 2 as soon as possible.

17. Do not add the proteinogenic amino acids whose codons are reprogrammed by non-proteinogenic acyl-tRNAs.

18. Since the FIT system includes T7 RNA polymerase, i.e., a transcription-translation coupled system, the DNA template is in situ transcribed into mRNA that is translated to the designated peptide.

Acknowledgments

We thank Dr. Hiroshi Murakami for the contributions to the development of the methods presented in this study. We thank Dr. Takayuki Katoh for assistance with manuscript preparation. We also thank Mr. Dai Shimomai for proofreading. This work was supported by grants of Japan Society for the Promotion of Science Grants-in-Aid for Scientific Research (S) (16101007), Specially Promoted Research (21000005), and a research and development projects of the Industrial Science and Technology Program in the New Energy and Industrial Technology Development Organization (NEDO) to H.S., and grants of Japan Society for the Promotion of Science Grants-in-Aid for Young Scientists (B) (22750145) and PREST, Japan Science and Technology Agency to Y.G..

References

1. Forster, A. C., Tan, Z., Nala, M. N., Lin, H., Qu, H., Cornish, V. W., and Blacklow, S. C. (2003) Programming peptidomimetic syntheses by translating genetic codes designed de novo *Proc. Natl. Acad. Sci. USA.* **100**, 6353–6357.

2. Josephson, K., Hartman, M. C., and Szostak, J. W. (2005) Ribosomal synthesis of unnatural peptides *J. Am. Chem. Soc.* **127**, 11727–11735.

3. Murakami, H., Ohta, A., Ashigai, H., and Suga, H. (2006) A highly flexible tRNA acylation method for non-natural polypeptide synthesis *Nat. Methods* **3**, 357–359.

4. Niwa, N., Yamagishi, Y., Murakami, H., and Suga, H. (2009) A flexizyme that selectively charges amino acids activated by a water-friendly leaving group *Bioorg. Med. Chem. Lett.* **19**, 3892–3894.

5. Sako, Y., Morimoto, J., Murakami, H., and Suga, H. (2008) Ribosomal synthesis of bicyclic peptides via two orthogonal inter-side-chain reactions *J. Am. Chem. Soc.* **130**, 7232–7234.

6. Goto, Y., Murakami, H., and Suga, H. (2008) Initiating translation with D-amino acids *RNA* **14**, 1390–1398.

7. Goto, Y., Ohta, A., Sako, Y., Yamagishi, Y., Murakami, H., and Suga, H. (2008) Reprogramming the translation initiation for the synthesis of physiologically stable cyclic peptides *ACS Chem. Biol.* **3**, 120–129.

8. Kawakami, T., Murakami, H., and Suga, H. (2008) Messenger RNA-programmed incorporation of multiple N-methyl-amino acids into linear and cyclic peptides *Chem. Biol.* **15**, 32–42.

9. Kawakami, T., Murakami, H., and Suga, H. (2008) Ribosomal synthesis of polypeptoids and peptoid-peptide hybrids *J. Am. Chem. Soc.* **130**, 16861–16863.

10. Ohta, A., Murakami, H., Higashimura, E., and Suga, H. (2007) Synthesis of polyester by means of genetic code reprogramming *Chem. Biol.* **14**, 1315–1322.

11. Goto, Y., and Suga, H. (2009) Translation initiation with initiator tRNA charged with exotic peptides *J. Am. Chem. Soc.* **131**, 5040–5041.

12. Xiao, H., Murakami, H., Suga, H., and Ferre-D'Amare, A. R. (2008) Structural basis of specific tRNA aminoacylation by a small in vitro selected ribozyme *Nature* **454**, 358–361.

13. Kawakami, T., Ohta, A., Ohuchi, M., Ashigai, H., Murakami, H., and Suga, H. (2009) Diverse backbone-cyclized peptides via codon reprogramming *Nat. Chem. Biol.* **5**, 888–890.

14. Milligan, J. F., Groebe, D. R., Witherell, G. W., and Uhlenbeck, O. C. (1987) Oligoribonucleotide synthesis using T7 RNA polymerase and synthetic DNA templates *Nucleic Acids Res.* **15**, 8783–8798.

15. Saito, H., Kourouklis, D., and Suga, H. (2001) An in vitro evolved precursor tRNA with aminoacylation activity *EMBO J.* **20**, 1797–1806.

16. Olins, P. O., Devine, C. S., Rangwala, S. H., and Kavka, K. S. (1988) The T7 Phage Gene 10 Leader Rna, a Ribosome-Binding Site That Dramatically Enhances the Expression of Foreign Genes in Escherichia-Coli *Gene* **73**, 227–235.

17. Martinis, S. A., and Schimmel, P. (1992) Enzymatic aminoacylation of sequence-specific RNA minihelices and hybrid duplexes with methionine *Proc. Natl. Acad. Sci. USA.* **89**, 65–69.

18. Kung, H. F., Redfield, B., Treadwell, B. V., Eskin, B., Spears, C., and Weissbach, H. (1977) DNA-directed in vitro synthesis of beta-galactosidase. Studies with purified factors *J. Biol. Chem.* **252**, 6889–6894.

19. Shimizu, Y., Inoue, A., Tomari, Y., Suzuki, T., Yokogawa, T., Nishikawa, K., and Ueda, T. (2001) Cell-free translation reconstituted with purified components *Nat. Biotechnol.* **19**, 751–755.

20. Schagger, H. (2006) Tricine-SDS-PAGE *Nat. Protoc.* **1**, 16–22.

INDEX

Jörg S. Hartig (ed.), *Ribozymes: Methods and Protocols*, Methods in Molecular Biology, vol. 848,
DOI 10.1007/978-1-61779-545-9, © Springer Science+Business Media, LLC 2012